Design of Industrial Information Systems

Design of Industrial Information Systems

Thomas O. Boucher
Professor
Industrial & Systems Engineering
Rutgers University
New Brunswick, New Jersey

Ali Yalçın
Associate Professor
Industrial & Management Systems Engineering
University of South Florida
Tampa, Florida

AMSTERDAM • BOSTON • HEIDELBERG • LONDON
NEW YORK • OXFORD • PARIS • SAN DIEGO
SAN FRANCISCO • SINGAPORE • SYDNEY • TOKYO
Academic Press is an imprint of Elsevier

Academic Press is an imprint of Elsevier
30 Corporate Drive, Suite 400, Burlington, MA 01803, USA
525 B Street, Suite 1900, San Diego, California 92101-4495, USA
84 Theobald's Road, London WC1X 8RR, UK

This book is printed on acid-free paper. ♾

Library of Congress Cataloging-in-Publication Data
Application Submitted

British Library Cataloguing-in-Publication Data
A catalogue record for this book is available from the British Library.

ISBN 13: 978-0-12-370492-4
ISBN 10: 0-12-370492-8

For information on all Academic Press publications
visit our Web site at www.books.elsevier.com

Transferred to Digital Printing 2009

Contents

Chapter 7
Executing an Information System Design Project: A Case Study

Chapter 8
E-business and Web-Enabled Databases

Chapter 9
Unified Modeling Language

Preface

Long term growth in productivity is critically important to improving the standard of living of our society. Over the past ten years (1996–2005), average annual productivity growth in the non-farm business sector has been twice as high as it was in the prior fifteen years (1981–1995). In the manufacturing sector, it has been even more spectacular, averaging about 4.4% since 1995.

There is much debate about the sources of productivity growth in the industrial sector from the 1990s to the present. However, there is general agreement among economists and business analysts that capital investment in information technology (IT) was one of the major factors. A significant amount of time elapses between the development of a new technology and its diffusion throughout industry. The development of the microprocessor and its implementation in computers and industrial controllers began in the 1970s and early 1980s. The resulting productivity gains occurred with a time lag of two decades.

The 1990s saw an enormous increase in the number of industrial firms that implemented enterprise-wide information technology software. By some estimates, the number of companies investing in such systems had quadrupled when compared to the mid 1980s. These investments have had a significant impact on the efficiency of resource use within companies, the relationship between companies and their suppliers and customers, and the business strategies adopted by companies that use information technology as a competitive tool. For all these reasons, information technology has become an important subject in the academic training of engineers and industrial managers.

In order to place the current importance of IT into perspective, it is useful to review a little history. Thirty years ago, information technology was looked upon as just another labor saving tool, like any other form of mechanization. Early adoption of information systems was in the bookkeeping and payroll functions, where computers were substituted for accountants in maintaining and summarizing financial records. In production operations, automatic control of processes based on analog electronics was being replaced by digital controllers, such as programmable logic controllers, which were more flexible in supporting system maintenance and functional change.

By the 1980s, industrial companies became aware of the need to integrate operations. Integration first took place by expanding business applications beyond accounting into order entry, purchasing, production planning and inventory control. In so doing, company managements began to focus on company "processes"; i.e., interrelated groups of activities across which information could be shared. On the manufacturing shop floor, developments in shop floor local area networks provided the technology for gathering data from digital controllers and sensors and transmitting it to factory databases for analysis of factory operations. It was during this period that the concept of

Computer Integrated Manufacturing (CIM) was popularized as a vision for the integration of IT throughout all functions of the industrial enterprise.

During the 1990s, industrial managers began to view information systems as a tool to improve their processes and eliminate functions that did not add value to the enterprise. The emphasis was placed on organizational redesign in order to rationalize the functions of the enterprise in relation to information technology strategy. During this period several companies emerged with software solutions for managing industrial enterprises. The application of these products crossed the boundaries between management functions, shop floor functions, and warehousing and distribution functions, such as supply chain management systems. The Enterprise Resource Planning (ERP) software industry was born and many firms reorganized their modes of operation around the possibilities that were enabled by these software. For example, the most widely used ERP system, SAP R/3, was installed in over 15,000 locations by the year 2000.

Since the mid 1990s these software have been more Internet-enabled, giving industrial firms the ability to easily integrate customer relations and supplier relations with the overall enterprise database. Examples of this are numerous. Some packaged food manufacturers now monitor their grocery store chain customers' inventory positions in real-time. As checkout counters automatically debit inventory using bar code scanning, inventory levels can be monitored by suppliers who can then anticipate the need for a new shipment of product to their customer. Apparel manufacturers typically produce an entire line of garments at the beginning of each season. Some apparel firms now electronically monitor the retail sales of their garments at the beginning of the season and determine which styles are catching on. Production schedules are then modified to support those styles that are expected to do well, based on early sales. This strategy also requires flexible manufacturing operations, which is part of the change in enterprise organization that is motivated by the information technology possibilities. Everyone is familiar with the new make-to-order environment, such as that of Dell Computer, where the Internet allows the customer to tailor the product to his or her own specification on-line. In 2001, Internet business was responsible for over $700 billion in sales for manufacturing companies. This was about 18% of total manufacturing sales.

As the foregoing discussion has indicated, industrial companies have been leaders in the adoption of IT. This textbook addresses the subject of the application of information systems within the context of industrial firms. It combines the teaching of the principles of information systems and database design with examples and case studies in industrial companies. This text is especially relevant to students and professionals who will be working with information technology in industrial organizations. The teaching of IT topics provides an excellent opportunity to introduce engineering and industrial management students to applications of this technology in the industrial environment.

INTENDED AUDIENCE

This textbook is intended for use in a one-semester course on information systems at the advanced undergraduate or first-year graduate level. The student does not require any specialized background except a course in a high-level programming language. It specifically addresses the needs of students of industrial and systems engineering, engineering management, and industrial management. It provides the relevant material in the design of information system architectures, database design and usage, and Internet applications. Special attention is given to the integration of these technologies in industrial enterprises.

A typical textbook in information systems and database design usually focuses on three design layers: 1) conceptual, 2) logical, and 3) physical. The conceptual design is the process of constructing an information model independent of all physical implementation considerations. The logical database design is the process of modeling the information of the conceptual level using a specific data model. Current textbooks emphasize the relational model, as do we. The logical model is independent of any particular database management system. The physical design describes the storage structures and access methods used to store and access the data on secondary storage. It focuses on such topics as file structure, indexing and optimizing queries, managing transaction concurrency, data backup and recovery. This book focuses on the conceptual and logical design layers. The relevant audience for this book will be involved with information systems as part of design teams or user groups. As design team members, they will focus on design architectures and design methods as well as specifying the use of the information system throughout the enterprise. As professional users they are mostly involved in the development of decision support systems that must interface with the enterprise database in using information. They are not frequently involved in the details of file structures, indexing, or database security, tasks that are usually done by computer science professionals and database administrators. For that reason, less emphasis is placed on physical layer design and more emphasis is placed on the application of the technology to solve operational problems of the enterprise.

DISTINGUISHING FEATURES

- Numerous realistic case studies in industrial enterprises that illustrate how information systems solve industrial problems.
- Demonstration of how to determine system specifications from user requirements by way of interview processes and analyzing paper records of the enterprise.
- Hands-on exercises in key chapters that allow the student to put a learned concept to immediate use in a relational database application.
- Detailed coverage of HTML, ASP, and XML, with numerous application examples.
- A chapter on the Unified Modeling Language, with comparisons to traditional design methods.
- A chapter on Workflow Management Systems that shows the integration of database systems with the management and control of enterprise operations.

GUIDELINES FOR ALTERNATIVE COURSE STRUCTURES

There are alternative ways to structure a course based on this book. The authors have used it to teach a traditional approach to determining information system requirements, designing a functional and informational architecture, and implementing forms and reports. The course coverage is as follows:

- Chapters 1–3: Introduction and fundamentals of database design.
- Chapters 4, 5: Architecture design methods.
- Chapters 6–8: Implementation in stand-alone and Web applications.

In our approach we have put a great deal of emphasis on the design and execution of a real problem by the students. Students are required to execute a project in teams based on some organizational need with which they are familiar through an internship in a company, an on-campus organization with which they are familiar, an on-campus facility needing a database application, a family business, or similar environment. A surprising array of projects has emerged over the years. The coverage of the eight chapters listed above has proven to be more than enough material for a one-semester course.

An alternative approach to using this book is to focus on architecture design principles and to leave teaching the details of implementation to student self study. In this approach, lectures could be based on the following coverage:

- Chapters 1–3: Introduction and fundamentals of database design.
- Chapters 4, 5, 9: Architecture design methods.
- Chapter 10: Information technology and management of industrial workflows.

If an instructor wants to include database implementation outside of lecture time, Chapters 6 and 8 have been carefully designed for self study. Throughout these two chapters students are asked to complete exercises using Microsoft Access DBMS, as well as coding and executing HTML and ASP Web page applications. With the occasional help of a teaching assistant as a resource person, there should be no problem integrating these chapters into the course without formally covering them in lectures.

SUPPLEMENTARY MATERIALS

Students and instructors can access supplementary materials online. The Web site that supports this book is *http://books.elsevier.com/companions/0123704928*. The available supplementary materials are as follows.

Students:

- Databases used in the chapters and in the end-of-chapter review exercises.
- HTML, ASP, and XML files for in-chapter Internet application examples.
- *Visio Lab Manual*, which illustrates how to use Microsoft Visio for creating activity and information design models as shown in the chapters of this book. A free trial version of Visio 2003 can be downloaded form the Microsoft Web site.

Instructors:

- Databases for all in-chapter and end-of-chapter exercises completed with all the exercise requirements.
- A special populated database titled "University Food Master Database," which integrates many of the data models of the book into a single master database.
- The answer book for all end-of-chapter review exercises.
- All HTML, ASP, and XML files for in-chapter Internet applications and end-of-chapter review exercises.
- Additional case studies and problem sets not published in the textbook.

ACKNOWLEDGMENTS

The origins of this book began with the authors' experience with a project to design the automated control systems and the information system of a demonstration factory to be used in the production of packaged food products. That factory now exists as a food industry extension service of Rutgers

University and an engineering R&D facility of the Defense Logistics Agency, where it supports R&D, new product prototyping, and production services for various clients. The reader will note that the hypothetical company that is used for illustration throughout this book, University Food Company, is loosely based on this demonstration facility. Thus, our first debt of gratitude is to Don O'Brien and Russ Eggers, managers of the contract for the Defense Logistics Agency, the funding agency for this project. Thanks also to the principals who managed this project at Rutgers, particularly Jack Rossen and John Coburn, both of whom served as Senior Associate Directors of the Center for Advanced Food Technology before their retirements. Special thanks also to our colleague and partner in the information system design component, Professor Nabil Adam, Director of the Center for Information Management, Integration and Connectivity at the Rutgers University Graduate School of Business. And special thanks to Rieks Bruins, currently Associate Director of the Center for Advanced Food Technology, who worked closely with us in defining technical specifications for key software modules.

The design of this book greatly benefited from colleagues who reviewed earlier versions of chapters and made substantial recommendations. Particular thanks goes to Richard E. Billo of the University of Texas at Arlington, Reiks Bruins of Rutgers University, Gary P. Moynihan of the University of Alabama, and Tsuta Tai of State Street Corp.

Several students have made contributions to case studies and end of chapter exercises. Dr. Boucher would like to thank Andrew Balin, John Fischer, Carolina Zonensein Horner, Stéphane Ricard, Dustin Runkle, Sameer Shah, Varun Sharma, and Özgecan Uluscu. Dr. Yalçın would like to thank Amin Mokrivala, Bharath Natarajan, and Rami Salhab Al-Tamimi.

We are also grateful to our acquisitions editor, Joel Stein, for his support, advice, and encouragement during the preparation of the book, and Carl M. Soares, our production manager, for his professionalism in working with us to realize the quality we wanted in the appearance of the final product.

Finally, we would like to acknowledge the encouragement and support of our families and friends, who were very patient in our absence while working on this book.

Chapter 1

Introduction

1.1 INTRODUCTION

Information is an important asset in the management of industrial enterprises, and ***information technology (IT)*** helps manage that asset. IT enables firms to integrate the decision functions that exist in the myriad of subsystems required to manufacture and distribute a product. These subsystems include sales, purchasing, production planning, quality control, process control, and supply chain logistics.

An industrial firm is a web of activities, or processes, that interact with each other. At one level of abstraction, that interaction can be viewed as a process of creating and exchanging information. For example, before production personnel can use incoming raw material, it often must be examined and approved by quality control personnel. The fact that quality control has given final approval to use the material is "information" passed on to production before production personnel can process the material. In many ways the quality control organization operates within its own processes, executing its rules of inspection for raw materials. Concurrently, production personnel operate within their own set of processes, executing production according to the rules that govern the planning and execution of production. However, quality control personnel and production personnel are coupled together by the information transfers that enable each other's work.

An organization operates more efficiently and better manages its resources when activities are clearly coordinated among the subsystems of the organization. If the information flows are unavailable for the decision-making processes, poor coordination develops. This results in lagging responses to changing conditions on the factory floor and lost productivity. One response to this need in organizations has been the implementation of computer technology in the form of information systems, databases, local and wide area networks, and middleware. In the information technology arena, this is known as ***enterprise integration***.

In manufacturing industries, specific standard solutions have been developed since the 1970s to address particular parts of the overall enterprise integration problem. One of the first attempts at integration occurred in the late 1960s through the 1970s. Known as ***material requirements planning (MRP)***, it combined database management systems with application software in order to manage inventory levels and plan production schedules. In the classic MRP approach, a master production schedule for finished products is developed for a finite time horizon. This master production schedule is a plan of when specific products should be completed by production. The plan is derived from outstanding sales orders or finished goods inventory levels that need to be replenished. The master production schedule for end products is used to compute the requirements to build intermediate subassemblies and components. This is called the "bill of materials explosion," in which the product bill of materials is used to convert the end product demand into the demand for subassemblies and individual components. Finally, the raw materials required for production are determined from the individual components that have to be built.

If the material requirements that are necessary to execute the master production schedule are not in inventory, requisitions are sent to purchasing, and purchasing places the materials on order. MRP is basically a production/inventory management system that provides information to the production department so that it can rationally plan to produce products, knowing it will have the proper materials on hand to do so.

MRP II, which was built on the basic MRP model, extended MRP to include capacity planning. When MRP provides a schedule of manufacturing activities required to meet a master plan, MRP II takes over and balances the available manufacturing resources, such as machines and workers, to the required production activity. MRP II answers the question of whether or not sufficient week-by-week plant capacity exists to meet the planned production schedule. In the case of insufficient capacity, MRP II allows the planner to evaluate alternative plans by rescheduling the required work within the overall plant capacity limits. The general MRP/MRP II schema is shown in Figure 1.1.

The standard MRP system has been expanded to include much more functionality within a concept known as ***enterprise resource planning (ERP)***. In addition to the traditional MRP production activities, ERP has added support for some of the following functions:

- Quality management
- Sales and distribution
- Human resource management
- Project management

More recently, ERP was extended beyond the factory and the firm to include functions that link the company to its customers and suppliers, such as the following:

- Logistics supply chain management
- Intercompany communications
- Electronic commerce

MRP/MRP II/ERP are generally thought of as "planning" systems. They are responsible for supporting the planning of production, but they are not very well integrated into the execution of production. This void in available software solutions on the shop floor has led to the development of the ***manufacturing execution system (MES)***. The MES is an attempt to manage resources, including materials, machines, and personnel, on a daily or even hourly basis. Typical MES functions include the following:

- Dispatching and monitoring production—that is, controlling the release of work orders to the shop floor and tracking work-in-process inventory.
- Detailed scheduling associated with specific production units in order to meet specific performance criteria.
- Data collection from factory floor operation to provide a history of factory events.
- Quality data analysis—that is, providing real-time analysis of measurements collected from manufacturing, notifying production personnel of out-of-tolerance conditions, and sometimes analyzing data to recommend corrective action.
- Product history recording—that is, providing a history of product manufacture for genealogy or regulatory purposes and tracking the specific lot of materials and specific machines used

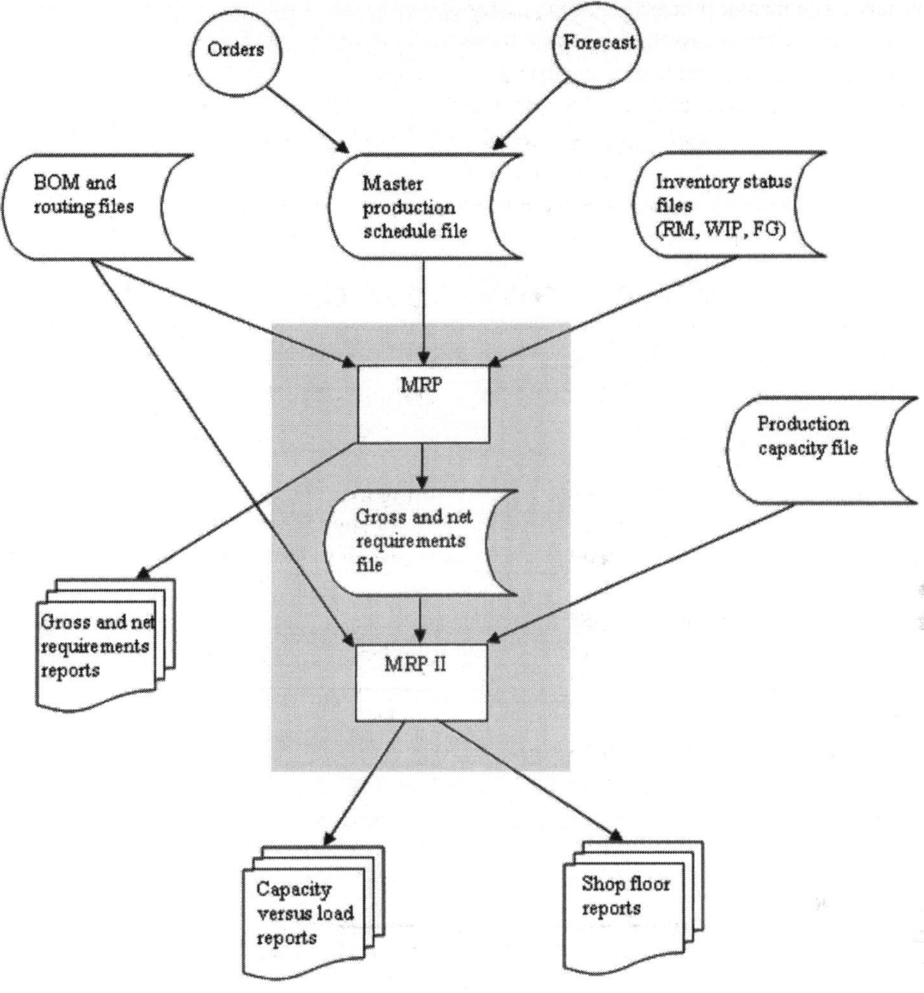

Figure 1.1 Typical structure of MRP planning system.

in the manufacture of a specific lot of finished product. This is useful in isolating the source of problems in cases of product recalls.

In effect, the MES provides a functionality closer to the execution of production than that provided by a typical MRP system. The MES also provides actual results that can be fed back to planning systems for updating plans over the next planning horizon. An MES provides functions that are just above the level of actual real-time control of machines and processes.

At the very bottom of the manufacturing information system is the electromechanical ***controls layer***. The tasks required to fabricate a part are a series of physical operations performed on materials. The control system is programmed to execute the decision process for these manufacturing tasks. Typical examples are the computer numerical control (CNC) parts program that controls motors and other actuators that move a machine tool through its cutting process, or a process controller that regulates the temperature of a heating kettle in a chemical processing plant. The components of the control layer usually include programmable logic controllers, robots, conveyors, computer numerically controlled machines, operator interfaces, and display devices. Unlike the higher levels of the information hierarchy, the software solutions for information management at the control level are not very standardized. In fact, individual machines and processes are usually highly engineered for specific functions and applications. However, the computer programs that enable them to execute a particular sequence of instructions are considered to be part of the overall repository of information on plant operations.

1.2 ERP/MES/CONTROL: A HIERARCHY OF INFORMATION

The standard software solutions used in industrial information systems reflect the fact that a hierarchy of decisions must be made in manufacturing, from the machine, or unit operation control level, up through the overall planning of plant operations. This hierarchy is illustrated in Figure 1.2.

The ***machine control level*** is responsible for ensuring that the sequence of machine operations corresponds to the planned sequence necessary to fabricate the part. Typically, the sequence of operations is carried out as prescribed by the program resident in the machine controller, and there are few, if any, decisions to be made.

At the ***production line or work cell level,*** the objective is to supervise the interactions between a group of related machines or processes. This level of decision making is not concerned with the operation of the machine or process. The emphasis here is on the release and delivery of materials to a machine in the work cell at the correct time. Examples of decisions at this level include routing of material among machines fabricating a component and the decision to extract out-of-specification components while they are being processed on a production line or within a manu-

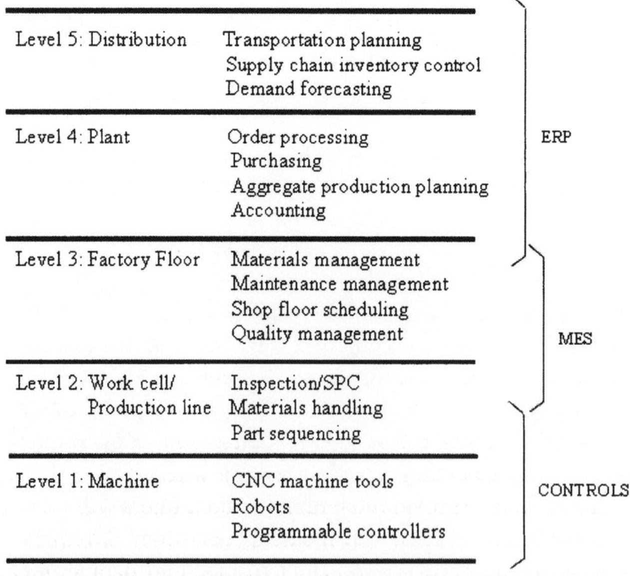

Figure 1.2 Decision hierarchy of industrial information system.

facturing cell. The work cell or production line level is typically considered part of the MES level, but there is some overlap with the controls level.

At the ***factory floor level***, decisions are made that affect groups of production lines or work cells. For example, several production lines or work cells may be serviced by the same materials management system that requests the movement of raw materials from storage to production so that they can be manufactured into a finished product. Since this resource is shared among production lines and work cells, there must be a supervisory level of decision making that decides when to bring material to the shop floor, particularly when conflict occurs (e.g., when it is required to service two production lines at the same time). Other examples include the scheduling of which production line or work cell will fabricate a part when there is a choice and the management of common facilities, such as storage areas for raw materials and finished products. There is some overlap between MES and ERP responsibilities at this level of the hierarchy.

At the ***plant level***, decisions are less concerned with the daily operation of the factory and are more closely related to the business planning objectives of the firm. A typical plant-level production control decision is *aggregate production planning*, which refers to the process of planning the use of the plant's production capacity to meet customer demands over a period of months or a year. The output of this plan is a master schedule of what products will be produced during each period of time going forward over the planning horizon of the plan. Decisions and functions at this level of the hierarchy are considered part of MRP/ERP functions.

Finally, at the ***distribution system level*** the emphasis is on coordinating the supply of finished product to the end customer. Maintaining appropriate and cost-effective inventory levels, as well as managing the transportation of product between warehouse locations in the supply chain, is a problem addressed at this level.

The decision hierarchy illustrated in Figure 1.2 starts from the top and works its way down. For example, the supply chain replenishment policy creates demand at the plant level. In turn, the plant-level aggregate production plan sets the overall boundaries of which products will be produced and the time interval in which they will be produced. This provides a constraint on the shop floor level, which must then allocate the required production to machining cells or other production processes in the most effective manner in order to meet that overall plan. Once a specific machining cell or set of production processes is allocated its production schedule for a specific day, it is the responsibility of the work cell/production line level to coordinate the manufacture of the product through the related machines and processes it requires. Finally, when the machine is assigned its role in partially fabricating the product, it is the responsibility of the machine-level controller to execute the correct steps of the fabrication process. This paradigm has led to modeling the industrial production problem as a hierarchy of decisions, where the upper levels of the hierarchy place constraints on each succeeding lower level. The objective is to assign each control decision to the lowest possible level in the hierarchy at which that decision can be made. The complete integration of all of these levels of decision processes, supported by computer information systems, is the domain of an industrial information system.

The interaction of the hierarchy of functions and standard software solutions is illustrated in Figure 1.3. From the planning level, ERP provides the MES level with an overall plan of what is to be produced during the current planning horizon, typically a couple of weeks in duration. The manufacturing execution level is then responsible for detailed production operations on the factory floor. As production is executed, actual results concerning what was produced are fed back to the planning level. The MES level also provides the machine controllers with the information concerning how to produce a particular part by downloading controller programs to the machines as required. The MES level monitors real-time actual results, and data summaries are logged for storage in factory databases.

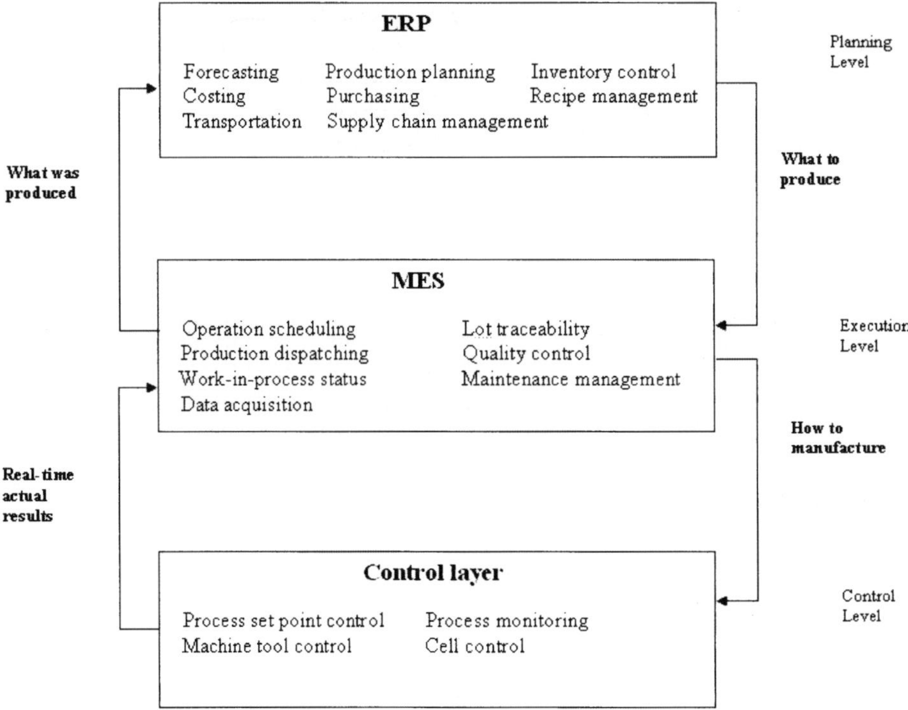

Figure 1.3 Coordinating the layers of the information/control system.

Implementing the model of Figure 1.3 within any specific industrial enterprise requires tailoring its functionality to the particular system environment. Specific cases will be examined in later chapters.

1.3 NETWORK ARCHITECTURE

Enterprise integration is about the integration of functional areas through information sharing. To realize efficient information sharing, it is desirable to network the levels of the hierarchy shown in Figures 1.2 and 1.3. The ***network architecture*** is a description of how the various layers of the decision hierarchy will communicate with one another. The network architecture is typically implemented with the use of local area networks. A ***local area network (LAN)*** is a communication network that is implemented over a limited area and is usually owned by one organization. It is a common medium that allows several computers or several machine controllers to be connected, and, as long as each computer uses the protocol convention of the LAN, communication can take place at high data rates.

Figure 1.4 shows a typical example for a modern industrial company. This architecture shows three distinct layers. At the bottom, the physical processes of manufacturing on the shop floor are arranged in stations along a production line or in manufacturing cells. Controllers that are dedicated to those physical processes control the operations at each station. These controllers may be computers, but they are more likely special-purpose computers, called ***programmable logic controllers (PLC)***, which have been specifically designed to control and collect formation from machinery.

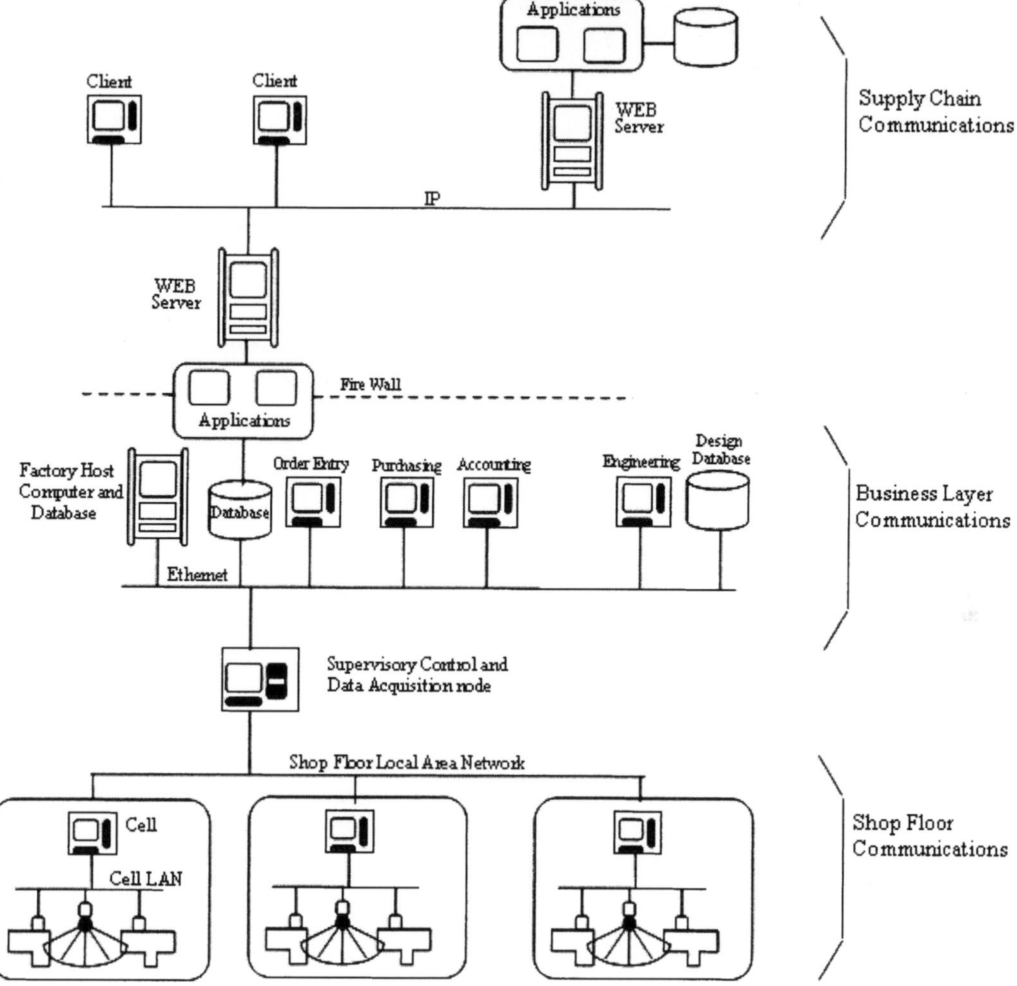

Figure 1.4 Typical network architecture.

Historically, manufacturers of machine controllers have implemented communication requirements using local area networks designed specifically for their own controllers. In Figure 1.4, we are assuming that the controllers have the capability of communicating with each other over a common local area network, sometimes called a ***shop floor data highway***. Such communication capability allows individual machines to "talk" to one another as peers. This is known as "peer-to-peer" communication capability. So, for example, if a machine at one station has a failure, the controller at that station can transmit that information to other stations or to the supervisory controller, which may respond by stopping the processes and signaling for assistance.

Another distinct layer of the architecture is the business layer, at which functional departments are connected to one another via computers and a local area network. The factory host computer is the repository of data for factory operation. The individual functions shown are the high-level business and planning functions of the factory. In Figure 1.4, these functions are interconnected with each other and the factory database over their own local area network. The distinction between the shop floor LAN and the business-level LAN is one that has developed in practice over time because of differences in the communication performance requirements at each level as well as available LAN options on the factory floor.

The middle layer connecting the shop floor and the factory host database at the business layer is there to provide upward communication of actual data from the operations on the factory floor and downward communication of manufacturing instructions and production requirements to the factory floor. A common implementation of this connecting layer is the ***supervisory control and data acquisition (SCADA)*** node. A SCADA node is a computer that has communication drivers for both the factory LAN and the business-level LAN, allowing it to communicate in both directions. Technically, it is a communication ***gateway***, which means it provides a way for two dissimilar networks to communicate with each other at the application layer.

Finally, beyond the local area networks of the plant there is the Internet, connecting the plant electronically with other units of the firm and with other firms. Through the use of the Internet, the plant can share information with customers and with other businesses that are suppliers or partners of the firm. The technical details of the use of the Internet for connecting firms to customers and suppliers will be covered in a later chapter on the subject of e-business.

Figure 1.4 illustrates only one way in which the factory information can be shared among functions. Other architectures are available and have been implemented in factories. However, the common feature of these architectures is to link the shop floor and business layers in a seamless information system, giving them access to the same information held in one or more databases. We refer to the combination of networks, databases, and application programs at the ERP, MES, and control layers as an ***industrial information system (IIS)***.

1.4 SOME KEY APPLICATION AREAS OF AN INDUSTRIAL INFORMATION SYSTEM (IIS)

There are several vendors of "generic" database applications for industry. Some of the leading vendors are SAP AG, Oracle Corporation, PeopleSoft, Baan Corporation, and Manugistics Inc.

These generic database designs were developed to be broadly applicable to a number of industry types. In most cases, the implementation at a specific company involves tailoring the application to the specific environment in which the company operates, including peculiarities of the enterprise organization. These generic database applications are often preferred to developing the application in-house, for a variety of reasons:

- The applications are designed as modules that can be expanded over time with guaranteed system integration.
- They allow the enterprise to forego the high cost of the software development process since the development cost is implicitly shared among the purchasing enterprises.
- They are supported technically by vendor companies, allowing the user enterprise to maintain a relatively small information system support group.

These database application modules are targeted to the key management processes of the enterprise. In this section, we discuss some of the processes currently supported and the role that the IIS plays. These processes include the following:

- Customer relationship management
- Order fulfillment management

- Warehouse management
- Quality management
- Human resource management
- Accounting and financial management
- Distribution and supply chain management

1.4.1 Customer Relationship Management (CRM)

Most industrial enterprises are customer centric. The retention and satisfaction of customers are necessary to maintain customer loyalty and profits. A CRM system is composed of a database and application software that help the firm maintain contact with customers, automate the sales and marketing functions, and support the customer in using the product after sales. For example, through the use of the Internet and a database of customer electronic contact information, regular new product announcements and updates, as well as special sales offers, can reach the customer in a timely manner. Using a database to keep track of the frequency of customer visits by sales personnel, as well as the forward scheduling of such visits, can help ensure that the firm has good information about the future needs of its customers. A technical help hotline that manages customer complaints and maintains a database about the frequency of reporting of specific problems helps the enterprise to target problem areas and improve customer satisfaction more quickly. In addition, CRM can support order status monitoring so that customers can be provided with real-time status information about the delivery schedule of their orders.

The use of information systems to manage the customer relationship has proven to be cost-effective and necessary in cases where real-time information from and to the customer is an important competitive advantage in the industry. It would be virtually impossible to implement this function without information systems, backed up by information technology and the use of the Internet.

1.4.2 Order Fulfillment Management

Order fulfillment management lies at the heart of the industrial process, and it integrates several enterprise subsystems into a single overall function. Some of the key subsystems are shown in Figure 1.5. They include the following:

- Order entry
- Production control
- Purchasing
- Inventory, or material management
- Shipping

The process begins when an order is taken via the order entry system, which includes the appropriate database tables for registering the order. Unfilled orders from the order entry database are the inputs to the production control system, which includes the databases and software modules that plan and monitor the execution of production. Production control is perhaps the most complex set of modules in the enterprise system and has many details. Some of these components were illustrated in Figure 1.1.

The production control function interacts with the purchasing function to ensure that materials are acquired in order to produce and fill orders. It also interacts with the inventory database in the scheduling of the processing of materials and the monitoring of the levels of finished products.

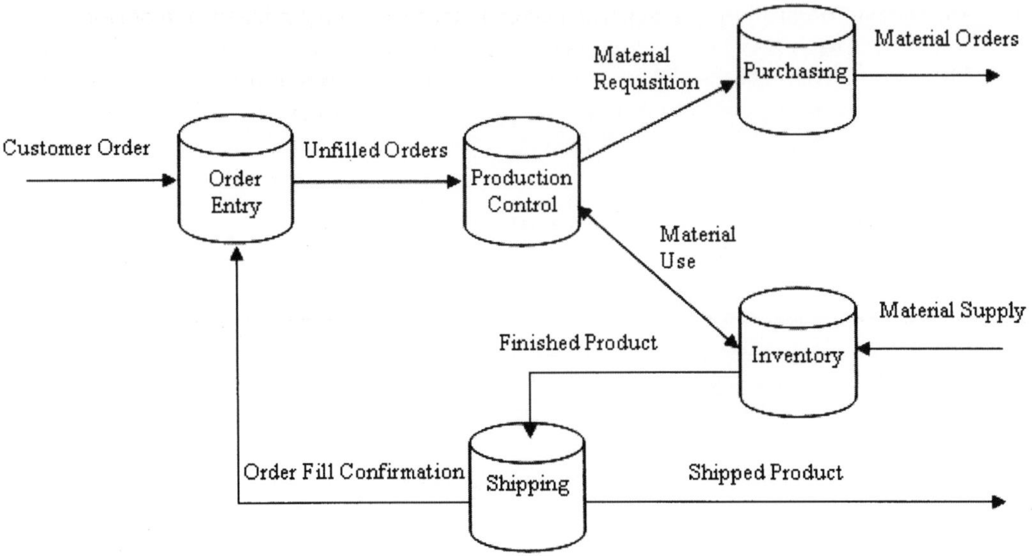

Figure 1.5 Order fulfillment process.

Finally, order fulfillment includes the shipping function, which delivers the order to the customer while informing the order entry database that the order is filled.

The integration of these functions into one process illustrates the power of the underlying enterprise database and information system. The information system makes important contributions toward executing this process:

- *Speed.* It reduces the time for functions involved in the process to communicate with each other.
- *Accuracy.* It ensures an accurate count of materials involved in the process needed to complete the order.
- *Coordination.* It avoids wasted time by coordinating work for more efficient resource utilization.
- *Process performance improvement.* It retains a chronology and history of events that can be studied to improve future performance.
- *Supports decision making.* The underlying database management system can generate timely and user-friendly reports that can be used in management decision making.

The order fulfillment process is central to industrial enterprises. It also lies at the boundaries of enterprise systems and execution systems because it supports both planning and execution. For this reason, examples from this area are used for illustration in the text and in the end-of-chapter exercises in this book.

1.4.3 Warehouse Management System (WMS)

Inventory management involves managing the levels of inventory by material, as well as their reorder points and reorder status. Warehouse management systems add the capability to manage inventory lots by storage bin location. The material manager can optimize the use of storage bins and monitor storage areas in complex systems. It enables material management and automated systems to direct transport devices to specific locations for the retrieval of materials for use in production. Pure inventory management enables the management of levels of inventory, but not the location. Some other activities of a WMS are the following:

- Keeping track of changes in status, such as the location of material lots on hold waiting for inspection or material lots released from inspection and available for use in production
- Directing the picking of finished goods by warehouse location for retrieval for shipping
- Through the report writing facilities of the database management system, automating the publication of reports of materials and finished product by location for management review
- Keeping track of material management activities in order to put a dollar value on the functions of storing and retrieving warehoused products and inventories

1.4.4 Quality Management

Quality management information integrates with order fulfillment and warehouse management because materials and finished products are regularly tested and classified for use. Quality management systems incorporate quality documentation, such as inspection and testing procedures, laboratory instruments that are used in testing, and documents of the test results, including audit and defect tracking. These systems also track the calibration and maintenance of laboratory equipment and supplies used in quality tests, referred to as laboratory information management systems (LIMS). Software modules are provided that interface to the database to download data and provide facilities for the statistical analysis of data.

Quality management information systems are widely used in all industries, but they have become essential in industries regulated by the Food and Drug Administration (FDA), such as the food and pharmaceutical industries. They provide automated checking to ensure that proper testing has occurred, and they provide ready access to test results taken during production should a defect problem occur in customer use.

1.4.5 Human Resource Management (HRM)

In large enterprises, the management of information about the firm's employees is a significant task that is made easier through the use of databases. The activities under HMR are related to the following:

- Administration of employee records
- Administration of personnel development programs
- Providing self-help information facilities for personnel

Employee records include compensation packages, payroll, travel expense reimbursement, and medical programs and cost, among others. Personnel development programs have to do with worker training and skills development. This may include the scheduling of in-house courses and training sessions and managing personnel assignments by skill levels. Self-help facilities are information services, such as company web sites where information is made available to the employee. This may include company forms and applications, information on company benefits, descriptions of corporate retirement programs, and similar items of interest to the employee.

Unlike customer relationship management, order fulfillment management, and warehouse management systems, HRM is a supporting activity of the enterprise and is not often thought of as generating profits. However, it is an important activity because it can affect employee morale and help ensure a proper match between employee skills and job allocation. Information systems help companies manage this important function with minimum staffing and overhead.

1.4.6 Accounting and Financial Management

The first corporate applications of database systems were in accounting and financial reporting. Accounting and financial reporting are strictly governed by rules from the Financial Accounting Standards Board (FASB), and financial records of publicly traded corporations are audited annually by certified public accounting firms for compliance to those rules. All companies today depend on computerized financial records in databases.

Financial and accounting responsibility is usually divided between the corporate finance and corporate treasury departments. Corporate finance is responsible for capital budgeting and general ledger accounting, including audits. Treasury is responsible for cash management, corporate funds that are invested outside the firm (for example, in the corporate retirement fund), and for raising capital through the issuance of stocks and bonds. The corporate accounting and financial functions are the most standardized across industry types of any functions performed within an enterprise. This is so because of the oversight provided by law, by FASB, and by the rules of the exchange on which the firm's stock is traded. Therefore, there is a realistic expectation that, with respect to database and information system design, one size fits all. The subject of corporate financial information systems is vast and not central to the purposes of this text, which focuses on the operations side of the enterprise. Therefore, we only mention it in passing and refer the interested reader to other references (Perry and Schneider, 2004; Romney and Steinbart, 2005).

1.4.7 Distribution System and Supply Chain Management

Supply chain management is concerned with the cost-effective integration of supplier coordinating and product distribution with enterprise operations. Figure 1.6 characterizes a hypothetical enterprise with several layers of stocking points: a central distribution center (CDC) that feeds regional distribution centers (RDC), which, in turn, feed local distribution centers (LDC). The enterprise may or may not own all the locations in a distribution chain, so other configurations are possible. The supply chain includes the distribution of product ultimately to the customer and the upstream procurement of materials from the supplier base.

Information technology and information systems are enablers for managing these functions. Typical operational objectives are to reduce inventories, reduce reorder lead times, and increase

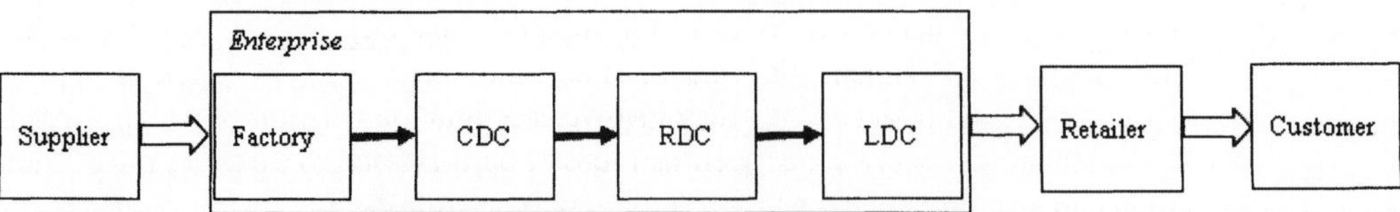

Figure 1.6 Supply chain spans beyond the enterprise.

the customer service levels while reducing cost. Information flows between companies, their customers, and their suppliers are critical for achieving these objectives. Some typical applications are as follows:

- Tracking of orders through distribution
- Managing multiechelon inventory levels
- Point-of-sale product tracking
- Data interchange with supply chain partners
- Use of workflow technologies

A key contribution of information systems is visibility throughout the supply chain that allows tracking of orders in the physical distribution system. It is important to know the whereabouts of product and orders so the information can be available to customers on demand. Technologies such as global positioning systems (GPS) and wireless communication can be used to bring real-time information about the location of a product in transit to the database system. Bar coding and radio frequency identification (RFID) facilitate easy maintenance of location data in intermediate storage, such as regional and local distribution centers.

Information system visibility also assists in managing multiechelon inventory levels. This refers to the quantities of a stock-keeping unit (SKU) held at various levels of the system; for example, in central, regional, and local warehouses. Depending on lead times between stocking points and the cost of holding inventory at each level, there are improved policies for moving lots of SKUs through the system. System-wide visibility helps in the application of these policies by providing real-time information about the inventory status at each level. Also, if there is a condition that will require a change in production scheduling at the plant, it is necessary to communicate this information to production in advance.

Point-of-sale product tracking extends data gathering beyond the firm into its customers' businesses. For example, a clothing manufacturer may have arrangements with its customer, a retailer, to supply daily information about the sale of its clothing lines at the retail level. Such information will enable the manufacturer to gauge its production of specialty or seasonal items to the demand as it is happening, thus reducing the likelihood of overproduction or shortages during the prime selling season.

When it comes to sharing information by supply chain partners across enterprise information systems, as in the case of point-of-sale data, the issue of standards for data interchange becomes important. One solution for getting data from one computer system to another is to send the data in simple text files. However, this does not solve the problem of automating computer-to-computer understanding of what the text files mean. Extensible markup language (XML) is a meta language that allows users to design and format documents to standardize their meaning. Vendors of information systems and database management systems support XML specifications and enable the automatic insertion of XML file data into databases.

When work is shared by two or more enterprises or more than one department of the same company, each controlling a subset of tasks, coordination of activities becomes important. Workflow systems are used to define a set of tasks and their precedence relationships for completing a job. Once a workflow is incorporated within an information system, it will manage the activities of the users such that the precedence of tasks is enforced and the successful completion of the workflow will result in the appropriate entries into the enterprise database. For example, completing a reservation with a Web-based travel agency may involve coordination with partner enterprises in air travel, hotel, and car rental services, as well as a credit card service to hold the reservation. Before a reservation can be completed, confirmations are needed from each partner service in some order, requiring the completion of reservation tasks by each enterprise. Managing these interactions to a successful completion is a good application for workflow software. Major vendors of database management systems now offer a facility for designing workflow systems as part of the information system application.

1.5 INFORMATION SYSTEMS AND DECISION SUPPORT SYSTEMS

In the previous sections, we described two components of industrial information systems. One component is the information system design, and management of information for enterprise functions, such as order fulfillment management or customer relationship management. The second component is the decision making that can be accomplished using the data from these systems. This was indicated in the description of MRP II and the coordination hierarchy of Figure 1.3. The first component has to do with information system design, and the second component has to do with information use. The second component comes under the general heading of decision support systems.

A ***decision support system (DSS)*** supports management decision making by using data from the information system in conjunction with sophisticated algorithms programmed in software modules for the purpose of analyzing how best to operate the enterprise. For example, forecasting models, simulation models, corporate planning and scheduling models, and the like are decision support systems that utilize data from the enterprise database.

In this book, we describe the technologies and modeling frameworks used in designing and implementing the information system itself. The primary purpose of this book is to explain how to use these technologies and design frameworks to make useful information available to workers and managers in order for them to make the decisions necessary to execute their functions. Hence, we focus our attention on the first component as described earlier. The design of decision support systems is not a subject of this text and we refer the interested reader to Hanna et al., 2003; Pol and Ahuja, 2003; Turban et al., 2005, for more on this topic.

1.6 PRODUCTION SYSTEM CLASSIFICATIONS AND INFORMATION REQUIREMENTS

At the heart of linking business processes with production processes are the unique manufacturing system characteristics of different industries. There is a formal structure for describing the basic information for building a product that is related to the characteristics of the product and the production system organization and design. We distinguish broadly between the manufacture of mechanical parts, including electrical assembly, and the industries whose manufacturing processes and products are based largely on producing chemical changes in the raw materials, usually referred to as the process industries. This section provides an overview of the information related to the product structure and production processes for these two categories of industries.

1.6.1 MECHANICAL FABRICATION INDUSTRIES

In the fabrication and assembly of mechanical products, the traditional manufacturing system is the ***functional***, or ***job shop***, organization of production, illustrated in Figure 1.7. In this design, machines that perform the same manufacturing processes are grouped together within the same department. Hence, there is a department for lathes, which remove the metal from cylindrical work

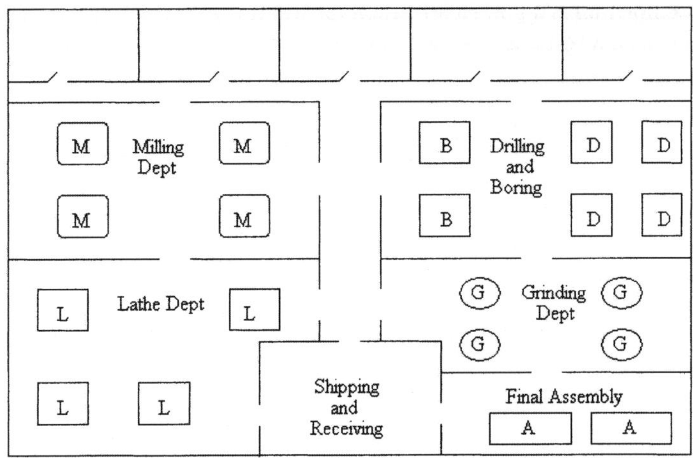

Figure 1.7 Functional or job shop design.

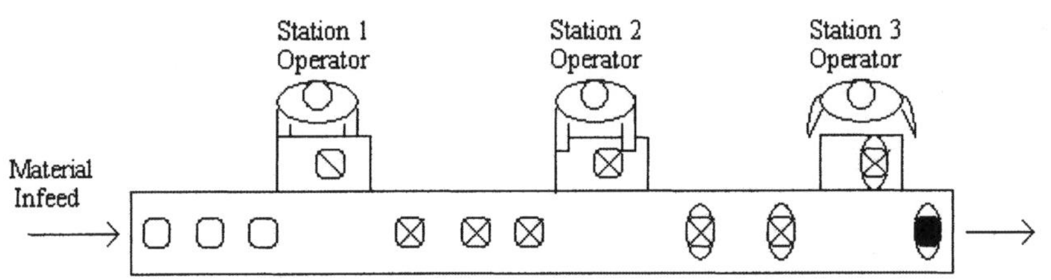

Figure 1.8 Flow line design.

pieces, a department for milling machines, which remove metal from prismatic parts, and so forth. Workpieces are routed among departments in a prescribed sequence, called a *routing plan*. They emerge from that sequence as a finished component. From there they may be assembled with other components to make a finished product.

This manufacturing system design is typically used for the manufacture of components in batches. It is a very unfavorable organization of production because it is difficult to manage the movement of batches of components through the plant without incurring high work-in-process inventory levels. To ensure an even balance of work in the various departments, an investment in large quantities of work-in-process inventory is often required. In addition, this system does not lend itself very well to factory automation, except at the machine level.

The ***production flow line***, illustrated in Figure 1.8, is a method of organizing production in the mechanical industries and electrical assembly industry such that individual fabrication operations are arranged along the steps necessary to manufacture the product. The automotive assembly line is an instance of flow line production, as is the manufacture of metal containers such as cans. These manufacturing system designs tend to be composed of closely coupled unit operations linked by conveyors. They are dedicated to a narrow range of products and are relatively high speed. A high degree of automation along the production line is not unusual.

A form of production organization in the mechanical parts industries that is intermediate between flow line and job shop organization is called ***cellular manufacturing***. Cellular manufacturing takes advantage of some of the favorable aspects of job shop production and flow line production. It is based on the group technology manufacturing philosophy. *Group technology* identifies components of the product mix that are similar in design and that require roughly the same kinds of manufacturing operations. Hence, components requiring primarily external turning operations, such as shafts, are collected in one group, while components requiring surface grinding operations and drilling operations, such as plates, are assigned to a different group. These groups become the basis on which production engineers can reorganize a traditional job shop into a plant design with machining cells in which machines are arranged such that each machining cell can complete the fabrication of one or a few groups of parts. Despite the fact that each individual component may have a small annual demand, when components are grouped together they add up to a large enough annual production quantity to utilize the capacity of the dedicated manufacturing cell.

Figure 1.9 illustrates the difference between the flows of products in a job shop versus cellular manufacturing. Through the simplification of routings, the cellular design eliminates the difficult routing and control problems that exist in a job shop. In addition, there is a dramatic reduction in setup times on each machine because the fixtures and work holding devices for each machine are redesigned to accommodate the components in the group without major changeover. With respect to automation, the cellular design makes it economic to load and unload the machining operations automatically—for example, using robots. Also, the manufacturing cell has been shown to be economic in manufacturing components in small batch sizes.

Figure 1.10 shows just such an automated cell. Here a robot is attending several machining operations. Incoming parts are placed in an input buffer and finished parts are put into an output buffer. Parts that have undergone some machining operations are in a work-in-process buffer. The program to control the fabrication steps for each machine resides in the controller of the machine.

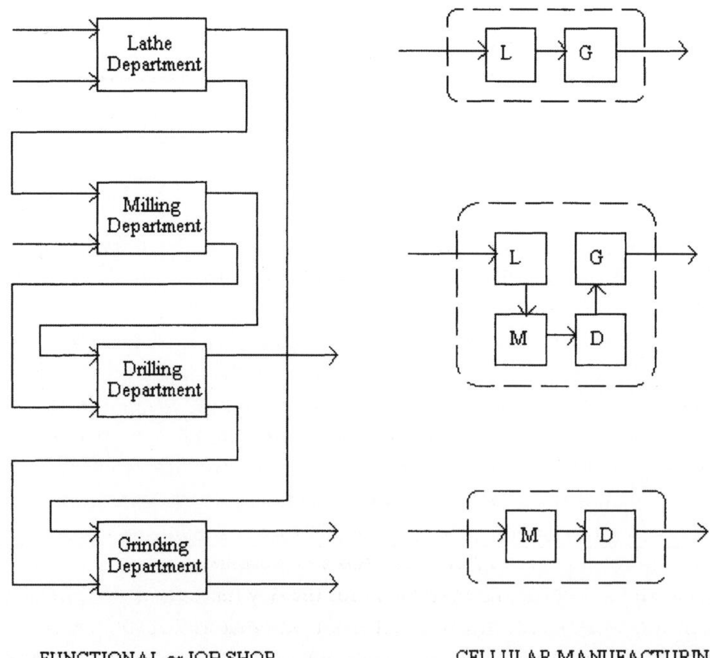

Figure 1.9 Product flow under functional and cellular manufacturing plant designs.

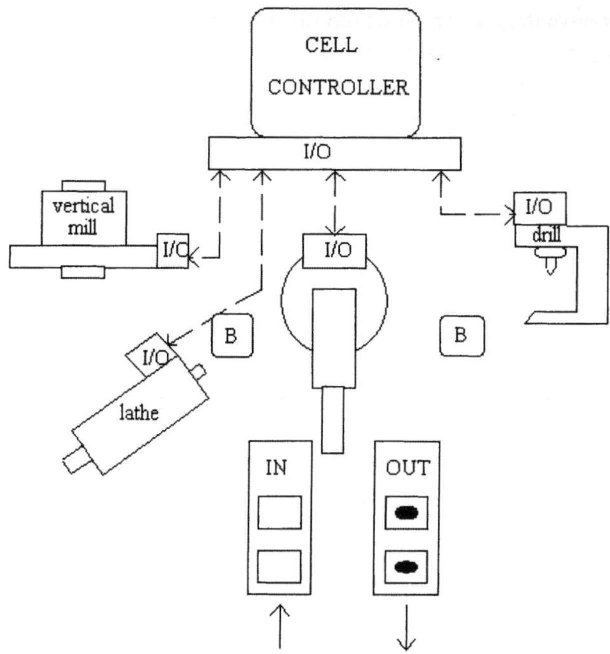

Figure 1.10 Automated machining cell design.

The program to load and unload components to and from machines resides in the robot controller. Overall supervision of the cell is the responsibility of the cell controller. If more than one component is being manufactured at a time, it is necessary to identify the incoming component and download the appropriate program to the machine tool before directing the machine to begin its operation.

The group technology philosophy has been extended to large-scale manufacturing systems composed of machines and cells linked together by automated materials handling. The design is usually referred to as a ***flexible manufacturing system*** **(FMS)**. An example is shown in Figure 1.11. Here we show manufacturing cells that are multipurpose machining centers serviced by a loading/unloading machine, sometimes a robot. These machining centers are capable of performing more than one machining operation on a single machine. They are linked together by an automated transport system. This may be an ***automated guided vehicle*** **(AGV)** that is programmed to follow a fixed path and carry components to be manufactured between cells, or it may be a pallet transport system that moves parts along an asynchronous conveyor on fixtures, called pallets. These pallets are of standardized sizes and they can be directly loaded onto the multipurpose machine tool when they arrive at the cell. Workers in a loading/unloading area do the mounting of workpieces on pallets. Hence, once a workpiece is mounted on a pallet in the loading area, it stays on that pallet through its machining operations and transport back to the unloading area.

There is a hierarchy of supervisory control problems associated with the operation of an FMS. Assuming that a schedule of parts to be produced for the day has been determined and downloaded to the FMS controller, workpieces are shipped from raw material inventory into the FMS to be machined. The FMS supervisory controller calls for the mounting of the workpiece when its scheduled time in the sequence occurs. When a workpiece is released to production, the FMS

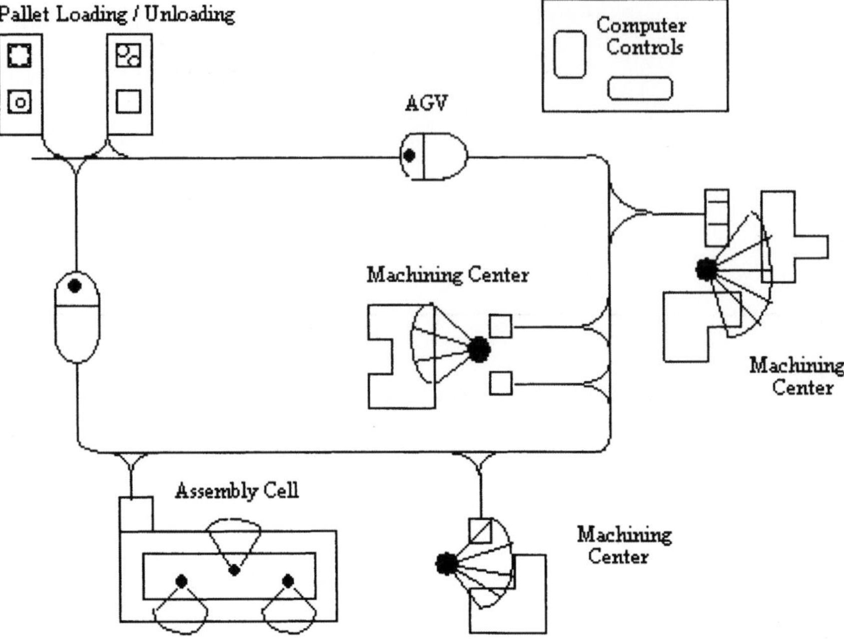

Figure 1.11 FMS design.

supervisory controller must evaluate the operational requirements of the component, which are stored in the supervisory controller's database. The status of machines that can perform those operations as well as the availability of the required tooling at the machining centers must then be determined. If there is only one machine cell that can perform the operation, the pallet is sent to the queue in front of that cell. If a choice among cells exists, the FMS supervisory controller evaluates the workload already assigned to each cell and selects a cell based on available machining time and the efficiency of using that machining cell for that particular component.

There are other resource sharing considerations that are managed by the FMS supervisory controller. In introducing workpieces into the system, the controller must consider the availability of the fixtures appropriate to mounting the workpiece. When directing the transport of workpieces among stations, the controller must consider the availability of a transporter and, if more than one route exists between stations, which route to take.

Regardless of whether the manufacturing system is designed for functional layout, as stand-alone machining cells, or as an FMS, a common set of product and process information is required in manufacturing the product. This information is also central to the information system.

The important product-related information is known as the product ***bill of materials (BOM)***. The bill of materials is a master list of the components, purchased parts, and subassemblies required to produce a complete product. It is derived from the design drawing of the product. This document provides the engineer with information on all of the components that must be brought together in order to deliver one unit of the final product. The bill of materials is also the key to the raw materials that will be required to make the components for the product. Figure 1.12 shows the structure of a typical bill of materials.

Design engineering makes the engineering drawings of the product and describes the components to be manufactured but does not define how the components will be manufactured. Manu-

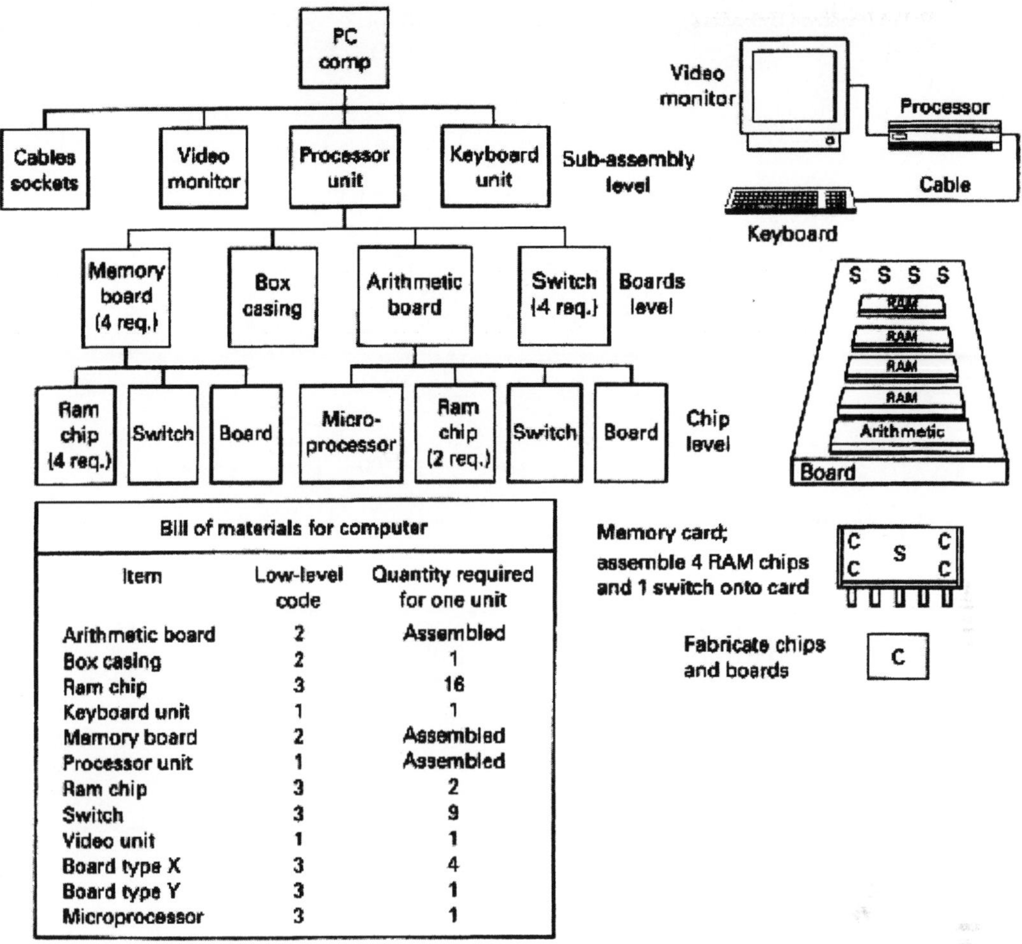

Bill of materials for computer		
Item	Low-level code	Quantity required for one unit
Arithmetic board	2	Assembled
Box casing	2	1
Ram chip	3	16
Keyboard unit	1	1
Memory board	2	Assembled
Processor unit	1	Assembled
Ram chip	3	2
Switch	3	9
Video unit	1	1
Board type X	3	4
Board type Y	3	1
Microprocessor	3	1

Figure 1.12 Example of a product and its bill of materials. (Materials and Processes in Manufacturing, E.P. Degarmo, J.T. Black and R.A. Kohser, 2003. Reprinted with permission of John Wiley & Sons, Inc.)

facturing engineering does this. Based on the design features of a component and the material it is to be made from, the manufacturing engineer defines one or more ***process plans***. A process plan is a sequence of machining operations that take a raw material (workpiece) and transform it into a component usable in the final product. In the case of assembly, the process plan defines the set of operations for assembling components and subassemblies into a final product. The process plan is usually incorporated into a ***routing sheet***. Since the machining processes are selected from among those available in the factory, the routing sheet describes the specific machine type to be used in the performance of the machining. Figure 1.13 shows a routing sheet that describes the process plan for a component, indicating the specific machine types used in each step of the manufacturing process. When multiple process plans are defined for a component, there will be more than one routing sheet.

The bill of materials and the routing sheets of each component, subassembly, and final assembly form the backbone of the information needed in the manufacture of mechanical parts, as well as electronic assemblies. Together they describe what components are needed for building the product

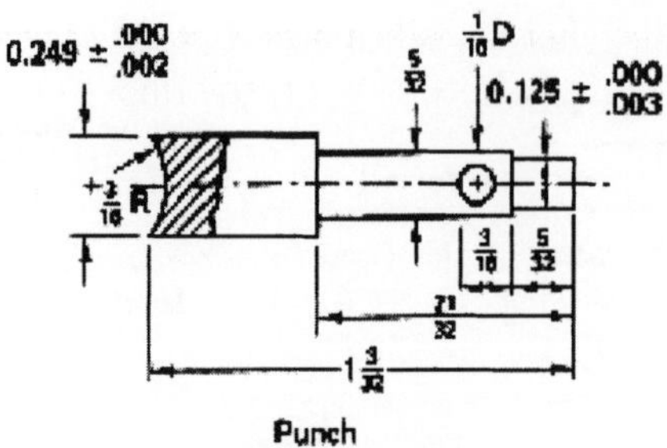

Punch

Matl. – 0.250 dia. AISI 1040
H.T. to 60 R.C. on 0.249 dia.

DARIC INDUSTRIES

ROUTING SHEET

NAME OF PART Punch PART NO. 2

QUANTITY 1,000 MATERIAL SAE 1040

OPERATION NUMBER	DESCRIPTION OF OPERATION	EQUIPMENT OR MACHINES See Fig 41-4	TOOLING
1	Turn, $\frac{5}{32}$, 0.125, and 0.249 diameters	J & L turret lathe	#642 box tool
2	Cut off to 1$\frac{3}{32}$ length	"	#6 cutoff in cross turret
3	Mill $\frac{3}{16}$ radius	#1 Milwaukee	Special jaws in vise $\frac{3}{16}$ form cutter × 4" D
4	Drill $\frac{1}{16}$ hole	Turret Drill	
5	Heat treat. 1,700° F for 30 minutes, oil quench	Atmosphere furnace	
6	Grind (cutting edges)	Surface grinder	$\frac{3}{16}$ end radius on wheel
7	Check hardness	Rockwell tester	

Figure 1.13 Engineering drawing and part routing sheet. (Materials and Processes in Manufacturing, E.P. Degarmo, J.T. Black and R.A. Kohser, 2003. Reprinted with permission of John Wiley & Sons, Inc.)

and how to manufacture each component. When manufacturing involves automatic machining, as in the case of computer-controlled machine tools, the information on how to fabricate the product is incorporated in computer code. This software, known as a ***CNC parts program***, contains the detailed instructions of how to move the axes of the machine tool in order to produce the component. When the processes are not automated, it is left to the machinist to use the part drawings in conjunction with process plan information in order to semiautomatically move the machine tool through the steps to make the part.

1.6.2 Process Industries

The process industries, such as the chemical, pharmaceutical, and food industries, have a common set of product and process information that is used to guide production. In this section we discuss this information in relation to continuous and batch processes.

The ***continuous process design***, illustrated in Figure 1.14, has several production steps, or processes, linked together to provide continuous inflow and outflow of processed materials. Petroleum refineries and most chemical plants are large-scale production facilities that use the continuous process design model. In general, these production systems provide high production rates, but they are dedicated to the production of a narrow range of products. The primary control problem in continuous processes is to maintain the set points of the process. These plants usually have multiple inputs and multiple outputs, and the control of each variable, such as temperature or pressure, is interactive with other variables. These interactive effects lead to the development of quite sophisticated control models, which require in-depth treatment of the process thermodynamic and transport phenomena.

The ***batch process design*** is characterized by a sequence of several unit operations that accept and dispatch material in discrete batches as opposed to continuously. In general, the batch process facility tends to be more flexible in terms of the number of products it can produce, but the throughput rate is lower than that of a continuous operation.

Typical batch process designs include those used in the food and pharmaceutical industries, as well as with fine chemicals. Figure 1.15 shows the batch processing steps for a typical packaged food manufacturing plant. Materials are first processed in a kettle, where they are mixed and precooked. A packaging or canning operation, where the mixed and processed material is filled into containers, usually follows this. Containers are then thermally processed in a retort (thermal

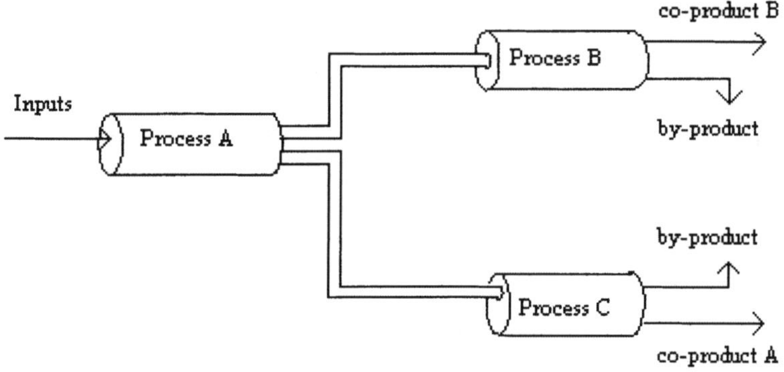

Figure 1.14 Illustration of a continuous process design.

sterilization process) in order to kill microbes that may reside in the sealed container. After sterilization, containers are labeled and packaged for shipment.

Just as in the case of mechanical parts, there are descriptors of the product and its processing requirements for continuous and batch processes. With regard to the product structure, the ***formula*** (product formulation) is the process industry equivalent of the bill of materials. The formula describes process inputs, process parameters, and process outputs. The process input is the quantities of raw materials required to make the product. The process input includes the name of the input and the quantity required for the specified quantity of output. The process output is a specification of the amount of output that is expected to result from one execution of the formula. An example of a formula typical for a food industry product appears in Figure 1.16, which lists the component materials for a 300-gallon batch of chicken broth.

The formula is a subcategory of a more general specification called a ***recipe***. The recipe is the master plan for producing the product. It includes the formula, the equipment requirements, and the detailed procedures of manufacturing. It contains product-specific process information necessary to manufacture a particular batch of product. It contains all of the operating parameters of each piece of equipment used. Therefore, the recipe, which includes the formula, is analogous to the requirements of bill of materials, process plan, routing sheet, and detailed manufacturing steps that are present in the mechanical fabrication industries.

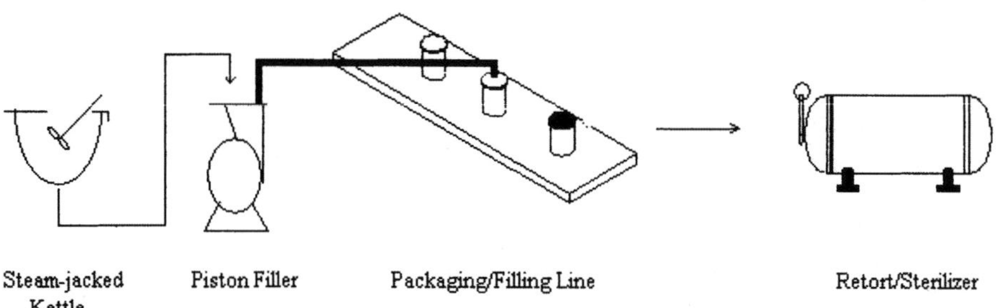

Figure 1.15 Batch processes in packaged food industry.

CODE: 1020 **ITEM:** Chicken Broth **Gallons:** 300 **Yield:** 100 cans

Processing Time: 50 min. **Initial Temp:** 75°F **Cook Temp:** 240-260°F

Percent	ITEM	Units	Amt	
4.2	Chicken Broth, 10%	lbs.	100	
2.0	Chicken Fat	lbs.	50	
1.5	Carrot Puree	lbs.	34	
1.1	Salt	lbs.	26	
0.01	All Spice	oz.	4	
90.99	Water	gal.		As needed to attain 300 gallons.

Figure 1.16 Example of a formula in the food industry.

Figure 1.17 shows a recipe for making cheese sauce. Note that it combines the formula for the specification of materials along with the processing steps, including the settings of critical process parameters such as temperature. The recipe serves as both a bill of materials and a process plan.

1.6.3 SERVICE INDUSTRIES

Service industries are those that do not produce capital or consumer goods. Instead, they supply services, which include financial services (banks, stock brokerages), medical services (hospitals, clinics, doctor practices), travel services (agencies, airlines, hotels, car rental agencies), and so forth. Unlike the manufacturing industries, which have some common structures to them, the service industries present a more unique set of functions and information requirements. In this text, we take most of our examples from the manufacturing sector. However, some end-of-chapter exercises are drawn from examples in the service sector.

Product: Cheese Sauce **Revision Date: 9/09/05**
Material ID Number: 74 **Supersedes: 10/01/03**

Sauce preparation procedure: (50 Gallons)

1. Place water in the Groen Kettle (No. 1).
2. Turn on agitator at highest speed.
3. Add dry skim milk and reconstitute.
4. Add ingredients (corn starch, flour, salt, mustard powder, and paprika powder). Mix at room temperature until all ingredients are dissolved.
5. Add margarine and heat with agitation to 45-50 degree Celsius (115 – 125 degrees Fahrenheit), until all margarine is melted.
6. Add cheddar cheese, mix thoroughly, and add annatto cheese color.
7. Heat with agitation to a temperature of 85 – 90 degrees Celsius (180 – 195 degrees Fahrenheit) and hold this temperature for a minimum of 10 minutes.
8. Take sample to quality control for viscosity measurement.

FORMULA (50 Gallon Batch)

INGREDIENTS	In Lbs	
100 Water	308 (38.5 gal.)	
66 Cheddar Cheese	32.0	
152 Skim Milk Powder	24.0	
40 Margarine	11.0	
90 Corn Starch	8.28	
71 Flour, all purpose	6.2	
70 Salt	5.57	
68 Mustard Powder	0.8	
81 Paprika Powder	0.4	
72 Annatto Cheese Color	0.28	

Figure 1.17 Example of a recipe in the food industry. (Courtesy Rutgers Center for Advanced Food Technology.)

1.7 ABOUT THIS BOOK

This book is about the application of information technology in industrial companies. The emphasis is on teaching principles of information system design through applications in typical industrial environments. These applications touch on all levels of the manufacturing decision hierarchy; however, special attention is paid to the MES level, which is the connecting link between general business functions and the control of physical processes. The reader of this text will come away with general knowledge about information systems design and insight into how manufacturing enterprises are managed.

The first part of this book addresses fundamentals of information system/database design. Chapter 2, "The Relational Database Model," describes the most common database model used in industry today. This chapter includes instruction on the use of structured query language, which is the standard programming language for manipulating data in relational databases.

Chapter 3, "Data Modeling," introduces the entity-relationship model, which is the classic modeling tool for describing data and the semantics of its relationship with other data.

Chapter 4, "Structured Analysis and Functional Architecture Design," addresses the development of models of the business processes that are to be integrated by the information system design. This is the first step in understanding the information requirements of an enterprise.

Chapter 5, "Informational Architecture and logical Database Design," addresses the logical specification of data models using modeling tools that are commonly used in industry. This chapter is an extension of the data modeling material introduced in Chapter 3. The output of the information architecture design process is a data model that can support the functional requirements of the business processes and that can be directly implemented in a relational database.

The second part of this book addresses implementation. We use Microsoft Access and Microsoft Internet Information Server to demonstrate implementation issues because of their ease of use, availability to students, and portability for stand-alone applications. This part emphasizes hands-on exercises for the reader.

Chapter 6, "Design of a User Interface," shows how the functional and informational architectures are combined and implemented in a relational database management system. Here the reader learns how to design and implement forms and reports based on user requirements.

Chapter 7, "Executing an Information System Design Project: A Case Study," presents an example of how an information system design project is executed using the design tools discussed in Chapters 4 and 5.

Chapter 8, "E-business and Web-Enabled Databases," introduces the reader to the role of the World Wide Web as a tool of e-commerce. In particular, the design and implementation of dynamic web pages for database applications will be covered here.

The third part of this book covers advanced topics. With the advent of Internet applications, the direction of IIS design is changing. In particular, there is a closer coupling of software modules that interface the user with the database management system. This has led to new design paradigms and software tools.

Chapter 9, "Unified Modeling Language," presents a relatively new modeling formalism for designing both software and database applications. Unified Modeling Language offers an alternative to the design methods presented in Chapters 4 and 5. It is based on object-oriented design principles. Here, its application to design projects using relational databases, as well as object-oriented databases, is described.

Chapter 10, "Workflow Management Systems," covers a relatively new technology in the use of information systems to control business processes. A workflow is a sequence of tasks that must be performed in a specified order to fulfill the requirements of a particular business process. Applications in e-business are illustrated.

1.8 SUMMARY

Information is an important asset in the management of industrial companies. Today, industrial companies are using information technology to coordinate the activities of the many functions involved in running the enterprise. Traditional categories of functions have included planning functions, shop floor execution functions, and machine and process control functions. These functions are hierarchically linked because the upper level planning functions typically constrain the activities that are allowed at the lower levels. In firms with advanced enterprise integration, these levels are coupled through the network architecture and a set of common databases that hold the most current information and make it available to all functions requiring it.

Chapter 2

The Relational Database Model

2.1 INTRODUCTION

In this chapter, we introduce the basic concepts of the relational database. Early attempts at developing database systems were based on the ***hierarchical*** and ***network*** models. However, these have been largely displaced in industry by the ***relational*** model, which is the foundation of modern database management systems. Therefore, it shall be the relational model that we focus on exclusively in this chapter.

Learning to use a database is greatly enhanced by hands-on experience. For that reason, we provide many examples and encourage the reader to try them on a database management system. We have chosen to illustrate the examples on Microsoft Access because of its wide availability on personal computers and its ease of use. The programming instructions are based on Access 2003.

2.2 THE DATABASE MANAGEMENT SYSTEM (DBMS)

A database can be thought of as a computerized filing cabinet. This filing cabinet electronically stores data defined and "filed" by users within the organization that maintains the database, usually referred to as the "enterprise."

The database system has both hardware and software components. Hardware is the physical storage medium for the data (hard disk, tape, etc.). The software is the medium through which the user accesses the physically stored data. This software is called the ***database management system (DBMS)***. The DBMS allows the user to store, retrieve, and update data without having particular knowledge about the physical location of data or how related data are stored. In effect, the user is provided a *view* of the data that makes it easy for him or her to access and use.

The most widely used DBMSs are relational (RDBMSs). There are three classes of database systems with different levels of complexity and sophistication: enterprise databases, workstation databases, and personal databases. An ***enterprise database*** is a large database that runs on one or more servers and may have client users spread throughout many locations. It must be capable of handling a large quantity of transactions and the execution must be in real-time. For example, a transaction involving an ATM debit should be recorded in the time frame of seconds. It uses sophisticated security measures and can allow different levels of access by client users. Database management systems such as Oracle (Oracle Corporation) and DB2 (IBM) are typically used for these applications. A ***workgroup database*** typically runs on one server and distributes information to several client machines running on the same local area network. The level of transaction processing is much lower than that of an enterprise database, but the DBMS must be capable of handling multiple clients who are independently generating transactions that change the contents of one or more databases running concurrently on the DBMS. Microsoft's SQL Server, which supports client-server architecture, is a popular choice for workgroup applications. A ***personal database*** runs on a single personal computer. It has a lower transaction handling rate and is not designed with sophisticated administrative tools for setting levels of security. The Access DBMS is a good example of a personal database. Access is used in exercises throughout this book because it is suitable for demonstrating all the concepts presented and it is easily available to PC users.

The relational database uses the concepts of *attribute*, *domain*, *relation*, *tuple*, and *primary key*. An ***attribute*** is a name, or label, for a set of data that refer to the same thing. For example, "employee_last_name" could be a label for a set that contains the last names of the employees of an enterprise. A ***domain*** is the smallest unit of data in the database (i.e., it is an individual value). Therefore, if the enterprise has three employees (Joseph Smith, John Doe, and Mary Murphy), then the three values "Smith," "Doe," and "Murphy" are the domain of the attribute "employee_last_name." A ***relation*** refers to a set of related attributes as defined by a user. Consider, for example, the following three attributes: "employee_SS_no," "employee_last_name," and "employee_first_name." If the user defines these three attributes as a relation, then the following data set is defined as a relation:

036-27-5192, Smith, Joseph
357-19-9921, Doe, John
142-36-1529, Murphy, Mary

A ***tuple*** is a set of related data from within a relation. The previous example has three tuples, each consisting of a value for employee_SS_no, employee_last_name, employee_first_name. A ***primary key*** is an attribute in which the domain value is unique (i.e., not repeated in any tuple of the relation). In the previous example, employee_SS_no corresponds to a primary key. The attributes employee_last_name and employee_first_name do not correspond to primary keys because it may be possible at some future time to have two or more employees with the same last or first names. However, social security numbers are uniquely assigned to individuals. In the following sections, we will build on these concepts in describing the relational database and its use.

2.3 THE RELATIONAL DATABASE VIEWED AS A SET OF TABLES

The actual storage of data in a database is fairly complex from a physical point of view. However, the relational model allows the user to view the data in a simple intuitive tabular structure, called a ***table***. This structure allows the user to model the logic of the data manipulation such that it is independent of the physical storage of the data.

A ***table*** is a logical view of related data. The table is defined by the entity set and attributes that it represents. An ***entity*** is a person, place, event, concept, or thing about which information is to be kept in the database tables. For example, a particular employee, a specific material, or a manufacturing process plan for a particular component could qualify as entities. A related group of entities, the information about which is maintained in the same table, is called an ***entity set***. Each entity set has unique characteristics, which is the set of information that is kept on that entity set.

These unique characteristics are called ***attributes***. An attribute consists of an ***attribute name***, and no two attributes of the database may have the same name.

All of these concepts are captured in the structure of a table. Figure 2.1 shows a set of tables that contain information useful to the purchasing department of an enterprise in tracking purchase orders. Figure 2.2 shows a typical purchase order. It consists of a header that identifies the vendor and shows the purchase order number. A description of the individual items that are being ordered

Entity Set: VENDOR

VENDOR_ID	V_NAME	V_STREET	V_CITY	V_STATE	V_ZIP
V110	Jersey Vegetable Co.	2 Main St.	Patterson	NJ	07055
V25	General Provisions	125 Common St.	Boise	ID	44830
V250	Spices Unlimited	25 Salty Lane	East Hampton	NY	10027
V75	Pasta Supply, Inc.	34 Henry St.	Philadelphia	PA	09098

Entity Set: PURCHASE_ORDER

PO_NUMBER	RELEASE_DATE	PO_STATUS	PO_AMT	VENDOR_ID
2591	2/10/06	CLOSED	$4,300.00	V110
2592	2/10/06	OPEN	$505.50	V25
2593	2/11/06	OPEN	$4,000.00	V110
2594	2/12/06	OPEN	$3,280.00	V250
2595	2/15/06	OPEN	$500.00	V250
2596		HOLD	$1,000.00	V75

Entity Set: PO_DETAIL

PO_NUMBER	PO_LINE_IT	MATERIAL_ID	UNITS	QUANTITY	BALANCE	PROMISED_DEL_	UNIT_COST	STATUS
2591	1	RM201	LB	1000.0	0	2/20/06	$2.00	CLOSED
2591	2	RM202	LB	1000.0	0	2/20/06	$2.00	CLOSED
2591	3	RM205	LB	300.0	0	2/20/06	$1.00	CLOSED
2592	1	RM805	GAL	800.5	0	2/25/06	$0.50	CLOSED
2592	2	RM810	GAL	210.5	210	3/10/06	$0.50	OPEN
2594	1	RM310	LB	4000.0	4000	3/12/06	$0.50	OPEN
2594	2	RM311	LB	2000.0	2000	3/12/06	$0.25	OPEN
2594	3	RM318	LB	2000.0	2000	3/12/06	$0.25	OPEN
2594	4	RM340	LB	560.0	560	3/20/06	$0.50	OPEN
2593	1	RM210	LB	1000.0	500	2/25/06	$2.00	OPEN
2593	2	RM211	LB	2000.0	2000	3/10/06	$1.00	OPEN
2595	1	RM305	LB	400.0	400	2/27/06	$0.50	OPEN
2595	2	RM308	LB	1200.0	1200	2/27/06	$0.25	OPEN
2596	1	RM502	LB	5000.0	5000		$0.20	OPEN

Figure 2.1 Tables for entity sets VENDOR, PURCHASE_ORDER, and PO_DETAIL.

UNIVERSITY FOOD INCORPORATED
1776 NEW ENGLAND AVENUE
PISCATAWAY, NJ 08855

VENDOR: Spices Unlimited
25 Salty Lane
East Hampton, NY 10027

PO Number: 2595
Date: 2/15/06
Amount: $500.00

Line Item	Material_ID	Unit Price	Quantity	Del. Date
1	RM305	$0.50/lb	400 lbs	2/27/06
2	RM308	0.25/lb	1200 lbs	2/27/06

Figure 2.2 Information on a purchase order.

is shown in the body of the purchase order. In a later chapter we shall show that the purchase order presented in Figure 2.2 is, in a practical sense, deficient in many ways. However, it will be useful for illustrating several points in this chapter.

The tables in Figure 2.1 are used to store the information on the purchase order. Consider the table "VENDOR" shown in Figure 2.1. This table maintains information on the suppliers of materials to a company, called the company's vendors. The entities are the individual vendors themselves. When grouped in a table as a related set of entities (the entity set), the table is titled "VENDOR." The title or name given to the group of entities in a table is synonymous with the concept of an entity set. The unique characteristics, or attributes, of the entities have the attribute names: VENDOR_ID, V_NAME, V_STREET, V_CITY, V_STATE, and V_ZIP. The meaning of these attributes is obvious. No two attributes have the same name. The same interpretation can be given for the tables PURCHASE_ORDER and PO_DETAIL.

A table is a ***relation*** because it contains a set of attributes about a group of related entities, the entity set. As we build our knowledge of database models, you will see that a database contains many such tables and that these tables are related to one another. For example, the table VENDOR is related to the table PURCHASE_ORDER by the attribute that is common to both tables (i.e., the VENDOR_ID). However, the term *relational* in relational database does not refer to the relationships among tables, but to the fact that the entities in the entity set of a table are related by a common set of attributes.

The attribute columns of the entity sets PURCHASE_ORDER and PO_DETAIL are a tabular representation of the data that appear on the physical document called a purchase order. The reader will note that PURCHASE_ORDER tabulates data in the header and PO_DETAIL is the body of the purchase order.

Tables are constructed of rows and columns, and we usually use these terms in referring to a table structure. A ***row*** represents a single entity, or instance of the entity set. A row is sometimes referred to as a ***record***, which is a term carried over from a period when computer information was maintained in a file structure. A ***column*** represents the attributes of the entity set. Sometimes columns are referred to as ***fields***. The VENDOR table contains four rows (entities or records) and six columns (attributes or fields).

Relational database management systems support a variety of data types. Typical data types are *numeric*, *character* or *text*, *date*, and *currency*.

Numeric data types are classified as integer, floating point, or decimal. Integer types are either SMALLINT (also referred to as Integer in Microsoft Access) or INTEGER (also referred to as Long Integer in Microsoft Assess). Floating point includes Single precision (4 bytes of data storage) and Double precision (8 bytes of data storage). Single and Double data types use an internal storage format that can handle very large and very small numbers, but is somewhat imprecise. The Decimal data type is a formatted data type in which the DBMS stores the number, including fractional parts, as an integer with up to 12 bytes of data storage. It separately stores a value for the number of significant digits to the right of the decimal point. The Decimal data type insures absolute fractional precision up to 28 decimal places. The numerical ranges for Numeric data types are listed in Table 6C.4 of Appendix 6C.

Character or **text data** are represented as an alphanumeric string in the range of 1 to 254 characters. The length of the string is indicated in parentheses following the type designation, as in CHAR(10). All of the attributes in the table VENDOR could be character data types. Although V_ZIP (zip code) could be a SMALLINT, it is not a number that would usually be subject to a mathematical calculation. Therefore, from that perspective, there is no need to assign it as a number type. In Access, CHAR data type is also referred to as TEXT.

The **DATE data type** tells the DBMS to interpret the field as a date. The attribute RELEASE_DATE in the table PURCHASE_ORDER and PROMISED_DEL_DATE in the table PO_DETAIL are examples of the DATE data type. This data type is stored as a sequential number (Julian date) within the memory of the computer, thus allowing computations to be made on the DATE data type. For example, it may be of interest to compute the number of days from the current date to the promised delivery date for a certain material in order to estimate the time over which the current inventory of the material must satisfy production needs.

The **CURRENCY data type** tells the DBMS that the numerical value is a monetary value. The attribute PO_AMT in PURCHASE_ORDER and UNIT_COST in PO_DETAIL are examples of the CURRENCY data type.

When the attribute fields of a table are defined, the user also defines the data type. When values are entered into the table, the DBMS will not allow the user to enter data into an attribute field that does not conform to the data type of that attribute. This will be clarified later when we create tables and populate them with data.

The information that appears on a purchase order such as that shown in Figure 2.2 has been divided among three tables in Figure 2.1. It is reasonable at this point to ask why all the data are not placed in one table, for example, a table named PURCHASE_ORDER_DATA. For the time being, the reader should accept the three-table model as a proper representation of the data for a relational database. A more complete justification will be given in the following chapters.

2.4 KEY ATTRIBUTES AND LINKING TABLES

To maintain the integrity of a database, certain attributes of each table are designated as ***key attributes***. The two most important key attributes are the ***primary key*** and the ***foreign key*** attributes.

In every table, one (or a combination) of the attribute fields must be used to uniquely identify each row (entity). This attribute is called the ***primary key*** attribute. Its value in each row is only allowed once—that is, the value can exist in one and only one row of the table. For example, in the VENDOR table, the VENDOR_ID is unique. The VENDOR_ID attribute, assigned to the vendor by the database user, will always refer to one and only one vendor. It may occur that two vendors will have the same names, causing confusion when the computer must differentiate between them. No such confusion exists when a unique number or alphanumeric code is assigned to each vendor in order to differentiate between them. The primary key for the PURCHASE_ORDER table is the PO_NUMBER. Only one purchase order number is assigned to a purchase order, and that number is unique to that purchase order.

At times the row of a table may not be distinguished by a single attribute value. For example, in the PO_DETAIL table, the PO_NUMBER is repeated for each line item on the purchase order. However, each purchase order contains only one line item 1, one line item 2, and so forth. Therefore, the combination of the attributes PO_NUMBER and PO_LINE_ITEM represents a unique combination for each row. In such a case, one can declare a ***composite primary key attribute***.

When a table is created, it is necessary to declare the attribute(s) that will be the primary key. When data are retrieved from the table, the primary key can be used to ensure that the retrieved data set is unique.

Previously we pointed out that related tables are linked to each other by having attribute fields in common. For example, the table PURCHASE_ORDER is related to the table PO_DETAIL by the attribute PO_NUMBER. The attribute field that relates one or more entities in one table to the primary key attribute in another table is called a ***foreign key***. In the table PO_DETAIL, the attribute

PO_NUMBER serves as a foreign key. Unlike a primary key, a foreign key may have duplicate instances. However, the foreign key in one table must be related to a primary key in another table. Otherwise, the reference between tables could not be guaranteed to be unique. This is referred to as ***referential integrity*** and its importance shall become clear in later sections of this chapter. In a similar manner, VENDOR_ID is a foreign key in the PURCHASE_ORDER table, which provides a reference to the primary key VENDOR_ID in the VENDOR table.

When a column is designated as a key attribute, any row of the table must have an entry in that column to ensure the integrity of the table and its references. This means that no "null" values are allowed. A *null* value refers to the absence of a value; it does not refer to a value of zero. For example, the RELEASE_DATE of PO_NUMBER 2596 in the PURCHASE_ORDER table is a null value. This could occur, for example, because the purchase order has been made up but the purchasing agent has not yet released it. Null values may be allowed in nonkey attribute columns, but null values are not allowed to exist in key attribute columns. Alternatively, the database designer may enforce a "NOT NULL" requirement on the database user by specifying to the DBMS that null values are not allowed for a specific nonkey attribute. In any case, the DBMS will not allow a null value to be entered into a NOT NULL column of a table. Any attempt to do so will result in an error message. The user declares which attributes are NOT NULL when the table is created.

2.5 STRUCTURED QUERY LANGUAGE (SQL)

The relational database community has defined a standard language for manipulating data in a database called ***Structured Query Language*** **(SQL)**. The American National Standards Institute (ANSI) has standardized SQL. SQL describes a basic set of keywords and their meaning. Although the standard is largely followed, not all commercial RDBMSs use all the keywords of the standard, and many RDBMSs have extended the standard by adding keywords.

SQL is a nonprocedural language. In a procedural language, the programmer must define the steps by which a program executes. Examples of procedural languages include Basic, Pascal, and C. In a nonprocedural language, you are not concerned with the details of how the work gets done, you only have to define what you want to have done. In SQL, the RDBMS decides how to get it done. The "what" is defined by the programmer using the SQL command set.

There are about 30 standard instructions in the basic SQL command set. These instructions allow the user to perform operations for the following purposes: (1) to create a database and its table structure, (2) to manage the data in the database tables, and (3) to summarize the data into useful information for decision making. In the sections that follow, we describe some of the commands and their meaning. This will be followed by the syntax as it is used in the Access RDBMS and instructions on how to execute examples.

2.5.1 SQL: CREATING THE DATABASE AND TABLE STRUCTURE

The CREATE command keyword is used for creating databases and tables. The ANSI syntax for database creation is

```
CREATE DATABASE <database name>;,
```

which is entered at the command prompt in RDBMSs complying with the ANSI standard for this command. When this command is executed, a database with the given name is created. The programmer must then populate the database with its table structure.

The Access RDBMS does not support the CREATE DATABASE command. Instead, it provides a Windows menu-driven route to establishing the database. The following exercise assumes the reader is using Microsoft Access 2003.

Access Exercise 2.1:

We wish to establish a database for the tables of Figure 2.1. We will use the database title ***MATERIAL_MANAGER***. Establish the database by following these steps:

1) From the start menu, launch Microsoft Office Access 2003.
2) From the ***File*** Menu, select ***New***. From the New File selection panel, select ***Blank Database***. The ***File New Database*** dialog window will open.
3) Type in the database name: ***MATERIAL_MANAGER***. Click the ***CREATE*** command button and a database by that name is established.

When the database is created, the Access RDBMS puts the user into the *database window* shown in Figure 2.3. Access provides a highly intuitive menu-driven environment for creating tables and manipulating data. However, it is our purpose in this chapter to illustrate the use of SQL commands, and we shall bypass these utilities in what follows.

The following conventions will be used to describe SQL commands. The command will be written in capital letters. Commands are usually followed by arguments. Arguments are placed within angle brackets, < >. If an argument is optional, it will be bounded by square brackets, [< >].

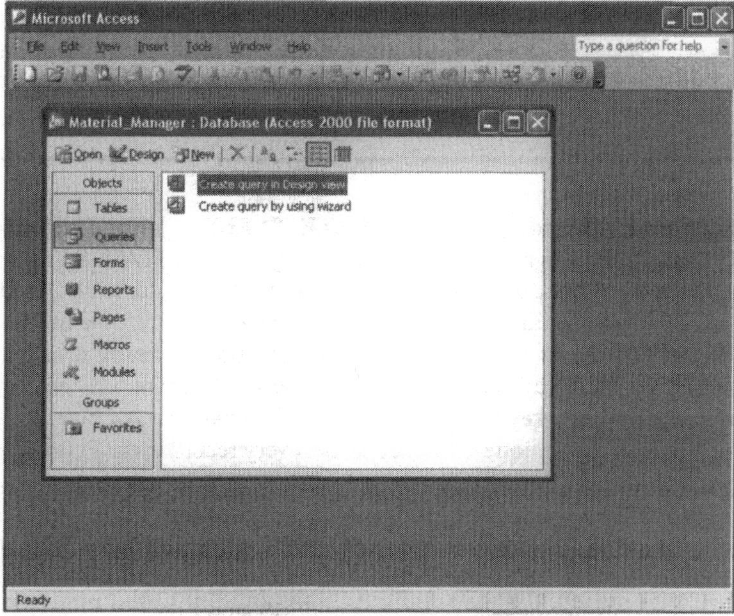

Figure 2.3 Access database window.

The syntax for creating a table within a database uses the CREATE keyword as follows:

```
CREATE TABLE <table name>
(<[attribute1 name]> <data type>,
<[attribute2 name]> <data type>,
...           ...        ...
<[attributeN name]> <data type>);
```

When the table is created, the columns will be ordered by the attribute name order given in the **CREATE TABLE** command. We will illustrate by establishing the table for PURCHASE_ORDER.

Access Exercise 2.2:

In order to establish a database table using the SQL CREATE command, follow these steps:

1) From the database window for ***MATERIAL_MANAGER***, click on ***Queries*** and then double click on ***Create query in Design View.*** The New Query window opens.
2) The Show Table window will now be open. Close the window by clicking on ***Close.***
3) From the Toolbar, click on the ***SQL*** menu item. This will open a Query window that will allow you to enter a query in SQL.
4) Delete any existing query in the query window and enter the following query:

```
CREATE TABLE PURCHASE_ORDER
([PO_NUMBER]      SMALLINT,
[RELEASE_DATE]    DATE,
[PO_STATUS]       CHAR(6),
[PO_AMT]          CURRENCY,
[VENDOR_ID]       CHAR(5),
CONSTRAINT [INDEX1] PRIMARY KEY (PO_NUMBER));
```

5) Click the run button on the Toolbar **!** to execute the query. The table is created.
6) Close the query window (do not save the query) and click the ***Table*** tab on the database window.
7) Confirm that a table with the name ***PURCHASE_ORDER*** now exists. The empty table can be opened by double clicking on the table name.

There are some points to note about the **CREATE TABLE** command statement in Access Exercise 2.2. By limiting the length of the character strings for PO_STATUS and VENDOR_ID, we save computer memory storage space. If we did not limit the character length, the DBMS would have defaulted the field to 50 characters. The largest text field size is 255 characters. Note the use of the **CONSTRAINT** clause in the last line of the query. The constraint clause is used to add constraints on any of the attribute fields of the table. In Access Exercise 2.2, the constraint clause is used to constrain the primary key to PO_NUMBER. Therefore, the attribute PO_NUMBER is constrained to be **Not Null** and **Unique**. This means that every record in the PURCHASE_ORDER table must have a value for PO_NUMBER and that value cannot be repeated.

Within the constraint clause, an index object has also been defined. When a table is established, it is a good idea to create an index on one or more fields that uniquely identify a record. It is not necessary to create indexes for tables, but it speeds up the processing times because Access uses the index to process records in a defined order. Each DBMS stores data in a manner unique to the

DBMS. When an index is created, an auxiliary data structure is created that allows the DBMS search algorithm to locate a particular record in a table faster than would be possible without the index. This can be helpful in databases containing many large tables.

Readers who are using Microsoft Access at this time should follow Access Exercise 2.2, presented earlier, and establish the tables for VENDOR and PO_DETAIL. The appropriate command structure is as follows:

```
CREATE TABLE VENDOR
([VENDOR_ID]          CHAR(5),
[V_NAME]              CHAR(20),
[V_STREET]            CHAR(20),
[V_CITY]              CHAR(20),
[V_STATE]             CHAR(2),
[V_ZIP]               CHAR(5),
CONSTRAINT [INDEX2] PRIMARY KEY (VENDOR_ID));

CREATE TABLE PO_DETAIL
([PO_NUMBER]          SMALLINT,
[PO_LINE_ITEM]        SMALLINT,
[MATERIAL_ID]         CHAR(10),
[UNITS]               CHAR(4),
[QUANTITY]            SINGLE,
[BALANCE]             SINGLE,
[PROMISED_DEL_DATE] DATE,
[UNIT_COST]           CURRENCY,
[STATUS]              CHAR(6),
CONSTRAINT [INDEX3]
PRIMARY KEY (PO_NUMBER, PO_LINE_ITEM));
```

2.5.2 SQL: Managing the Data in the Database Table

Once created, the tables must be populated in order to be useful. In this section we shall discuss and illustrate several keyword commands for populating and manipulating data in the database; in particular, the keywords **INSERT**, **SELECT**, **UPDATE**, and **DELETE**.

2.5.2.1 INSERT Keyword

The tables that were created in Section 2.5.1 are now empty. Records can be placed into these tables using the SQL **INSERT** command. The syntax is as follows:

```
INSERT INTO <table name>
([<attribute1 name>], <[attribute2 name]>,...)
VALUES (<value1>, <value2>,...);
```

When all attribute columns of a table are being populated, it is unnecessary to define the attribute name. This is needed only when some specific attribute(s) of the table is being inserted.

Access Exercise 2.3:

Return to the query window. Follow the steps of Access Exercise 2.2 if needed.

1) Delete any existing query in the query window and enter the following query:

INSERT INTO PURCHASE_ORDER
VALUES (2591, "02/10/06", "CLOSED", 4300.00, "V110");

2) Click the run button on the Toolbar [!] and ***YES*** on the warning message to execute the query. The first record is now inserted into the table. You can view this record by closing the query window, returning to the database window, selecting the table name from the ***table objects*** window and clicking on ***Open***.
3) Following the same procedure, enter the remaining data for the ***PURCHASE_ORDER*** table.

Note, in Access Exercise 2.3, that TEXT and DATE data types are enclosed in quotations, whereas numerical and CURRENCY data types are not. Also, only one record at a time can be inserted into a table.

In addition to the use of SQL, Access also provides a highly intuitive route to establishing a database and populating tables. Instead of using SQL and the INSERT keyword, the user can open a table and directly insert data into the rows and columns of the table. Each new row of data requires a unique primary key in order for the data to be accepted by Access. Readers who are using Microsoft Access at this time should open the VENDOR table and insert the data from Figure 2.1 into the appropriate cells of the table. Similarly, the reader should populate the PO_DETAIL table.

2.5.2.2 SELECT Keyword

The workhorse in retrieving and manipulating data in a table is the **SELECT** keyword, which can be used in many command variations. In this section we describe its use in many examples. The syntax of a **SELECT** command is as follows:

```
SELECT [DISTINCT] <attributes/*>
FROM <table name>
WHERE <condition>
ORDER BY <attribute name> ASC/DESC;
```

The functions of the various clauses shall be illustrated in the following examples.

When the **SELECT** command is used, the minimum requirement is to identify the attributes to be selected and the table from which they are selected. The wildcard * identifies all attribute fields of the table. Thus, the command to retrieve the entire contents of the table **PURCHASE_ORDER** would be as follows:

```
SELECT *
FROM PURCHASE_ORDER;
```

On execution of this command, the entire contents of the table **PURCHASE_ORDER** is retrieved and displayed on the screen. Users can be more specific about the columns they wish to retrieve by specifying the column's attribute name in the SELECT clause.

Access Exercise 2.4:

1) Open the Query window and retrieve the following data from the table ***PURCHASE_ ORDER***

```
SELECT PO_NUMBER, PO_STATUS, PO_AMT
FROM PURCHASE_ORDER;
```

2) After execution, the table of retrieved data should appear as shown in Figure 2.4.

PO NUMBER	PO STATUS	PO AMT
2591	CLOSED	$4,300.00
2592	OPEN	$505.50
2593	OPEN	$4,000.00
2594	OPEN	$3,280.00
2595	OPEN	$500.00
2596	HOLD	$1,000.00

Figure 2.4 Result of Access Exercise 2.4.

The record of retrieved data in Figure 2.4 is called a "recordset." A ***recordset*** is a view of the data from one or more tables, selected and sorted as specified by the query. Sometimes it is desirable to arrange the retrieved tabular information in a particular order. This can be done by using the **ORDER BY** clause, followed by the attribute name of the column on which you wish to impose the order. If the column is a numeric, date, or currency data type, the order will be determined by the magnitude of the number. The default is to use ascending order. For descending order, the keyword **DESC** is used following the attribute name. If the column is a text data type, the order is determined by ASCII equivalent. The computer stores text data by its numeric equivalent in ASCII. Thus, the letter *A* is equivalent to the decimal value 65 in ASCII, *B* is 66, *C* is 67, and so forth. The lower case letter *a* has a value of 97, *b* is 98, *c* is 99, and so forth. Therefore, in sorting alphabetic strings is ascending order, *A* precedes *B*, and *B* precedes *a*, which precedes *b*, and so on.

Access Exercise 2.5:

1) Open the Query window and retrieve the following data from the table ***PURCHASE_ ORDER***

```
SELECT PO_NUMBER, PO_STATUS, PO_AMT
FROM PURCHASE_ORDER
ORDER BY PO_AMT;
```

2) After execution, the table of retrieved data should appear as shown in Figure 2.5.

PO NUMBER	PO STATUS	PO AMT
2595	OPEN	$500.00
2592	OPEN	$505.50
2596	HOLD	$1,000.00
2594	OPEN	$3,280.00
2593	OPEN	$4,000.00
2591	CLOSED	$4,300.00

Figure 2.5 Result of Access Exercise 2.5.

Note that, by default, the order is ascending. Also, all other columns have been reordered to correspond to the order of PO_AMT. The referential integrity of the table is conserved.

Sometimes it is desirable to have an ordering within an ordering. For example, it may be desirable to order purchase orders by VENDOR_ID and, within each vendor group, to order the PO_AMT. The ORDER BY clause has the flexibility to provide multiple sorting of the data in the order in which the attribute names of columns are listed. This is illustrated in Access Exercise 2.6.

Access Exercise 2.6:

1) Open the Query window and retrieve the following data from the table ***PURCHASE_ORDER***

```
SELECT VENDOR_ID,PO_AMT,PO_NUMBER, PO_STATUS
FROM PURCHASE_ORDER
ORDER BY VENDOR_ID, PO_AMT DESC;
```

2) After execution, the table of retrieved data should appear as shown in Figure 2.6.

VENDOR_ID	PO_AMT	PO_NUMBER	PO_STATUS
V110	$4,300.00	2591	CLOSED
V110	$4,000.00	2593	OPEN
V25	$505.50	2592	OPEN
V250	$3,280.00	2594	OPEN
V250	$500.00	2595	OPEN
V75	$1,000.00	2596	HOLD

Figure 2.6 Results of Access Exercise 2.6.

Note the way VENDOR_ID has been sorted. VENDOR_ID is a text field and is sorted by ASCII equivalent. The first entry in the field, "V," is identical for all VENDOR_IDs. The next entry in the field is to be sorted as the ASCII equivalent of a numeric symbol. The ASCII equivalent of 1 is 49, of 2 is 50, and of 7 is 55. Therefore, the second entry of the field has determined that V75 will follow V250 and V110 even though 75 is a number smaller than 250 or 110. Remember that we are sorting a text data type, not a number. At the fourth entry of the field, it is determined that V25 shall precede V250.

Note also that the order of PO_AMT is descending. This occurs because we used the keyword DESC following the attribute name of that column.

The **DISTINCT** keyword in the SELECT clause allows the user to sort a single column of a table and to return a list of the unique entries in that column. In effect, the resulting list does not contain redundancies. For example, the following command will return the unique VENDOR_IDs in the PURCHASE_ORDERS table:

```
SELECT DISTINCT VENDOR_ID
FROM PURCHASE_ORDER;
```

The command returns:

VENDOR ID
V110
V25
V250
V75

It is often of interest to select out specific rows from a table based on a criterion. This is done using the **WHERE** clause of the SELECT command. For example, you may want to retrieve information that is pertinent to one and only one vendor. A SELECT command that accomplishes this is as follows:

```
SELECT *
FROM PURCHASE_ORDER
WHERE VENDOR_ID="V110";
```

The command returns:

PO NUMBER	RELEASE DATE	PO STATUS	PO AMT	VENDOR ID
2591	2/10/06	CLOSED	$4,300.00	V110
2593	2/11/06	OPEN	$4,000.00	V110

Note the use of the equal (=) sign to indicate the row selection criteria. The use of ***comparison operators*** is a common method of indicating the selection criteria. The comparison operators are as follows:

= equal to
!= not equal to
< less than
<= less than or equal to
> greater than
>= greater than or equal to

Access Exercise 2.7:

1) Open the Query window and retrieve the following data from the table ***PURCHASE_ORDER***

```
SELECT VENDOR_ID,PO_NUMBER, PO_AMT
FROM PURCHASE_ORDER
WHERE PO_AMT > 1000
ORDER BY PO_AMT;
```

2) After execution, the table of retrieved data should appear as shown in Figure 2.7.

VENDOR ID	PO NUMBER	PO AMT
V250	2594	$3,280.00
V110	2593	$4,000.00
V110	2591	$4,300.00

Figure 2.7 Results of Access Exercise 2.7.

There may be more than one selection criterion for a retrieval. This can be handled by extending the WHERE clause using logical operators. The ***logical operators*** are AND, OR, and NOT. For example, suppose we wish to look at all open orders that are above $500. This would require the logical AND, since we want the orders that are open AND greater than $500. The WHERE clause is extended using logical operators as follows:

```
SELECT *
FROM PURCHASE_ORDER
WHERE PO_STATUS="OPEN"
AND PO_AMT>500;
```

The command returns:

PO_NUMBER	RELEASE_DATE	PO_STATUS	PO_AMT	VENDOR_ID
2592	2/10/06	OPEN	$505.00	V25
2593	2/11/06	OPEN	$4,000.00	V110
2594	2/12/06	OPEN	$3,280.00	V250

Access does not support != (not equal to). Instead, the NOT operator is used to return the complementary set of an operation. For example,

```
SELECT *
FROM PURCHASE_ORDER
WHERE NOT PO_STATUS = "OPEN";
```

This will return all purchase orders that are not open as follows:

PO_NUMBER	RELEASE_DATE	PO_STATUS	PO_AMT	VENDOR_ID
2591	2/10/2006	CLOSED	$4,300.00	V110
2596		HOLD	$1,000.00	V75

With creative use of the logical operators it is possible to sort data based on a variety of criteria. Remember, text fields are ordered based on their ASCII values. This means that a sorting by text fields is possible. Furthermore, the WHERE clause can be a compound statement of logical operators. These points are illustrated in the following example:

Access Exercise 2.8:

1) Open the Query window and retrieve the following data from the table ***PURCHASE_ORDER***

```
SELECT *
FROM PURCHASE_ORDER
WHERE PO_STATUS="OPEN"
AND (VENDOR_ID>"V150" AND PO_AMT>500)
ORDER BY PO_AMT;
```

2) After execution, the retrieved data should appear as shown in Figure 2.8.

PO NUMBER	RELEASE DATE	PO STATUS	PO AMT	VENDOR ID
2592	2/10/06	OPEN	$505.50	V25
2594	2/12/06	OPEN	$3280.00	V250

Figure 2.8 Results of Access Exercise 2.8.

Note that the compound logical clause must be enclosed in rounded brackets (parentheses). The first condition of selection eliminates all orders that are not open. The next condition eliminates VENDOR_ID "V100." Recall that this is done by evaluating the ASCII values of a text field. Therefore, "V25" is retrieved because ASCII "2" > ASCII "1" where the second entry of the field is evaluated. Finally, the third condition eliminates the PO_AMTs less than $500. The reader should examine this result closely against the PURCHASE_ORDER table shown in Figure 2.1.

Another useful operator is the **BETWEEN** keyword. This allows the user to specify a range of values over which the data will be retrieved. It can be used in sorting numeric, date, currency, and text fields. Consider the following command, which sorts records by numerical and date data type fields:

```
SELECT *
FROM PURCHASE_ORDER
WHERE ((PO_AMT BETWEEN 505 AND 4000)
AND (RELEASE_DATE BETWEEN #2/11/06# AND #2/13/06#));
```

This query returns the following result:

PO NUMBER	RELEASE DATE	PO STATUS	PO AMT	VENDOR ID
2593	2/11/06	OPEN	$4,000.00	V110
2594	2/12/06	OPEN	$3,280.00	V250

Note that the BETWEEN operator is inclusive—that is, it includes the values at the boundaries of the query range. Also, the complement of BETWEEN is NOT BETWEEN, which will return the complement set from the table. Also note the use of the pound sign (#) around the DATE data type values. In Access SELECT command statements, the # sign is used to indicate a DATE data type.

SQL also provides a predicate operator that allows a search to be done on strings that are a partial match to the predicate. This is done using the **LIKE** keyword. There may be times when a user is not quite sure of the name or spelling in the attribute field of an entity. The LIKE operator allows the user to search for a match with partial information. For example, suppose the user wishes to identify a vendor, but has forgotten the name of the company. However, she or he knows that the name begins with the word "spice." It's "spice" something. Using the LIKE operator and the wildcard (*), the following query will return the desired information:

```
SELECT *
FROM VENDOR
WHERE V_NAME LIKE "SPICE*";
```

This query returns the following:

VENDOR ID	V NAME	V STREET	V CITY	V STATE	V ZIP
V250	Spices Unlimited	25 Salty Lane	East Hampton	NY	10027

The wildcard (*) in the WHERE clause indicates that the user does not care what follows the *e* in "spice." Hence, with partial information, it is possible to locate a record in the database.

So far we have searched for records using attribute fields in which data existed. It is also possible to search for records having null values. This is done using the **NULL** keyword. For example, if we wanted to know what purchase orders have been made up but not yet released, we could use the following query:

```
SELECT *
FROM PURCHASE_ORDER
WHERE RELEASE_DATE IS NULL;
```

This query returns the following:

PO NUMBER	RELEASE DATE	PO STATUS	PO AMT	VENDOR ID
2596		HOLD	$1,000.00	V75

The SELECT command has been illustrated in some of its forms in this section. The reader will become more familiar with its use by attempting the exercises at the end of this chapter.

2.5.2.3 UPDATE Keyword

The **UPDATE** keyword allows the user to replace existing values in a table with new values. For example, if a vendor were to change the promised delivery date of an order, it would be desirable to reflect this change immediately in the appropriate table of the database.

The syntax of the UPDATE command is as follows:

```
UPDATE <table name>
SET <attribute name> = <value/expression> [...]
[WHERE <condition>];
```

The updated field value may be a new value or any admissible expression, typically an arithmetic expression. So, for example, if the attribute is the price of a product and the company raises all its prices by 2%, the SET clause could read as follows:

SET PRICE = PRICE * 1.02

Access Exercise 2.9:

Vendor V250 has just informed us that all items on purchase order number 2594 will not be delivered until 03/22/06.

1) Open the Query window and perform the following data update in the table ***PO_DETAIL***.

```
UPDATE PO_DETAIL
SET    PROMISED_DEL_DATE = "03/22/06"
WHERE  PO_NUMBER=2594;
```

2) After execution, open the table ***PO_DETAIL*** and confirm that the changes have occurred.

2.5.2.4 DELETE Keyword

The **DELETE** keyword allows the user to remove one or more rows from a table (i.e., delete one or more records). The syntax for the DELETE command is as follows:

```
DELETE FROM <table name>
WHERE <condition>;
```

So, for example, if it was determined that purchase order number 2596 would not be released because the requisitioning department reconsidered the purchase, we may want to delete that record from the database. This can be accomplished with the following commands:

```
DELETE FROM PO_DETAIL
WHERE PO_NUMBER = 2596;

DELETE FROM PURCHASE_ORDER
WHERE PO_NUMBER = 2596;
```

Note that the DELETE command will only allow one table to have its row(s) deleted at a time. It is incumbent on the user to specify all the tables involved in a DELETE operation. Note also that the order in which deletions are made in tables is important. To preserve relational integrity, the last table on which the deletion is performed must be the table that contains the primary key attribute. Relational integrity requires that any instance of a foreign key in one table (called the child table) be related to an instance of a primary key in another table (called a parent table). If the order in which the preceding deletions were performed began with the table PURCHASE_ORDER, there would have been instances of the foreign key PO_NUMBER in the child table PO_DETAIL that no longer existed in the parent table. This condition would violate relational integrity. This point will be discussed further in later chapters.

2.5.3 SQL: Converting Data into Information

The purpose of having data is to be able to summarize them in ways that support decision making. In Chapter 6, we shall be describing the use of ***forms*** and ***reports*** as ways of presenting such information in an interactive manner. Often, making the data useful involves making computations on the fields within the tables. If the computation is particularly complex, it may be necessary to write a separate program in a procedural language that takes data from the database and computes the desired result. However, some basic arithmetic and logical functions are built into SQL that enable relatively simple calculations. In addition, compiling information for summary purposes often involves retrieving information from more than one table at the same time. In our examples thus far, we have shown queries that access one table at a time. In this section, we shall look into these issues of accessing multiple tables and performing computations.

2.5.3.1 Aggregate Functions in SQL

The ***aggregate functions*** allow the user to specify a summary mathematical operation with a keyword. The basic aggregate functions of the SQL are ***AVG***, ***SUM***, ***MIN***, ***MAX***, and ***COUNT***. The syntax of the aggregate functions is as follows:

```
SELECT AGGREGATE FUNCTION ([DISTINCT] <attribute name>)
FROM <table name>
WHERE <condition>
GROUP BY <attribute name> [HAVING <condition>];
```

The following examples illustrate the use of the aggregate functions. Suppose we are interested in the average dollar value of open purchase orders.

```
SELECT AVG(PO_AMT)
FROM PURCHASE_ORDER
WHERE PO_STATUS="OPEN";
```

which yields the following result:

Expr1000
$2,071.38

Exprl000 is a default sequential number assigned by Access to indicate the requested computation and refers to the expression AVG(PO_AMT), which is the argument of the SELECT clause. SQL assigns a default column label to the attribute if the programmer does not specify a label. However, a label can be specified by the programmer using the AS clause. Consider the use of the **AS** clause in the following example. The command

```
SELECT MIN(PO_AMT) AS MIN_AMT
FROM PURCHASE_ORDER;
```

yields

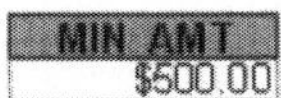

MIN_AMT
$500.00

The result is the minimum entry in the PO_AMT column. The column label "MIN_AMT" is defined in the AS clause.

The **COUNT** keyword is used to return a count of the number of rows in a column having entries that satisfy the WHERE clause of the command:

```
SELECT COUNT(PO_STATUS)
FROM PURCHASE_ORDER
WHERE PO_STATUS="OPEN";
```

This command returns a count of the number of purchase orders in the database that are still open.

Expr1000
4

Access Exercise 2.10:

Try the following combination search.

1) Open the Query window and perform the following query from the table ***PURCHASE_ORDER***

```
SELECT MIN(PO_AMT), MAX(PO_AMT), SUM(PO_AMT)
FROM   PURCHASE_ORDER
WHERE  PO_STATUS="OPEN";
```

2) The result should appear as in Figure 2.9, below.

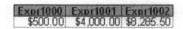

Expr1000	Expr1001	Expr1002
$500.00	$4,000.00	$8,285.50

Figure 2.9 Result of Access Exercise 2.10.

It is also possible to embed an arithmetic operation in an aggregate function SELECT clause or to use a logical operator for multiple criteria in the WHERE clause of an aggregate function. For example, suppose the Spices Unlimited Company informs us that a 10% freight (shipment) charge will be added to the purchases on purchase orders 2594 and 2595, which are now scheduled to be shipped together. We wish to know the total combined amount to be paid on those purchase orders. These total amounts can computed by using the following query:

```
SELECT    SUM(PO_AMT)*1.10
FROM      PURCHASE_ORDER
WHERE     VENDOR_ID="V250"
AND       (PO_NUMBER=2594 OR PO_NUMBER=2595);
```

Answer:

Expr1000
4158

2.5.3.2 Grouping Data

The **GROUP BY** clause can be used to group data from an aggregate function by a column attribute. This allows you to display the results of several aggregates by some meaningful summary. For example, suppose you want to summarize aggregates of data from the PO_DETAIL table by PO_NUMBER. Consider the following example:

```
SELECT     PO_NUMBER, MIN(QUANTITY), MAX(QUANTITY)
FROM       PO_DETAIL
GROUP BY   PO_NUMBER;
```

which returns the following:

PO_NUMBER	Expr1001	Expr1002
2591	300	1000
2592	210.5	800.5
2593	1000	2000
2594	560	4000
2595	400	1200

A subclause of the GROUP BY clause is the **HAVING** clause. This clause allows the programmer to place a condition or filter on the group. For example, suppose we are interested in the maximum and minimum quantities of those purchase orders that have a maximum quantity greater than 1000. This can be accomplished by applying the HAVING clause as follows:

```
SELECT     PO_NUMBER, MIN(QUANTITY), MAX(QUANTITY)
FROM       PO_DETAIL
GROUP BY   PO_NUMBER HAVING MAX(QUANTITY)>1000;
```

The following recordset is returned:

PO_NUMBER	Expr1001	Expr1002
2593	1000	2000
2594	560	4000
2595	400	1200

2.5.3.3 Subqueries in SQL

Understanding how to write subqueries is important for getting useful information from tables. Subqueries allow the user to condition one query on the results of another query from a table and also allow the user to retrieve information in a table based on the results of a query in another table.

As an example of a subquery on a single table, suppose that you want to list all purchase order numbers that have amounts that are greater than the average amount of all purchase orders. One way to attack this problem is to find the average amount for all purchase orders and then to use that information in the WHERE clause of another query as follows:

```
Query 1:    SELECT  AVG(PO_AMT)
            FROM    PURCHASE_ORDER;

Query 2:    SELECT  PO_NUMBER, PO_AMT
            FROM PURCHASE_ORDER
            WHERE PO_AMT > ( result of query 1);
```

The problem with this approach is that you have to execute two queries. A command that uses a subquery structure incorporates both queries in one command. Such a command defines a ***main query*** (query 2) and a ***subquery*** (query 1). The main query uses the result of the subquery as a criterion for its command. The syntax is as follows:

```
MAIN QUERY  SELECT     <attribute name(s)>
            FROM      <TABLE NAME>
            WHERE     <column name> <criterion> / IN>

   SUBQUERY       (SELECT      <column name>
                  FROM        <table name>
                  [WHERE      <condition>]);
```

> ***Access Exercise 2.11:***
>
> Execute the two part query above as a subquery.
>
> 1) Open the Query window and perform the following query from the table ***PURCHASE_ORDERS***.
>
> ```
> SELECT PO_NUMBER,PO_AMT
> FROM   PURCHASE_ORDER
> WHERE  PO_AMT > (SELECT AVG(PO_AMT)
>                  FROM PURCHASE_ORDER);
> ```
>
> 2) The result should appear as in Figure 2.10, below.

A subquery can also be used to retrieve related data in more than one table. For example, suppose you are interested in seeing the details of all the purchase orders that have been placed with vendor V250. These details are listed in the table PO_DETAIL. However, the table PO_DETAIL does not

PO_NUMBER	PO_AMT
2591	$4,300.00
2593	$4,000.00
2594	$3,280.00

Figure 2.10 Result of Access Exercise 2.11.

contain any reference to VENDOR_ID V250. On the other hand, V250 appears *in* the table PURCHASE_ORDER, and that table is related to PO_DETAIL by the attribute they have in common: PO_NUMBER. Therefore, if the linking attribute is retrieved in a subquery, the results of the subquery can be used in a main query to retrieve the open purchase orders.

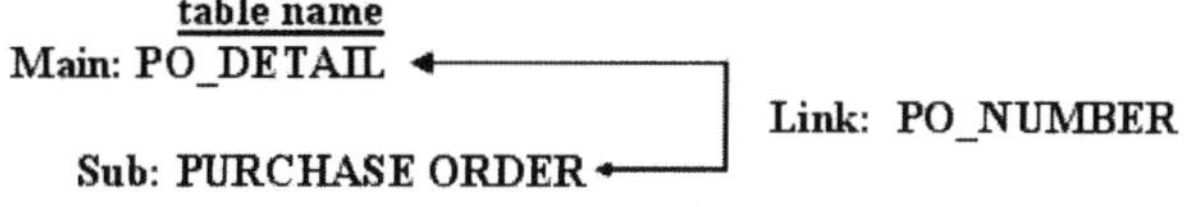

Access Exercise 2.12:

1) Open the Query window and perform the following subquery.

```
SELECT *
FROM   PO_DETAIL
WHERE  PO_NUMBER IN (SELECT PO_NUMBER
                     FROM PURCHASE_ORDER
                     WHERE VENDOR_ID = "V250");
```

2) The result should appear as in Figure 2.11, below.

PO_NUMBER	PO_LINE_ITEM	MATERIAL_ID	UNITS	QUANTITY	BALANCE	PROMISED_DEL_DATE	UNIT_COST	STATUS
2594	1	RM310	LB	4000	4000	3/22/06	$0.50	OPEN
2594	2	RM311	LB	2000	2000	3/22/06	$0.25	OPEN
2594	3	RM318	LB	2000	2000	3/22/06	$0.25	OPEN
2594	4	RM340	LB	560	560	3/22/06	$0.50	OPEN
2595	1	RM305	LB	400	400	2/27/06	$0.50	OPEN
2595	2	RM308	LB	1200	1200	2/27/06	$0.25	OPEN

Figure 2.11 Result of Access Exercise 2.12.

The **IN** operator states that the main query is conditioned on the PO_NUMBER(s) that are returned *in* the subquery.

2.5.3.4 Appending Tables Using Joins

Note that the subquery in the previous section used the IN operator to return information from one table as criterion for retrieving from another table. However, the displayed information came only from the table referenced in the main query. There are times when it is desirable to display information from more than one table on the same retrieval. For example, one might want to display the VENDOR_ID and RELEASE_DATE from the PURCHASE_ORDER table along with the related details from the PO_DETAIL table. This is the purpose of a table ***join***. The syntax for a join is as follows:

```
SELECT    <table1 name.attribute name>, <table2 name.attribute name>,...
FROM      <table1 name>, <table2 name>,...
WHERE     <join condition>
ORDER BY  <column name>
```

As indicated in the SELECT clause, in order to use a join it is necessary to indicate the table from which the attribute is to be retrieved. Calling the attribute name with the table name as a prefix does this.

Access Exercise 2.13:

1) Open the Query window and perform the following query which joins data from ***PURCHASE_ORDER*** and ***PO_DETAIL*** tables.

```
SELECT PURCHASE_ORDER.VENDOR_ID,
       PURCHASE_ORDER.RELEASE_DATE,
       PURCHASE_ORDER.PO_NUMBER,
       PO_DETAIL.*
FROM   PURCHASE_ORDER,PO_DETAIL
WHERE  PURCHASE_ORDER.PO_NUMBER =
       PO_DETAIL.PO_NUMBER
 AND   PURCHASE_ORDER.VENDOR_ID="V250";
```

2) The result should appear as in Figure 2.12, below.

VENDOR_ID	RELEASE_DA	PO_NUMBER	PO_NU	PO_LIN	MATE	UNITS	QUANTITY	BALANC	PROMIS	UNIT_CO	STATUS
V250	2/12/06	2594	2594	1	RM3	LB	4000	4000	3/12/06	$0.50	OPEN
V250	2/12/06	2594	2594	2	RM3	LB	2000	2000	3/12/06	$0.25	OPEN
V250	2/12/06	2594	2594	3	RM3	LB	2000	2000	3/12/06	$0.25	OPEN
V250	2/12/06	2594	2594	4	RM3	LB	560	560	3/20/06	$0.50	OPEN
V250	2/15/06	2595	2595	1	RM3	LB	400	400	2/27/06	$0.50	OPEN
V250	2/15/06	2595	2595	2	RM3	LB	1200	1200	2/27/06	$0.25	OPEN

Figure 2.12 Result of Access Exercise 2.13.

In Figure 2.12, the first three columns were retrieved from PURCHASE_ORDER and the remaining columns came from PO_DETAIL. The foreign key PO_NUMBER was used to make the match by the criterion statement in the WHERE clause. The most common cases in which a join is used involve two tables having a 1:M relationship. The foreign key of the table on the M side of the relationship is used to join to the primary key of the table on the 1 side of the relationship. The foreign key in PO_DETAIL that joins to PURCHASE_ORDER is PO_NUMBER.

It is possible to have more complex joins when more than two tables are involved. For example, let us say we want to display the following information: VENDOR_ID, V_NAME, PO_NUMBER, RELEASE_DATE, PO_LINE_ITEM, MATERIAL_ID, and QUANTITY. This requires that we retrieve information from all three tables. One solution is as follows:

```
SELECT  VENDOR.VENDOR_ID, VENDOR.V_NAME,
        PURCHASE_ORDER.PO_NUMBER,
        PURCHASE_ORDER.RELEASE_DATE,
        PO_DETAIL.PO_LINE_ITEM, PO_DETAIL.MATERIAL_ID,
        PO_DETAIL.QUANTITY
FROM    VENDOR, PURCHASE_ORDER, PO_DETAIL
WHERE   VENDOR.VENDOR_ID = PURCHASE_ORDER.VENDOR_ID
AND     PURCHASE_ORDER.PO_NUMBER=PO_DETAIL.PO_NUMBER
AND     VENDOR.VENDOR_ID="V250";
```

This query returns the following:

VENDOR_ID	V_NAME	PO_NUMBER	RELEASE	PO_LINE	MATERIAL_ID	QUANTITY
V250	Spices	2594	2/12/06	1	RM310	4000
V250	Spices	2594	2/12/06	2	RM311	2000
V250	Spices	2594	2/12/06	3	RM318	2000
V250	Spices	2594	2/12/06	4	RM340	560
V250	Spices	2595	2/15/06	1	RM305	400
V250	Spices	2595	2/15/06	2	RM308	1200

2.6 SUMMARY

The relational database model can be viewed by a user as a set of tables. The tables are defined by an entity set name and contains the values of attributes for instances of that entity set. Each instance in a table is uniquely identified by a primary key. Tables are related to each other through the use of foreign keys. The programming language of relational databases is Structured Query Language. In this chapter, we demonstrated the use of SQL to give the reader some concept of how tables are related to each other in the relational design. We will return to the use of SQL in Chapter 6, when we discuss the design of forms and reports.

An information system design project begins by establishing the requirements of the design project. The next three chapters focus on this aspect of the design problem by demonstrating the use of design methodologies at the conceptual level.

REVIEW EXERCISES

2.1 Consider the tables in Figure 2.1. Write the SQL code for the queries that will return the following:

(a) The PO details associated with Purchase Order 2594.

PO_NUMBER	PO_LINE_ITEM	MATERIAL_ID	UNITS	QUANTITY	BALANCE	PROMISED_DEL_DATE	UNIT_COST	STATUS
2594	1	RM310	LB	4000	4000	3/12/2006	$0.50	OPEN
2594	2	RM311	LB	2000	2000	3/12/2006	$0.25	OPEN
2594	3	RM318	LB	2000	2000	3/12/2006	$0.25	OPEN
2594	4	RM340	LB	560	560	3/20/2006	$0.50	OPEN

(b) The list of purchase order numbers, without duplication, that have an "OPEN" line item.

PO_NUMBER
2592
2593
2594
2595
2596

(c) The following PO details from purchase orders placed with Vendor V250: PO_NUMBER, PO_LINE_ITEM, MATERIAL_ID, QUANTITY, UNIT_COST, TOTAL_COST (QUANTITY × UNIT_COST).

PO_NUMBER	PO_LINE_ITEM	MATERIAL_ID	QUANTITY	UNIT_COST	TOTAL_COST
2594	1	RM310	4000	$0.50	2000
2594	2	RM311	2000	$0.25	500
2594	3	RM318	2000	$0.25	500
2594	4	RM340	560	$0.50	280
2595	1	RM305	400	$0.50	200
2595	2	RM308	1200	$0.25	300

(d) Repeat part c, but order the result by TOTAL_COST increasing.
(e) Return the number of purchase orders whose PO_AMT is greater than 3000.

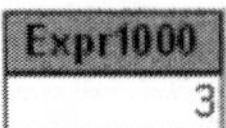

Expr1000
3

(f) Return the vendor ID, without duplication, of any vendor having an open PO line item where quantity times unit cost is more than $500.

VENDOR_ID
V110
V250
V75

(g) Return V_NAME, PO_NUMBER, PO_LINE_ITEM, TOTAL_ COST=(QUANTITY*UNIT_COST) from any vendor with an OPEN purchase order having a PO_AMT greater than $2000.

V_NAME	PO_NUMBER	PO_LINE_ITEM	TOTAL_COST
Jersey Vegetable	2593	1	2000
Jersey Vegetable	2593	2	2000
Spices Unlimited	2594	1	2000
Spices Unlimited	2594	2	500
Spices Unlimited	2594	3	500
Spices Unlimited	2594	4	280

2.2 USF Rent-a-Car keeps track of its vehicle fleet using a database with the following tables:

LOCATION—These are the locations that maintain and rent the cars.
CLASS—This is a grading system for rental rates.
MODEL—This is a description of the model types that USF Rent-a-Car has in its fleet.
CAR—These are the actual vehicles that are owned and rented by USF Rent-a-Car.

All attributes are text data types except YEAR, which is numeric and the rental rates, which are currency data types.

LOCATION				
LOCATION_ID	STREET	CITY	STATE	ZIP
L100	200 Fowler Ave.	Tampa	Florida	32401
L102	45 Beach Place	St. Petersburg	Florida	33567
L103	32 Airport Road	Orlando	Florida	23453

CAR			
VIN	MODEL_ID	LOCATION_ID	COLOR
1HGCC1650HA062999	M301	L100	Blue
1HGCC230HA9866554	M301	L100	Green
1HGCC450JA9987334	M301	L103	Green
CVC39439JK324927384	M100	L102	Brown
CVC39439JK324987790	M100	L103	Black
CVC39493JK233820900	M100	L100	Blue
CVI39493JK233822222	M300	L102	Brown
CVI39493JK233822654	M300	L102	Brown
CVI39493JK233829922	M300	L103	Green
CVM39493JK233820900	M200	L103	Blue
CVM39493JK233820999	M200	L100	White
CVM39493JK233822222	M200	L100	Black
DOD87E67364528346	M101	L102	Red
DOD87E67364555680	M101	L103	Blue
DOD87E67364575775	M101	L100	Green
DOS87E67364528346	M202	L100	Blue
DOS87E67364528676	M202	L103	Green
DOS87E67364577669	M202	L102	Red
FOD39493JK233820900	M201	L103	Yellow
FOD39493JK233820989	M201	L103	Green
FOD39493JK233822345	M201	L100	Red
KIA87E67364522294	M102	L102	Grey
KIA87E67364528346	M102	L102	Grey
KIA87E67364529899	M102	L102	Red
WVWFE83A5SY913999	M302	L103	Red
WVWFE83A5SY989983	M302	L100	Red
WVWFE84A5SY889343	M302	L100	Green
WVWFE88A9SY998883	M302	L103	Green

MODEL				
MODEL_ID	MAKE	MODEL	YEAR	CLASS
M100	Chevrolet	Cobalt	2007	Compact
M101	Dodge	Neon	2007	Compact
M102	Kia	Rio	2007	Compact
M200	Chevrolet	Monte Carlo	2007	Standard
M201	Ford	Mustang	2007	Standard
M202	Dodge	Stratus	2007	Standard
M300	Chevrolet	Impala	2007	Full Size
M301	Honda	Accord	2007	Full Size
M302	Volkswagon	Passat	2007	Full Size

CLASS		
CLASS_ID	DAILY_RATE	WEEKLY_RATE
Compact	$23.00	$115.00
Full Size	$40.00	$200.00
Standard	$33.00	$165.00

(a) For each table, identify the primary key(s) and foreign key(s).
(b) Create the database and populate the tables.
(c) Write the SQL commands for the following:
  (1) List the vehicle identification number (VIN) of all Chevrolets.

VIN
CVC39493JK233820900
CVC39439JK324927384
CVC39439JK324987790
CVM39493JK233820900
CVM39493JK233820999
CVM39493JK233822222
CVI39493JK233822222
CVI39493JK233822654
CVI39493JK233829922

  (2) List the VIN and year of all Chevrolet Impalas located at St. Petersburg.

VIN	YEAR
CVI39493JK233822222	2007
CVI39493JK233822654	2007

  (3) List all Volkswagon Passat cars by location and VIN as follows:

LOCATION_ID	STREET	CITY	STATE	ZIP	VIN
L103	32 Airport Road	Orlando	Florida	23453	WVWFE83A5SY913999
L100	200 Fowler Ave.	Tampa	Florida	32401	WVWFE83A5SY989983
L100	200 Fowler Ave.	Tampa	Florida	32401	WVWFE84A5SY889343
L103	32 Airport Road	Orlando	Florida	23453	WVWFE88A9SY998883

(4) List all red standard cars by location_ID and VIN.

LOCATION_ID	VIN
L103	WVWFE83A5SY913999
L100	WVWFE83A5SY989983
L102	DOD87E67364528346
L102	KIA87E67364529899
L102	DOS87E67364577669
L100	FOD39493JK233822345

(5) How many full-size cars are owned by USF Rent-a-Car?

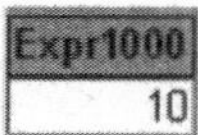

Expr1000
10

(6) How many compact cars are located at the Tampa location?

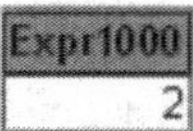

Expr1000
2

(7) What is the cost savings by class for a customer who pays the weekly rate for a 7-day rental instead of the daily rate?

CLASS_ID	Cost Saving
Compact	$46.00
Full Size	$80.00
Standard	$66.00

(8) List the different classes that are located at each location as follows:

LOCATION_ID	CLASS
L100	Compact
L100	Full Size
L100	Standard
L102	Compact
L102	Full Size
L102	Standard
L103	Compact
L103	Full Size
L103	Standard

2.3 The human resources department of an enterprise maintains a database on the company's employees in order to track their qualifications at various skill levels. The tables are as follows.

EMPLOYEE			
EMPL_ID	LAST_NAME	FIRST_NAME	DEPT_ID
1001	Brown	George	20
1002	Lee	Harry	20
1003	Kirk	Bernard	30
1004	James	William	30
1005	James	Harry	40

SKILL	
SKILL_ID	SKILL_DESC
110	C Programmer
111	Systems Analyst
112	Project Manager
113	Java Programmer
114	Database Administrator

DEPARTMENT			
DEPT_ID	DEPT_NAME	DEPT_LOCATION	DEPT_EXTENTION
20	ERP Development	Room 101	5-2007
30	Technical Assistance	Room 103	5-2004
40	Programming Support	Room 105	5-2008

EMPLOYEE_SKILL_SET		
EMPL_ID	SKILL_ID	DATE_QUALIFIED
1001	110	9/1/1990
1001	114	7/10/1995
1002	111	5/10/1992
1002	112	6/12/1995
1003	113	6/6/2000
1004	110	9/10/1999
1004	113	6/8/2004
1005	110	1/15/1996
1005	111	2/1/1999
1005	112	3/3/2001

All attributes are text data types except DATE_QUALIFIED, which is a date data type. The database can be downloaded from the web site supporting this book.

(a) Identify the primary key(s) and foreign key(s) of each table.

(b) Write the SQL commands to return the following data from the database:

  (1) The first and last names of the employees who work in Department 30

  (2) The first and last names of employees with the skill ID of 110

  (3) The first and last names of the employees who are "C Programmers"

  (4) The number of "Java Programmers" in Department 30

  (5) The last and first names of the "Java Programmers" in Department 30 listed alphabetically by last name

  (6) The department ID and department name, listed without repetition, of any departments that do not have "Java Programmers"

2.4 The ACME Machine Shop is a small job shop that does machining of components for customers that assemble finished products. The customers give ACME engineering drawings of parts and request production of parts from ACME as needed. The manager of the machine shop has used a database to try to organize some basic information on the customers' parts and their machining requirements as well as scheduling information based on orders. The tables and their design requirements are shown here. The meaning of the tables is as follows:

PART—The parts produced by ACME
PROCESS_PLAN—The operations that must be performed to manufacture the parts
MACHINE—The machines used to manufacture the parts
SCHEDULE—A set of production schedules for part manufacture

```
CREATE TABLE PART
([PART_NO]              CHAR(4),
[PART_DESC]             CHAR(20),
[MATERIAL_ID]           CHAR(10),
[PROCESS_PLAN_ID]       CHAR(10),
[DRAWING_NO]            CHAR(10),
CONSTRAINT [INDEX1] PRIMARY KEY (PART_NO));

CREATE TABLE PROCESS_PLAN
([PROCESS_PLAN_ID]          CHAR(10),
[OP_NO]                    SMALLINT,
[OP_DESC]                  CHAR(20),
[MACHINE_ID]                CHAR(5),
[TOOLING_ID]                CHAR(5),
[STD_SU_MIN]               SMALLINT,
[STD_UNIT_MACH_MIN]          SMALLINT,
CONSTRAINT [INDEX2] PRIMARY KEY (PROCESS_PLAN_ID, OP_NO));

CREATE TABLE MACHINE
([MACHINE_ID]            CHAR(5),
[MACHINE_DESC]           CHAR(20),
CONSTRAINT [INDEX3] PRIMARY KEY (MACHINE_ID));

CREATE TABLE SCHEDULE
([SCHEDULE_ID]            CHAR(5),
[PART_NO]                CHAR(4),
[S_LOT_SIZE]             SMALLINT,
[S_RELEASE_DATE]          DATE,
[S_COMPLETION_DATE]        DATE,
CONSTRAINT [INDEX4] PRIMARY KEY (SCHEDULE_ID));
```

MACHINE

MACHINE_ID	MACHINE_DESC
M1	J&L Turret lathe
M2	#5 Milwaukee
M3	#4 Cross Turret
M4	Bridgeport Saw
M5	J&L Profiler
M6	Cincinnati Grinder
M7	Yamaguchi Drill
M8	Bridgeport Mill

PART

PART_NO	PART_DESC	MATERIAL_ID	PROCESS_PLAN_ID	DRAWING_NO
1002	Punch	RM201	PP250	402
1010	Rod	RM302	PP125	467
1011	Cam	RM210	PP101	219
1015	Plate	RM501	PP280	109
1020	Shaft	RM302	PP180	345
1022	Mounting Bracket	RM501	PP105	203

SCHEDULE

SCHEDULE_ID	PART_NO	S_LOT_SIZE	S_RELEASE_DATE	S_COMPLETION_DATE
S210	1011	20	4/1/06	4/2/06
S211	1002	15	4/1/06	4/3/06
S212	1015	25	4/2/06	4/5/06
S213	1020	10	4/2/06	4/5/06
S214	1022	12	4/2/06	4/6/06

PROCESS_PLAN

PROCESS_PLAN_ID	OP_NO	OP_DESC	MACHINE_ID	TOOLING_ID	STD_SU_MIN	STD_UNIT_MACH_MIN
PP101	1	cut round stock	M4	T210	2	2
PP101	2	profile cam	M5	T50	5	15
PP101	3	drill pin hole	M7	T101	3	2
PP105	1	cut bar stock	M4	T400	2	2
PP105	2	plane slot	M2	T25	5	20
PP105	3	drill mounting	M7	T102	3	2
PP125	1	cut round stock	M4	T210	2	2
PP125	2	turn round stock	M1	T202	10	20
PP125	3	finish grind	M6	T40	8	15
PP180	1	cut round stock	M4	T210	2	2
PP180	2	turn round stock	M1	T215	10	20
PP250	1	cut round stock	M4	T210	2	2
PP250	2	turn stock	M1	T215	10	25
PP250	3	taper angle	M3	T30	12	30
PP250	4	finish grind	M6	T40	8	15
PP280	1	cut plate stock	M4	T250	4	6
PP280	2	mill surface	M8	T602	12	25
PP280	3	drill holes	M7	T101	3	2

(a) Create the database and tables in Access DBMS.
(b) Write the SQL code that will generate answers to the following questions:
   (1) List all the entries from the parts table.
   (2) List the instances of parts that use raw material RM302.
   (3) List the instances of all parts, alphabetically ordered by part description.
   (4) List the instances of all schedules with lot sizes greater than 15.
   (5) List the instances of all parts that require machine M1.
   (6) List the machines and their average, minimum, and maximum unit machining times.
   (7) For each schedule, list the schedule ID, part number, operation number, machine ID, setup time, total machining time, and total time (setup plus machining). Be sure that each attribute column has a title.

(8) List the tooling that is used with each raw material without repetition in the tuple.
(9) List the schedule release date, part number that will be machined, and machine ID that will be used by schedule release date descending.
(10) List the part number and machine ID that will be used in production on schedule release date 4/1/06.
(11) List just the machine ID (without duplication) that will be used by schedules released on 4/1/06.
(12) Using a query/subquery, list all the part numbers that have operations with a setup time greater than 8 minutes.
(13) List the material ID and the number of pieces of raw material that will be required for all the scheduled job releases on 4/1/06.

2.5 Bikes-R-US maintains a database of products sold to customers, and the invoices for those sales are used to bill customers. The Access database can be downloaded from the web site supporting this book. The following tables are kept in the database:

PRODUCT—This is the list of products sold by the company.
CUSTOMER—These are customers to whom the company sells its products.
INVOICE—When a sale is made, the company sends an invoice to its customer.
INVOICE_LINE—A sale may be for one or more products. These are the line items that appear on the invoice.

INVOICE

INVOICE_NO	CUSTOMER_NO	SALE_DATE
1015	123	6/3/2006
1016	124	6/3/2006
1017	125	6/4/2006
1018	123	6/5/2006
1019	126	6/5/2006

PRODUCT

PRODUCT_ID	PRODUCT_DESC	PRICE
B1	Kids Bike	$55.00
B10	Mountain Bike	$160.00
B25	Road Bike	$110.00
B33	Tandem Bike	$180.00

CUSTOMER

CUSTOMER_NO	LAST_NAME	FIRST_NAME	STREET	CITY	STATE
123	Harris	Catherine	13 East St.	Montclair	NJ
124	Smith	Joan	24 Shaw St.	Caldwell	NJ
125	Levi	Jerome	45 Tappan St.	Glen Rock	NJ
126	Smithson	Richard	14 West St.	Long Branch	NJ

INVOICE_LINE			
INVOICE_NO	LINE_NO	PRODUCT_ID	QTY_SOLD
1015	1	B1	1
1015	2	B10	2
1016	1	B10	3
1017	1	B1	2
1017	2	B10	1
1017	3	B25	1
1018	1	B10	2
1019	1	B33	2

Assume that attributes are text data types except for SALE_DATE (date type), PRICE (currency type), and attributes with "NO" in their name (number types).

(a) Write the SQL statement that produces the following recordset:

INVOICE_NO	LINE_NO	PRODUCT_ID	QTY_SOLD
1017	3	B25	1
1017	2	B10	1
1017	1	B1	2

(b) Write the SQL statement that produces the following recordset:

INVOICE_NO	SALE_DATE	LAST_NAME	STATE
1015	6/3/2006	Harris	NJ
1018	6/5/2006	Harris	NJ

(c) Write the SQL statement to list customer rows for customers whose last names begins with "Smith" (returns both Smith and Smithson).

(d) Write the SQL statement that will add one product row to the PRODUCT table. The row to be added is as follows:

PRODUCT_ID	PRODUCT_DESC	PRICE
B101	Custom Mountain Bike	$225.00

(e) Write the SQL command that summarizes total sales by invoice number as follows:

INVOICE_NO	SALE_DATE	PRODUCT_DESC	Unit_Price	QTY_SOLD	Total_Amt
1015	6/3/2006	Mountain Bike	$160.00	2	$320.00
1015	6/3/2006	Kids Bike	$55.00	1	$55.00
1016	6/3/2006	Mountain Bike	$160.00	3	$480.00
1017	6/4/2006	Road Bike	$110.00	1	$110.00
1017	6/4/2006	Mountain Bike	$160.00	1	$160.00
1017	6/4/2006	Kids Bike	$55.00	2	$110.00
1018	6/5/2006	Mountain Bike	$160.00	2	$320.00
1019	6/5/2006	Tandem Bike	$180.00	2	$360.00

(f) Write the SQL statement to return the customer numbers of customers who have purchased product ID B10.

2.6 The ACME metalworking Machine Shop has implemented a highly automated cellular manufacturing system. In this system, machines are grouped into autonomous machining cells, each attended by a robot for material handling. Parts to be machined are moved to the cells by an automated guided vehicle (AGV). When a part is taken to a specific cell, it is completely manufactured within that cell. Some parts can be produced in either cell. ACME has implemented two cells, as shown here:

Manufacturing Cell 2 Manufacturing Cell 1

ACME has designed a database and implemented some tables that describe the relationships between cells, machines, and the parts that they produce. The Access database can be downloaded from the web site supporting this book. The tables are shown here and have the following meaning:

CELL—Identifies the different cells
MACHINE—Identifies the different machines
PART—Identifies the different parts manufactured in the cells
PROCESS_PLAN—Identifies the different possible process plans that can be used to manufacture each part (some parts can be manufactured by more than one process plan)
PROCESS_PLAN_STEP—Defines each operation of each process plan and the machines used in the operation

CELL

CELL_ID	CELL_DESCRIPTION
1	Cell for mostly round stock
2	Cell for mostly prismatic stock

MACHINE

MACHINE_ID	MACHINE_NAME	MACHINE_DESCRIPTION
M1	B&S Lathe	18 in. engine lathe
M2	Bridgeport Mill	Horizontal milling machine
M3	W&S Drill	Vertical drill
M4	H&Z Lathe	20 in turret lathe
M5	Mazak machining center	Drill/Boring/Milling

MACHINE_CELL_XREF

CELL_ID	MACHINE_ID
1	M2
1	M3
1	M4
2	M1
2	M5

PART

PART_ID	PART_DESCRIPTION
1	Part 1
2	Part 2
3	Part 3
4	Part 4

PROCESS_PLAN

PROCESS_PLAN_ID	PART_ID
1.1	1
2.1	2
2.2	2
3.1	3
3.2	3
4.1	4

PROCESS_PLAN_STEP

PROCESS_PLAN_ID	OPERATION_NO	MACHINE_ID	MACHINE_TIME
1.1	1	M2	100
1.1	2	M3	52
2.1	1	M4	210
2.1	2	M2	20
2.2	1	M1	10
2.2	2	M5	100
3.1	1	M2	25
3.1	2	M3	30
3.2	1	M5	205
4.1	1	M1	304
4.1	2	M5	310

(a) For each table identify the primary key(s) and foreign key(s).

(b) Write the SQL statements for the following:

(1) A list of machines by cell.

CELL_ID	MACHINE_ID	MACHINE_NAME
1	M2	Bridgeport Mill
1	M3	W&S Drill
1	M4	H&Z Lathe
2	M1	B&S Lathe
2	M5	Mazak machining center

(2) The minimum, maximum, and average processing time on a machine.

MACHINE_ID	MIN	MAX	AVG
M1	10	304	157
M2	20	100	48.3333333333333
M3	30	52	41
M4	210	210	210
M5	100	310	205

(3) The total machining time by cell and by part.

CELL_ID	PART_ID	TOT
1	1	152
1	2	230
1	3	55
2	2	110
2	3	205
2	4	614

(4) Cells having a machine with a machining time >300.

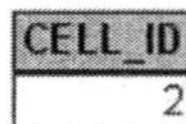

CELL_ID
2

2.7 Flotsum Pharmaceutics Ltd. manufactures a variety of prescription drugs. By government regulation, each batch of manufactured product is subjected to laboratory testing before it can be sold. This is done at the Flotsum quality assurance lab. The lab was recently reorganized in an effort to make its operation more efficient. Previously, lab technicians would withdraw bulk chemicals from inventory at the time of testing. Using the bulk chemicals, they would prepare reagents for the tests and use the reagents and other chemicals to test a sample of the product brought to the lab from production. Under the new system, "testing kits" are prepared in advance of testing by chemists and are stored in cabinets in the lab. Each testing kit contains all the reagents and chemicals necessary for completely performing a specific laboratory test. This eliminates the withdrawal and mixing of bulk chemicals at the time of testing.

The following tables are used in the quality assurance laboratory database:

TEST_KIT

TEST_KIT_ID	KIT_SPEC_ID	CABINET_NO
TK1	S1	5
TK2	S2	1
TK3	S3	2
TK4	S1	5

KIT_SPEC_ID—A description of chemicals and reagents in a kit of the particular specification
TEST_KIT_ID—The identification of an actual kit that is built to the associated specification
CABINET_NO—The location where the kit is stored

QA_TEST

TEST_NO	LAB_PROC_SPEC_ID	BATCH_NO	ACTUAL_VALUE	TEST_RESULT	TEST_DATE	TECHNICIAN_ID
1	LAB_2	2445	0.5	P	11/10/2006	036-26-0000
2	LAB_3	2445	1.2	P	11/10/2006	036-26-0000
3	LAB_3	2446	1.25	P	11/11/2006	036-26-0000
4	LAB_1	2446	0.7	P	11/12/2006	036-26-0000
5	LAB_1	2447	0.8	F	11/13/2006	036-26-0000

LAB_PROC_SPEC_ID is the unique ID for a testing procedure. The TEST_NO identifies an actual test in which the procedure was used on a BATCH_NO of product. The ACTUAL_ VALUE is the numerical result from the test, which yields a TEST_RESULT of either pass or fail.

The following table keeps track of the usage of the test kits. Kits are withdrawn or replaced (TRANS TYPE) during each usage transaction (TRANS_NO).

USAGE_TRANSACTION

TRANS_NO	TEST_KIT_ID	TEST_NO	TEST_DATE	TRANS_TYPE
1	TK3	1	11/10/2006	W
2	TK3	1	11/10/2006	R
3	TK1	2	11/10/2006	W
4	TK1	2	11/10/2006	R
5	TK1	3	11/11/2006	W
6	TK1	3	11/12/2006	R
7	TK2	4	11/12/2006	W
8	TK2	4	11/12/2006	R
9	TK2	5	11/13/2006	W
10	TK2	5	11/13/2006	R
11	TK4	6	11/13/2006	W

LAB_PROCEDURE_SPEC

LAB_PROC_SPEC_ID	LP_DESCRIPTION	TEST_ACCEPT_VALUE_RANGE	KIT_SPEC_ID
LAB_1	Acid reflux.	0.6-0.7	S2
LAB_2	PH	0.4-0.6	S3
LAB_3	Granular course	1.2-1.3	S1
LAB_5	Granular fine	1.18-1.22	S1

A lab procedure specification is defined by a LAB_PROC_SPEC_ID, which uniquely defines the kit to be used, the description of the test, and the acceptable range of the outcome in order for the batch of product to pass the test.

The Access database can be downloaded from the web site supporting this book. Write the SQL commands required to return the following recordsets:

(a) A list of quality assurance (QA) tests performed, the lab procedure used, and the test results.

TEST_NO	LAB_PROC_SPEC_ID	RESULT
4	LAB_1	P
5	LAB_1	F
1	LAB_2	P
2	LAB_3	P
3	LAB_3	P

(b) A comparison of actual test results with acceptance ranges.

TEST_NO	LAB_PROC_SPEC_ID	TEST_ACCEPT_VALUE_RANGE	ACTUAL_VALUE	TEST_RESULT
4	LAB_1	0.6-0.7	0.7	P
5	LAB_1	0.6-0.7	0.8	F
1	LAB_2	0.4-0.6	0.5	P
2	LAB_3	1.2-1.3	1.2	P
3	LAB_3	1.2-1.3	1.25	P

(c) The number of times there was a withdrawal of a test kit from Cabinet 5.

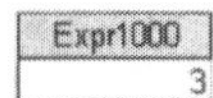

Expr1000
3

(d) The first and the latest date that test kit TK1 was withdrawn from the cabinet.

MIN	MAX
11/10/2006	11/11/2006

APPENDIX 2A QUERY BY EXAMPLE

SQL is the common language of relational databases. However, another standard that has been developed and is available in DBMSs is **query by example (QBE)**. QBE is a visual method of establishing a query within or among tables. It was developed as a user-friendly way of designing queries without having to write an SQL statement—in fact, without having to know SQL at all.

In what follows we shall describe QBE as it is implemented in Microsoft Access RDBMS. As we create a query in QBE, Microsoft Access develops an equivalent SQL command in the background. We will show some queries that appear in Chapter 2 as they would be designed in QBE.

We begin with a SELECT query example. The following steps are involved in getting to the appropriate screen.

Step 1. In the MATERIAL_MANAGER database window, select the **Query** tab and click on **New**.

Step 2. From the Main Menu bar, select **Design View** and click **OK**.

Step 3. At the Show Table window, select the PURCHASE_ORDER table and click **Add**. Then **close** the Show Table window.

At this point the window that you see should look like Figure 2A.1.

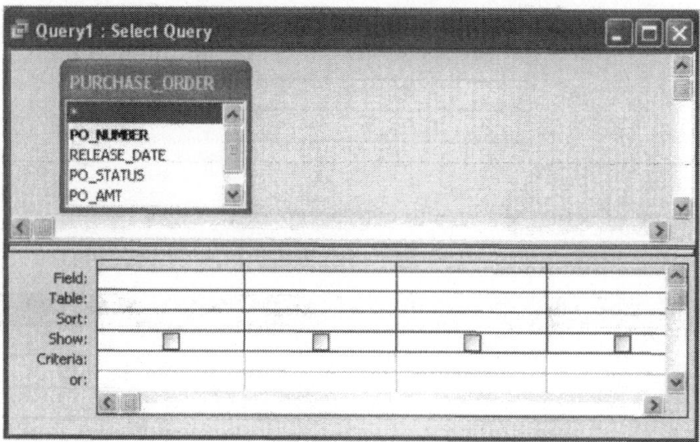

Figure 2A.1 Initial query window.

ACCESS EXERCISE 2.4

```
SELECT PO_NUMBER, PO_STATUS, PO_AMT
FROM PURCHASE_ORDER;
```

The QBE equivalent is shown in Figure 2A.2. At the Field row, pull down the menu and add the attributes of the PURCHASE_ORDER table in the order: PO_NUMBER, PO_STATUS, and PO_AMT. The checked boxes in the Show row indicate that these attributes will be retrieved and shown in the results of the query. Execute the query, and observe that the appropriate rows are retrieved.

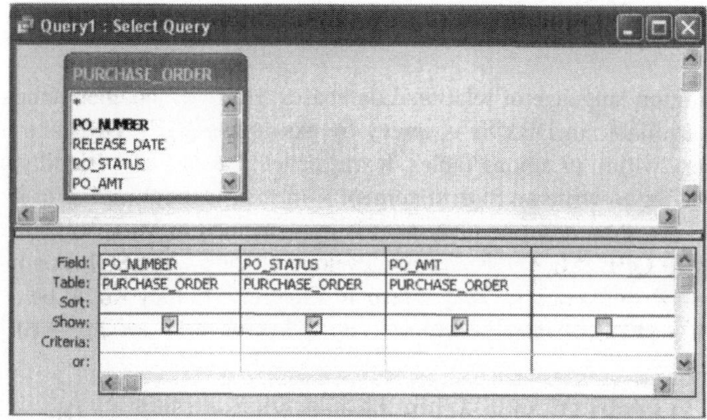

Figure 2A.2 Access Exercise 2.4.

You can see that Access has converted the QBE into an SQL statement. Go to the View menu item on the toolbar. Pull the menu down and choose "SQL View." A window should appear with the appropriate SQL statement.

The following QBE screens are shown to illustrate how to replicate exercises in the chapter.

ACCESS EXERCISE 2.5

```
SELECT PO_NUMBER, PO_STATUS, PO_AMT
FROM PURCHASE_ORDER
ORDER BY PO_AMT;
```

Note that, in Figure 2A.3, the ORDER BY clause is implemented in the Sort row of the QBE screen.

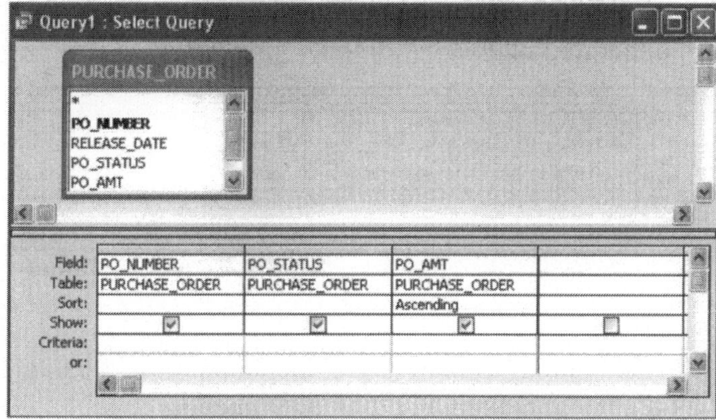

Figure 2A.3 Access Exercise 2.5.

ACCESS EXERCISE 2.6

```
SELECT VENDOR_ID, PO_AMT, PO_NUMBER, PO_STATUS
FROM PURCHASE_ORDER
ORDER BY VENDOR_ID, PO_AMT DESC;
```

Note that, in Figure 2A.4, multiple orderings are implemented in the Sort row.

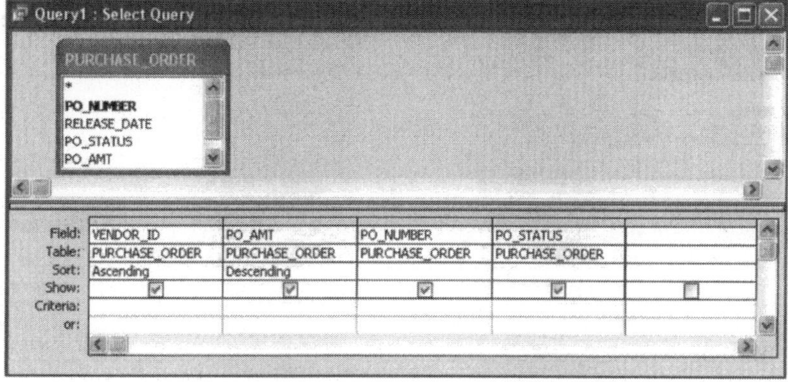

Figure 2A.4 Access Exercise 2.6.

Access Exercise 2.7

```
SELECT VENDOR_ID, PO_NUMBER, PO_AMT
FROM PURCHASE_ORDER
WHERE PO_AMT > 1000
ORDER BY  PO_AMT;
```

Note that, in Figure 2A.5, WHERE clause criteria are implemented in the Criteria row.

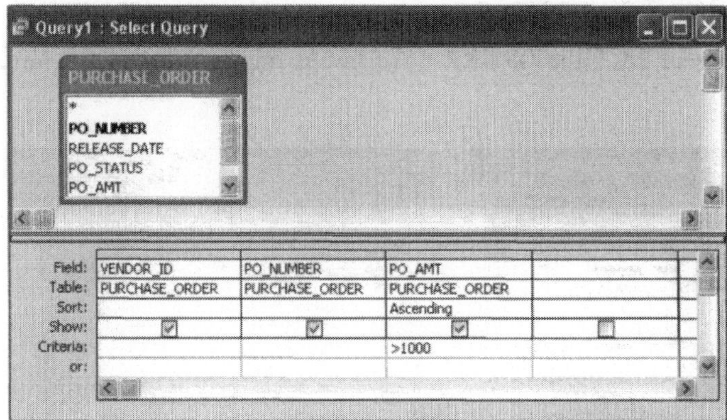

Figure 2A.5 Access Exercise 2.7.

Access Exercise 2.8

```
SELECT *
FROM PURCHASE_ORDER
WHERE PO_STATUS=“OPEN”
AND (VENDOR_ID>“V150” AND PO_AMT>500)
ORDER BY  PO_AMT;
```

Figure 2A.6 illustrates the use of multiple criteria on numeric and text attributes.

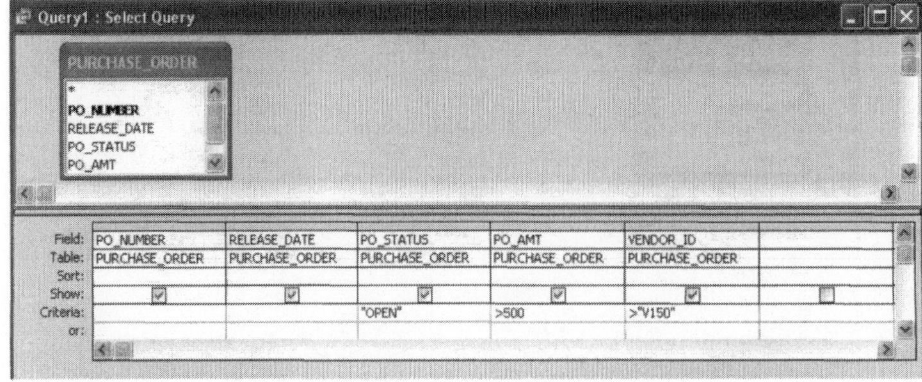

Figure 2A.6 Access Exercise 2.8.

Access Exercise 2.9

```
UPDATE   PO_DETAIL
SET      PROMISED_DEL_DATE = "03/22/06"
WHERE    PO_NUMBER=2594;
```

An UPDATE query is executed in two stages. First, a SELECT query is used to select the attributes to be updated. Following that, an UPDATE is executed on the attributes. The SELECT query is shown in Figure 2A.7.

Step 1. Begin a new query based on the table PO_DETAIL.
After selecting the record by executing the SELECT query of Figure 2A.7, the SELECT query is converted into an UPDATE query.

Step 2. Click on the "Query" menu item on the tool bar. Select "Update Query" from the pull-down menu.

Step 3. In the row titled "Update to," add the new data as shown in Figure 2A.8.

Go to the PO_DETAIL table and check the results.

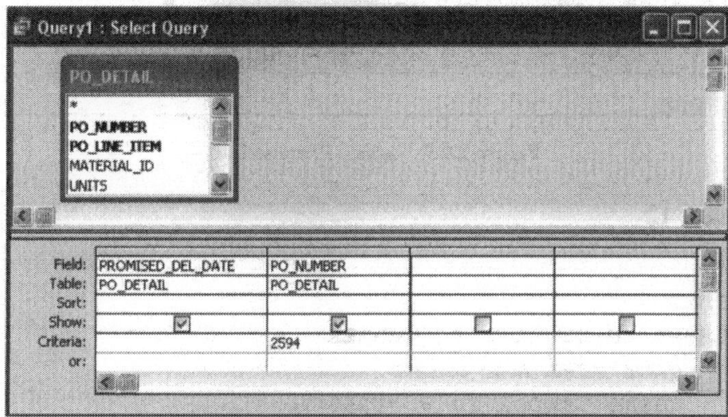

Figure 2A.7 Access Exercise 2.9, Step 1.

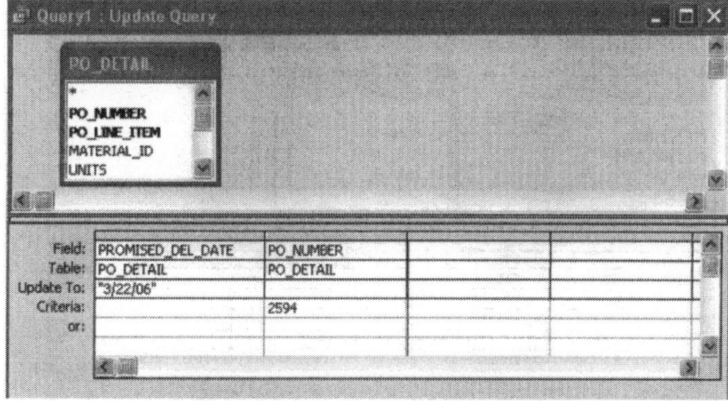

Figure 2A.8 Access Exercise 2.9, Step 2.

Access Exercise 2.10

```
SELECT   MIN(PO_AMT), MAX(PO_AMT), SUM(PO_AMT)
FROM     PURCHASE_ORDER
WHERE    PO_STATUS="OPEN";
```

Starting from the Query window, choose "View" on the toolbar and click on "Totals" on the pull-down menu. This will add a new row to the query window titled "Total." The QBE is shown in Figure 2A.9. Notice that the check box under PO_STATUS is not checked so that it will not show in the tuple that is returned.

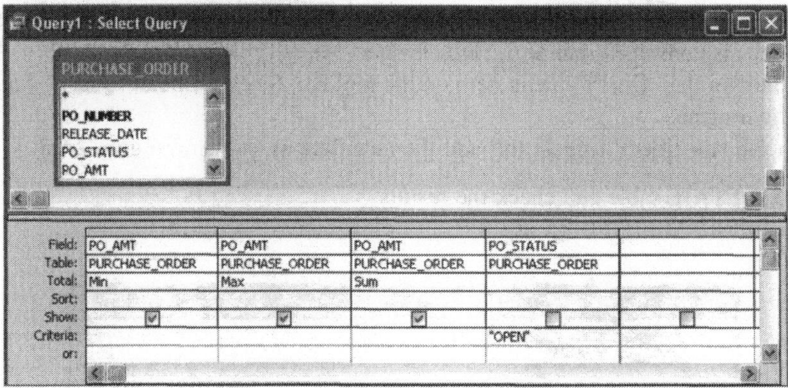

Figure 2A.9 Access Exercise 2.10.

Access Exercise 2.12

```
SELECT   *
FROM     PO_DETAIL
WHERE    PO_NUMBER IN (SELECT PO_NUMBER
                 FROM PURCHASE_ORDER
                 WHERE VENDOR_ID = "V250");
```

In this exercise, we require two tables. From the Show Table window, add the tables PURCHASE_ORDER and PO_DETAIL. A line is drawn automatically that connects the two Tables based on the common attribute PO_NUMBER that exists in both tables. Close the Show Table window. The exercise can now be completed as shown in Figure 2A.10.

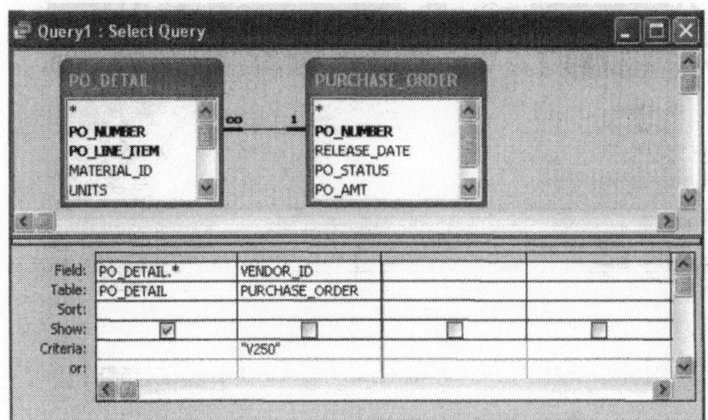

Figure 2A.10 Access Exercise 2.12.

ACCESS EXERCISE 2.13

```
SELECT    PURCHASE_ORDER.VENDOR_ID,
          PURCHASE_ORDER.RELEASE_DATE,
          PURCHASE_ORDER.PO_NUMBER,
          PO_DETAIL.*
  FROM    PURCHASE_ORDER, PO_DETAIL
  WHERE   PURCHASE_ORDER.PO_NUMBER =
          PO_DETAIL.PO_NUMBER
  AND     PURCHASE_ORDER.VENDOR_ID="V250";
```

Their Join query is shown in Figure 2A.11.

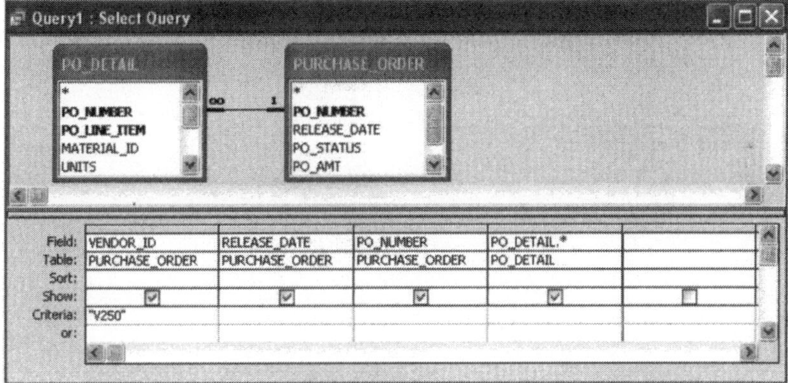

Figure 2A.11 Access Exercise 2.13.

Chapter 3

Data Modeling

3.1 INTRODUCTION

An important task during the design of a database application is for the design team to provide themselves with an abstract representation of the data on which the application is to be based. This abstract representation should be independent of how the data are stored and manipulated in the computer. In 1975, a general architecture for data representation was developed by the Standards Planning and Requirements Committee (SPARC) of the American National Standards Institute (ANSI). The ANSI/SPARC architecture is a three-level architecture based on three views of the data in the database: the (1) external, (2) conceptual, and (3) internal levels, as shown in Figure 3.1.

The ***external level*** addresses the way in which different users view the database. It includes those entities and attributes that the user sees and interacts with. There may be other entities and attributes that the user does not see and that are not part of the external level. The actual implementation at the external level involves the user interfaces that are used to interact with the database. These interfaces are typically designed as "forms" and "reports," a topic that is discussed in Chapter 6. The user view can also include representations of the data that are not stored in the database. For example, summary statistics on attributes (such as the average dollar value of purchase orders) can be part of the user view but not stored as a data summary in the database tables.

The ***conceptual level*** is a holistic view of the database at the level that defines entities, their attributes, and their relationships. It describes *what* data are stored in the database, but not *how* they are stored. The conceptual level is a logical description of the database without saying anything about its implementation. Therefore, it is independent of any specific hardware or software platforms. The conceptual level is the foundation for the external level design. That is to say, the data provided to the user views at the external level must be represented at the conceptual level. A query or summation of data to be performed for the external level depends on the entities, attributes, and relationships defined at the conceptual level. This chapter (as well as Chapters 4 and 5) addresses the conceptual-level design problem.

The ***internal level*** addresses the data structures and the file organization that are used to store data within the computer. Whereas the conceptual level defines *what* data are stored, the internal level defines *how* data are stored. This includes data compression, data encryption, use of indexes, hashing schemes, and other details. The internal level is dependent on the operating system and physical components of the computer system on which the database resides. Given that the focus of this book is on information systems as opposed to database technology, the details of the internal level are not covered here. Interested readers are referred to Lewis et al. (2002).

Design of Industrial Information Systems

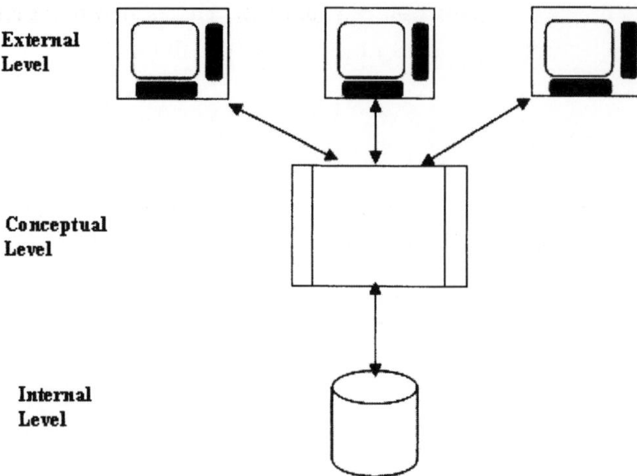

Figure 3.1 ANSI/SPARC three-level architecture.

In this chapter, we focus on the conceptual, or logical, design of the database. We refer to this as the design of a data model. The data model provides a representation of the entities in the enterprise, the attributes of those entities, and the relationships that exist among entities. Several related modeling formalisms are used to develop relational data models. They are derivatives of the seminal modeling approach developed by Peter Chen in the late 1970s (Chen, 1976). In this chapter, we describe Chen's entity-relationship (E-R) modeling technique as the foundation of principles for data modeling. In Chapter 5, we shall also describe the Integrated Computer-Aided Manufacturing Definition 1, extended (IDEF1X), which is a data modeling formalism specifically designed for manufacturing applications.

3.2 ENTITY-RELATIONSHIP (E-R) MODELING

The purpose of E-R modeling is to design a conceptual schema of entities and their relationships in order to (1) facilitate the process of communication among analysts, managers, and users of data; (2) design a common data model that will accommodate the different needs of individuals and organizations within the enterprise; and (3) establish a logical model that can be implemented in a database management system (DBMS). The E-R model is a conceptual model and does not depend on any particular hardware or software. E-R modeling is an interactive process that is carried on between the designer of the database application and the individuals in the organization (enterprise) that is the subject of the data model. In this section, we describe the fundamentals of E-R modeling.

3.2.1 E-R MODELING PRIMITIVES

The E-R modeling approach focuses on three aspects of data: entity sets, attributes, and relationships between entity sets. To provide a consistent example throughout, we will illustrate using the entity sets and attributes of the tables from Chapter 2.

The E-R diagram uses a box to represent an entity set. This is shown in Figure 3.2 as the entity sets PURCHASE_ORDER, PO_DETAIL, and VENDOR. In the E-R modeling formalism, the term "entity" is used to describe what is referred to in Chapter 2 as "entity set." Also, the term "entity instance" is used to describe what we refer to in Chapter 2 as an "entity" or "record." Henceforth, we will use this terminology associated with E-R diagramming.

E-R diagrams distinguish between weak and strong entities. An entity is ***weak*** if its existence is dependent on the existence of another entity. An example of this occurs in the case of PO_DETAIL. PO_DETAIL is dependent on the existence of PURCHASE_ORDER. A weak entity is sometimes referred to as a *child* in a *parent-child* relationship. The primary key of the weak (child) entity is at least in part derived from the parent entity. A weak entity is shown in an E-R diagram by using a double-sided rectangle for the entity. A ***strong entity*** is an entity that is not existence-dependent on some other entity. A strong entity is represented by a single-sided rectangle.

In the E-R modeling formalism, attributes are represented by ovals with a line joining them to the entity to which they apply. Primary keys are indicated by underlining the attribute name. Thus, PO_NUMBER is the primary key attribute of the entity PURCHASE_ORDER, and VENDOR_ID is the primary key for VENDOR.

In E-R models, foreign keys are not shown. For example, VENDOR_ID is the foreign key that links PURCHASE_ORDER to the entity VENDOR. It is understood that when tables are implemented from an E-R model, a table must contain, as a foreign key, the primary key of any table to which it is related. Therefore, it is unnecessary to show the attribute VENDOR_ID with the

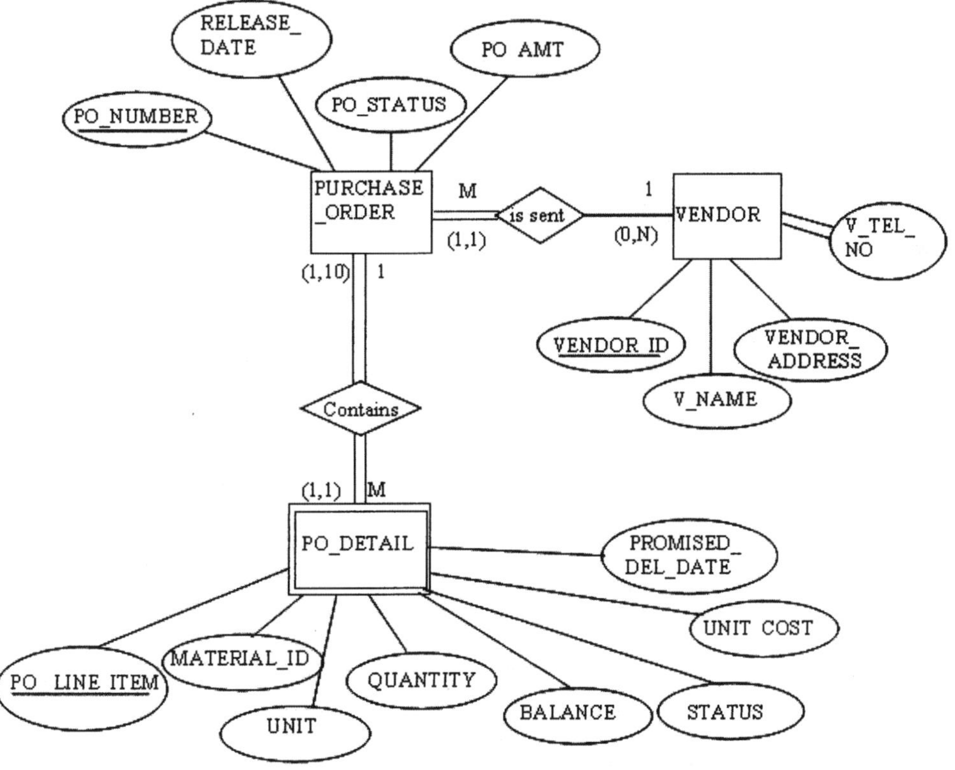

Figure 3.2 Entity relationship diagram.

entity PURCHASE_ORDER. VENDOR_ID will be inherited by PURCHASE_ORDER when the tables are created. Note that the entity PO_DETAIL will inherit the attribute PO_NUMBER from PURCHASE_ORDER. However, PO_DETAIL is a weak entity and the inherited foreign key, PO_NUMBER, will become part of the composite primary key of PO_DETAIL.

In the conceptual modeling stage, attributes are sometimes introduced without consideration as to how they would be implemented in the database tables. An example of this is introducing ***composite attributes*** into the model. A composite attribute, not to be confused with a composite key, is an attribute that can be further broken down into more specific attributes. As an example, we have shown the attribute VENDOR_ADDRESS in the entity VENDOR. This attribute can be subdivided into street, city, state, and zip code, as we did in Chapter 2. To make it possible to query the database for vendor by city, state, or zip code, such a breakdown is necessary. At the initial conceptual design stage, it may be efficient to use composite attributes in descriptive models, but these should eventually be broken down before database implementation into tables.

An attribute may be described as being either single valued or multivalued. A ***single-valued attribute*** will have only one entry in each instance of that attribute. For example, there is only one purchase order number to be entered in each entry of the attribute PO_NUMBER. On the other hand, there may be attributes that have more than one value. Consider the attribute V_TEL_NO of the entity VENDOR. The vendor may have more than one primary telephone number that should be kept track of in the database. Therefore, it may be useful to define a primary and secondary telephone number in the final data tables. However, at the early design stage it is sufficient to indicate that this is a multivalued attribute by using the double lines to connect the attribute with the entity, as shown in Figure 3.2.

The ***relationship*** is a description of the association between entities. Relationships have three components: *connecting arcs*, *diamonds*, and *cardinality*.

Arcs connect entities that are related to each other. In Figure 3.2, VENDOR is related to PURCHASE_ORDER and PURCHASE_ORDER is related to PO_DETAIL. A verb or verb phrase is used to express the general relationship. This is placed in a diamond on the arc. Thus, the phrases "VENDOR *is sent* PURCHASE_ORDER" and "PURCHASE_ORDER *contains* PO_DETAIL" describe how the entities are related. The appropriate verb or verb phrase is at the judgment of the analyst.

Cardinality expresses the number of entity occurrences associated with one occurrence of the related entity. Cardinality is indicated by numbers or symbols written at the ends of the relationship arc. Three cardinalities are common on E-R diagrams:

One-to-many: (1 : M)
One-to-one: (1 : 1)
Many-to-many: (M : N)

The cardinality 1 : M means that each instance of one entity is associated with zero, one, or many instances of the entity on the M side of the relationship. The relationships and their cardinalities in Figure 3.2 form concise expressions:

"One vendor is sent 0,1 or many purchase orders."
"One purchase order contains 0,1 or many purchase order details."

In general, the one-to-many relationship is common in a relational database table implementation. A one-to-one relationship often indicates that the two entities involved can be combined into a single entity. This is so because a record in one table will be related to one and only one record in another table to which it has a one-to-one relationship. Therefore, no increase in the number of rows will occur if both entities are combined in one table.

The many-to-many relationship, which is difficult to handle in a database query, is usually expanded into a series of one-to-many relationships in the final database tables. We will look at these cases more closely in later sections.

A useful graphical tool that assists in visualizing the cardinality of a relationship is called a ***semantic net diagram***. In a semantic net diagram, examples of the instances of two entities are connected by relationship lines. If instances of one entity connect to many instances of another entity, it indicates the existence of a 1:M relationship. Figure 3.3 shows such a diagram for the relationship between PURCHASE_ORDER and PO_DETAIL. The examples used are from Figure 2.1.

Note that in Figure 3.3 the diagrams are drawn such that each connection between instances of the entities requires a separate instance of a relation. If, in a semantic net diagram, there are more instances of a relationship than there are entity instances on one side and an equal number of instances of the entity on the other side of the relationship, a 1:M relationship exists.

For another example, consider a relationship between the entities MANAGER and DEPARTMENT. If each department has one, and only one, manager and each manager manages one, and only one, department, this situation is a 1:1 relationship. It would be drawn in a semantic net diagram like that shown in Figure 3.4. Here we see that the number of instances of a relationship is equivalent to the number of instances in each entity set. This is typical of a 1:1 relationship.

In an M:N relationship, the number of instances of a relationship is greater than the number of related instances in an entity set. Consider the relationship between two entities: VENDOR and MATERIAL. A vendor *supplies* material to a company. One vendor may supply more than one material. Also, a specific material may be supplied by more than one vendor. Hence, an M:N relationship exists. This is captured in the semantic net diagram shown in Figure 3.5. Note that the instances of the entity VENDOR have multiple relationships and the instances of the entity MATERIAL also have multiple relationships.

It is often desirable to add ***cardinality limits*** to the E-R diagram in order to be more specific about the domain of the cardinality relationship. Consider the VENDOR/PURCHASE_ORDER relationship in Figure 3.2. The 1:M relationship allows M to take on any integer value. Is this correct? This cardinality means that if you were to observe any instance of the entity

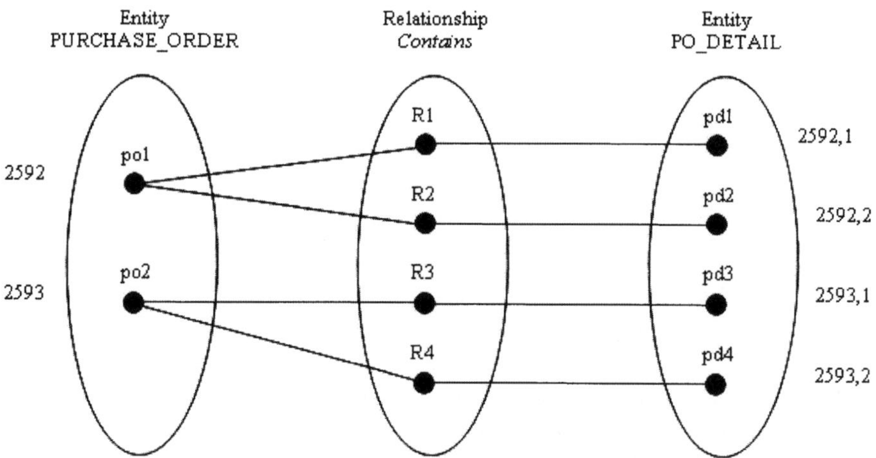

Figure 3.3 A 1:M relationship in a semantic net diagram.

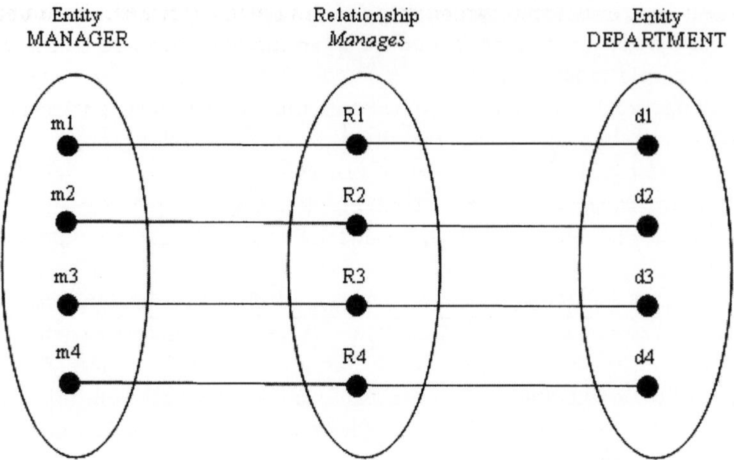

Figure 3.4 A 1 : 1 relationship in a semantic net diagram.

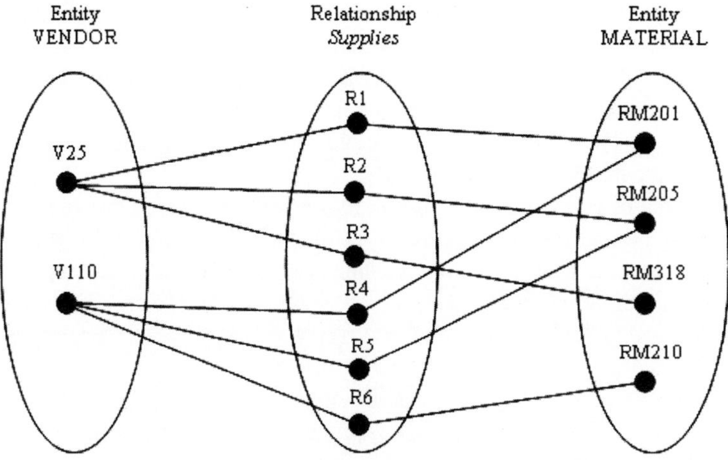

Figure 3.5 M : N Relationship in a semantic net diagram.

VENDOR and look for its occurrence in the table PURCHASE_ORDER, you may find it in zero, one, or many instances. If you consider the way in which most businesses operate, this is a very reasonable result. In most industries, the vendors to a company are first approved by management based on criteria such as product quality and financial stability. There may very well be a vendor that has been approved (thus, existing in the VENDOR table) but has not yet been given a purchase order. Thus, 0 is a possible cardinality. On the other hand, there may be one or many purchase orders that are being filled by the same vendor. From this we conclude that the cardinality 1 : M reflects the way in which the business is being operated.

On the other hand, consider the relationship between PURCHASE_ORDER and PO_DETAIL. When the purchasing agent writes a purchase order, it is surely a purchase order for something.

Therefore, it must have line items that indicate what is being purchased. A cardinality of 0 does not exist. In addition, the company may place an upper limit on the number of line items allowed on a single purchase order. Since purchase orders are often mailed on paper forms, this limit may simply reflect the physical limits of space on the form. On the other hand, it may simply be a policy the company adopts in order to avoid large purchase orders. We assume that such a limit has been placed at 10 and reflect that limit in Figure 3.2.

Cardinality limits are shown in Figure 3.2 as numbers in parentheses. In the relationship between PURCHASE_ORDER and PO_DETAIL, the cardinality limits for PO_DETAIL are (1,1), which means that each PO_DETAIL will be associated with one purchase order. Similarly, the cardinality limits on PURCHASE_ORDER is (1,10), which means that there will be a minimum of 1 and a maximum of 10 line items associated with one purchase order.

Cardinality limits are imposed based on the way in which the business is operated. Therefore, they are said to reflect the ***business rules*** of the enterprise. Whenever possible, identifying business rules and making them explicit in the E-R diagram helps to sharpen the understanding of how the database tables are to be implemented for data integrity.

Business rules also affect the degree of an entity's participation in a relationship. There are two categories of entity participation: optional and mandatory. The participation is ***optional***, or ***partial***, if one entity's occurrence does not require the participation of the other entity in the relationship. Consider an enterprise that has projects and that has employees who work on these projects. One relationship that exists in the enterprise is shown in Figure 3.6. The entity EMPLOYEE has a relationship to the entity PROJECT named *manages*. It is a 1 : M relationship. Each employee is allowed to manage up to two projects, and each project is managed by one employee. However, some employees of the enterprise are not hired to be project managers, and those employees are not allowed to have management responsibility in a project. Therefore, those employees do not have a mandatory relationship with the entity PROJECT. Therefore, the entity EMPLOYEE has a partial involvement in the relationship "manages." The optional entity is shown by placing a single-line arc near the optional entity. On the other hand, a PROJECT requires the participation of an EMPLOYEE, who manages it. Therefore, a double-lined arc is placed next to PROJECT to indicate that the participation is mandatory in the relationship.

A relationship is ***mandatory*** if one entity's occurrence requires the participation of the other entity in the relationship. Consider the relationship between PURCHASE_ORDER and PO_DETAIL. Each occurrence of PO_DETAIL requires an occurrence of PURCHASE_ORDER and vice versa. This means that when a purchase order is created, a detail must also be created at the same time. If PO_DETAIL were optional to PURCHASE_ORDER, then a purchase order could be created without creating the detail simultaneously. Therefore, the definition of which relationships are mandatory and which are optional has much to do with the way in which the business operates (i.e., what the business rules are).

Figure 3.2 shows an optional participation between the entity VENDOR and the relation "is sent." This implies that it is possible to have some vendors in the database for which a purchase

Figure 3.6 Optional participation.

order does not exit. As previously mentioned, most firms operate this way. They approve vendors to be potential suppliers of materials and enter these approved vendors into the database without any immediate intention of buying from them. Therefore, VENDOR has optional participation in the relationship. On the other hand, there may be firms that enter new vendors into the database only when they write a purchase order to buy materials from them. If this were the business rule used by the firm, then VENDOR would be mandatory to the relationship and a double line would be shown on the arc. Also, the cardinality limits on the VENDOR side would be (1,N) instead of (0,N). The degree of participation reflects the business rules used by the firm.

3.2.2 The Degree of a Relationship

The ***degree of a relationship*** is a measure of the number of entities sharing the same association. There are four cases: binary relationships, ternary relationships, n-ary relationships, and unary relationships.

The *binary* relationship is most common, and it exists when two entities have an associated relationship. For example, in Figure 3.2, VENDOR and PURCHASE_ORDER share a binary relationship, as does PURCHASE_ORDER and PO_DETAIL.

A *ternary* relationship exists when three entities share a common relationship. Consider, for example, Figure 3.7, which models certain entities involved in the inspection process for raw materials received by an enterprise. For each type of MATERIAL to be inspected by quality assurance, there is a QUALITY_TEST for that specific material. The test requires that a sample is taken from the lot of delivered material and subjected to the test. However, the QUALITY_TEST may be a variable even for the same material. This is because different vendors may require different inspection levels. The inspection level can be viewed as a set of state assignments given to vendors that indicates the severity of the inspection of their shipments to the enterprise. Terms such as "minimum," "normal," and "restricted" could be used to describe three levels. For vendors who have historically shipped superior materials, a minimum inspection standard may be required, or they may not require inspection at all. A vendor with whom the company has had a bad experience in the past may be put on a higher (restricted) inspection level. In any event, for this situation, these three entities (VENDOR, MATERIAL, and QUALITY_TEST) are interactive, as indicated by the ternary relationship. If the quality tests for a material were independent of the vendor, then the appropriate form of the relationship would be binary (i.e., two M:N relationships). VENDOR would have a relationship with MATERIAL, and MATERIAL would have a relationship with QUALITY_TEST. It is the particular practice of the company that determines whether the relationship is binary or ternary.

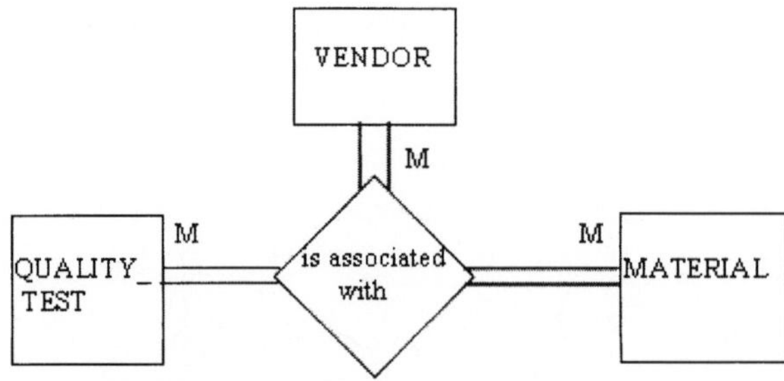

Figure 3.7 Example of a ternary relationship.

When a ternary relationship appears in an E-R diagram, it is often an M:N relation. It is desirable to find a way to reduce it to a series of binary 1:M relationships, which is possible in this case. This will be left as an end-of-chapter exercise for the reader.

When more than three entities share a relationship, it is known as n-ary. This situation seldom occurs and can be ignored for our purposes.

A *unary* relationship exists when an association is maintained within a single entity (i.e., the entity has a relationship with itself). A candidate for a unary relationship occurs in the area of project management and is shown in Figure 3.8. A *project* is defined as a sequence of activity steps. These activity steps are sequential tasks that must be performed in order to complete the entire project. Therefore, a project has one or more activity steps. Another characteristic of a project is that the activity steps have a precedence relationship. Some activity steps must be performed before other activity steps. Hence, each activity step has a relationship to other activity steps called "immediate predecessors." The cardinality of that relationship can be 0 (for the first activity step of the project, which has no predecessor), 1, or many. This is reflected in the unary relationship of ACTIVITY_STEP with itself, which is also called a ***recursive relationship***.

The cardinality of the unary relationship will depend on the specific network structure of the projects themselves. Here we show a M:N relationship for illustration. This is based on a project network as shown in Figure 3.9. The arcs indicate the individual project steps, each having a certain duration. The directions of the arcs indicate the precedence relationship of the steps. Note that S1 is the initial step of the project, and S6 is the final step.

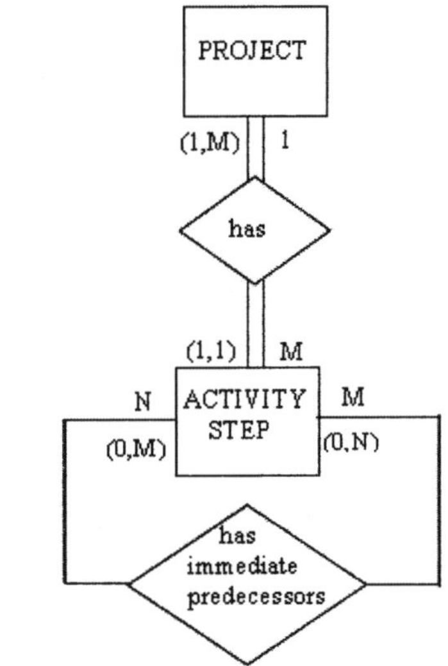

Figure 3.8 Example of a unary relationship.

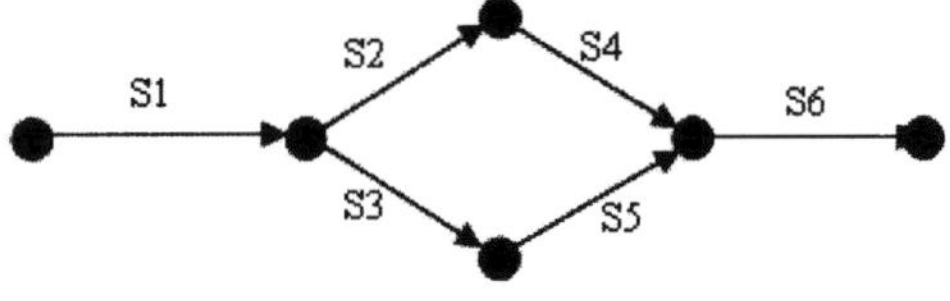

Figure 3.9 Typical project network.

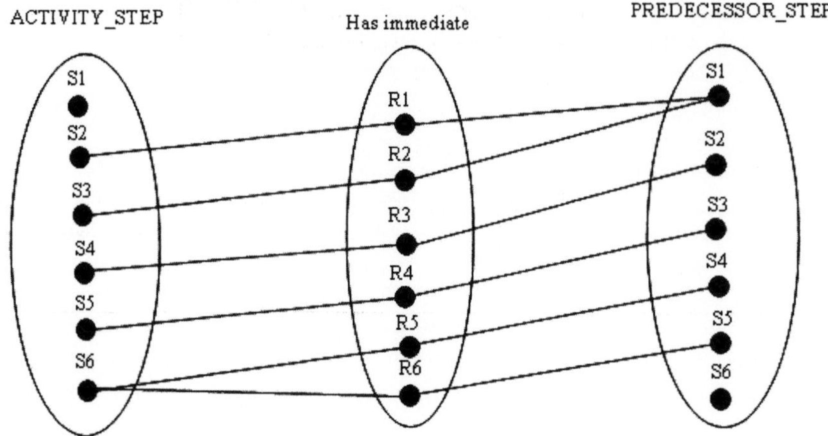

Figure 3.10 Semantic net diagram for Figure 3.8.

We create a semantic net diagram between activities and their predecessors, as shown in Figure 3.10. Step S1 on the activity step side does not participate in the relationship since it has no predecessors. Also, step S6 on the predecessor side does not participate in the relationship because it is not a predecessor of any other step. This is reflected in the optional participation in Figure 3.8. Step S6 on the activity step side has two relationships (S4, S5), and step S1 on the predecessor side has two relationships (S2, S3). Hence, the cardinality is M:N.

3.2.3 Composite Entities

The relational model requires the use of 1:M relationships. This enables queries to be designed that use the foreign key of the table on the M side of the relationship to retrieve corresponding unique rows on the 1 side of the relationship. In the course of developing an E-R diagram, M:N relationships are often encountered. When this happens, the analyst can break down the M:N relationship into a series of 1:M relationships using a ***composite entity***.

Consider the situation shown in Figure 3.11. It is clear that the vendors to the enterprise supply the materials. However, any particular vendor may supply more than one material and any material may be supplied by more than one vendor. In effect, there is a many-to-many relationship between VENDOR and MATERIAL.

Since the VENDOR_ID is the primary key for VENDOR, it violates the rules of the relational database to add the attribute MATERIAL_ID to VENDOR and duplicate the VENDOR_ID instance for each material that the vendor supplies. It can be argued that such a table would still be valid if you changed the primary key to the composite key VENDOR_ID/MATERIAL_ID. However, this would result in many more entries in the table, with the street, city, state, and zip code duplicated for each instance of VENDOR_ID/MATERIAL_ID.

On the other hand, it can be argued that it is possible to add multiple material ID attributes to VENDOR (MATERIAL_ID1, MATERIAL_ID2, etc.). However. since the number of such attributes is determined by the vendor that can service the most materials, this will result in numerous NULL fields when the table is populated.

The solution is the use of a composite entity, as shown in Figure 3.12. The entity VENDOR_MATL_XREF is a bridge between the entities VENDOR and MATERIAL. Note that each instance

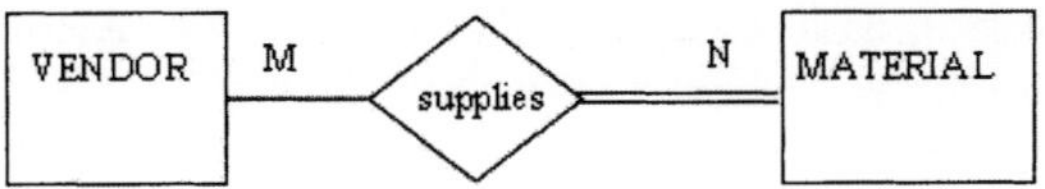

Entity Set: VENDOR

VENDOR ID	V NAME	V STREET	V CITY	V STATE	V ZIP
V110	Jersey	2 Main St.	Patterson	NJ	07055
V25	General	125 Common	Boise	ID	44830
V250	Spices	25 Salty Lane	East Hampton	NY	10027
V75	Pasta Supply,	34 Henry St.	Philadelphia	PA	09098

Entity Set: MATERIAL

MATERIAL ID	MATL DESCRIPTION
RM201	Carrots, whole
RM202	Carrots, diced, 1/4 inch
RM205	Potatoes, Eastern,
RM210	Peas, shelled
RM211	Tomatoes, whole
RM310	Garlic, whole
RM311	Garlic powder
RM318	Salt, iodized
RM308	Onion salt
RM305	Paprika
RM340	Sugar, bulk
RM805	Olive oil
RM810	Vinegar, white

Figure 3.11 Relationship between VENDOR and MATERIAL.

VENDOR M Supplies N MATERIAL

Entity Set: VENDOR

VENDOR ID	V NAME	V STREET	V CITY	V STATE	V ZIP
V110	Jersey	2 Main St.	Patterson	NJ	07055
V25	General	125 Common	Boise	ID	44830
V250	Spices	25 Salty Lane	East Hampton	NY	10027
V75	Pasta Supply,	34 Henry St.	Philadelphia	PA	09098

Entity Set: MATERIAL

MATERIAL ID	MATL DESCRIPTION
RM201	Carrots, whole
RM202	Carrots, diced, 1/4 inch
RM205	Potatoes, Eastern, whole
RM210	Peas, shelled
RM211	Tomatoes, whole
RM310	Garlic, whole
RM311	Garlic powder
RM318	Salt, iodized
RM308	Onion salt
RM305	Paprika
RM340	Sugar, bulk
RM805	Olive oil
RM810	Vinegar, white

Entity Set: VENDOR_MATL_XREF

VENDOR ID	MATERIAL ID
V25	RM201
V25	RM205
V25	RM318
V25	RM340
V110	RM201
V110	RM202
V110	RM205
V110	RM210
V110	RM211
V250	RM310
V250	RM311
V250	RM318
V250	RM340
V250	RM308
V250	RM305
V25	RM805
V25	RM810

Figure 3.12 Use of the composite entity.

in VENDOR is associated with M instances of VENDOR_MATL_XREF, and each instance of MATERIAL is associated with M instances of VENDOR_MATL_XREF. Therefore, the M:N relationship is converted to two 1:M relationships, and the referential integrity is preserved. The tuple VENDOR_ID/MATERIAL_ID serves as the (composite) primary key of VENDOR_MATL_XREF, while VENDOR_ID is a foreign key referencing VENDORS and MATERIAL_ID is a foreign key referencing MATERIAL.

When M:N relationships are encountered during the data modeling design process, the requirement for a composite entity can be indicated by placing a box around the relationship diamond, as shown in Figure 3.12. When it is time to implement the data model in database tables, a cross reference table is established as a bridge. Thus, the existence of a 1:M relationship is ensured in the implemented database.

3.2.4 Recursive Entities

Recursive entities are entities with unary relationships. Such entities are naturally occurring in some important manufacturing problems. The typical example of this is the product bill of materials (BOM), in which parts are made of other parts. Consider the hypothetical product structure shown in Figure 3.13. There are two end products: 1 and 2. Each end product is composed of sub-

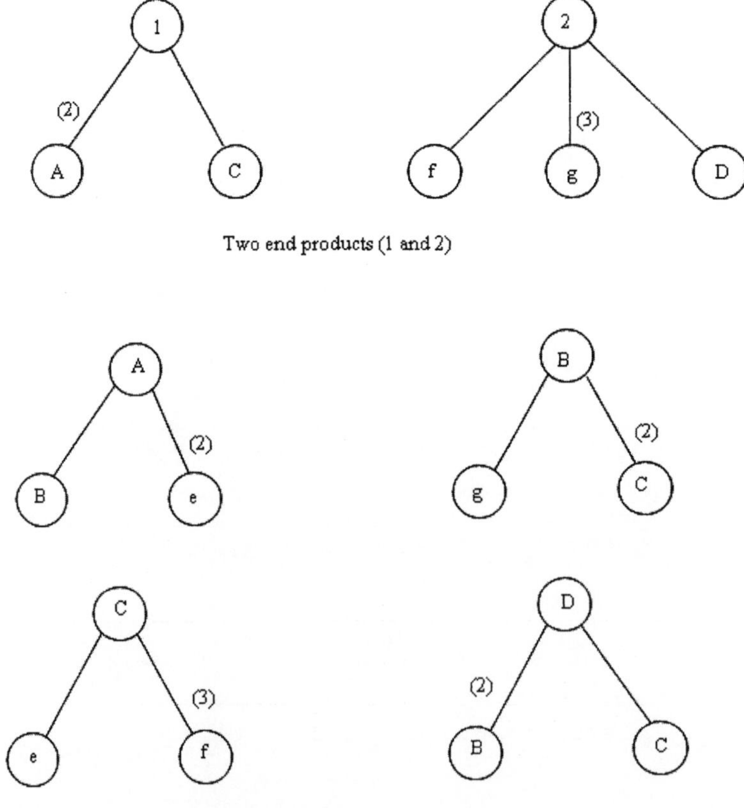

Figure 3.13 Product structure for two hypothetical end products.

assemblies and components. There are four subassemblies: A, B, C, and D. Each subassembly is composed of other subassemblies and components. There are three components: e, f, and g. The numbers next to the subassemblies and components indicate the quantity required. Where no quantity is shown, a quantity of 1 is implied. The BOM shows all the subassemblies and parts that are required to build one unit of an item.

Whether the item is a product, a subassembly, or a component, the database design will be more parsimonious if a generic term is used to describe these items. A term such as PART could describe any and all materials kept by the enterprise. Since parts are made up of parts, PART will have a relationship with itself in the data model, as shown in Figure 3.14.

A compact way of expressing the product structure shown in Figure 3.13 is to use a bill of materials matrix, which is shown in Table 3.1. The rows of Table 3.1 are the *how-constructed files* of the BOM; reading across, we can determine exactly what is required to go directly into an end product or subassembly. For example, product 1 requires two units of A and one unit of C. The columns of Table 3.1 are the *how-used files*. For example, C is used directly in the production of product 1 and subassemblies D and B.

The bill of materials matrix is organized as an upper triangular matrix. It is organized with products first, followed by subassemblies and then components. Also, within subassemblies and components, the order is based on the *level* of the subassembly or component. It is common for subassemblies and components to be numbered by levels. The level numbers describe the relative position of assemblies and parts to one another and to the final assembly. Level numbers are assigned on the basis of the maximum number of stages of assembly required to get the assembly or component into the end product.

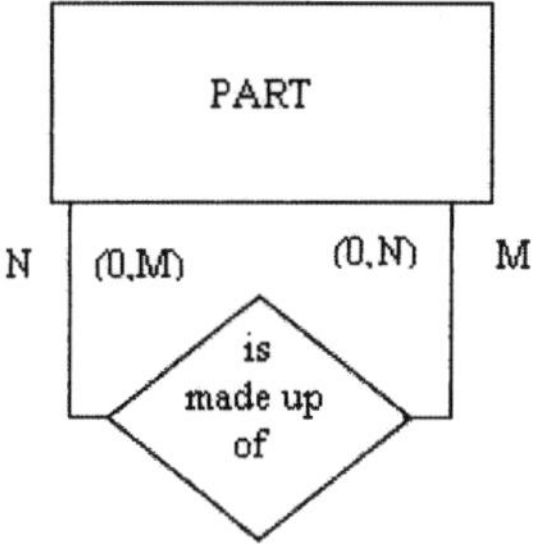

Figure 3.14 A recursive entity.

Table 3.1
Bill of Materials Matrix

	1	2	A	D	B	C	g	e	f
1			2			1			
2				1			3		1
A					1			2	
D					2	1			
B						2	1		
C								1	3
g									
e									
f									

As an example of the level numbering system, consider Figure 3.13 in relation to Table 3.2. Figure 3.13 shows that subassembly A goes directly into product 1 and is not used anywhere else. Hence, subassembly A is one level removed from an end product and is assigned to level 1, as shown in Table 3.2. By similar reasoning, it can be shown that subassembly D is a level-1 item.

Subassembly B goes directly into product 2. However, subassembly B is also used to produce level-1 subassemblies A and D. Hence, the maximum number of steps of assembly required to reach the end product is two. Subassembly B is a level-2 item.

By continuing the same reasoning process, level assignments are made as shown in Table 3.2. These level assignments correspond to the ordering of items within the bill of materials matrix given in Table 3.1.

The recursive relationship of Figure 3.14 describes the fact that PART has a relationship to itself. However, it does not describe how to implement that relationship in a table. When a recursive relationship exists, its implementation in tables depends on the particular relationship and how it is to be employed when using the database. Therefore, we shall discuss some possibilities with respect to the BOM.

In production planning, the bill of material information is used to determine the total requirements of production to meet end product demand. The total requirements of production include all the subassemblies and components that have to be built in order to meet the demand for the end products. So, from Table 3.1, demand for a unit of end product 1 requires building two units of subassembly A and one unit of subassembly C; but in order to build a unit of C, one unit of component e and three units of component f must be manufactured. Each time a demand of n units of a product is called for, it is necessary to *explode* this demand into the demand for all the subassemblies and components that must be fabricated.

For this reason, it is useful to convert the bill of materials matrix into a *total-requirements matrix*. With regard to this, we note that components and subassemblies enter the final product both directly and indirectly. To account for the complete requirements of an nth-level component, we must sum its relationship to subassemblies and final products. This can be done starting from the end product in Table 3.1 and tracing each subassembly and component that goes into it, or by using matrix algebra to compute total requirements. For a complete description of the matrix algebra approach, the reader is referred to Elsayed and Boucher (1994). The total-requirements matrix is shown in Table 3.3.

Table 3.2

Item Level Assignments

		Level		
0	**1**	**2**	**3**	**4**
1	A	B	C	e
2	D		g	f

Table 3.3
Total Requirements Matrix

	1	2	A	D	B	C	g	e	f
1	1		2		2	5	2	9	15
2		1		1	3	7	6	7	21
A			1		1	2	1	4	6
D				1	2	5	2	5	15
B					1	2	1	2	6
C						1		1	3
γ							1		
α								1	
β									1

Consider the entries of the total-requirements matrix. Each row shows the total units of an item required in order to build one unit of the item indicated in the row heading. For example, building 1 unit of product 1 requires 15 units of f. This requirement can be traced from Table 3.1. In Table 3.1, 1 unit of product 1 requires 2 units of A and 1 unit of C. The row for subassembly C indicates that producing C requires 3 units of f. Therefore, this relationship accounts for 3 units of f, leaving 12 units still to be accounted for. This remaining 12 units can be found by tracing the 2 units of A. From the subassembly A row, 2 units of A requires 2 units of B. From the subassembly B row, 2 units of B requires 4 units of C. From the subassembly C row, 4 units of C requires 12 units of f. Hence, we have accounted for the 15 units of f in the total requirements for end product 1. The reader can review the remainder of the total-requirements matrix to trace the source of these entries.

The recursive relationship in the E-R diagram shown in Figure 3.14 is many to many. Therefore, it is necessary to decompose it to obtain the desired 1:M relationship when creating tables. However, unlike the situation of MATERIAL and VENDOR in Figure 3.12, the case of a simple bridge entity is not appropriate because there is only one table (entity) involved. In the case of an M:N unary relationship, one approach is to create two tables as shown in Figure 3.15. Figure 3.15 shows a simple structure that may be considered. The PART table is designed for basic information and a table titled PART_COMPONENT is used to reference the BOM structure of Table 3.1. The primary key of the PART_COMPONENT table is the composite key (PART_NO, PART_COMP_NO), where the component is from the BOM matrix. Although this is a simple 1:M structure, upon reflection it is not easy to query the table and return the total requirements for the production of, say, *n* items of product 1 and *m* items of product 2. Remember, this is the kind of information we wish to retrieve from the database. However, by writing a high-level program, for example, in C language, it would be possible to retrieve the level-1 requirements and, using that information, return to the table and retrieve the level-2 requirements, and so on until the total requirements are returned.

A table structure that makes the query for total requirements easier is shown in Figure 3.16. Here, a 1:M relationship is established between a parent table PART and a child table PART_COMPONENT. PART_COMPONENT is designed from the total-requirements matrix of Table 3.3. It includes the composition of all the end items and subassemblies. Because subassemblies are used

TABLE PART

PART_NO	PART_DESC	DRAWING_NO
1	Product 1	102-23
2	Product 2	110-20
A	Subassembly A	290-10
B	Subassembly B	220-05
C	Subassembly C	256-01
D	Subassembly D	245-90
e	Component e	335-23
f	Component f	304-20
g	Component g	356-90

TABLE PART_COMPONENT

PART_NO	PART_COMP_NO	PART_COMP_QTY
1	A	2
1	C	1
2	D	1
2	f	1
2	g	3
A	B	1
A	e	2
B	C	2
B	g	1
C	e	1
C	f	3
D	B	2
D	C	1

Figure 3.15 PART tables using the BOM matrix.

TABLE PART

PART_NO	PART_DESC	DRAWING_NO
1	Product 1	102-23
2	Product 2	110-20
A	Subassembly A	290-10
B	Subassembly B	220-05
C	Subassembly C	256-01
D	Subassembly D	245-90
e	Component e	335-23
f	Component f	304-20
g	Component g	356-90

TABLE PART_COMPONENT

PART_NO	SUB_PART_NO	SUB_PART_QTY
1	A	2
1	B	2
1	C	5
1	e	9
1	f	15
1	g	2
2	B	3
2	C	7
2	D	1
2	e	7
2	f	21
2	g	6
A	B	1
A	C	2
A	e	4
A	f	6
A	g	1
B	C	2
B	e	2
B	f	6
B	g	1
C	e	1
C	f	3
D	B	2
D	C	5
D	e	5
D	f	15
D	g	2

Figure 3.16 PART tables using total-requirements matrix.

in various end products and components are used in various subassemblies and end products, the table is quite large. In fact, it is carrying all the possible combinations of parts that exist in the BOMs of the enterprise. Figure 3.15 is efficient in terms of data storage. Figure 3.16 is efficient in terms of query performance. If storage is not an issue and BOM queries are made often, it may be more effective to store data using the total-requirements format. It is a matter of choice on the part of the user to decide how data will be stored to maximize performance in the database.

3.2.5 Superclass and Subclass Entity Types

To simplify the data model as much as possible while maintaining useful information, it is desirable to aggregate people, places, and things into as few entities as possible, consistent with the semantic requirements of the problem. Thus, raw materials, subassemblies, final assemblies, office materials, repair and maintenance materials, and so forth could all reasonably be part of an entity called MATERIAL. Similarly, all employees, whether managers, engineers, technicians, or production workers, could be part of an entity called EMPLOYEE. Although there are distinctions to be made among the people, places, and things being aggregated, the members within a group do have certain attributes in common. For example, all instances of the MATERIAL entity have in common a material number, a description, a unit of measure (gallons, pounds), and so on. All instances of the EMPLOYEE entity have common attributes such as social security number, name, and address. However, there are attributes of MATERIAL and EMPLOYEE that are unique to subgroups. For example, some employees may be salaried workers while others are paid hourly This gives rise to the use of superclass and subclass entity types.

A ***superclass*** is an entity type that represents a set of similar people, places, or things that have common attributes but represent distinct subclasses. The ***subclass*** is an entity that is a member of a superclass but has distinct attributes that are not in common with all other members of the superclass. Figure 3.17 illustrates this point. EMPLOYEE is a superclass entity. All employees share the attributes of social security number, date of birth, name, and address. One could also include other shared attributes such as gender and telephone number. On the other hand, there are attributes that are not shared by all employees. Some employees are paid on a salary basis, while other employees are paid on an hourly basis. Machinists and other skilled workers are assigned a skill classification, while only sales personnel are assigned a sales district. To carry all of these attributes within the superclass EMPLOYEE would result in many NULL entries in the table. Therefore, a relationship is defined between the superclass and each set of its subclasses.

A superclass-subclass relationship is modeled using the entity rectangle to describe both superclass and subclass entities. The arcs connecting the superclass to subclasses enters and exits a specialization circle, which is coded to indicate the degree of specialization in the relationship. The code "d" indicates *distinct.* This means that any instance of the superclass entity may be in one and only one of the subclass entities. In Figure 3.17, the subclass entities FULL_TIME and PART_TIME are distinct. An employee is either working full time or part time, but not both. The code "o" indicates "overlapping." This means that any instance of the superclass entity may be in more than one subclass entity. In Figure 3.17, For example, a person may have dual roles as both a sales person and the sales manager. Therefore, he or she has a sales district *and* is required to manage other sales personnel.

There is a 1 : 1 relationship between the superclass and its subclasses. Every member of a subclass has to be a member of the superclass. However, not every member of the superclass has to be a member of a subclass. When every member of the superclass must be a member of a subclass, a double line is drawn between the superclass entity and the specialization circle. In Figure 3.17,

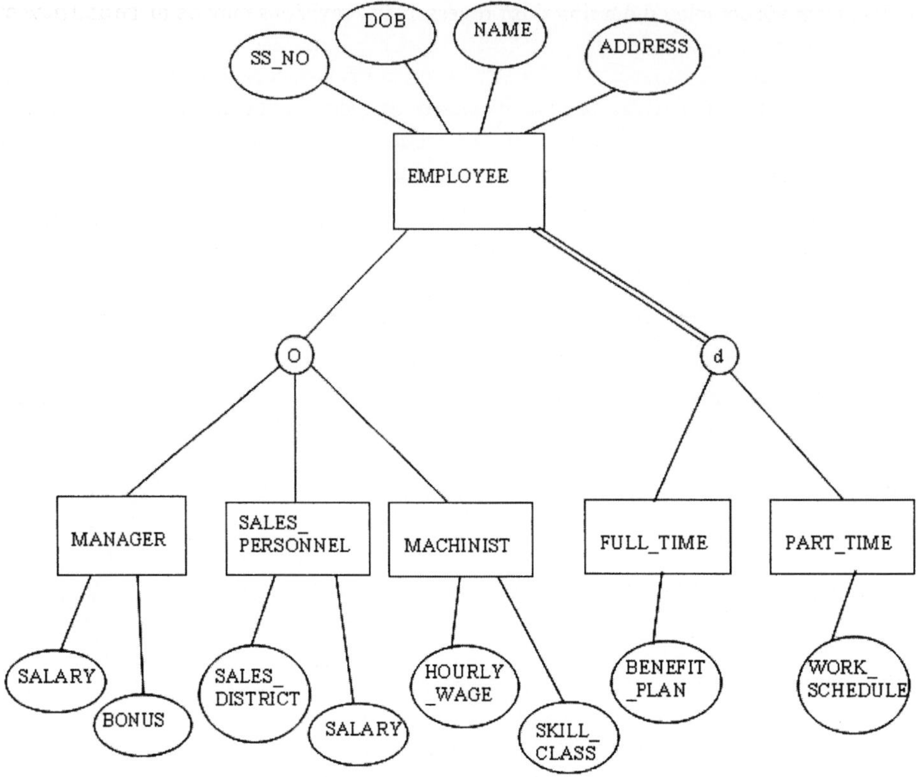

Figure 3.17 Superclass and subclass hierarchy.

this is the case for the relationship between EMPLOYEE and the subclass entities FULL_TIME and PART_TIME. If every instance of the superclass is not required to be a member of a subclass, a single line is drawn between the superclass entity and the specialization circle.

During the process of building the data model for an enterprise, the designer may find that several entities that have been defined have many attributes in common. This is an indication that a superclass entity may be appropriate. On the other hand the designer may begin with large classifications of things into entities, only to find that there are significant differences in the attributes required to completely describe them. If significant common attributes exist, this is an indication that subclass entities should be created. Finding superclass-subclass relationships can work from the bottom up or top down. One of the roles of the designer is to be aware of the possibilities when he or she encounters them.

3.3 CASE STUDY IN DATA MODELING

Developing and E-R diagram usually involves the collaboration between the analyst and individuals within the organization who are familiar with the entities and attributes important to the organization. Of course, they probably do not think of them as entities and attributes. The analyst must lead in an iterative process of uncovering these entities and attributes and refining a data model as the process unfolds. The model is reviewed with individuals in the organization until a final E-R diagram is completed. During this process, many revisions may be made and new entities and relationships may be uncovered.

In this section, we illustrate the process of developing an E-R model using a hypothetical organization within the firm that is responsible for managing the materials inventoried by the firm. We will assume that the following information results from an interview process with key individuals in the organization:

1. The department is responsible for the inventory of three material types: raw material, finished goods, and supplies. Each material, identified by a material ID, belongs to one material type. Each material type classifies many different materials.

These statements indicate the existence of two entities. One entity is the MATERIAL_TYPE, which defines a class of materials. The other entity is MATERIAL, with its primary key of *material ID*. At this point, we will refrain from including attributes in the diagram and focus on entities and relationships. The E-R diagram is shown in Figure 3.18. At this point, it is useful to mention that MATERIAL_TYPE and MATERIAL are concepts, not tangible things. Material becomes tangible when it is delivered and placed into inventory. Prior to that it is just a classification scheme.

2. When a raw material or supply is ordered, it is done using a purchase order, where it appears on the purchase order detail. Purchase orders are also used to order capital equipment and services.

This statement introduces the PURCHASE_ORDER and PO_DETAIL, which were discussed earlier in this chapter. The issue is how the entity MATERIAL relates to these. With our prior knowledge of the structure of a purchase order, we know that the material identification appears on the detail, not on the header of the purchase order. Therefore, MATERIAL is related to PO_DETAIL, as shown in Figure 3.19.

Note that MATERIAL and PO_DETAIL have a 1 : M relationship. The entity material is the unique set of all the different materials purchased and inventoried by the enterprise. Each material appears once in this set and is defined by the unique material ID. Each of the materials may appear more than once on a PO_DETAIL.

MATERIAL is optional to the relationship because finished goods, one of the material types, is not ordered on a purchase order. Also, PO_DETAIL is optional in the relationship because some of the instances of PO_DETAIL are for capital goods and services, not material.

3. When the materials or supplies ordered on a purchase order arrive or when finished goods are produced, they are a tangible manifestation of the concept of MATERIAL. They are called *material lots*, and each is assigned a unique identification called a *material lot number*. For each material, there may be zero, one, or many material lots. Each material lot is the tangible manifestation of one and only one material.

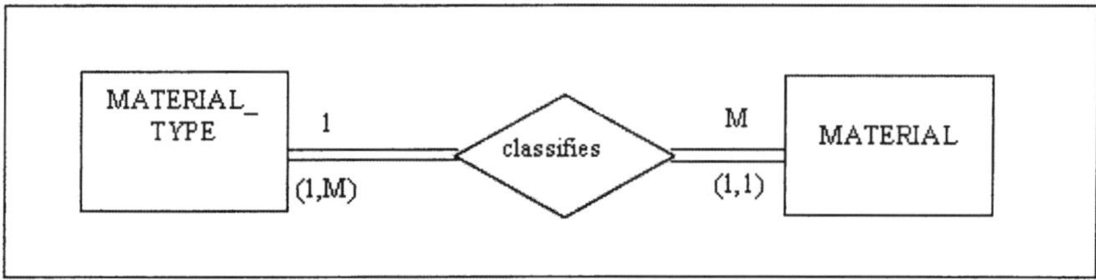

Figure 3.18 Relationship 1.

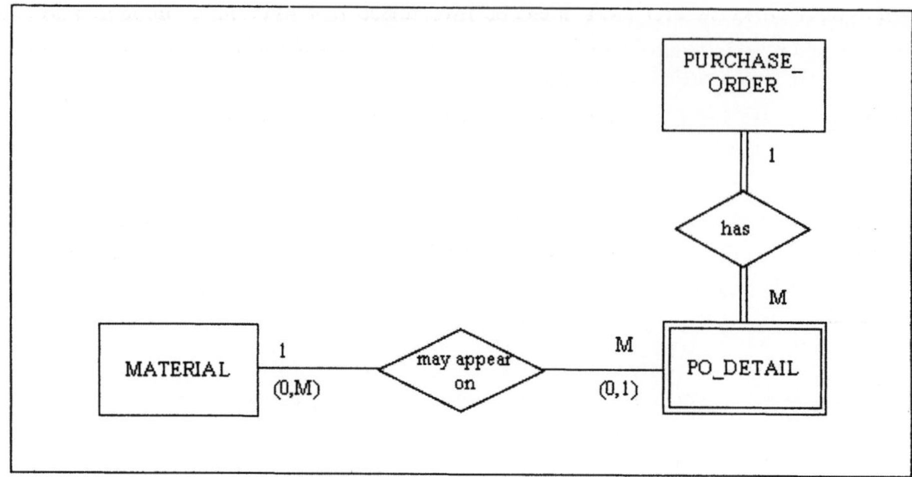

Figure 3.19 Relationship 2.

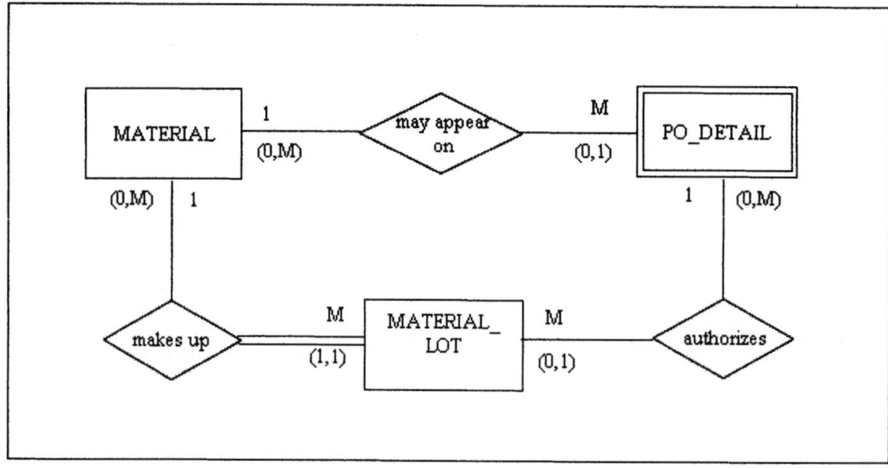

Figure 3.20 Relationship 3.

As Figure 3.20 shows, MATERIAL_LOT has a relationship with MATERIAL and with PO_DETAIL. We know from the description that the material lot is the physical manifestation of the concept of material. Since it is possible that some instance of material may have been defined but never purchased or produced, the entity MATERIAL has an optional relationship to MATERIAL_LOT. However, each material lot must participate in the relationship to material.

The MATERIAL_LOT is the thing that is delivered to the enterprise to satisfy the order that appears on the PO_DETAIL. Not all purchase order details are ordering material lots since, as previously mentioned, capital equipment and services are also ordered on purchase orders. Some material lots are finished goods, so MATERIAL_LOT is optional in the relationship to PO_DETAIL.

4. The entity MATERIAL_LOT includes inventoried raw materials, supplies, and finished goods. These three categories are nonoverlapping and each instance of a material lot must appear in one and only one category.

This is an example of a superclass-subclass hierarchy. There is good reason to decompose MATERIAL_LOT into these subclasses. These subclasses do have some attributes in common. For example, the attributes material ID, material description, quantity on hand, and so on, are common to all materials. However, there are clearly some attributes that are different. Purchased raw materials and supplies will be related to vendors and purchase orders on which they were purchased. This is not the case with finished goods. The finished goods are related to the production lines on which they were made, which is not relevant to raw material and supplies. In Figure 3.21, the mandatory and distinct relationship is shown because all the material lots must be represented in one and only one subclass.

5. A material lot is stored in a warehouse location. A warehouse location may store zero, one, or many material lots. If a material lot is too large to be stored in one warehouse location, it may have to occupy more than one location. A material lot may not be assigned to a warehouse location at some points in time. For example, when raw material is taken from storage and moved to the production area for use in making finished product, it still exists as a material lot but is not in storage.

The warehouse is a thing about which the company needs to keep information. Here we are interested in the material lots that occupy the locations in the warehouse. As Figure 3.22 shows, MATERIAL_LOT is optional in the relationship because it is not necessarily in warehouse storage. Some warehouse locations may not be used for storage; they may be empty. Therefore, WAREHOUSE_LOCATION is optional with respect to this relationship.

The cardinality of the relationship is M:N. A material lot may be stored in more than one location, and a warehouse location may contain more than one material lot.

6. If a material lot is finished goods, it was produced on a production line. A production line can produce more than one material lot of material, and each lot of finished goods will be produced on one and only one production line.

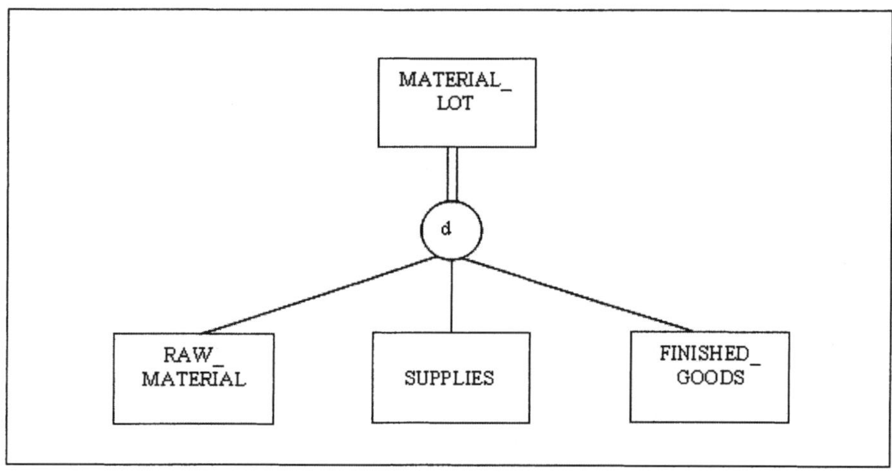

Figure 3.21 Relationship 4.

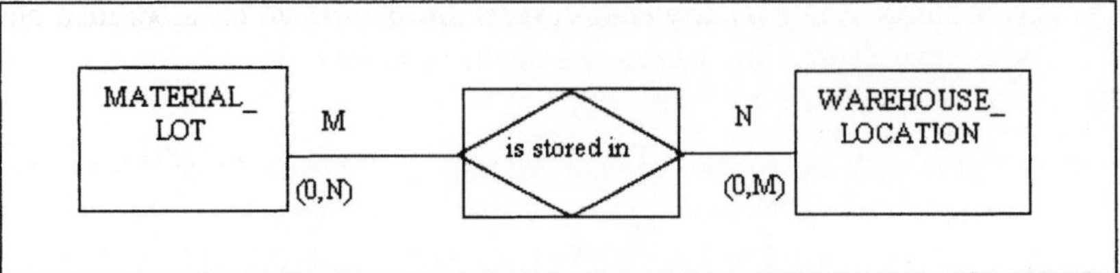

Figure 3.22 Relationship 5.

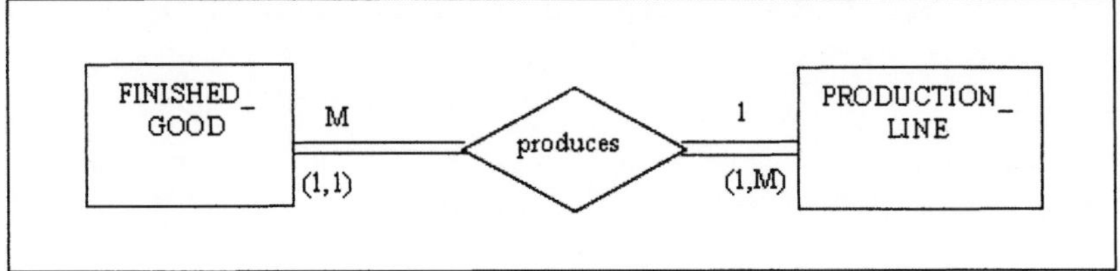

Figure 3.23 Relationship 6.

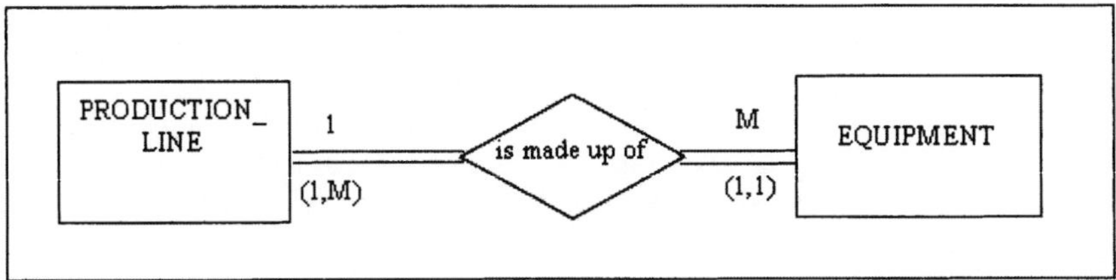

Figure 3.24 Relationship 7.

This relationship is between a subclass of the entity MATERIAL_LOT and a new entity, PRODUCTION_LINE. Figure 3.23 shows this relationship to the subclass FINISHED_GOOD instead of the superclass MATERIAL_LOT. When implemented in a table, the subclass FINISHED_GOODS will hold the foreign key *production line number.* Therefore, any query intended to retrieve the relationship between a material lot and a production line on which it was made will include the table based on the entity holding the foreign key.

Again, it is worth discussing the difference between a concept of a thing and its physical manifestation. A finished good lot is a physical manifestation of a product. It is possible that a product (the concept) may be produced on any one of several different production lines or assembly lines in the plant. However, when a finished good lot of that product is actually produced, it is produced on one, and only one, production or assembly line. Thus, a 1 : M relationship exists.

7. A production line is made up of one or more pieces of equipment, each with its own *equipment ID.* Each piece of equipment belongs to one and only one production line.

Figure 3.24 shows the relationship, which has cardinality 1 : M and is mandatory on both sides.

8. Materials may have quality control tests associated with them. This is true of some raw materials and all finished goods. Some raw materials and all supplies are not subject to any

quality control tests. Some quality control tests are performed on more than one material, and some materials require more than one quality control test.

Note that a quality control test is a description of a procedure used to perform the test. It is not the test itself. Actual tests are performed on material lots, the physical manifestation of material. Figure 3.25 shows the association between a description of a test procedure and the entity MATERIAL. MATERIAL is optional to the relationship because some material is not associated with a test. On the other hand, QUALITY_CONTROL_TEST is mandatory in the relationship because it requires MATERIAL to which the test is associated.

A summary view of the relationships described in this section is shown in Figure 3.26. There is still much work to be done on specifying attributes and creating detailed tables. We will revisit these issues in Chapter 5, where we will implement more detailed specifications.

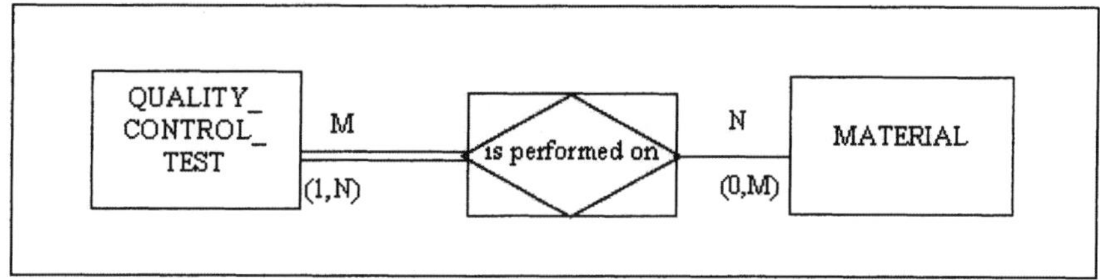

Figure 3.25 Relationship 8.

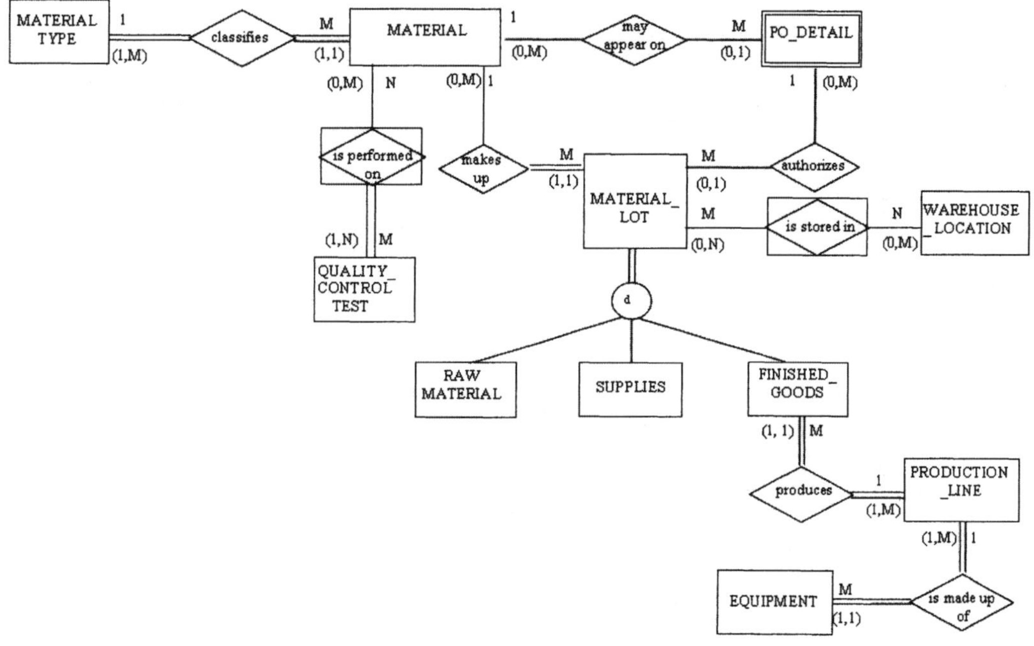

Figure 3.26 Summary E-R diagram.

3.4 NORMALIZATION

As previously mentioned, when designing a data model, attribute field redundancy must be avoided as much as possible. This is so that the implementation in tables is not burdened with the storage of unnecessary data. ***Normalization*** is the name given to the process of evaluating a model in order to control data redundancies.

Data redundancy causes problems with the integrity of data when the usual database operations of insert, delete, and update are performed on the data. These problems are called anomalies, most commonly referred to as insertion anomalies, deletion anomalies, and update anomalies. To illustrate, we refer to Figure 3.27. The Vendor and Purchase Order tables of Figure 2.1 are shown in Figure 3.27a. The structure of these tables conform to a 1 : M relationship in which Vendor_ID is a foreign key of the table Purchase Order. We can write the table names and their corresponding sets of relations as follows:

```
Purchase Order (PO Number, Release_Date, PO_Status, PO_Amt, Vendor_ID)
Vendor (Vendor ID, V_Name, V_Street, V_City, V_State, V_Zip)
```

When using this notation the primary key attribute(s) in the relation is underlined.

Figure 3.27b is a join of the Purchase Order and Vendor tables. It is not inconceivable that a person might construct such a table to hold the purchase order header information, thinking that it is unnecessary to have a separate table for the vendors. It is obvious from what we have learned thus far that Figure 3.27b is a poor model for a relation because of the redundant values in the attribute columns. Each time a V_Name is repeated, all the address information is repeated as well. We name the entity set "PO_Vendor" and write the entity set as follows:

Table: Purchase_Order

PO NUMBER	RELEASE DATE	PO STATUS	PO AMT	VENDOR ID
2591	2/10/06	CLOSED	$4,300.00	V110
2592	2/10/06	OPEN	$505.50	V25
2593	2/11/06	OPEN	$4,000.00	V110
2594	2/12/06	OPEN	$3,280.00	V250
2595	2/15/06	OPEN	$500.00	V250
2596		HOLD	$1,000.00	V75

Table: Vendor

VENDOR ID	V NAME	V STREET	V CITY	V STATE	V ZIP
V110	Jersey Vegetable Co.	2 Main St.	Patterson	NJ	07055
V25	General Provisions	125 Common St.	Boise	ID	44830
V250	Spices Unlimited	25 Salty Lane	East Hampton	NY	10027
V75	Pasta Supply, Inc.	34 Henry St.	Philadelphia	PA	09098

a) 1:M Relationship

Table: PO_Vendor

PO NUM	RELEASE DAT	PO STAT	PO AMT	V NAME	V STREET	V CITY	V STATE	V ZIP
2591	2/10/2006	CLOSED	$4,300.00	Jersey Vegetable Co.	2 Main St.	Patterson	NJ	07055
2592	2/10/2006	OPEN	$505.50	General Provisions	125 Common St.	Boise	ID	44830
2593	2/11/2006	OPEN	$4,000.00	Jersey Vegetable Co.	2 Main St.	Patterson	NJ	07055
2594	2/12/2006	OPEN	$3,280.00	Spices Unlimited	25 Salty Lane	East Hampton	NY	10027
2595	2/15/2006	OPEN	$500.00	Spices Unlimited	25 Salty Lane	East Hampton	NY	10027
2596		HOLD	$1,000.00	Pasta Supply, Inc.	34 Henry St.	Philadelphia	PA	09098

b) Redundant data entries

Figure 3.27 A comparison between table structures for the same data.

```
PO_Vendor (PO_Number, Release_Date, PO_Status, PO_Amt, V_Name, V_Street,
           V_City, V_State, V_Zip)
```

Note that PO_Number is the primary key attribute of the relation PO_Vendor. All the remaining attributes of the tuple are dependent on PO_Number.

3.4.1 Insertion Anomalies

Different kinds of anomalies can occur when inserting new records into a table structured like Figure 3.27b. For example, if the enterprise qualifies a new vendor and wishes to record vendor information in the database, the new record in PO_Vendor will start from V_Name. The first five fields will be NULL values until a purchase order is assigned to the vendor. In fact, it will not be possible to enter the vendor record because PO_Number is the primary key attribute and a primary key cannot be NULL. Therefore, it will not be possible to retain vendor information in PO_Vendor without first assigning a purchase order.

Consider a different possibility in which a new purchase order is being assigned to an existing vendor. When the data are entered for the new purchase order, there is always the possibility that an error will occur in entering an attribute value. For example, the wrong vendor street number or zip code could be entered. Then there will be two different descriptions of the same vendor in the database. This introduces ambiguity into the vendor information.

Note that these anomalies do not pertain to the structure of Figure 3.27a. In the first case, a new vendor can be entered into the database by inserting a record into the Vendor table. In the second case, there is only one record of vendor information for each Vendor_ID. Since all purchase orders having the same V_Name are related to the same record in the Vendor table, there is no chance of having ambiguous or conflicting information about a vendor.

3.4.2 Deletion Anomalies

The deletion anomaly results in the complete loss of information when a related record is deleted from the database. Again consider Figure 3.27b. Assume that purchase order number 2596 is canceled and the purchasing agent wishes to delete it from the database. The vendor for this purchase order is Pasta Supply Inc. In fact, this is the only purchase order to which that vendor is assigned. If the record for PO number 2596 is deleted, all information in the database concerning Pasta Supply Inc. will be lost. Note that this does not happen in Figure 3.27a. Any purchase order can be deleted without affecting the records in the Vendor table.

3.4.3 Update Anomalies

Consider the situation in which a vendor changes its address. For the table structure of Figure 3.27a, there is one update to the set of attribute values required. For the PO_Vendor table structure of Figure 3.27b, multiple records will need to be changed depending on the number of purchase orders assigned to the vendor. In fact, in large databases this may become very unwieldy and introduces a risk of missing values that have to be updated or of introducing different values for the same data element due to data entry errors.

3.4.4 Normal Forms

Normalization is the process of controlling data redundancy and eliminating the possibility of the anomalies previously described. If an analyst finds a document used by the enterprise and considers putting all the data elements into one table, what rules can she or he use to systematically determine whether or not redundancy can be eliminated? The process of normalization tries to uncover *functional dependencies* between attributes that are being considered for inclusion in a relation and provides the rules to be applied.

A functional dependency between two attributes (A and B) exists if a value of A uniquely determines a value of B. Functional dependency can be identified from a data set by identifying a value of A, "a," and, for each instance of that value, confirming that it is associated with the same value of B, "b." It is clear in Figure 3.27b that for the V_Name attribute with a value "Jersey Vegetable Co.," the corresponding value of the attribute V_Street is "2 Main St." We say that V_Name functionally determines V_Street. It can also be said that V_Name is functionally dependent on V_Street since each unique value of V_Street determines a unique value of V_Name. At this point it is clear that the tuple (V_Name, V_Street) belongs in the same relation.

By similar logic, the attributes V_City, V_State, and V_Zip are functionally dependent on V_Name and V_Street. That is to say, for each value of V_Name or V_Street, the same values of V_City, V_State, and V_Zip occur. However, the inverse is not true in general. Because of the limited data set presented in Figure 3.27, we do not see that there are other vendors who reside in the same city, state, or zip code as the vendors shown. With additional data it would be obvious that V_Name and V_Street functionally determine V_City, V_State, and V_Zip, and not vice versa.

From a systematic evaluation of the data of Figure 3.27b, the subset of attributes having functional dependencies can be isolated as a separate relation and given a suitable entity name.

```
Vendor (V_Name, V_Street, V_City, V_State, V_Zip)
```

Since the Vendor relation is jointly determined by V_Name and V_Street, it is unclear at this point which attribute should serve as a primary key attribute. We can immediately discount V_City, V_State, and V_Zip, since other vendors may have these attribute values. Although it is unlikely that two vendors will have the same name and street address, a vendor may change these attributes at some point in time. It is better to supply a more permanent unique attribute, such as Vendor_ID. By this reasoning we arrive at the model shown in Figure 3.27a.

```
Vendor (Vendor_ID, V_Name, V_Street, V_City, V_State, V_Zip)
```

A data model is said to exhibit a normalization state, called a *normal form*. Three such states shall be discussed here: first normal form (1NF), second normal form (2NF), and third normal form (3NF). One characteristic of a good data model is that it is in third normal form. One of the steps in refining a data model is to try to revise entities in first or second normal forms to third normal form. In this section, we will motivate the subject of normalization by introducing some poorly defined entities and improving their definition through the normalization process.

In moving from 1NF to 3NF, two types of dependencies must be eliminated:

- *Partial dependency.* Dependency based on part of a composite primary key
- *Transitive dependency.* Dependency among nonkey attributes

Transitive dependency exists in Figure 3.27b. Finding and eliminating this dependency resulted in the data model shown in Figure 3.27a, which is in third normal form. What follows is a description of a formal process for moving from 1NF to 3NF.

3.4.4.1 A Table in First Normal Form

Consider an analyst who is trying to model the information on a purchase order in tabular form. He collects a group of purchase orders, all having the general format of Figure 2.2. He designs a simple report that could be used to list certain key components of the purchase order as follows:

Po_no	Date	Ven ID	Vendor	Line item	Mat'l ID	Description	Amt	Del date
3502	5/28/06	V25	Gen. Provisions	1	RM805	tomato paste	4760	6/25/06
3503	5/29/06	V110	Jersey Vegetables	1	RM201	carrots, whole	1000	6/20/06
				2	RM202	carrots, diced	1400	6/20/06
				3	RM201	carrots, whole	1300	7/25/06
3504	5/29/06	V250	Spices Unlimited	1	RM305	paprika	800	6/30/06
				2	RM340	sugar, bulk	1000	6/30/06

Based on the requirements of the report, the analyst defines an entity set called PO_REPORT, with a table designed to match the requirements of the report. The table, shown in Figure 3.28, would appear to be the easiest way to maintain the data in table form for generating the required report. However, it does not conform very well to the relational model. The table has considerable unnecessary data redundancy.

From the table it is clear that each row is uniquely identified by the two attributes: PO_NUMBER and PO_LINE_ITEM. This is the composite primary key for the table. The values of all the other attributes are said to be *dependent* on this composite primary key. However, other subdependencies exist within the table structure. These can be seen by looking at a ***dependency diagram*** of the table. This is shown in Figure 3.29.

PO NUMBER	RELEASE	VENDOR	V NAME	PO LINE I	MAT'L I	MAT'l DESC	AMT	DEL DAT
3502	5/28/06	V25	Gen.	1	RM805	Tomato paste	4760	6/25/06
3503	5/29/06	V110	Jersey	1	RM201	Carrots,	1000	6/20/06
3503	5/29/06	V110	Jersey	2	RM202	Carrots, diced	1400	6/20/06
3503	5/29/06	V110	Jersey	3	RM201	Carrots,	1300	7/25/06
3504	5/29/06	V250	Spices	1	RM305	Paprika	800	6/30/06
3504	5/29/06	V250	Spices	2	RM340	Sugar, bulk	1000	6/30/06

Figure 3.28 Table in first normal form.

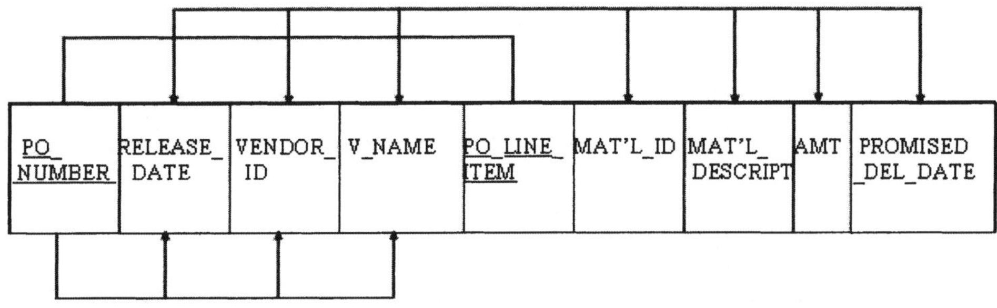

Figure 3.29 Dependency diagram of table in Figure 3.28.

The dependency diagram shows all the attributes in the table. The primary key attributes are indicated by underlining them and connecting them with a line. The line with arrows above the boxes points to the attributes that are dependent on the primary key. The line with arrows below the boxes shows the dependencies based on part of the composite primary key. This is called a *partial dependency.*

The table PO_REPORT is said to be in first normal form (1NF). First normal form has the following characteristics:

1. The key attributes are defined.
2. All attributes are dependent on the primary key(s).
3. No composite attributes exist (i.e., each row/column intersection contains only one value, rather than a set of values).

A table in first normal form can contain partial dependencies. A table that contains partial dependencies will have data redundancy. For example, each time a user enters PO_NUMBER, she or he must also enter RELEASE_DATE, VENDOR_ID, and V_NAME, all of which are uniquely dependent on PO_NUMBER. Data redundancy based on primary key attributes is eliminated when going from first normal form to second normal form.

3.4.4.2 Conversion to Second Normal Form

A table is converted to second normal form by eliminating partial dependencies. There are two steps in the process as follows:

1. Write each key component that has partial dependency on a separate line and the original key on the last line:

   ```
   PO_NUMBER
   PO_NUMBER, PO_LINE_ITEM
   ```

2. Define an entity name for each primary key (or composite key), and write the primary key attribute and dependent attributes of the entity.

   ```
   PURCHASE_ORDER(PO_NUMBER, RELEASE_DATE, VENDOR_ID, V_NAME)
   PO_DETAIL(PO_NUMBER, PO_LINE_ITEM, MAT'L_ID, MAT'L_DESCRIPT,
             AMT, PROMISED_DEL_DATE)
   ```

The resulting tables are said to be in second normal form. A table is in 2NF if

1. It is in 1NF.
2. There are no partial dependencies based on the primary key(s).

We encounter a problem when we reach second normal form. This problem is illustrated in the dependency diagrams shown in Figure 3.30. Although there are no partial dependencies based on primary key attributes, there are dependencies within the nonkey attributes. In particular, VENDOR_ID implies V_NAME and MAT'L_ID implies MAT'L_DESCRIPTION. This dependency among nonkey attributes is known as *transitive dependency.*

3.4.4.3 Conversion to Third Normal Form

When transitive dependencies are eliminated, a table has been converted to third normal form. A table is in 3NF if

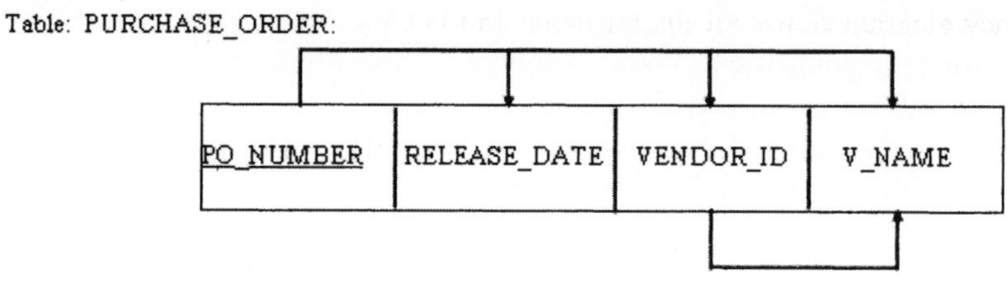

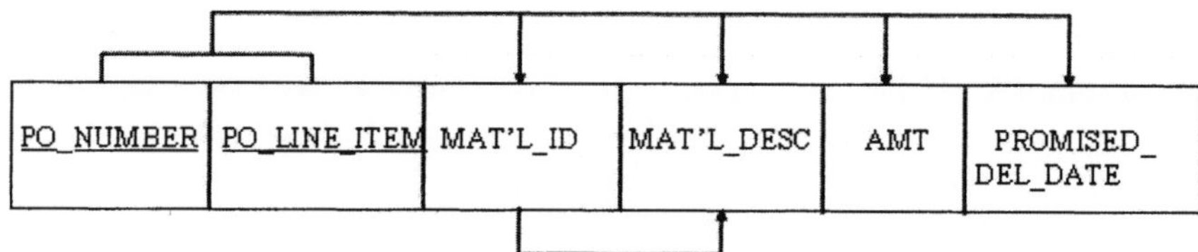

Figure 3.30 Dependency diagram for tables in 2NF.

1. It is in 2NF.
2. It contains no transitive dependencies.

The conversion process is a direct application of the steps followed in going from 1NF to 2NF, except that we are eliminating transitive dependencies instead of partial dependencies. The conversion process results in the following table definitions:

```
PURCHASE_ORDER(PO_NUMBER, RELEASE_DATE, VENDOR_ID)
VENDOR(VENDOR_ID, V_NAME)
PO_DETAIL(PO_NUMBER, PO_LINE_ITEM, MAT'L_ID, AMT,
        PROMISED_DEL_DATE)
MATERIAL(MAT'L_ID, MAT'L_DESCRIPTION)
```

Note that the new primary key attributes are retained in their original tables as foreign keys because they will be needed for referential integrity.

3.4.4.4 Denormalization

Sometimes a well-designed database in 3NF will not be the best design for a specific application. Many applications include functions that require quick response times, and the only way to ensure those response times is to carry all the data in a redundant table. The process of denormalization proceeds in two steps: (1) identify the functions requiring quick response times, and (2) once identified, examine the SQL code requirements to retrieve data for those functions. Potentially slow response times are indicated by the number of tables to be joined in a search and the number of derived data values to be calculated.

Normalization reduces the complexity of maintaining data integrity by eliminating as much duplicate data as possible. It is best to establish an initial design at the highest level of normalization possible. Denormalize only when it is deemed necessary for performance purposes.

3.5 SUMMARY

In this chapter we introduced the entity-relationship approach to developing a data model of the enterprise. The E-R modeling methodology as discussed in this chapter is the basis for other formats that are used for designing data models in industry. In industry, *computer-aided software engineering (CASE)* tools are widely used to document information system design projects and to help the designer to be more productive in executing those projects. In the next two chapters, we examine a specific set of CASE tools within the context of a design project. When discussing the data model tool set, it will be apparent that it is related to the entity-relationship modeling approach.

REVIEW EXERCISES

3.1 Consider the ternary relationship shown in Figure 3.7. Show how you would reduce this to a series of 1 : M relationships when you implement the tables in a database.

3.2 Create a semantic net diagram for the unary relationship of Figure 3.14 using the data of Table 3.1.

3.3 The following is a description of some data requirements for a chain of pharmacies. Draw the appropriate entity-relationship (E-R) diagram. Clearly show all cardinality constraints, cardinality limits, and existence dependencies.

(a) A pharmaceutical company manufactures one or more drugs, and each drug is manufactured and marketed by exactly one pharmaceutical company.

(b) Drugs are sold in pharmacies. Each pharmacy has a unique identification. Every pharmacy sells one or more drugs, but some pharmacies do not sell every drug.

(c) Drug sales must be recorded by prescription, which are kept as a record by the pharmacy. A prescription clearly identifies the drug, physician, and patient, as well as the date it is filled.

(d) Doctors prescribe drugs for patients. A doctor can prescribe one or more drugs for a patient and a patient can get one or more prescriptions, but a prescription is written by only one doctor.

(e) Pharmaceutical companies may have long-term contracts with pharmacies and a pharmacy can contract with zero, one, or more pharmaceutical companies. Each contract is uniquely identified by a contract number.

3.4 Bikes-R-Us sells standard and customized bicycles over the Web. The company buys components from various vendors in its supply chain and assembles the components into bicycles. Standard bicycles are produced to inventory, while custom bicycles are only made to order. Bikes-R-Us wants to develop a database for certain parts of its business. The company needs an E-R diagram to use as a basis for a database design. Develop the E-R diagram from the following statements:

(a) Bikes-R-Us will keep track of materials. The materials are of two kinds: components and bikes.

(b) Materials are related to each other by their bill of material structure. Each bike requires M components. Each component may go into 1 or more bikes.

(c) Components are provided by vendors. A component must be ordered from one or more vendors. A vendor must provide one or more components.

(d) Inventoried material is known as a "material lot." A lot is a grouping of the same material either supplied by the same vendor (components) on a particular shipment or produced in the same production run (bikes).

(e) A material lot may be provided by a vendor. When a material lot is provided by a vendor, it is associated with one and only one vendor. A vendor may have provided zero, one, or many material lots. Some material lots (e.g., bikes) are not provided by vendors.
(f) A warehouse location is a place where a material lot is stored. A material lot may be stored in more than one location. A warehouse location may have zero, one, or many material lots.
(g) A material lot may be produced on an assembly line. Each assembly line is associated with one or more material lots. Some material lots (e.g., components) are not produced on an assembly line.
(h) An assembly line is composed of stations. Each assembly line has one or more stations, and each station is associated with exactly one assembly line.
(i) An assembly process plan describes the steps by which a bike is assembled. Each bike has one assembly process plan, but the same assembly process plan may be used by more than one bike.
(j) Each assembly process plan has several steps. A step is associated with one process plan.
(k) Each process plan step is associated with a station, where that step is executed. Each step is associated with one and only one station. However, a station may be used in many process plan steps.

3.5 USF Rent-a-Car wishes to implement a database to control all aspects of its operations, including tracking car inventories, rental contracts, and billing. The company previously implemented a partial database as indicated by the tables shown in Review Exercise 2.2. The following statements of business rules and relationships are used to construct an E-R model:
(a) A model is defined as a specific manufacturer, model type, and year (e.g., 2007 Honda Accord). Each model is a member of one and only one car class (compact, standard, or full size). A class has many models.
(b) A model is associated with one or more cars.
(c) Cars are assigned to locations, and each location has one or more cars.
(d) A customer who wants to rent a car makes a reservation. The reservation is made for pick-up of a particular class of car at a specific location. The same customer may make more than one reservation over time.
(e) In the normal course of events, a reservation results in a rental agreement, which is established when the customer comes to the location to pick up the car. However, this is not always the case, since a reservation may be canceled or the customer with a reservation may not show up.
(f) A rental agreement is for a specific vehicle. At any point in time, a specific vehicle may have participated in zero, one, or more rental agreements.
(1) Construct an E-R diagram based on the preceding statements.
(2) The following discussion of the reservation process will help you identify some attributes. Assign the attributes to the appropriate entities. Indicate primary key and foreign key attributes.

Car rental rates are determined by the class of the car. USF Rent-a-Car has two rental rates for each class: daily and weekly. The car model includes a make (Ford, Honda, etc.), the year of the model, and the model name. Each car is uniquely identified by a vehicle identification number (VIN). The branch to which the vehicle is assigned has an address and a location ID.

The process of renting a car is as follows. Typically, a customer first makes a reservation with a location by telephone prior to arriving at the branch location to pick up the car. The USF Rent-a-Car service representative takes the customer name and address and the class of vehicle and period of rental (date and time in and out) that the customer desires. The customer is informed of the rental rate.

When the customer arrives at the branch location to pick up the car, the service representative first checks for a reservation and, if a reservation exists, she draws up a rental agreement. At that time the service representative obtains other customer information, such as his operator's license number and state that issued it and the customer's credit card type and number, including the expiration month and year. If the customer has made a reservation, then the reservation information is used to assign a specific vehicle to the rental agreement. If the customer is a walk-in (no reservation), the service representative fills out the reservation information first as part of the process. All rentals must be associated with a reservation. The rental agreement has a contract number that uniquely identifies it, the VIN number of the vehicle that is being rented, the current date and time for the rental to start, and a current odometer reading. The customer is given copy of the rental agreement along with the keys to the car. This ends the activities at the time the vehicle is picked up.

After use, the car is returned to the branch location. Information that will be filled in when the car is returned is the date and time at which the rental ends and the ending odometer reading. When the rental agreement is completed, the actual cost of the rental is computed using the class rental rate and the cost is charged to the customer's credit card. No other form of payment is accepted.

3.6 Pants-R-Us is an apparel industry innovator, designer, and producer of high-quality branded and retailer private label pants. It sells its products to major retailers who distribute through their own stores in major cities and shopping malls. The primary product lines feature dress and casual pants for men and women. The products are marketed under widely recognized national brands.

Pants-R-Us is in the process of designing a database system. The following information is provided to design the entity-relationship model for the database.

(a) Develop the entities and relationships that correspond to the following rules:

(1) Pants-R-Us classifies its product line as "Styles." Each style is a design classification for a stock-keeping unit (SKU) where the SKU is the basic style in a particular color, waist, and length. Each style may have zero, one, or many SKUs, but an SKU is a member of only one style category.

(2) Pants-R-Us has created several unique colors for its pants. Each color is uniquely identified by a color_id. A color may be used in zero, one, or more SKUs, but each SKU has only one color. Some colors have been created that have not been assigned to an SKU and may never be used.

(3) The retailers who sell the products are the customers of Pants-R-Us. Customers will contract with Pants-R-Us based on styles it will distribute. A customer may distribute one or more styles for Pants-R-Us, and a particular style may be distributed by more than one customer.

(4) Customers have stores, which is the place where they sell the pants and other products. A customer may have one or more stores. To identify a store in the Pants-R-Us database for shipping purposes, all customers have agreed to provide their store_id to Pants-R-Us.

(5) SKUs are inventoried at the stores of the customers. A store will inventory one or more SKUs and each SKU may be inventoried in zero, one or more stores.

(b) The following is a list of attributes. Assign each attribute to the appropriate entities. If a composite entity has to be introduced in order to assign an attribute, name that composite entity. Indicate which attributes are key attributes.

STYLE_ID—A unique identifier of a style of pants
STYLE_DESCRIPTION—A description of a style (e.g., "Loose Fit Denim," or "Pleated Cuffed Pants")
CUST_ID—Unique identifier of a customer of Pants-R-Us
CUST_NAME—The name of the company that is the customer
CUST_ADDRESS—The address of the company that is the customer
STORE_ID—The identifier of the customer's store as provided by the customer
STORE_ADDRESS—The address of the store
COLOR_ID—Unique identifier of a color
COLOR_DESC—Description of a color (e.g., navy, indigo, stone)
UPC—Universal product code, which is a unique identifier of an SKU
WAIST—The waist dimension of an SKU
LENGTH—The length dimension of an SKU

Pants-R-Us will receive actual point-of-sales data by SKU from the customers' stores and keeps track of the stores' inventory positions.

STD_SALES—Customer sales to date from the store location
STD_RETURNS—Cumulative returns to date from the store
ON_HAND—The amount of On_Hand units of inventory of an SKU in a store
IN_TRANSIT—The number of units of inventory of an SKU that has been shipped to a store but not yet received
ON_ORDER—The number of units of an SKU ordered by a store but not yet shipped
PERPETUAL_INV—ON_HAND + IN_TRANSIT + ON_ORDER

3.7 Flotsum Pharmaceutics Ltd. manufactures a variety of prescription drugs. By government regulation, each batch of manufactured product is subjected to laboratory testing before it can be sold. This is done at the Flotsum quality assurance lab. The lab has recently been reorganized in an effort to make its operation more efficient. Previously, lab technicians would withdraw bulk chemicals from inventory at the time of testing. Using the bulk chemicals, they would prepare reagents for the tests and use the reagents and other chemicals to test a sample of the product brought to the lab from production. Under the new system, "testing kits" are prepared in advance of testing by chemists and stored in cabinets in the lab. Each testing kit contains all the reagents and chemicals necessary for completely performing a specific laboratory test. This eliminates the withdrawal and mixing of bulk chemicals at the time of testing.

Management wants to design a database to support the operation of this lab. It will manage the use of the kits, the testing procedures, and the work of the quality assurance technicians. Develop an E-R diagram based on the following description of the business rules governing the lab.

(a) Reagents are made up of more than one chemical. A chemical may be used in more than one reagent. Some chemicals are not used in reagents.
(b) A "kit specification" is a description of reagents and chemicals that are in each kit. Each chemical and reagent may be specified for more than one kit, and a kit can require more than one reagent or chemical.
(c) "Chemical lots" are the physical realizations of the material classification "Chemical." A chemical may have zero, one, or many chemical lots associated with it. Each chemical lot is associated with only one chemical.

(d) Reagent lots are made from more than one chemical lot. A chemical lot may be used in zero, one, or many reagent lots. Some chemical lots are not used in reagent lots.
(e) A "test kit" is the physical realization of a kit specification. Each kit specification may have zero, one, or more test kits associated with it, but each test kit is based on only one kit specification.
(f) Test kits are composed of reagent lots and chemical lots. Reagent lots and chemical lots may be used in one or more test kits, and each test kit may contain more than one reagent lot or chemical lot number.
(g) A "laboratory procedure specification" describes a quality control test. A "kit specification" is associated with one or more laboratory procedure specifications. However, a laboratory procedure specification notes the use of one and only one kit specification.
(h) A laboratory procedure specification involves more than one "procedure steps," which are the sequential steps used to carry out the procedure. Each procedure step is noted in one and only one laboratory procedure specification.
(i) "Test kits" are stored in laboratory cabinets. A laboratory cabinet will store more than one test kit. A specific Test Kit ID is stored in one and only one cabinet.
(j) When a technician uses a "laboratory procedure specification" to perform a test, it is called a "QA_Test" and is given a unique ID each time it is performed. In other words, the QA_Test is the actualization, or use, of the specification. A QA_Test uses one and only one laboratory procedure, but the same laboratory procedure can be used in more than one QA_Test.
(k) When test kits are withdrawn from a cabinet to be used in a QA_Test, or replaced back into a cabinet after the test, a "usage transaction" is recorded. This keeps track of test kit usage. A test kit will be involved in more than one usage transaction, but each usage transaction involves only one test kit. A QA_Test will normally be involved in two usage transactions: (1) the withdrawal and (2) the replacement. Each usage transaction involves only one QA_Test.

CASE STUDIES

3.8 University Food Company Case (A)

The University Food Company wants to develop a data model that represents the entities and relationships in its manufacturing operations. As discussed in Chapter 1, food products are manufactured in batch operations like the ones described in Figure 1.15. First, raw ingredients are combined in a kettle, where they are mixed and may be preheated. The batch is then transferred to a filling machine, which dispenses the ingredients into packages along a conveyor line. Sealed packages are then subjected to sterilization by heating them in a retort.

The management of University Food Company wants to develop a relational database that can be used in recipe management. The database should bring together all the information that is needed to manufacture the product in the plant. For each product this includes the materials that make up the product formula, the operations performed during manufacturing of the product, and the instructions required for manufacture of the product. The combination of all of this information is what is commonly called the product "recipe." Management wants to begin by developing an appropriate data model that reflects the business rules under which the enterprise operates.

The following list gives information on various entities and their relationships. Attributes will be mentioned, but you will only be required to draw E-R diagrams showing the entities and their relationships. Some entities are already indicated in capital letters; however, others will have to be defined to make the model complete. Figure CS3.1 may be a helpful reference in completing this exercise.

1. Each PRODUCT has one or more recipes. Each RECIPE can be used in one or more products. For example, the recipe for cheese sauce can be used in the product macaroni and cheese, as well as the product cheese enchiladas.
2. Each recipe involves one or more PRODUCTION STEPS. For example, Figure CS3.1 shows the recipe for making cheese sauce, which involves eight production steps.
3. The production steps have a production step sequence that describes the precedence relationships among production steps.
4. A recipe includes one or more FORMULAS. The details of the formula describe the INGREDIENTS used in the recipe and their quantities. Each recipe may have more than one formula, which are different in quantity aspect (e.g., a 50-gallon batch versus a 100-gallon batch).
5. Each detail of a formula contains one ingredient, and ingredients can be used in one or more formula details.
6. The PRODUCTION_STEP of a recipe may include one or more equipment settings—for example, temperature settings for a cooking or mixing kettle. In the industry, a type of equipment setting is often referred to as a "tag." The TAG has a name given to the setting, such as "Temp_Setting_1."

Product: Cheese Sauce **Revision Date: 9/09/05**
Material ID Number: 74 **Supersedes: 10/01/03**

Sauce preparation procedure: (50 Gallons)

1. Place water in the Groen Kettle (No. 1).
2. Turn on agitator at highest speed.
3. Add dry skim milk and reconstitute.
4. Add ingredients (corn starch, flour, salt, mustard powder, and paprika powder). Mix at room temperature until all ingredients are dissolved.
5. Add margarine and heat with agitation to 45-50 degree Celsius (115 – 125 degrees Fahrenheit), until all margarine is melted.
6. Add cheddar cheese, mix thoroughly, and add annatto cheese color.
7. Heat with agitation to a temperature of 85 – 90 degrees Celsius (180 – 195 degrees Fahrenheit) and hold this temperature for a minimum of 10 minutes.
8. Take sample to quality control for viscosity measurement.

FORMULA (50 Gallon Batch)

INGREDIENTS	In Lbs	
100 Water	308 (38.5 gal.)	
66 Cheddar Cheese	32.0	
152 Skim Milk Powder	24.0	
40 Margarine	11.0	
90 Corn Starch	8.28	
71 Flour, all purpose	6.2	
70 Salt	5.57	
68 Mustard Powder	0.8	
81 Paprika Powder	0.4	
72 Annatto Cheese Color	0.28	

Figure CS3.1 Cheese sauce recipe sheet.

7. The value of the tag setting is referred to as the "Tag Value," such as 120 degrees Fahrenheit. Production steps in recipes use tag values. Tags may be used for minimum and maximum values also.
8. A production step of a recipe may use equipment. Each piece of EQUIPMENT must have its setting defined by a tag value. For example, in Figure CS3.1, time and temperature settings are provided.
9. The enterprise considers PRODUCTS and INGREDIENTS to be subclasses of MATERIALS, which also includes spare parts for equipment and office supplies.

(a) Show the E-R diagram for the situation described.
(b) Using Figure CS3.1 and some intuition, define a set of attributes for each of the following entities: RECIPE, PRODUCTION_STEP, FORMULA, and FORMULA_DETAIL. Indicate the primary key and foreign key attributes.

3.9 ACME Machine Shop Case (A)

The ACME Machine Shop is a small job shop that does machining of components for customers that assemble the components into finished products. The customers give ACME engineering drawings of parts and request production from ACME as needed. A typical drawing and the routing plan for manufacturing the part at ACME is shown in Figure 1.13. The manager of the machine shop wants to set up a database that can be used for maintaining records of the machine shop, including the equipment and human resources involved. He has given the following information about the parts manufactured by ACME and the resources of the shop:

1. Each part has one or more drawings. A drawing is specific to a part.
2. Each part has one or more routing plans. A routing plan is specific to a part.
3. Each part is machined from a workpiece, which is a material purchased by the company. A workpiece may be used to produce more than one part.
4. Each routing plan is composed of a number of operations. Operations always use equipment and sometimes require tooling as well.
5. Each operation of a routing plan uses one piece of equipment. However, each piece of equipment can be used in more than one routing plan operation.
6. Each operation may use tooling, but each tooling is used in one or more routing plan operations.
7. Each piece of equipment is operated by an equipment operator of the skill class appropriate to the operation of that piece of equipment.
8. An operator is an employee. Operators may have achieved more than one skill class, and a particular skill class may be assigned to more than one employee. Some employees are not operators and do not require a skill class designation.
9. Some operations in the routing plan are inspections, or quality control tests, such as operation 7 in Figure 1.13. Hardness testing is a quality control procedure. A particular quality control procedure may apply to more than one operation.
10. Each quality control procedure has a set of details, a step-by-step description of how to do the test. There are usually several steps in each procedure.

(a) Show the E-R diagram for the situation described.
(b) Four routing sheets from ACME machine shop, like that shown in Figure 1.13, were sampled. The attributes were listed in a single table, as shown next. Using the procedure for normalization, show the steps required to normalize the data to third normal form.

PART_NO	PART_NAME	QUANTITY	MATL_NAME	OPER_NO	DESCRIPTION	EQPT	TOOLING
2	Punch	1000	SAE 1040	1	turn......	J&L turret lathe	#642 box tool
2	Punch	1000	SAE 1040	2	cut....	J&L turret lathe	#6 cutoff
2	Punch	1000	SAE 1040	3	mill...	#1 Milwaukee	3/16 vise
2	Punch	1000	SAE 1040	4	heat treat...	Furnace	
2	Punch	1000	SAE 1040	5	degrease	degreaser	
2	Punch	1000	SAE 1040	6	check hardness	Rockwell	
6	Shaft	50	SAE 1040	1	turn...	J&L turret lathe	#642 box tool
6	Shaft	50	SAE 1040	2	drill...	#2 drill	#10 box tool
6	Shaft	50	SAE 1040	3	heat treat	Furnace	
10	Cam	1000	SAE 4050	1	mill...	#1 Milwaukee	5/16 vise
10	Cam	1000	SAE 4050	2	bore...	#2 Cincinnati	#40 box tool
10	Cam	1000	SAE 4050	3	heat treat...	Furnace	
10	Cam	1000	SAE 4050	4	check roundness	#3 CMM	
10	Cam	1000	SAE 4050	5	check hardness	Rockwell	

3.10 Regal Foreign Car Repair Shop Case (A)

Regal Foreign Car Repair Shop is currently keeping track of its repair orders using the work order history form shown in Figure CS3.2. Typically, a customer brings his or her car to the garage with a specific complaint. The customer service manager places the customer information in section A of the work order and the specifics of the vehicle in section B of the work order. The customer complaints are written in the comments area of section B. A unique repair order number is assigned, as shown in section B.

Based on the information provided by the customer, the service manager assigns a mechanic to investigate the complaint. The mechanic ID is placed in column C4. Every attempt is made to assign one mechanic to an automobile, but it is possible that, because of different skill sets, different complaints will require different mechanics. Thus, there may be different mechanics involved with different items on the work order. However, only one mechanic is assigned to each item. After investigating the complaint, the mechanic assigns a trouble code in column C2. The "trouble code" is a code that covers all types of repair work done on foreign cars. A sample list of trouble codes and their meanings are shown in the exhibit on the next page.

Each trouble code has a standard number of hours to complete the work (called flat hours, column C3). The service manager quotes the cost of the repairs to the customer based on flat hours multiplied by the labor rate of the mechanic skill required, plus an estimate of the cost of parts involved. The mechanic labor rate is placed in column C5. Finally, when the job is done, the mechanic lists actual hours worked on the task in column C6. Regal wants to record actual hours worked in order to rate the efficiency of mechanics and possibly revise the flat hours required for different jobs. If parts are involved in the service, the part numbers and descriptions are placed in section C7.

When all the work is done, the service manager summarizes the cost of the service. This is done at the bottom of section C. The invoice for the total cost is presented to the customer for immediate payment by cash or credit card. The customer is then given a copy of the paid bill, which includes a copy of the work order describing the work done.

Trouble_Code	
Trouble_Code_ID	**TC_Description**
1	Tune up
2	Engine Seal Leak
3	Transmission Replacement
4	Fan Belt Replacement
5	Starter Replacement
6	Battery Replacement
7	Muffler Replacement
8	Exhaust Replacement
9	Electrical Problem

Est. 1947

Regal

Foreign Car Repairs

55 Frelinghuysen Road
Piscataway, NJ 08854

A. Customer's Name, Mailing Address, and telephone no.

B. General Information

Repair Order no.

Chassis no.
Engine no.
Registration no.
Make/Model/Yr.
Milage
Date & Time In:
Date & Time Out:

Comments:

Section C

1 Item No.	2 Trouble Code	3 Flat Hrs.	4 Mechanic ID	5 Labor Rate	6 Act. Hrs.	7 Parts and Part Cost
1.						
2.						
3.						
4.						

Repair Cost Summary:
Labor Cost:
Parts Cost:
Total Cost:

Form of Payment:
☐Cash ☐Credit Card
Card Type:
Card Number:

Figure CS3.2 Repair order.

(a) Create an E-R data model based on the following description:
   1. A customer may own one or more automobiles that are repaired at Regal.
   2. An automobile is serviced using a repair order. Over its life cycle, an automobile may have several repair orders.
   3. Each repair order contains one or more repair details, each detail based on a specific customer complaint.
   4. A trouble code is associated with zero, one, or more repair details.
   5. A mechanic is assigned to each repair detail. A mechanic will work on many repair details.
   6. A part detail lists all the parts associated with a repair detail. The part detail includes the part IDs and quantities used in the repair. Some repair details have no part detail because no parts are involved in the repair.
   7. A part is associated with zero, one, or many part details.
   8. A repair order has a repair cost summary, which includes the labor costs, part costs, and total cost.

(b) Using the information in Figure CS3.2 and elsewhere in the description of this case, assign attributes to entities, indicating the primary and foreign keys.

Chapter 4

Structured Analysis and Functional Architecture Design

4.1 INTRODUCTION

In this chapter, we focus on the first step in the design of an information system for an industrial enterprise. The design proceeds from a definition of a business model of the enterprise. The business model is a description of the functions of the business, the data requirements, and the interactions between the functions and data requirements. The term ***functional or activity architecture*** is often used to describe a conceptual model of the activities that operate the business and the relationships between those activities. The word *architecture* denotes the fact that the model has a layered structure.

A related conceptual model of the firm is the ***informational or data architecture***. This is a model of the information requirements needed to perform the functions of the business. The functional and informational architectures, when taken together, form a high-level blueprint for the implementation of computer integration in the enterprise.

At a minimum, there are three layers of system design to consider in an information system project, as shown in Figure 4.1. The *conceptual layer* consists of the logical design of function and data requirements, as previously mentioned. When a conceptual design is complete, the next step is to implement the blueprint in hardware and software. The *implementation layer* requires the selection of a database management system, hardware platforms, and a communication medium. At this level, specific commitments are made that place limitations on the actual realization of the concept models. At the *execution layer,* the concept models are coded in software in terms of forms and reports that are to interface with the individuals performing the functions defined in the functional architecture.

In this chapter, we focus on the conceptual layer and, in particular, the functional architecture design methodology. We shall look at the other layers of the design process in later chapters.

4.2 FUNCTIONAL ARCHITECTURE AND BUSINESS PROCESS REDESIGN

The functional architecture conceptual design has a lot in common with the practice called *business process redesign* (BPR). BPR is a term given to the study of the existing functionality

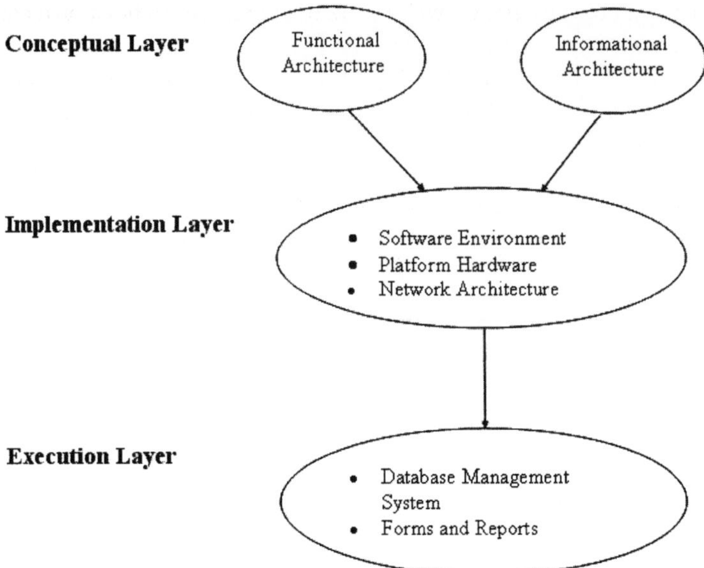

Figure 4.1 Layers of the information system design process.

and information used by the enterprise with the intention of redesigning the system to meet the same business objectives at higher performance or lower costs.

As in BPR, the process of designing a functional model follows a stepwise procedure. The first step is to study the existing enterprise and to deduce the business rules that govern its functions. The resulting model of the enterprise as it currently exists is referred to as the **as is** model. With an understanding of the business rules governing the functions, the objective is to eliminate redundancy and functions that are not required to meet the business objectives. In other words, reorganize the functionality (business processes) of the enterprise to reflect a more desirable model. The resulting model of the enterprise is called the **to be** model.

Business process redesign focuses on the *process*, which is a set of logically related functions that are organized to achieve a goal. These processes have customers that they are trying to service. Those customers may be outside the enterprise, as in the case of consumers of the final product, or they may be within the enterprise, as in the case of internal consumers of information or services generated by the process. The process may or may not be synonymous with an organizational unit of the enterprise. Most likely it is not, since processes usually cross organizational boundaries. As an example of a process in manufacturing, consider the replenishing of raw materials. It may begin in production planning with a requisition, flow through purchasing, go to an outside vendor, returning materials to receiving.

An important consideration in redesigning the process is a recognition that all organizations involved in the process should have access to the same data. Thus, an information system design consideration is that the databases and platforms are compatible. In an ideal case, all participants in the process may be operating off the same multiuser database.

The design of the functional and informational architectures may or may not be part of a business process redesign. However, if one is going to document the functions and data requirements of the enterprise, it presents an opportunity to review the way the processes of the enterprise are performed and how the information system can help to reshape those processes.

There are two methodologies that we will examine for designing a functional architecture. Both of these methodologies are known as "structured analysis" techniques. When an analyst approaches an information system project, he or she is often unfamiliar with the "business context" wherein the system will be implemented. The business context includes the functions performed by employees and the relevant data that are used to perform those functions. There is a learning process that the analyst must go through during which the functions are documented. The procedures and methods used in the documentation process are called "structured analysis" methods.

One application of structured analysis is the use of ***data flow diagrams***, which have been promoted in industry by Yourdon et al. (1979). This method is widely used by information system professionals in all industries. An alternative method is ***structured analysis and design technique (SADT)*** (Marca et al., 1988), which has been adapted for manufacturing enterprises under the name ***integrated computer-aided manufacturing definition 0 (IDEF0)***. Both of these methodologies are based on graphical notations used to describe information flows among processes of the enterprise being documented. There are differences in the graphical symbols being used in each case as well as some differences in emphasis on the kind of information to be captured in the documents. Next we will describe each method, beginning with IDEF0.

4.3 IDEF0 METHODOLOGY MODELING PRIMATIVES

IDEF0 is a modeling methodology for designing and documenting hierarchic, layered, modular systems. It is considered to be primarily relevant to manufacturing systems. The building blocks of IDEF0 are shown in Figure 4.2. The ***activity box*** is used to describe a function being performed in the enterprise. Typically, this is either a material conversion function, such as machining a part, or an information conversion function, such as processing a requisition for ordering materials. The same generic activity box primitive can model any function.

Inputs are shown at the left of the activity box. ***Inputs*** are those things that are transformed by the function. As an example of a material input, it could be a workpiece to be machined. As an example of an information input, it could be the information on a requisition to be transformed into a purchase order.

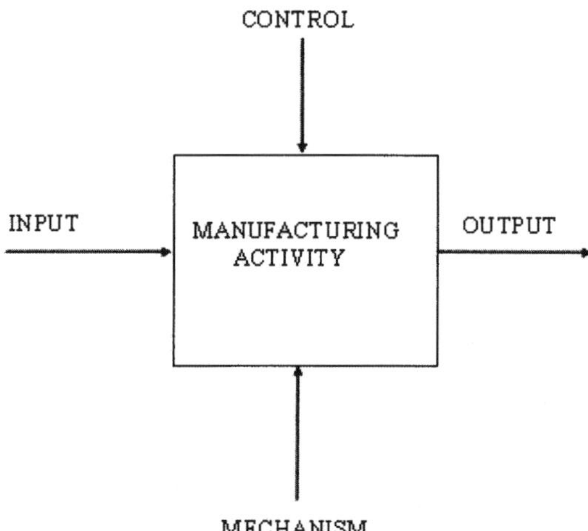

Figure 4.2 IDEF0 activity box and connecting arrows.

Outputs are shown at the right of the activity box. ***Outputs*** are the result of the transformation process provided by the activity. A material output could be the finished component after machining. The information output could be a purchase order for materials.

Mechanisms are shown entering an activity box at the bottom. ***Mechanisms*** are the means by which a function is realized. The material conversion of a workpiece to a finished component might require a lathe and lathe operator as mechanisms. The information conversion to process a requisition into a purchase order could involve a purchasing agent as the mechanism.

Controls are shown entering the activity box at the top. A ***control*** is a condition or set of conditions that guide or constrain the performance of the activity. For example, the machining activity may require a numerical control parts program if the machining is automatic or a set of mechanical drawings if the machining is done semi-automatically. For an example of information con-version, the requisition processing function may require adherence to a set of company rules or purchasing policy directives, such as purchasing only from approved vendors or obtaining high-level management approval for large purchases.

The activity box and four arcs provide a concise expression: an *input* is transformed into an *output* by an *activity* (function) performed by a *mechanism* and governed by a *control*. The specific activity, inputs, outputs, mechanisms, and controls are defined by the situation being modeled.

There is a grammatical convention used in naming activities and arcs. Activities represent actions being performed and are labeled with verb phrases. Inputs, outputs, mechanisms, and controls represent things and are labeled with noun phrases.

4.4 IDEF0 HIERARCHIC DECOMPOSITION

IDEF0 is a top-down modeling approach. The first layer is a single activity box that describes the overall function of the enterprise, organization, or process within the enterprise that is the subject of the model. This overall activity is then decomposed into its major subactivities at the second layer. Each of these subactivities are further decomposed at the next layer, and so forth.

Functions are related to each other by their material flows and information flows. For example, the output material or information of one activity may provide the input to another activity. These relationships are made explicit in the IDEF0 model. The hierarchic decomposition of IDEF0 models is conceptualized in Figure 4.3.

4.4.1 HIERARCHIC DECOMPOSITION ILLUSTRATED: NODE A0

Hierarchic decomposition is best illustrated with an actual case study. In this section, we introduce a hypothetical enterprise engaged in the manufacture of food products. In subsequent chapters and sections, we will return to this example as a pedagogical tool for illustrating important points.

Suppose an analyst were given the problem of developing an integrated model of an entire manufacturing enterprise. The place to begin would be with a definition of the overall activity of the enterprise. Figure 4.4 is a high-level view of the enterprise and its interactions with the outside world.

The inputs and outputs of interest at this level are the important input-output relationships at the boundary of the enterprise and the outside world that undergo the conversion process of the enterprise activity. We may highlight them as follows:

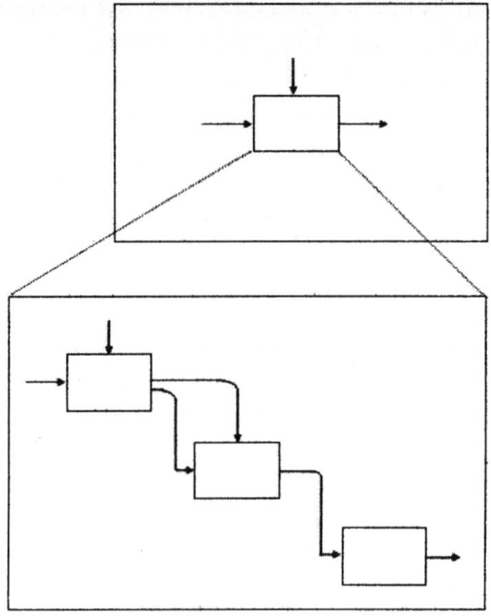

Figure 4.3 Relationship among levels in IDEF0 methodology.

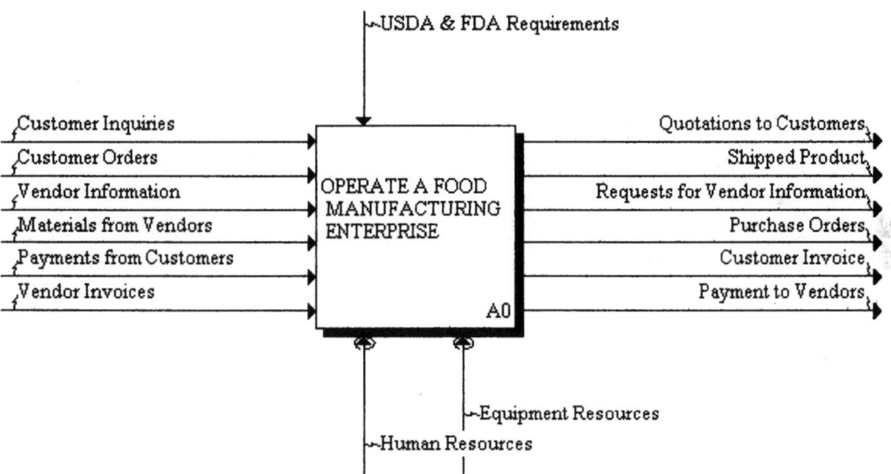

Figure 4.4 Top-level view of the enterprise: Node A0.

Customer inquiries (I)/quotations to customers (O). Before receiving an order and as part of servicing the customer, a company routinely receives inquiries from customers. For example, customers are interested in information on current pricing and delivery schedules (lead times). This information is routinely handled by the sales department, which may give informal or formal quotations to its customers.

Customer orders (I)/shipped product (O). Customer orders create the demand to which the manufacturing function responds eventually with shipped product.

Customer invoice (O)/payments from customers (I). This is the important financial transaction on the selling side. After an order is shipped to a customer, the enterprise bills the customer with an invoice. The customer responds by sending a payment to the enterprise.

Request for vendor information (O)/vendor information (I). This is the mirror image, on the buying side, of customer inquires. The enterprise is the customer of its vendors and requires information on pricing and delivery.

Purchase orders (O)/materials from vendors (I). This is the mirror image, on the buying side, of customer orders. The enterprise requires raw materials to produce its products. Those materials are bought on a purchase order, which eventually results in materials being delivered to the enterprise.

Vendor invoice (I)/payments to vendors (O). This is the mirror image, on the buying side, of receiving a payment. After material is delivered to the enterprise, the vendor presents an invoice for payment. This results in the enterprise sending a payment to its vendor.

To a large extent, the operation of a manufacturing enterprise can be viewed as three interrelated processes. First is the physical flow of materials. At node A0, this is represented by the input *materials from vendors,* which is transformed into the output *shipped product.* This physical conversion process is what we normally think of as the manufacturing process. The second process is the information flow. Internal to the organization, it is the information flow that leads to decisions that affect the physical flow system. At node A0, such items as *vendor information* and *quotation to customers* are typical of information flows. Finally, there is the financial flow system, which results in the collection and disbursement of wealth. *Payments from customers* and *payments to vendors* are obvious examples. There may be other important inputs and outputs that can be made explicit at this level of the IDEF0 diagram. This will depend on the purposes of the analyst. For this problem illustration, these inputs and outputs will be sufficient.

There are often many controls on the operation of an enterprise. When the entire enterprise is being modeled, we look for constraints at level A0 that are imposed on the organization from the outside. As the hierarchy is decomposed, it will emerge that various subsystems of functions within the organization have there own internal controls. There are obviously many outside controls that are placed on business enterprises, such as accounting standards, regulatory standards, and so forth. Here we are illustrating an important class of standards in the food industry, the Food and Drug Administration (FDA) and the United States Department of Agriculture (USDA) regulations. In a similar manner, we generalize the mechanism by which the activity of operating a food manufacturing enterprise converts inputs to outputs (i.e., human resources and equipment resources).

Node A0 is a starting point for modeling the enterprise, but it really does not tell us much. To understand more, we must look at a decomposition of Node A0.

4.4.2 Decomposition of Node A0

The decomposition of node A0 is illustrated in Figure 4.5. For our hypothetical enterprise, we identify four major activities at the next level: A1: *Manage Sales and Orders Process*, A2: *Plan for Manufacture*, A3: *Manufacture Product*, and A4: *Control Finished Goods*. For any particular enterprise, there may be other important functions at this level. However, for our illustration, we have chosen a generic set of functions that are common to most firms.

Decomposing node A0 illustrates several concepts about IDEF0 modeling. The IDEF0 model is a coordinated set of diagrams. For example, nodes A1 to A4 represent a decomposition of the parent node A0 into four major child activities. The decomposition of a parent activity is determined

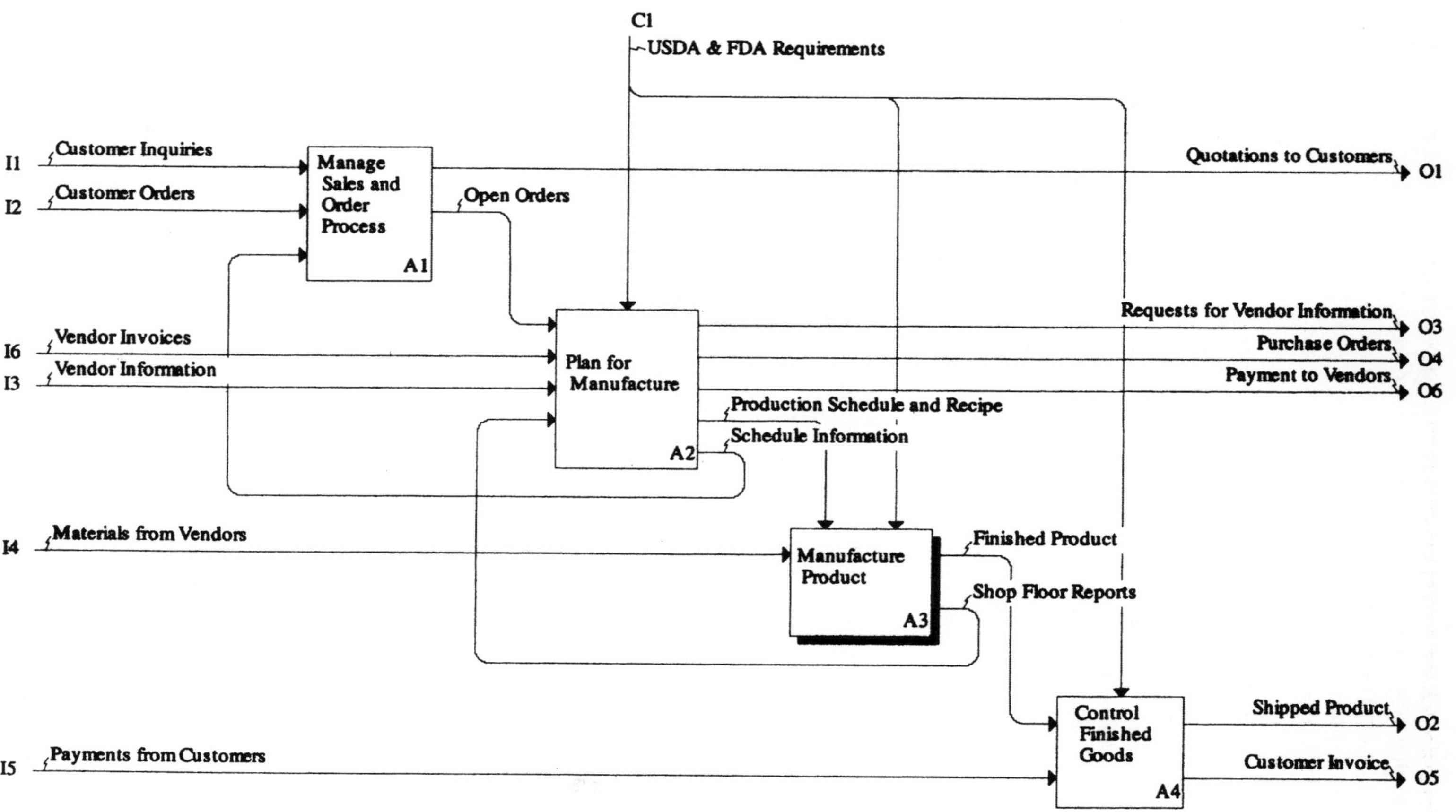

Figure 4.5 Decomposition of node A0.

by the analyst in conjunction with enterprise personnel. Another analyst may feel that one of these child activities is better modeled if it is broken into two distinct functions at this level, or may wish to combine two of the activities shown in Figure 4.5 into one activity. The important factor is that, when taken as a set, these A1 to A4 activities are a complete representation of the components of activity A0. The breakdown structure of an activity into its main child activities is usually kept track of in an indented list. The following indented list applies at this point:

A0—Operate a Food Manufacturing Enterprise
- A1—Manage Sales and Orders Process
- A2—Plan for Manufacture
- A3—Manufacture Product
- A4—Control Finished Goods

The higher level activity from which a decomposition occurs is called the *parent* activity, and the decomposed activities are called *child* activities. Inputs and outputs of the parent activity are also inputs and outputs of the child activity (i.e., they are inherited by the child). Thus, the *customer inquiries* and *customer orders* are handled by the sales organization and are inputs to node A1. The relationship with vendors is usually handled by the purchasing department of a company. The purchasing activity has been incorporated into the activity A2, Plan for Manufacture. Thus, *vendor information* and *vendor invoice* are shown as inputs to activity A2. Raw materials are input to the manufacturing process, so this is shown as an input to activity A3. Finished product leaves the enterprise from finished goods inventory; this is shown as an output of activity A4. The reader should compare the remaining inputs and outputs in Figure 4.5 to the inputs and outputs in Figure 4.4 to see that all inputs and outputs are accounted for.

As a model is decomposed, the relationships between activities begins to emerge. Figure 4.5 shows several examples of relationship primitives, which we shall explain. First note the *activity flow relationship.* When several activity boxes are drawn at the same level of decomposition, they are not placed randomly on the diagram. They are ordered by sequential order or some other dominance relationship. Thus, in Figure 4.5, the "Manage Sales and Order Processes" activity precedes the "Plan for Manufacture" activity, which precedes "Manufacture Product," ending with the "Control of Finished Goods." For ease of diagramming, the IDEF0 methodology recommends using a *staircase pattern* whenever possible. It is also recommended that, at each level of decomposition, from three to six child activities be created from each parent.

Connecting arcs are used to illustrate a number of output-input relationships among activities. The simple output-input relationship is shown in Figure 4.6. This occurs when the output of one activity is the input to one and only one succeeding activity. Examples are presented in Figure 4.5.

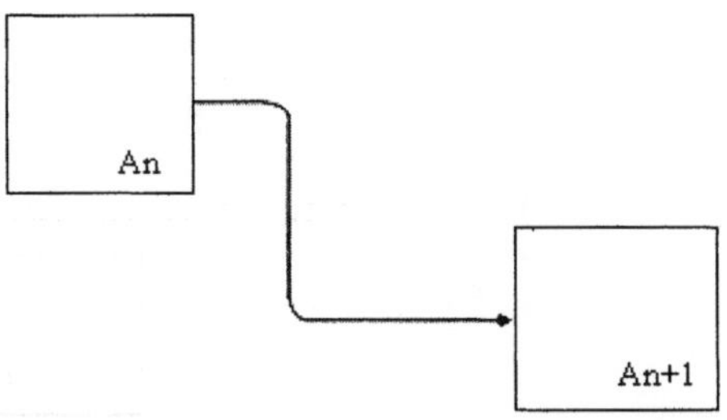

Figure 4.6 Simple output/input relationship.

For example, *open orders* are an information output of A1 that is input to the manufacturing planning function, A2. The planning process will convert this information into a planned schedule of manufacturing activities required to service those orders. Similarly, the arc *finished product* is a physical flow of material from A3 to A4.

Information or physical entities can flow from one or more activities to one or more other activities in various patterns. Figure 4.7 shows one such pattern. When the same information output of one activity is input to two or more subsequent activities, one arc leaving the source activity is divided into two or more input arcs to subsequent activities. This is what is referred to in some modeling practices as ***parallelism*** (i.e., both An + 1 and An + 2 are simultaneous recipients of the flow). This is more common with a flow of information than with a flow of physical entities.

Another possibility is where the output from an activity can be allowed to flow to either of two subsequent activities, as shown in Figure 4.8. This structure shows two separate arcs exiting the upstream activity with different arc labels.

One activity can provide inputs, controls, or both to other activities. Figure 4.9 illustrates a control relationship. The use of controls in the modeling methodology is a unique feature of IDEF0 and SADT. However, there is sometimes confusion in deciding whether an arc is either an input to an activity or a control. If the entity represented by the arc is converted into some other form

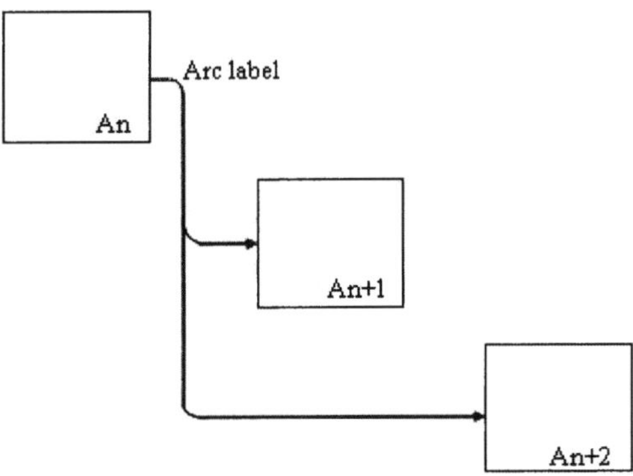

Figure 4.7 Simultaneous flow to more than one activity.

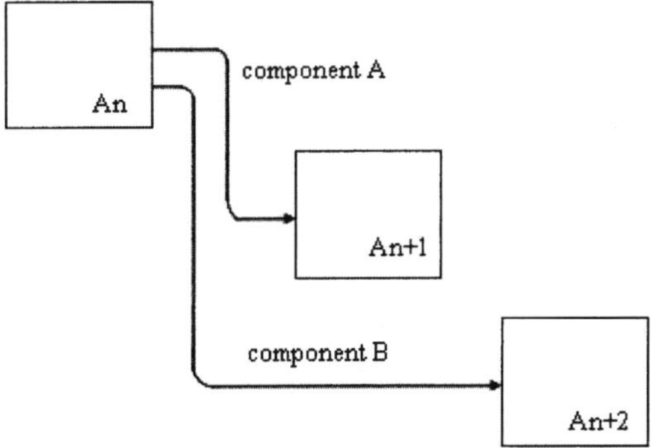

Figure 4.8 Distribution of flows to more than one activity.

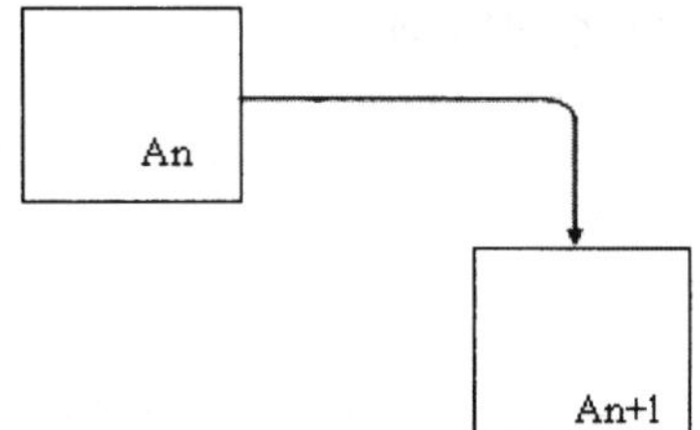

Figure 4.9 Activity An controlling activity An + 1.

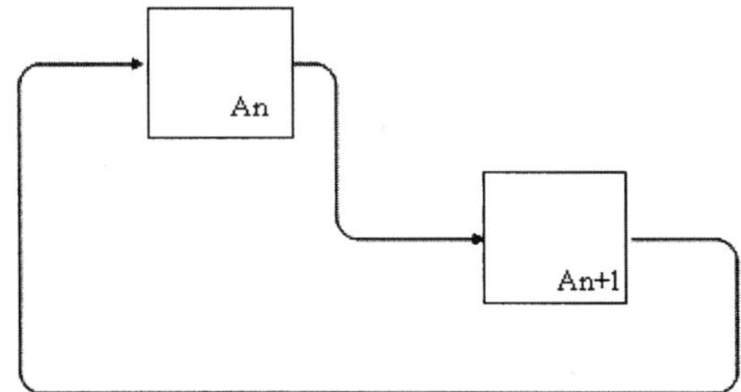

Figure 4.10 Activity An + 1 providing feedback to activity An.

by the activity, it is clearly an input. An example of this in Figure 4.5 is *finished product,* which activity A4 converts into *shipped product.* If the entity represented by the arc directs the activity as to how it will perform its function, it is clearly a control. An example of this in Figure 4.5 is *production schedule and recipe.* This information tells activity A3 what products will be produced on a specific day (production schedule) and how these products will be produced (recipe). Although the preceding rules should enable the analyst to distinguish between an input and a control, there will be cases where some judgment is required.

The output-input relationship between activities is not always in the forward direction. Data flow feedback can occur when information generated in a subsequent activity is used by a prior activity in the activity diagram. This is shown in Figure 4.10. An example shown in Figure 4.5 is the relationship between A1: *Manage Sales and Order Process* and A2: *Plan for Manufacture.* In order for activity A1 to inform customers regarding perspective delivery dates (quotations to customers), it is necessary to know the current planned schedule for the manufacturing facility in order to find the available time to produce the new order. Therefore, as plans are generated, this information is made available to sales. The model shows feedback from A2 to A1.

This example presents an interesting distinction between an input and a control. The production schedule is a control on the manufacturing process because it tells manufacturing *when* to produce. Scheduling information (which is similar) is an input to the sales function. It does not tell sales *how* or *when* to perform its activity. Rather, it provides the sales staff with information to assist them in providing lead time quotations for new orders. Therefore, essentially the same information can be a control on one activity and an input to another activity.

There are other modeling constructs that we will introduce later as we continue the decomposition. Note that we have not shown mechanisms in Figure 4.5. It is not necessary to identify mechanisms at every step of the modeling process. It is required at the ***elemental level*** (i.e., when a specific activity is identified at the lowest level of the hierarchy). This will be illustrated later.

4.4.3 Decomposition of Node A3

The continuation of the process of decomposition will be illustrated on node A3, Manufacture Product. Figure 4.11 is a decomposition based on our hypothetical enterprise, and it illustrates some new conventions. First, note the node numbering convention. The convention uses the parent node number with an extension. Thus, we have A31: *Control Incoming Materials,* A32: *Control Stored Materials,* and A33: *Control Production Processes.* At this point the indented diagram structure is as follows:

A0—Operate a Food Manufacturing Enterprise
  A1—Manage Sales and Order Process
  A2—Plan for Manufacture
  A3—Manufacture Product
    A31—Control Incoming Materials
    A32—Control Stored Materials
    A33—Control Production Processes
  A4—Control Finished Goods

An additional modeling concept, called *bundling,* is introduced in this diagram. You will note that, in Figure 4.5, activity node A2 provides a control for activity node A3 called *production schedule and recipe.* The production schedule allocates products to be produced to the equipment (production lines) by time periods. In this hypothetical enterprise, the production schedule is a daily schedule (i.e., production of a particular product is scheduled by day). The recipe is composed of many things. It includes the steps in the production process, the materials or ingredients that are used at each step to make the product, and the critical operating parameters of the production line (i.e., temperatures and time settings for cooking and sterilization). All are subsumed under the concept of a "recipe." In Figure 4.11 we have reached a level of decomposition where it is time to make these controls more explicit.

In our examples, the information entities that control aspects of the manufacturing processes do not explicitly appear on the diagram for node A3 (Figure 4.5). Instead, we had a more generic term, "production schedules and recipe." The production schedule and recipe in our enterprise is composed of four documents: (1) retort processing information, (2) cook sheet, (3) day production schedule, and (4) material move schedule.

Packaged food products are usually sterilized after packaging by putting the packages in a chamber of superheated water. This chamber is called a "retort." The *retort processing information* gives the time and temperature requirements for sterilization to the retort operator. These are the settings that he or she will use for products produced during the particular production day to which it applies. These settings vary by product and must be given to the retort operator each day.

The *cook sheet* contains the formula that must be used for each product. Each product has its own cook sheet. It describes which ingredient is to be added at each operation as well as other information about equipment settings, such as the time and rate for mixing ingredients.

The *day production schedule* tells the supervisor which production lines will be used to produce each of the products to be made that day as well as the order of production when a line is to produce multiple products. The *material move schedule* tells the forklift truck operator which lots of ingredients to transfer from storage to production on a particular production day.

Note that we have added a suffix to these information controls, (DB). This is being used to inform the reader that this is information that is currently derived from some data source, electronic

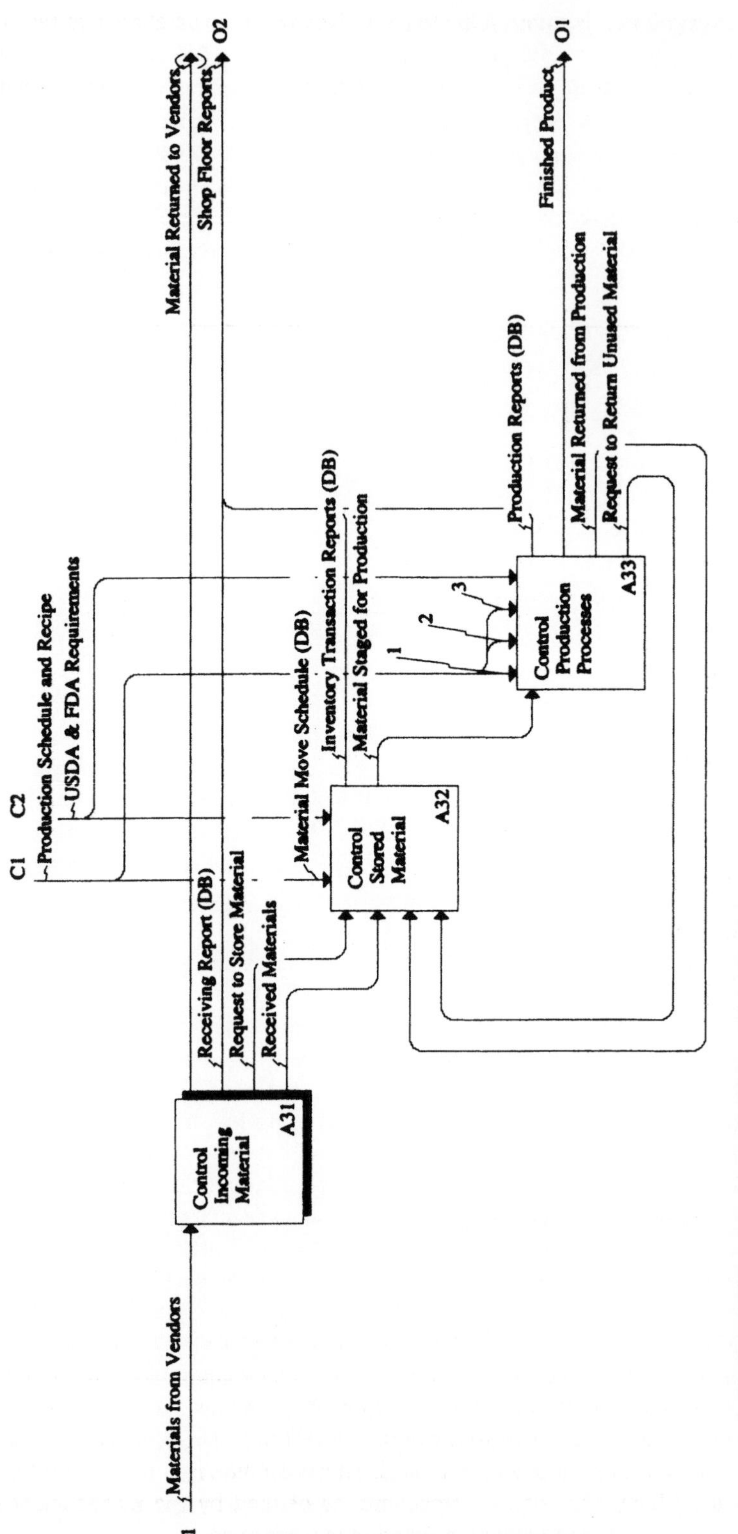

Figure 4.11 Decomposition of node A3.

or otherwise, and that we intend this information to be derived from the database that results from our later implementation of this concept model. This notation is not part of the IDEF0 methodology but is introduced here as a convenient way of identifying important information documents in the functional model.

Figure 4.11 also introduces the concept of *tunneling*. An example is shown by the output of activity A31, labeled "Material Returned to Vendors." Note the tunnel on the arrowhead of the arc. A tunnel arrow is used as a convenience. It can represent: (1) an external arrow that did not appear in the parent diagram (i.e., it has a hidden source) or (2) an arrow that goes to another activity but does not appear explicitly on the destination activity (i.e., a hidden destination). Tunneling is used when it is not convenient to show all inputs, outputs, controls, or mechanisms at every level of the hierarchy.

In completing the decomposition down to the elemental level, we will focus on node A31, Control Incoming Material. The elemental level is the most detailed level of analysis of functions. Node A32 will be decomposed in a later section, and the decomposition of node A33 will be the subject of a case study in Chapter 7.

4.4.4 Decomposition of Node A31

In Figure 4.12, node A31 has been decomposed into three child activities. At this point, the indented list of activities is as follows:

- A0—Operate a Food Manufacturing Enterprise
  - A1—Manage Sales and Order Process
  - A2—Plan for Manufacture
  - A3—Manufacture Product
  - A31—Control Incoming Materials
    - A311—Confirm Validity of Shipment
    - A312—Inspect Condition of Materials
    - A313—Receive Materials
  - A32—Control Stored Material
  - A33—Control Production Processes
- A4—Control Finished Goods

The elemental nodes should give a clear conceptual understanding of the processes that are taking place and the information requirements at each stage of the process. In effect, there should be a clear story (narrative) that goes with the diagram. This diagram and narrative are developed with those individuals within the enterprise who are actors in the process. For the reader to understand the details of activities A311 to A313, the narrative as developed by the analyst during the modeling process will be described as follows:

Node A311: Confirm validity of shipment. When a delivery truck arrives at the receiving dock, the receiving clerk is responsible for clearing the material for unloading. The clerk first compares the paperwork that comes with the shipment (bill of laden) with the enterprise's purchase order that is referenced on the bill of laden. Therefore, two required information inputs to this function are the bill of laden and the corresponding purchase order. If the material on the shipment bill of laden corresponds to the purchase order, the shipment is accepted. If no such purchase order exists, the shipment is refused. This procedure is defined as the "material receiving procedure." The steps of this procedure, as defined by the management of the enterprise, control the manner in which the function is performed.

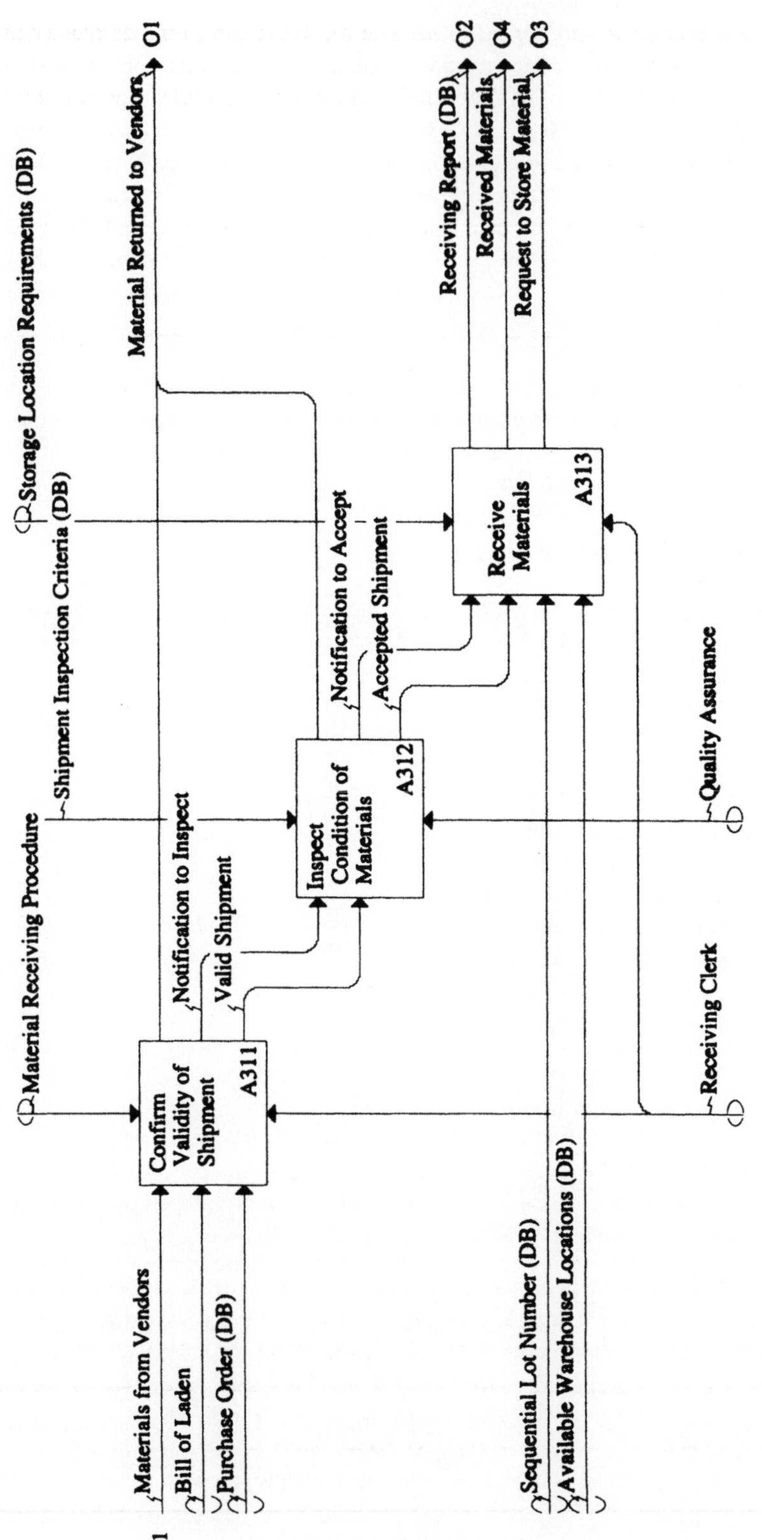

Figure 4.12 Decomposition of node A31.

Node A312: Inspect condition of material. The receiving clerk notifies quality assurance that material has arrived. Quality assurance personnel make a cursory examination of the truck and the condition of its contents. Broken containers or other conditions of transport can result in a partial or total rejection of the shipment. This is not an inspection of the quality of individual materials. Such testing is performed in the quality control lab and would be specified under another set of processes that occur after the material is stored in the warehouse and before it is used in production. When quality assurance clears the shipment for acceptance, the receiving clerk is notified.

Node A313: Receive materials. The enterprise keeps a record of shipments on a form called a *receiving report,* as shown in Figure 4.13. Upon accepting the shipment, the clerk makes out the top portion of the receiving report. The shipment is unloaded and made available for storage. The receiving clerk first assigns lot numbers to the accepted material. The lot number may identify an entire shipment, or the shipment may be divided into several lots, each with its own lot number. Lot numbers are assigned as sequential numbers and are obtained by the receiving clerk from a data source, such as a database or a list of sequential numbers kept at the receiving location. The receiving clerk also assigns the material to a storage location based on material location requirements and location availability. For example, some materials may require a freezer storage location while other materials can be stored at ambient temperature. The receiving clerk locates an available storage location, or locations, in the proper warehouse storage area. The location requirement (dry storage, refrigeration, etc.) serves as a constraint on the activity. The forklift truck operator is informed of the location to which the material should be moved, which is indicated by the output arrow labeled "Request to Store Raw Materials."

It is true that a great deal of detail is involved with each of these activities. Nothing prevents the analyst from making that detail explicit by further decomposing an activity. For the purposes of this illustration, we have chosen to stop here since enough detail about this operation has been uncovered at this point to provide a basis for defining the data requirements for the execution of these functions. These data requirements will be addressed in the next chapter.

RECEIVING REPORT

Supplier: **General Provisions**
125 Common St.
Boise, ID 44830

Purchase Order No.: **PO3502**
Date Received: **June 25 2006**

Quantity		Mfg. Lot No.	Item Code	Mat'L Lot No.	Description	Storage Location
accepted	not accepted					
1000		1275	RM805	97275	Tomato Paste, 1 gallon cans	Area A, Aisle 1 tier 1, bins 10-18
300		1283	"	97276	" " " "	Area A, Aisle 1 Tier 2, Bins 10-13
	100	"	"		" " " "	returned(1)

Comments: (1) returned due to case damage and badly dented containers.

Received by: J. Debbs

Figure 4.13 Receiving report.

4.5 THE PROCESS OF MODEL DEVELOPMENT AND VALIDATION

When an analyst begins the process of developing a functional model, she or he is entirely dependent on the cooperation of the individuals who work in the enterprise. They are the experts. They are the ones who know the most about how the individual functions are performed and how they relate to other functions. These experts will have different views of the organization, depending on their responsibilities. A general manager may have a rather global view of the important goals and interactions of business units to each other, whereas a purchasing agent will have detailed knowledge of the buying end of the business, and a receiving clerk may have yet another local view of how the enterprise works. The inputs from all of these people may be necessary at different stages of the modeling process, depending on the scope of the model to be developed.

Knowledge acquisition is largely an interactive review process. The IDEF0 modeling methodology is so easy to understand that it behooves the analyst to carefully explain the methodology to the experts and enlist their help in the graphical development of the model. It has been shown that, given the time and responsibility, these individuals are usually quite capable of taking the IDEF0 tool set and conducting their own model-building process with minimum support from the analyst (Young and Vesterager, 1991). In fact, this is most likely the preferred way to proceed because it ensures the contribution of good ideas in any attempt at redesign because these individuals will have to work with the redesigned system afterward. In any event, the graphic model should be refined in iterative interviews with the experts, presenting them with the most recent version of the model for comment at each interview. When there are no further revisions suggested, that portion of the model is final and the analyst moves on to the next process. For a detailed discussion of how to conduct and document the review process, see Marca and McGowan (1988).

Good documentation is essential in the model development process. Some of what goes on in running a manufacturing plant often relies on ad hoc and informal interactions that would never be apparent on the surface to an analyst. This is often referred to as the *informal organization*, and it is discovered during the detailed interactions with the experts. However, the formal organization usually runs on forms and reports that are used to authorize activities and record results. We saw some reports and schedules in the case study in this chapter. These should be carefully collected because they represent a wealth of information concerning data requirements that will be useful when the database is designed.

The IDEF0 diagrams developed in the previous sections are annotated with terms that describe physical entities and information entities. It is common for the same entity to be named by more than one term in different departments of the same enterprise. For example, the department that manages materials may refer to salt as a "raw material." However, on the factory floor, the operator of a sauce preparation kettle will refer to salt as an "ingredient." In fact, the forms used to make records of salt being stored in inventory (inventory record) and the forms used to record salt being added to a sauce (formula) may both refer to the same item (salt) by these different terms. This can be very confusing when it comes to establishing a database and information system. It is desirable to identify each entity by only one name to avoid confusion in the data management process. Therefore, when assembling existing forms and reports during the model development process, the analyst should take care to define each term used on a form or report to ensure complete understanding.

The narrative is important to the documentation process because it tells the story of what activities the information system is supposed to support. Over time, the nature of the business will change, and an activity and the procedures used by individuals performing the activity can change. The original narrative documentation is an important baseline to be able to return to when assessing the impact of those changes on information requirements.

4.6 DATA FLOW DIAGRAMS: AN ALTERNATIVE STRUCTURED ANALYSIS METHODOLOGY

There is an alternative to IDEF0 that is widely used in all industries, both manufacturing and service industries, called ***data flow analysis (DFA)***. Data flow analysis also models functions and the information that flows among functions. The focus of DFA is on the movement of data among those functions. It explicitly includes data repositories, called "data stores." Unlike IDEF0, it is not concerned with material flows or with the mechanisms that perform each activity. These latter considerations were incorporated into IDEF0 because they were determined to be quite important in the manufacturing arena. In the following sections, we examine the modeling conventions of DFA.

4.6.1 DFA MODELING PRIMITIVES

Data flow diagrams are constructed using four symbols, as illustrated in Figure 4.14. The ***process***, which is analogous to an IDEF0 activity, is shown as a circle. The process represents the people and procedures that transform data. In DFA, the emphasis is on how the data are used and transformed. Each process has data entering it and exiting it. If a process did not require a data input or produce a data output, it would not have a place in a data flow diagram.

An arrow is used to indicate ***data flow***. The arrow indicates the direction in which the data move. In DFA, "data" is a general concept. It can represent data sent to a computer file or information given from one process to another process by word of mouth, via e-mail, or by a telephone call. Note that arrows are not used to indicate physical flows of materials, as in IDEF0. The emphasis is on the flow of data.

The ***data store*** is represented by a rectangle with an open end. The data store is a place where data are preserved as a record. This could be a computer file or a paper filing cabinet. When the analyst finds that the process results in the storage of a document or a record, a data flow arrow goes from the process to a data store. Note that there is no explicit construct in IDEF0 that is analogous to a data store.

As in the case of IDEF0, the subject of the DFA is defined with reference to a specific system with boundaries. At the boundaries, the system interacts with outside people, processes, and organizations. These are indicated by a box, which represents ***sources and sinks*** for data. Sources are

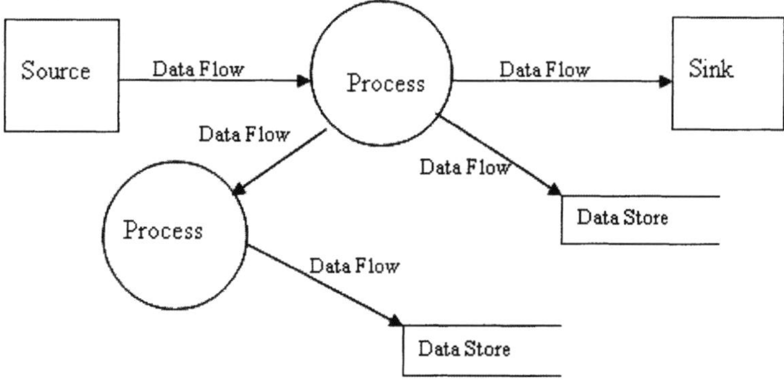

Figure 4.14 Generic data flow diagram.

entities outside the system that provide data input to the system. They are usually the entities that trigger events in the system. Sinks are entities outside the system that receive data. The same entity may be both a source and a sink if it both sends data to and receives data from the system.

4.6.2 Hierarchic Decomposition in DFA

Data flow diagrams follow the same principles of hierarchic decomposition as IDEF0. The highest level diagram is called the ***context diagram***. The context diagram includes the overall process and the sources and sinks that interact with the overall process. The context diagram is decomposed into a first-level diagram that shows more details of the process and data flow. The boundary relationships of the context level are maintained at each successive level of decomposition. Therefore, data flows from sources and to sinks that appear at the context level also appear at the first level of decomposition.

As the decomposition explores greater levels of detail, data stores not represented at the context level may be introduced. These data stores are internal to the process and, therefore, would not appear as external to the process at the context level. For example, the system under study may have a filing cabinet or local database in which it stores information locally.

As in the case of IDEF0, there is always a question of how much detail is required before the hierarchic decomposition stops. The analyst makes this judgement. In general, decomposition should be carried out to the degree necessary for the analyst to understand the details of the functions and data flows.

4.6.3 Hierarchic Decomposition Illustrated: Node A32

To demonstrate the use of data flow diagrams, we again refer to the example used in this chapter, Figure 4.11. Assume that the system of interest has been identified as all the activities involved in node A32, *Control Stored Materials*. A context diagram in the data flow diagram format would appear as shown in Figure 4.15.

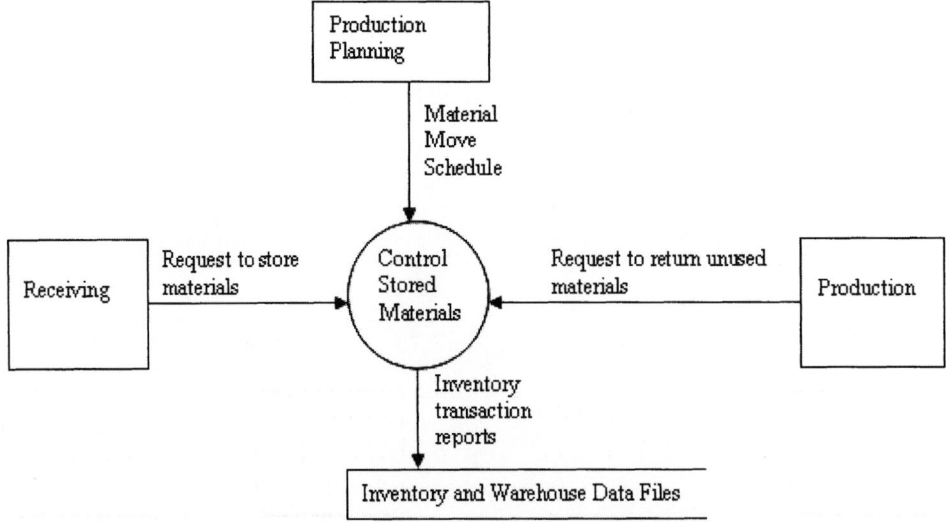

Figure 4.15 Data flow diagram context diagram for control stored materials.

The context diagram shows the process of interest as a circle with the appropriate label. This process defines the system to be documented. Note that this is the same as activity A32 in Figure 4.11. There are three source entities at the boundary of the system. They are receiving, production planning, and production. Receiving is the entity in charge of the process "Control Incoming Material," which we documented in Section 4.5. Receiving is a "trigger" for the process "Control Stored Materials" (i.e., it initiates an action in the process when it makes a "request to store materials"). This request is a data flow, which is made verbally to the forklift truck driver in our hypothetical example. The production planning department is the source of another trigger. The trigger to move raw material from the warehouse to work in process is the material move schedule. This is a daily schedule and is based on the products to be produced on a particular day. Similarly, a "request to return unused materials" from the production supervisor is another trigger to the process. Raw material that has been moved into production but not used must be returned to storage. Finally, the process sends an inventory transaction report to a data store. Compare Figure 4.15 with Figure 4.11 and note the similarities.

Information is generated within the context process. In particular, each movement of material requires that a transaction be recorded to indicate the debiting (removal) or crediting (addition) of inventory and the locations used. This must be done for both the inventory and warehouse files. The former maintains the status of the material, while the latter maintains the status of the warehouse locations.

4.6.4 DECOMPOSITION OF CONTEXT DATA FLOW DIAGRAM

In IDEF0 methodology, the overview of the decomposition of the highest level node is maintained by using an indented list. In a similar manner, the overall structure of the data flow diagram hierarchy is often shown in a ***process hierarchy chart***. The process hierarchy chart is a series of block diagrams that show the hierarchic relationship among processes that are documented in the data flow diagrams. An example is shown in Figure 4.16.

In Figure 4.16 the context process (Control Stored Materials) is composed of four level-1 processes: *Store Raw Materials, Move Raw Materials to WIP, Return Unused Raw Materials to Storage*, and *Transfer Daily Records*. In turn, the level-1 processes can be decomposed into level-2 processes. In Figure 4.16, we only show the decomposition to level 1.

The process hierarchy chart gives a picture of the relationship between processes without including the details of data flows. Each decomposed level of the process hierarchy chart is the subject

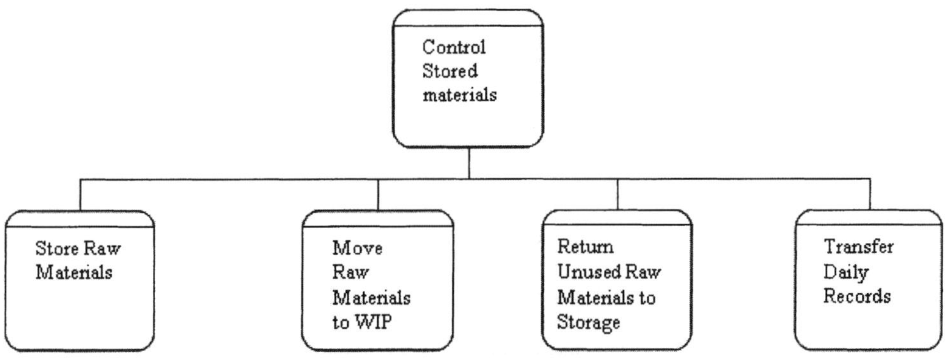

Figure 4.16 Process hierarchy chart.

of its own data flow diagram. Figure 4.16 provides an overall guide to the decomposition of processes.

The data flow diagram for the level-1 decomposition is shown in Figure 4.17. As in the case of IDEF0, interfaces at the boundary are conserved from the context-level diagram. That is to say, the external sources and sinks of data, both entities and data stores, that appeared in the context level diagram appear in Figure 4.17. The four processes are numbered from left to right. The numbering system does not indicate any particular precedence ordering in time since these processes may be going on simultaneously. The numbers are just labels that are used to trace the hierarchic decomposition through subsequent levels.

Each process must have an input data flow and an output data flow. If a process does not use data, it is not of interest. Remember, a data flow diagram is "data oriented." Thus, process 1.0 is triggered by the data flow "Request to Store Materials" and provides input to the inventory log by the data flow "Storage Transaction Details." The "Inventory Log" is a data store that is internal to the process. It is a temporary data store used by the forklift driver. At the end of the day, this record is transferred to the factory database files. As in the case of IDEF0, there is a narrative that accompanies each process, which we will now describe.

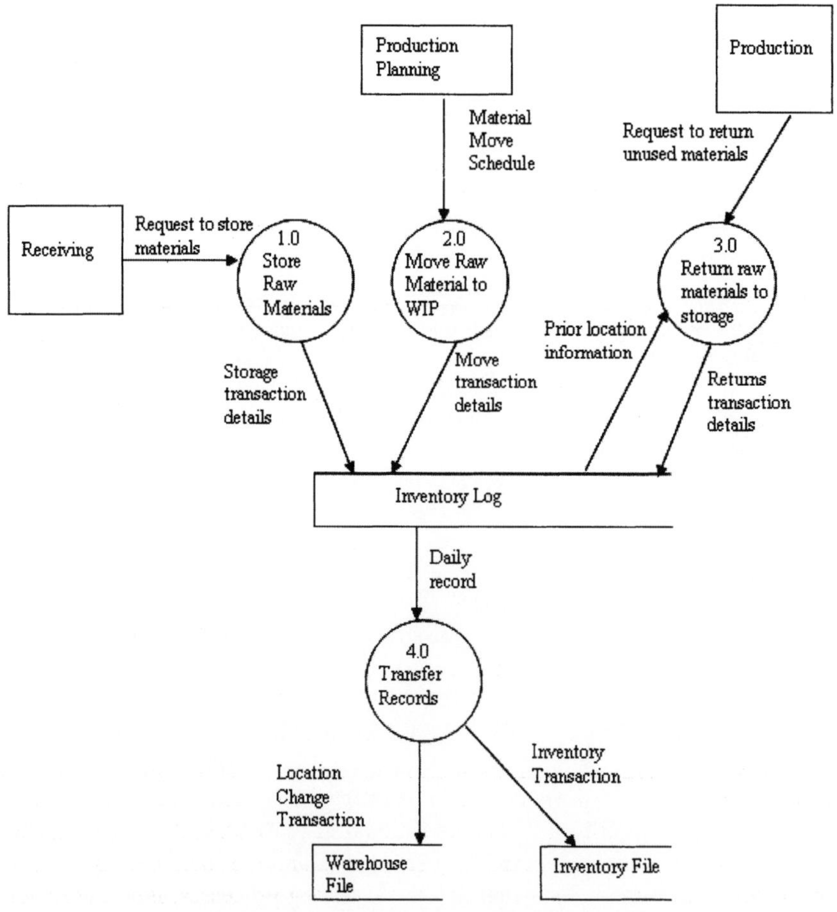

Figure 4.17 Decomposition of control stored materials.

1.0 *Store raw materials.* Receiving makes requests to forklift truck drivers to move material from the loading dock to inventory storage. The truck driver is informed of the location to use by the receiving clerk, who has indicated the location on the receiving report. The driver takes the material to the location. Each location in the warehouse is marked with a location number. The driver places the material in the location and then records the material, the location used, and the date and time of the transaction in the log. There is always the possibility that the receiving clerk was inaccurate and that the location is not available when the driver arrives. In the event that the location is not available, the driver goes to the next nearest location of the same type (refrigerated, dry storage, etc.) and stores the material there, recording the transaction.

2.0 *Move raw material to WIP.* At the beginning of the shift the forklift truck driver is given the schedule of material moves from storage to the factory floor. Each time the driver makes a move, raw material inventory is debited and the status of the warehouse location is updated. This is done by indicating a transaction to relieve inventory in the log, recording the material, location, date, and time.

3.0 *Return unused raw material to storage.* Some materials that are brought to the factory floor may be returned if they are not used. When production is completed for a particular product run, the unused materials are still in the staging area on the factory floor. Upon request from the production supervisor, the driver takes the material back to storage and logs the credit entry into the log.

4.0 *Transfer records.* The forklift truck driver's inventory log is used as the primary record for updating the warehouse and inventory records. This updating is done at the end of the shift. Materials management checks for any discrepancies between the receiving report and the actual location of material by comparing the log with the receiving report.

If Figure 4.17 were to be further decomposed, each process would have its own data flow diagram. The numbering system continues in the same manner as done in IDEF0. That is to say, a decomposition of process 2.0 would have labels 2.1, 2.2, 2.3, and so forth. A decomposition of process 2.2 would have labels 2.2.1, 2.2.2, 2.2.3, and so forth. Since Figure 4.17 clearly specifies how the system operates, we will go no further. Adding more detail will not uncover any additional data stores or data flows. The decomposition terminates here.

4.7 SUMMARY

The design of an industrial information system begins at the conceptual level with the development of two related architectures: the functional (activity) model and the informational (data) model. In this chapter, we focused on the functional model. Using the IDEF0 methodology, we illustrated the process of modeling based on activity blocks, inputs, outputs, controls, and mechanisms. We have shown that the modeling process requires a hierarchic decomposition of functions from parent activity to child activities. At each level of decomposition, consistency is maintained by having the child activity diagram inherit the inputs and outputs of the parent activity diagram. The process continues until the elemental level is reached at which enough detail is incorporated into each activity to understand the manner in which particular functions are performed and how they relate to other functions vis-à-vis material and information flows. We have also illustrated an alternative process for accomplishing a documentation of the processes that use data, the data flow analysis method. DFA is widely used in industry. It differs from IDEF0 in that it focuses exclusively on business processes and the information that flows among processes, ignoring material flows, mechanisms, and controls.

Diagramming tools are available for creating functional diagrams. The use of Microsoft Visio for diagramming is described in the Visio Lab Manual, which the reader can download from the Web site supporting this book. Professional IDEF software is available from several companies and can be downloaded in trial version, which may be useful in completing the end-of-chapter exercises. We refer the reader to selected Web sites listed at the end of the bibliography section at the end of this book. The IDEF0 standard, as published by the National Institute of Standards and Technology (NIST) can be obtained from http://www.itl.nist.gov/fipspubs/idef02.doc.

Although these structured analysis methods provide a standard set of graphic primitives and a set of rules to guide their use, different analysts often come up with different diagrams when employing these techniques (there is some latitude for interpretation, which causes this variation). However, their ease of use and their value as a communication tool among analysts and experts involved in the design process overcomes this shortcoming. The importance of the functional architecture becomes evident as analysts begin to define the specific information (data) needed to support these functions. In the next chapter, we focus on the information (data) model required to complete the conceptual modeling development process.

REVIEW EXERCISES

4.1 Describe the difference between an IDEF0 input and a control. For example, why is the production schedule and recipe a control instead of an input?

4.2 In Figures 4.11 and 4.12, the activity "Control Incoming Materials" was documented using the IDEF0 methodology. Using this example, document this process using the equivalent data flow diagram methodology.

4.3 In Figures 4.15 and 4.17, the activity "Control Stored Materials" was documented using the DFA methodology. Using this example, document this process using the equivalent IDEF0 diagram methodology. You will have to use the description of this case in the chapter to uncover the material flows involved, which are not explicit in Figures 4.15 and 4.17.

4.4 Review Exercise 3.5 provided a description of how a customer is processed by USF Rent-a-Car when she arrives at a branch location to pick up a car. From that description, create a set of data flow diagrams that describe the functional architecture for the process of picking up a car. The context diagram should be titled "Service Customer during Rental Pick Up."

CASE STUDIES

4.5 ACME Machine Shop Case (B)

Introduction

The ACME Machining Company is a small job shop that provides machining services. The company owns ten general-purpose computer numerical control machine tools and employs 15 people, 8 of whom are tool designers and machine tool operators. The company provides machining services to other manufacturers in the area who require machining services done that they are not interested in doing in-house. Therefore, ACME's orders from customers are usually for high-precision or special-purpose machined parts in small lots.

The most important assets of the business are the highly trained tool designers and machinists, and the very expensive precision tools owned by the company. The owner of the company has a simple view of his business: whenever a machinist or a machine is idle and there are jobs waiting to be processed, the company is losing money. Therefore, the owner has always put a high priority on machine maintenance, especially preventive maintenance, to avoid unnecessary downtime.

Maintenance is the responsibility of the plant supervisor, Mr. Bill Wrench. Bill has set up the entire preventive maintenance program for the plant, which consists of regularly performing certain preventive maintenance tasks based on the number of hours of operation on a machine. The manufacturers of each of the 10 machine tools have provided Bill with a list of preventive maintenance tasks and the frequency with which they are to be done.

The company owner has suggested to Bill that he maintain records of the preventive maintenance functions in a computer database. The owner has heard that there are inexpensive database systems that can be purchased and used for this purpose. Also, a nearby college has students that are knowledgeable about the use of databases, and they can be hired as interns to help to design the application. Bill decides to follow-up on his boss's suggestion, and he hires you to advise him on the design.

The Preventive Maintenance Function

Your first step is to meet with Bill to discuss the preventive maintenance function as it is now performed. The description that Bill provides you is given in the following paragraphs:

Bill: "First, let me show you my database; it's over here in the filing cabinet." [He opens a drawer.] "These documents are the most important information that I have. The machine tool vendors have given me the exact maintenance that I should follow on each machine. For example, look at this schedule [Exhibit CS4.1]. It's for the Mazak 2120 CNC Mill. See, every 100 hours of operation I have to check the slides for lubrication, and every 500 hours of operation I have to do a tear down of the milling head to check the condition of gears and belts. The other vendors have provided me with similar kinds of instructions for their machines."

Student: "So how do you know when the machine has been operated for 100 hours or 500 hours since the last maintenance?"

Mazak Corporation
122 Mill Street
Lexington, KY

PREVENTIVE MAINTENANCE SCHEDULE FOR: 2120 CNC Mill

PM TASK NO	FREQUENCY	DESCRIPTION	REFERENCE
1	Every 100 hours	Lubricate slides as needed	Manual M5.1, p. 4
2	Every 500 hours	Milling head: check gears and belt	Manual M5.1, p. 6
.	.	.	.
.	.	.	.
.	.	.	.

EXHIBIT CS4.1 Example Preventive Maintenance Schedule

Bill: "Oh, I keep records. Each machine has a meter on it that keeps track of operating hours. In the morning, the manufacturing department collects the meter readings and sends them to me. The machinists then reset the meters for the start of a new day. This way, I have a record of the hours of operation for each day. I also keep a sum column so that I know the total number of hours on the machine. See." [He holds up Exhibit CS4.2.]

Student: "Let me see if I understand. The first thing you do in the preventive maintenance function is that you receive the hours of operation data from manufacturing. Then you add those hours to your record. Then you look up the preventive maintenance schedule for each machine and compare it to the sum of the hours of operation in your record. If the prescribed number of hours of operation have elapsed since the last time a particular preventive maintenance task was done, you perform the task. One thing I don't understand. How do you know that the prescribed number of hours have passed since the last time you performed the task?"

Bill: "Oh, that's easy. I find that information in my file of completed maintenance records, which is this drawer containing completed maintenance work orders." [He opens another file drawer.] "Each time I request preventive maintenance on a machine, I make out a work order. I put the work order into this tray [points to tray on desk], which is my *open work order* file. The maintenance technician works off this tray. He takes each order in sequence, performs the maintenance, and then returns the work order to me when the maintenance is completed. He puts the time worked on the work order and returns it to me for filing. Here, look at this [Exhibit CS4.3]. This is one of

MACHINE OPERATING HOURS RECORD

	MAZAK 2120		CINCINNATI CNC LATHE	
DATE	DAY	TOTAL	DAY	TOTAL
2/1/99	4	90	-	-
2/2/99	8	98	-	-
2/3/99	6	104	5	5*
2/4/99	0	104	6	11
2/7/99	5	109	8	19
2/8/99	6	115	6	25
2/9/99	8	123	5	30
2/10/99	8	131	8	38
2/11/99	4	135	4	42
.	.	.	.	.
.	.	.	.	.
.	.	.	.	.

* first day put into service.

EXHIBIT CS4.2 Record of Operating Hours by Machine

PREVENTIVE MAINTENANCE WORK ORDER

PM WORK ORDER NUMBER: 00512
MACHINE ID: E41520
MACHINE NAME: Mazak 2120 CNC Mill
PM TASK: #1 – Lubricate slides as needed
MINUTES WORKED: 120

DATE WORK REQUESTED: 2/4/99
MACHINE HOURS: 104
DATE WORK DONE: 2/4/99

EXHIBIT CS4.3 Example Work Order

my records of preventive maintenance done in the past. The *work order number* is a unique number that I assign each time I request a specific task on a machine. Basically, it's just a rotating five-digit number. This one shows that this is the 512th preventive maintenance task I have performed on any of our machines. I put in the *machine ID* to identify the specific machine that is being worked on and the *date work requested* to show when the work was assigned. The machine ID is a number that comes off the manufacturer's data plate. It's basically his serial number, and it is unique for his machines. I also record the *machine hours*, which show how many total hours were on the machine when I did the maintenance. I get that number off this sheet [Exhibit CS4.2, again]. Finally, I record the preventive maintenance task that is to be performed, and the maintenance technician records how long it took to do the maintenance."

Student: "When the maintenance technician returns the completed work order to you, does he hand it to you? What actually happens?"

Bill: "No, he doesn't actually hand it to me. See this empty tray next to the one I use for open orders? He puts the completed work orders into this tray. I empty this tray at the end of the day. I add the last piece of information to each work order, the *date work done*, and then I put the completed work order into my completed maintenance record file. This is where I look to find out when a particular maintenance item was last done."

Student: "Okay, Bill, I think I understand what you are doing. Please do me a favor and provide me with any other information you think might be relevant [provided later as Exhibit CS4.4]. Also, since we are going to have your records on a database, try to describe the kinds of summary reports that you think would be desirable to print out for your own purposes or for your boss. We might as well plan to accommodate all the desired information in our design at this time." [Exhibit CS4.5 provided by Bill later.]

Requirements

1. Design a functional model using the IDEF0 methodology that shows how the preventive maintenance function is done.
2. Design a data flow diagram of the processes and information flows involved in the preventive maintenance process.

Machine name: Mazak 2120 CNC Milling Machine

Manufacturer: Mazak Corporation

Serial No.: E41520

First date in service: December 1, 1998

Warranty expiration date: December 1, 2003

EXHIBIT CS4.4 Machine Data

1. A list of warranty expiration dates by machine. We regularly extend our warranty dates by purchasing new contracts. I would like to be able to review them when I please.

MACHINE NAME	MACHINE SERIAL NO.	FIRST DATE IN SERVICE	WARRANTY EXP. DATE
2120 CNC Mill	E41520	12/1/98	12/1/03
Cincinnati CNC Lathe	C5001	2/3/99	2/3/01
Cincinnati CNC Lathe	C4010	4/1/96	4/1/01
.	.	.	.
.	.	.	.

2. I would also like to be able to obtain a preventive maintenance history by machine and by task.

MACHINE NAME	PM TASK NO.	DATE DONE	MACHINE HOURS	MINUTES WORKED
CNC Drill	1	1/10/98	205	30
CNC Drill	1	3/5/98	410	35
CNC Drill	1	7/2/98	605	40
CNC Drill	1	10/1/98	800	35
CNC Drill	1	2/1/99	1010	32
CNC Drill	2	6/2/98	500	150
CNC Drill	2	2/1/99	1010	130
.	.	.	.	.
.	.	.	.	.
.	.	.	.	.

EXHIBIT CS4.5 Requested Summary Reports

4.6 University Food Company Case (B)

The University Food Company wants to incorporate more information technology into its organization, particularly its shipping department. The sales department has already implemented a database for order entry functions, and the production planning department has an information system for tracking inventory, including finished goods. The shipping department currently uses information from these sources in the course of doing its daily tasks. The shipping supervisor has asked you to study the activities in the department and to recommend an information system design to support the operation. What follows is a description of the tasks performed by the shipping clerk and support personnel during a typical day.

At the beginning of each day, the sales department prints a report and sends it to the shipping department. This report, called the "Open Orders Report," is a list of orders taken by sales that have not been closed out. Orders are not closed out until the final shipment has been made against the order. An example of a typical open orders report is shown in Exhibit CS4.6. Open orders are listed in the ascending order of their promised delivery date.

University Food plans its production in two ways: produce to order and produce to inventory. In the first case, production is scheduled for a specific order. In the second case, production is done in the absence of a specific order. The second case occurs because a specific order is small and it is desirable to have a longer production run before changing over to another product. Changeovers require that the machinery is thoroughly cleaned before the next product can be made, thus incur-

University Food Company, Inc **Open Orders Report, 11/01/03**

Order #	Cust id	Name & Ship to	PO #	Product ID	Label	Qty(cases)	Price	Extension	Deliver by
03564	3950	Columbia University 4677 Broadway New York, NY	58900	243	Macaroni & Cheese	800	27.00	$21,600	10/29/03
03565	1000	Lehigh University 210 Packer Street Bethlehem, PA	03-100	430 350 235	Turkey Soup Beef Ravioli in Sauce Chili w/ Beans	600 300 1000	30.00 45.00 40.00	$18,000 $13,500 $40,000	11/01/03 11/01/03 11/01/03
03580	1050	New Jersey Institute 500 Martin Luther King Blvd. Newark, NJ	20300	350 440	Beef Ravioli in Sauce Corned Beef Hash	500 700	45.00 32.00	$22,500 $22,400	11/01/03 11/01/03
03566	873	University of Massachusetts 125 Springfield Ave. Amherst, MA	03-222	243 333	Macaroni & Cheese Chicken Noodle Soup	1000 3000	27.00 22.00	$27,000 $66,000	11/02/03 11/02/03
03564	3950	Columbia University 4677 Broadway New York, NY	58900	380	Fish Chowder	400	30.00	$12,000	11/03/03
03567	1200	New York University 10 Folly Square New York, NY	4999	430 905	Turkey Soup Beef Stew	500 600	30.00 50.00	$15,000 $30,000	11/03/03 11/03/03

EXHIBIT CS4.6 Open Orders Report

Finished Product Inventory Report, 11/01/03

Material ID	Description	Lot no.	On Hand	Location			
				area	isle	tier	bin
235	Chile w/ Beans	13180	500	5	1	1	6
		13210	2000	5	2	1	10
243	Macaroni & Beans	12920	400	5	1	1	8
		13010	1000	5	4	2	4
333	Chicken Noodle Soup	13220	1000	5	2	1	3
		13400	1000	5	2	2	1
350	Beef Ravioli in Sauce	12840	100	5	3	2	10
		13006	600	5	3	1	6
		13285	600	5	1	1	4
380	Fish Chowder	13401	600	5	4	2	2
430	Turkey Soup	13080	1000	5	1	1	5
.	.	.	.	.			
.	.	.	.	.			
.	.	.	.	.			

EXHIBIT CS4.7 Finished Product Inventory Report

ring a significant cost. The excess production is placed in inventory and is available for shipment the next time a customer orders the product. Also, sometimes production runs of the most popular food products are made to inventory during periods of slack demand.

The shipping clerk receives a report (printout) of the finished goods inventory from the production planning department, showing the location of the finished goods inventory by inventory lot. The report is ordered by material ID in the ascending order. The clerk examines each of the open orders and determines whether or not there exists a matching product inventory lot. If the match occurs, a shipment can be made. A typical finished product inventory report is shown in Exhibit CS4.7.

After identifying an inventory lot that can be shipped, the shipping clerk dispatches the forklift truck operator to the warehouse to fetch items to be shipped. He does this by giving verbal instructions to the driver concerning the lot number and location of the product. The driver collects the items and brings them on pallets to the shipping dock. He informs the shipping clerk that the pallets have been retrieved and gets his next assignment.

It is the responsibility of the shipping clerk to keep a daily record of everything that is transferred out of the warehouse for shipping and to indicate on the record the customer order against which it is shipped. The warehouse transfer report, shown in Exhibit CS4.8, is used for that purpose. The shipping clerk completes this record when pallets are brought to the loading dock. This document is transferred at the end of the day to the sales department where it is used as the source of information to close out sales orders.

After completing an entry on the warehouse transfer report, the shipping clerk directs his personnel to load the cases of product on to a truck. As cases are being loaded, he fills out another important report maintained by the shipping clerk, the shipping summary report. This report is kept daily and held in a file in the shipping department. It describes the quantity of cases that are shipped, the customers to whom they are shipped, and the truck number on which they are shipped. University Food maintains its own fleet of trucks, which is managed by the company's transportation department. The shipping department is only responsible for tracking the shipment to the truck on which it is dispatched. The transportation department is responsible for tracking deliveries to the final customer. A typical shipping summary report is shown in Exhibit CS4.9.

For each customer shipment on a truck, the shipping clerk prepares a bill of laden for the truck driver after the cases of product are loaded on the truck. Most of the information for the bill of laden is taken off the shipping summary and open orders reports. A copy of the bill of laden is sent to the transportation department at this time, and the other copies are given to the truck driver. No copies are kept in shipping. This bill of laden is presented to the customer when the driver arrives at the customer's receiving location. A signed copy of the bill of laden is turned in to the transportation department when the driver returns from a delivery. The transportation department matches this copy to the one sent by the shipping department and closes out the delivery in its records. A typical bill of laden is shown in Exhibit CS4.10.

Warehouse Transfer Report, 11/01/03

Order #	From Location	Product	Lot no.	Qty. (cases)
03564	5-1-1-8	Macaroni & Cheese	12920	400
"	5-4-2-4	"	13010	400
"	5-4-2-2	Fish chowder	13401	400
03567	5-1-1-6	Turkey soup	13080	500
"	5-5-1-1	Beef Stew	12604	600
03565	5-1-1-6	Turkey Soup	13080	500
"	5-3-2-10	Beef ravioli in sauce	12840	100
"	5-3-1-6	"	13006	400
.	.	.	.	.
.	.	.	.	.
.	.	.	.	.

EXHIBIT CS4.8 Warehouse Transfer Report

Shipping #	Customer	City & State	PO #	No. of Cases	Weight (Lbs.)	Truck No.
03550	Columbia Univ.	NewYork, NY	58900	1200	6000	3
"	NewYork Univ.	NewYork, NY	4999	1100	5500	3
03551	Lehigh Univ.	Bethlehem, PA	03-100	1900	9500	4
03552	New Jersey Inst.	Newark, NJ	20300	1200	6000	2

EXHIBIT CS4.9 Shipping Summary, 11/01/03

FREIGHT BILL OF LADEN

To: New Jersey Institute of Tech
500 Martin Luther King Blvd.
Newark, NJ 07050

From: University Food Company
1766 New England Ave.
Piscataway, NJ 08854

Date Shipped: 11/01/03
Truck Number: 2
B/L No.: 03552

Code	Quantity	Weight	Description	Reference
350	500 cases	2500 Lbs	Beef Ravioli in sauce	P.O 20300
440	700 cases	3500 Lbs	Corned Beef Hash	PO 20300
	Total Weight	6000 Lbs		

Received by:______________ **Date:**

Driver: ____________ **Lic. Number:** ______________

EXHIBIT CS4.10 Freight Bill of Laden

Requirements

1. Using the description presented earlier, develop a functional architecture based on the IDEF0 method. The context diagram should be labeled "Operate Shipping Department."
2. Draw a corresponding data flow diagram.

4.7 Regal Foreign Car Repair Case (B)

Case Study 3.10 includes a detailed description of the business procedures of Regal Foreign Car Repair Shop. The activities of the service manager and the mechanics are described in terms of their interface with the data requirements of the case. From this description, create a dataflow diagram that describes the processes and their interaction with sources, sinks, and data stores. Describe the data stores and data flows in terms of the entities described in the case. Title the contest diagram "Operate Car Repair Shop."

Chapter 5

Informational Architecture and Logical Database Design

5.1 INTRODUCTION

This chapter focuses on the conceptual, or logical, design of the database. We refer to this as the design of an informational architecture. Just as the functional architecture is developed to give a conceptual design of the operational requirements of the enterprise and the interaction of operations with information, there is a need for a conceptual model of the data requirements. This is the role of the informational architecture design. The informational architecture provides a model of the entities in the enterprise, the attributes of those entities, and the relationships that exists among entities. Several related modeling formalisms are used to develop relational data models. They are derivatives of the seminal modeling approach developed by Peter Chen in the 1970s (Chen, 1976) and described in Chapter 3. This chapter describes Integrated Computer-Aided Manufacturing Definition 1, extended (IDEF1X), which is the data modeling formalism associated with the IDEF CASE tools.

5.2 THE IDEF REPRESENTATION OF ENTITY-RELATIONSHIP MODELING

The rules of entity-relationship (E-R) modeling as they appear in the form originally developed for the representation of data were reviewed in Chapter 3. In E-R diagrams, the data model may contain several elements that cannot be directly mapped to database tables without further refinement. For example, the E-R model may contain composite attributes and relationships that are not 1 : M. Before the database tables are implemented, the data model should be in third normal form and composite attributes should be broken down into constitutive parts.

Over the years, some variations on E-R diagrams have appeared in commercial computer aided software engineering (CASE) tools. One of these alternative variations is called IDEF1X. IDEF1X is the data modeling methodology in the same family as IDEF0, the functional modeling methodology. IDEF1X was designed to complement IDEF0 by providing a representation of the information required to support the functions defined in IDEF0. It should be possible to directly compile the database tables from the IDEF1X model. In fact, CASE tools that support the IDEF1X formalism have the functionality to automatically create the tables from the model. Therefore, when we use the term "informational architecture," we are talking about a data model that is refined to this level and that complements a related functional architecture.

IDEF1X is based on relational theory concepts and entity-relationship modeling concepts. Therefore, it is a semantic data modeling technique that defines the meaning of data within the context of its interrelationship with other data. The methodology was designed to be used and understood by systems analysts and business professionals. In other words, it is designed to provide effective communication between designers and users during the conceptual design process. A number of software products are available to automate the generation of graphical designs using the IDEF1X methodology.

Figure 5.1 illustrates some of the fundamental concepts in IDEF1X modeling. A box represents the entity with the entity name above it. The entity name is a noun phrase that describes the set of things (persons, places, concepts, etc.) that the entity represents. The noun phrase is singular, not plural. When an entity name is more than one word, words can be separated by underscores, hyphens, or blanks.

Attributes are contained within the box. By containing attribute names within the entity box, IDEF1X modeling is more compact than generic E-R modeling. Primary key attributes are in the upper portion of the box. A dividing line separates the primary key attributes from nonprimary key attributes. Figure 5.1 shows only a subset of all the attributes of each entity for illustrative purposes.

IDEF1X makes a distinction between independent and dependent entities. An entity is ***independent*** if each instance of the entity can be uniquely identified without determining its relationship to another entity. An entity is ***dependent*** if the unique identification of an instance of the entity depends on its relationship to another entity.

In Figure 5.1, VENDOR and PURCHASE_ORDER are independent entities. Any instance of VENDOR is uniquely identified by VENDOR_ID, and no other entity is required to define VENDOR_ID. Each instance of PURCHASE_ORDER is uniquely defined by PO_NUMBER, and no other entity is required to define PURCHASE_ORDER. When an entity is independent, the box is drawn with right-angle corners.

The entity PO_DETAIL is dependent. Each instance of PO_DETAIL is uniquely identified by the tuple PO_NUMBER, PO_LINE_ITEM. However, PO_NUMBER depends on its relationship with PURCHASE_ORDER. Hence, PO_DETAIL is a dependent entity. Dependent entities are drawn with rounded corners. From this description it should be clear that if a child entity contains the foreign key of the parent entity as part of its primary key, that child entity is dependent.

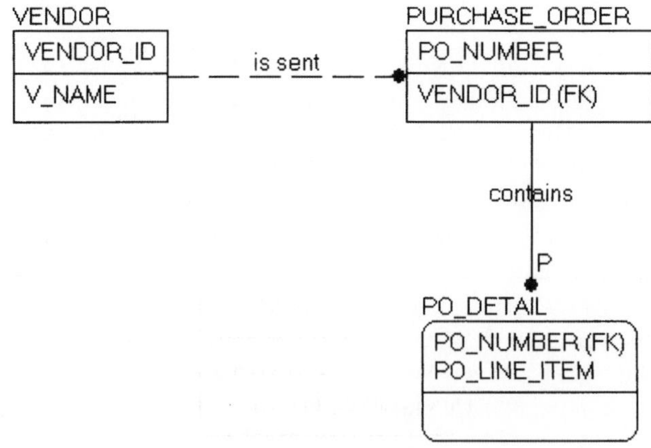

Figure 5.1 Basic Concepts in IDEF1X modeling.

A relationship (called a parent-child relationship) is an association between entities in which each instance of one entity, called the parent, is associated with zero, one, or more instances of another entity, called the child entity. Also, each instance of the child entity is associated with exactly one instance of the parent entity.

A parent-child relationship is shown as an arc from the parent to the child, with a dot at the child end of the arc. The cardinality is assumed to be 1:M unless a letter is affixed to the dot to indicate otherwise. A "Z" is placed beside the dot to indicate a cardinality of zero or one. A "P" is placed beside the dot to indicate cardinality of one or more. An "n" is placed beside the dot to indicate a cardinality of exactly n. In Figure 5.1, there are zero, one, or many purchase orders associated with each vendor. One or more purchase order details are associated with each purchase order.

A relationship is named with a verb phrase placed beside the relationship line. The relationship name is always expressed in the parent-to-child direction. Thus, "A vendor is sent zero, one, or more purchase orders" and "A purchase order contains one or more purchase order details" expresses the relationships in the model of Figure 5.1.

IDEF1X distinguishes between identifying relationships and nonidentifying relationships. An ***identifying relationship*** connects a parent entity to a child entity and *all* primary key attributes of the parent become primary key attributes of the child. In Figure 5.1, this occurs in the relationship between PURCHASE_ORDER and PO_DETAIL, where PO_NUMBER provides the identifying relationship. When an identifying relationship exists, the connecting arc is a solid line. However, if two entities are related and the foreign key that makes that relationship is not part of the primary key of the M side of the relationship, a ***nonidentifying relationship*** exists. In that case, the connecting arc between the two entities is dotted. In Figure 5.1, this occurs in the relationship between VENDOR and PURCHASE_ORDER. Another example is shown in Figure 5.2. Note that here the primary key for PO_DETAIL is now a single unique identifier that has replaced the composite primary key shown in Figure 5.1. It is a *surrogate* for the combination (PO_NUMBER, PO_LINE_ITEM), and it makes the relationship nonidentifying.

IDEF1X supports ***optional*** and ***mandatory*** relationships. If the dot on the child end of the relationship has no additional notation, it is assumed to mean 0, 1, or many. The "0" indicates that there may be instances of the parent entity that do not have a relationship to the child entity. Therefore, the parent is optional to the relationship. In Figure 5.1, this is the case in the relationship

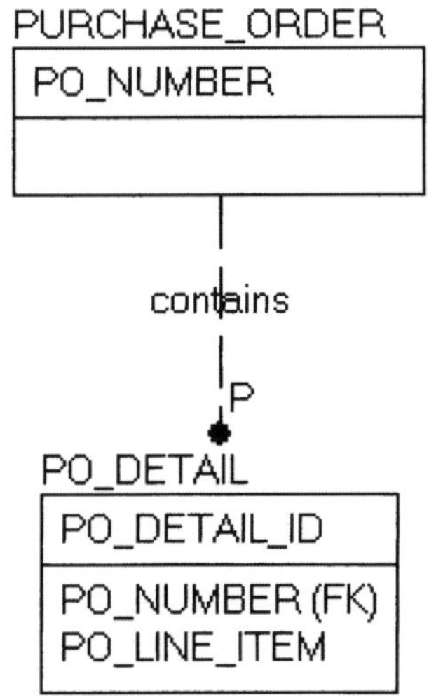

Figure 5.2 A Nonidentifying relationship.

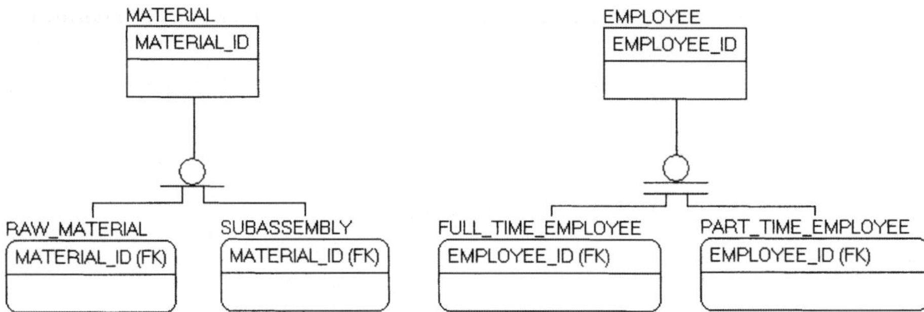

Figure 5.3 Categorization relationship.

between VENDOR and PURCHASE_ORDER. VENDOR is optional to the relationship. If a "P" is added next to the dot (1 or more), then the relationship is mandatory. This is shown in the relationship between PURCHASE_ORDER and PO_DETAIL in Figure 5.1. All purchase orders *must* have a detail.

Another way of indicating that an entity is optional to the relationship in IDEF1X is the placing of a diamond on the 1 side of a 1:M relationship. The diamond indicates that the M side may contain NULL values of the foreign key. This means that the relationship is optional to the child. This situation can only occur in a nonidentifying relationship, since, by definition, all the primary keys of a parent become part of the primary key attributes of the child in an identifying relationship, and primary key attributes cannot be NULL.

IDEF1X supports the definition of superclass/subclass entities, in which some entities are specific categories of a larger entity. In IDEF1X terminology these are called ***categorization relationships***. The superclass entity is called the *generic entity*, and the subclass entities are called the *category entities*. Such a relationship is shown in Figure 5.3, where the entity MATERIAL has a relationship to RAW_MATERIAL and SUBASSEMBLY. The category entities use the same primary key as the generic entity. Also, an instance of a generic entity is related to at most one instance of a category entity. In other words, an instance in MATERIAL may occur in either RAW_MATERIAL or SUBASSEMBLY, but not both.

A discriminator shows the nature of the relationship. The discriminator is either a single or double bar with a circle on the generic entity side. The cardinality of the relationship is not indicated, since it is implied to be at most one-to-one by the discriminator. When a single bar is used, the relationship is called an *incomplete categorization relationship*. This means that there are instances of the generic entity that do not appear in either of the category entities. For example, MATERIAL may include office materials, repair and maintenance materials, and so forth. These do not appear in any of the category entities, which describe production materials. When the instances of the category entities completely account for all the instances of the generic entities, a *complete categorization* is said to exist. In that case, the discriminator is shown as a double bar with a circle above it.

There are several other concepts in the IDEF1X modeling methodology. Here we have touched on the basic ideas that will allow us to model an actual problem. We will introduce other IDEF1X concepts as needed later in this chapter.

IDEF1X allows the modeler to design data models having 1:M, 1:1, and M:N relationships. In this chapter, we shall use the knowledge gained in Chapter 3 to directly design the data model using 1:M relationships.

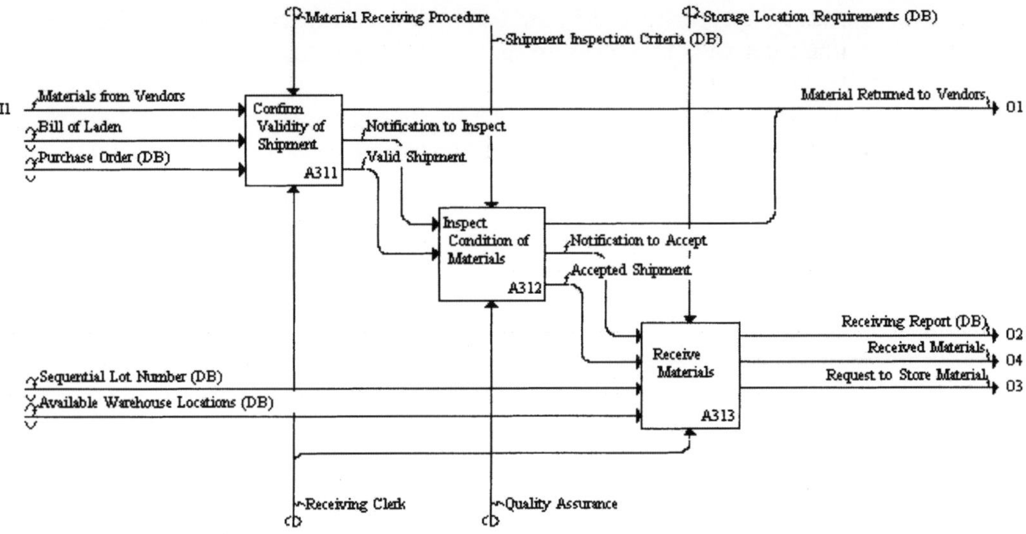

Figure 5.4 Decomposition of node A31.

5.3 A CASE STUDY IN DEVELOPING A DATA MODEL

The process of developing the data model of an enterprise is part art and part science. There are rules that govern the construction of the data diagram, and there are rules that govern the integrity requirements of the relational model. However, different analysts can look at the same organization and identify different entities to model, while still respecting the rules of the design. Moreover, more than one data model may be adequate for the intended purpose.

The process by which the analyst proceeds in the development of the model is usually iterative. The entities and relationships are discovered by interviewing the experts who do the work, by examining the paper reports and forms they use in performing their functions, and by being creative in extracting concepts out of these sources of information.

Entities can be concrete things, such as persons, places, equipment, and materials. Entities can also be abstractions, such as concepts or events. The best way to develop a sense of how to identify appropriate entities is through practice. In this section, we use the case study approach to emulate the discovery process that an analyst might go through in arriving at a suitable data model.

The case study will be a continuation of the discussion we began in Chapter 4—that is, the control of incoming material and the control of stored material. For the control of incoming material, we will utilize the narrative associated with the breakdown of node A31 as shown in Figure 4.12. For the control of stored material, we will use the narrative associated with the data flow diagram shown in Figure 4.17. What follows is a description of the kind of information that an analyst would have to extract out of multiple interviews. For convenience, Figure 4.12 has been reproduced as Figure 5.4. We will focus our attention on nodes A311 and A313, which are the activities handled by the receiving clerk. The reader may wish to review Section 4.4.4 before continuing.

5.3.1 ANALYSIS OF INFORMATION REQUIREMENTS AT NODE A311

The reader should refer to nodes A311 and A313. As described in Chapter 4, the receiving clerk is the initial contact when material arrives via truck at the receiving dock. The clerk is given a bill of laden by the truck driver, and he or she must confirm that the material being delivered is part of an open purchase order. Once it is determined that the material should be accepted, the truck is unloaded and the receiving clerk begins to account for the delivery by writing up a document called a receiving report. Therefore, three paper forms are being used at nodes A311 and A313: the bill of laden, the purchase order, and the receiving report. For clarity of illustration, these forms are shown in Figures 5.5 to 5.7.

FREIGHT BILL OF LADEN

To: The University Food Company
1766 New England Ave.
Piscataway, NJ 08855

From: General Provisions
125 Common St.
Boise, ID 44830

Date Shipped: Jun 20 06
Carrier: Northwest Trucking Co.
Hansford, WA
Car Truck Number: 5501
B/L No.: 06-225

UPC Code	Quantity	Weight	Description	Reference
20000 - 16605	1400	39,100	Gallon cans of Tomato paste	PO3502
	39	1,950	wood pallets	
	Total	41,050		

Received by: Jerry Dibbs   Date: June 25, 2006

Driver: Mack Truck   Lic. Number: E3589-002

Figure 5.5 Bill of laden.

RECEIVING REPORT

Supplier: General Provisions
125 Common St.
Boise, ID 44830

Purchase Order No.: PO3502
Date Received: June 25 2006

Quantity accepted	Quantity not accepted	Mfg. Lot No.	Item Code	Mat'L Lot No.	Description	Storage Location
1000		1275	RM805	97275	Tomato Paste, 1 gallon cans	Area A, Aisle 1 tier 1, bins 10-18
300		1283	"	97276	" " " "	Area A, Aisle 1 Tier 2, Bins 10-13
	100	"	"		" " " "	returned(1)

Comments: (1) returned due to case damage and badly dented containers.

Received by: J. Dibbs

Figure 5.6 Receiving report.

THE UNIVERSITY FOOD COMPANY
1766 NEW ENGLAND AVE.
PISCATAWAY, NJ 08855

VENDOR: General Provisions
125 Common St.
Boise, ID 44830
Attn: Dave Stauffer

P.O. Number: 3502
Date: 5/28/06
Amount: $4920.00

Line Item	Description	Unit Price	Amount	Del. Date
1	1400 gallons of tomato paste, Item # 2-414.	$3.40	$4760.00	6/25/06
	Freight (Est.)		160.00	
		Total	4920.00	

Ordered by: Randy Kim
Phone Number: 908-445-3657
Deliver to: The University Food Company
1766 New England Ave.
Piscataway, NJ 08855

Figure 5.7 Purchase order.

The bill of laden is a shipping document that is used to identify the contents of the shipment and the shipping authority. The carrier presents the bill of laden when it arrives with a delivery. The hypothetical bill of laden, shown in Figure 5.5, has header information regarding the sending and receiving organizations as well as the shipper and date shipped. The body of the form identifies what is being shipped and a reference to the billing authority, in this case a purchase order number. The bottom of the form requires a signature by the individual who accepts the shipment, usually the shipping clerk. A name and driver's license number also identify the truck driver.

The contents of a bill of laden will vary. However, we are going to assume that this is a generic form for our purposes in designing the data model for the function of receiving incoming materials.

The purchase order, which is shown in Figure 5.7, is similar to the one used in Chapter 2 (Figure 2.2). However, Figure 5.7 is more realistic. For one thing, the MATERIAL_ID is not shown on this purchase order. The MATERIAL_ID is an internal identifier assigned by the enterprise to keep track of its own inventory. This number means nothing to the vendors, who have their own identifier code for the products they sell. Thus, the purchasing agent uses the description field on the purchase order to specify the product using the vendor's own identification, in this case "Item# 2-414."

A company contact name and phone number is given at the bottom of the purchase order so that the vendor has someone to contact in the event she or he has to talk to a company representative. The contact is usually the purchasing agent initiating the purchase order. Also, a delivery address is given. A company may have more than one location that receives shipments; therefore, a shipping address is standard information on a purchase order.

At this point, enough new information has been introduced into the problem, and we should begin structuring it. There is a relationship between VENDOR, PURCHASE_ORDER, and PO_DETAIL, with which we have become familiar. However, we are now seeing new pieces of data on the purchase order: the line item amount, the estimated freight charge, the company contact, the contact's telephone number, and the vendor's item number for the material being ordered. What do we do with this new information? Let us take each item in its order.

5.3.1.1 Line Item Amount

The heading *Amount*, which appears in the purchase order detail, was not included in the original PO_DETAIL table shown in Figure 2.1. It was omitted because the information needed to compute it already existed in the table. That is to say, the QUANTITY and UNIT_COST attributes, when multiplied together, will yield the line item amount. The line item amount is an example of what is called a *derived attribute*, because its value can be derived from other attributes in the database. Therefore, it is not necessary to retain it in the table.

Whether or not an attribute is held in a table or treated as a derived attribute is a matter of judgment. It depends on how often the attribute is used and the difficulty (time) associated with computing it. There is no hard-and-fast rule on this; it is up to the database designer to decide whether or not to include an attribute in a table or to save the storage space by computing it each time it is required. For example, we have arbitrarily recorded PO_AMT in PURCHASE_ORDER (Figure 2.1), which could also have been derived.

5.3.1.2 Freight (Estimate)

The *estimated freight charge* is often quoted by the vendor to the purchasing agent at the time a purchase order is being written. It usually appears in the detail section, as shown in Figure 5.7. At first it may appear to be just another line item to be added to the PO_DETAIL table. However, some reflection should uncover problems with that solution.

If we make Estimated Freight an instance of PO_DETAIL, it would be given its own line item number, and it would have the unique primary (composite) key PO_NUMBER/PO_LINE_ITEM. However, it does not have a MATERIAL_ID, and most of the other attributes seem irrelevant as well. Upon reflection it would seem that freight (est.) is related to the entire purchase order and not the detail. Therefore, it seems appropriate to add this attribute to the entity PURCHASE_ORDER.

5.3.1.3 Ordered by, Phone Number

The name of an individual and his or her phone number appears on the purchase order: *Ordered by, Phone Number.* Only one name appears, and it is related to the entire purchase order, not a specific PO_DETAIL. Therefore, it would seem reasonable to carry this information in the table PURCHASE_ORDER.

However, some reflection on this set of attributes is needed. In particular, PURCHASE_ORDER is not the only place in the enterprise where these attributes are likely to be used. The purchasing agent Randy Kim is, after all, an employee of the enterprise and appears on the company's payroll as well as in its telephone directory. There is need to recall this information for purposes other than putting it on a purchase order. These attributes do, in fact, belong to a separate entity, which we shall call EMPLOYEE. This entity will have many different relationships within the enterprise data model, only one of which is to PURCHASE_ORDER.

5.3.1.4 Vendor Item Number

The *vendor item number* is a unique identifier of a material being offered for sale by a vendor. In our example, we are ordering gallons of tomato paste. However, the vendor may offer more than one grade of tomato paste for sale in gallon containers. The only way to positively identify the item being ordered is to use the vendor's unique code, which is specific to each item offered for sale. The number 2-414 in the description section of the purchase order identifies a specific tomato paste.

At first it may seem reasonable to keep this attribute (call it V_ITEM_CODE) in the entity PO_DETAIL. After all, it is required for the detail section of the purchase order and probably not used elsewhere. On the other hand, V_ITEM_CODE is actually a characteristic of a material and vendor combination. As we noted in Chapter 3, this is defined by the composite entity VENDOR_MATL_XREF. Therefore, this is the logical entity to retain this attribute.

5.3.1.5 A Preliminary Data Model

Based on the preceding observations, a tentative data model can now be developed, as shown in Figure 5.8. Note that VENDOR, MATERIAL, EMPLOYEE, and PURCHASE_ORDER are independent entities, and VENDOR_MATL_XREF and PO_DETAIL are dependent entities. The two dependent entities have foreign keys that are part of their primary key. Also, the relationship

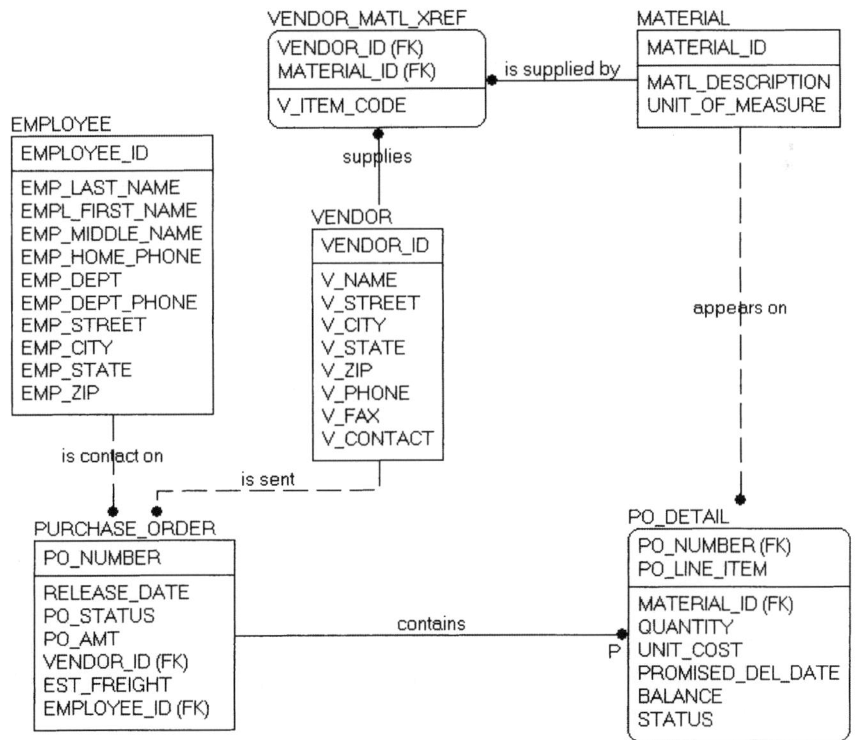

Figure 5.8 Data model after partial evaluation of node A311.

between MATERIAL and PO_DETAIL, for example, is nonidentifying because the foreign key MATERIAL_ID in PO_DETAIL is not part of its primary key.

It is useful at this point to reflect on how the entities of Figure 5.8 capture the data requirements necessary to construct or reconstruct a purchase order. It is true that a purchase order can be reconstructed most easily from a single table by simply carrying all the information regarding the purchase order in one table. However, when the same attributes are being used for other forms and reports, this will require duplication in other tables throughout the database. Aside from the storage problem, if the same attribute is duplicated in many tables throughout the database, a change in that attribute may cause the requirement that it be updated in many tables. In general, it is preferred that the instance of an attribute being updated exist in as few tables as necessary, consistent with being able to be retrieved when required using the relational model and SQL. The data model of Figure 5.8 is in 3rd normal form.

At this point, we have established three new tables for the MATERIAL_MANAGER database: MATERIAL, EMPLOYEE, and VENDOR_MATL_XREF. The reader should add those tables to the Access database MATERIAL_MANAGER, introduced in Chapter 2. The tables should be populated as shown next. All the attributes are text data types.

EMPLOYEE

EMPLOYEE_ID	EMP_FIRST	EMP_MI	EMP_LAST_N	EMP_STREE	EMP_CITY	EMP_STA	EMP_ZIP	EMP_DEPT	EMP_DEPT_PH
000-23-8900	Joseph	John	Swartz	233 East Ave	Fort Lee	NJ	06655	Purchasing	908-445-3654
042-38-6132	Randy	Randolf	Kim	23 Jay St.	Secaucus	NJ	06673	Purchasing	908-445-3657
098-67-9000	Thomas	Owen	Baker	34 Douglas F	Montclair	NJ	07043	Purchasing	908-445-9999

VENDOR_MATL_XREF

VENDOR_ID	MATERIAL_ID	V_ITEM_CODE
V110	RM211	765
V110	RM201	4562
V110	RM202	3423
V110	RM205	9899
V110	RM210	546
V25	RM810	F456
V25	RM201	A222
V25	RM805	F444
V25	RM340	C546
V25	RM318	C677
V25	RM205	A234
V250	RM318	3-212
V250	RM311	2-564
V250	RM340	3-675
V250	RM308	2-897
V250	RM305	2-312
V250	RM310	2-414

MATERIAL

MATERIAL_ID	MATL_DESCRIPTION	UNIT_OF_MEA
RM201	Carrots,whole	LB
RM202	Carrots,diced,1/4 in	LB
RM205	Potatoes,Eastern,whol	LB
RM210	Peas,shelled	LB
RM211	Tomatoes,whole	LB
RM305	Paprika	LB
RM308	Onion salt	LB
RM310	Garlic,whole	LB
RM311	Garlic powder	LB
RM318	Salt,Iodized	LB
RM340	Sugar,bulk	LB
RM805	Olive oil	GAL
RM810	Vinegar,white	GAL

Recall from Chapter 2 that the relational model allows a query to be made among multiple tables using foreign key attributes. Thus, the foreign keys MATERIAL_ID in the table PO_DETAIL and VENDOR_ID in the table PURCHASE_ORDER will allow the user to retrieve V_ITEM_CODE for a particular purchase order identified by PO_NUMBER.

Open the Access database MATERIAL_MANAGER, and enter the following query:

```
SELECT   V_ITEM_CODE
FROM     VENDOR_MATL_XREF
WHERE    VENDOR_ID IN (SELECT VENDOR_ID
                       FROM   PURCHASE_ORDER
                       WHERE  PO_NUMBER = 2594)
AND      MATERIAL_ID IN (SELECT   MATERIAL_ID
                       FROM   PO_DETAIL
                       WHERE  PO_NUMBER = 2594
AND    PO_LINE_ITEM = 1);
```

Executing the query should yield the vendor item code shown next. The readers should confirm to their own satisfaction that other data fields discussed in the previous sections can also be associated with purchase orders by a similar relational search.

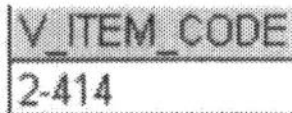

V_ITEM_CODE
2-414

5.3.2 Analysis of Information Requirements at Node A313

At this point, we have reviewed the data requirements at node A311. These were largely the data requirements associated with the purchase order, which had to be reviewed by the receiving clerk as a prerequisite to accepting the shipment. However, as shown in Figure 5.4, the receiving clerk must also provide data at node A313 by generating a *receiving report*, which is shown in Figure 5.6. This section addresses the new data requirements implied by the receiving report, beginning with a narrative based on the data fields shown on Figure 5.6.

The header information on the receiving report is by now quite familiar to us. The *supplier* is that which we have previously defined as the entity VENDOR. Note that the same organization, General Provisions, appears under the heading "VENDOR" on the purchase order form in Figure 5.7. It is quite common to find the same entity appearing under different headings on different forms used by the same enterprise. This occurs because forms have often been developed piecemeal over time, and no one has considered standardizing the terminology used. For this reason, an analyst can do the enterprise a great service by developing a data dictionary during the data-gathering phase so that the definition of terms for the same concept does not result in the definition of duplicate entities.

The *purchase order number* has been encountered before and it is the primary key attribute of the entity PURCHASE_ORDER. The data field *date received* is related to the attribute PROMISED_DEL_DATE, which is an attribute of PO_DETAIL. The date received is the actual delivery date as opposed to the promised delivery date. This is new information, and we must consider how it is to be handled.

The body of the receiving report contains information specific to the material being received. The *quantity accepted/not accepted* fields are used by the receiving clerk to indicate how much of the shipment of a particular material was accepted at the receiving dock. The quality assurance department is responsible for examining the condition of the shipment. Visibly identifiable damaged crates or damaged goods should be flagged immediately at the receiving station, and the material manager and purchasing agent should be informed. A joint decision is made whether or not to accept the damaged goods. If offloaded, damaged goods are accounted for separately, as they will be returned to the vendor and not paid for.

The *manufacturer's lot number* is a code placed on the material container by the manufacturing organization in order to indicate the date of manufacture and the facility at which the manufacture took place. Therefore, it identifies certain conditions of manufacture of the material. In many cases it uniquely identifies a production batch made in the supplier's plant. This is a relevant new data field, and we shall consider how it is to be handled.

The *item code* field is used to identify the material by a specific code. We have seen that the purchase order uses the vendor's item code (V_ITEM_CODE) so that the vendor will recognize the specific item being ordered. Ideally, the item code on the receiving report should refer to the MATERIAL_ID, which is how material in inventory is traced within the enterprise. However, in a database implementation, the receiving clerk can retrieve the MATERIAL_ID by a simple query of the table VENDOR_MATL_XREF shown in Figure 5.8.

The *material lot number* field is an important data field. The lot number is an identifier for a batch of material delivered to the enterprise. The enterprise assigns a lot number when it receives a new lot of material. In our hypothetical enterprise, the lot number is assigned as a sequential number. The characteristic of a lot is that it defines the following:

- a specific material (MATERIAL_ID),
- supplied by a specific vendor (VENDOR_ID),
- on a specific purchase order (PO_NUMBER),
- on a specific delivery date (DATE_RECEIVED),
- and, optionally, having a specific manufacturer's lot number.

The material lot number is the most specific identification of a material that is kept in the enterprise records. In many industries, such as the food and pharmaceutical industries, it is essential to keep track of the material of the enterprise by lot number. This is because there must be trackability from a finished product manufactured by the enterprise to the history of the materials that went into it. For example, the manufacture of an inferior (or defective) product may be the result of an inferior (or defective) material input. By keeping track of material by the lot number, and by keeping track of which material lots are used in which finished products, the enterprise can trace back from a defective final product to the input material and its source. In the food and pharmaceutical industries, this is a requirement of the U.S. Food and Drug Administration in the event of a product recall.

It is useful to discuss the distinction between the entity MATERIAL and MATERIAL_LOT. MATERIAL is the concept of materials. It is a categorization of the kinds of materials that the enterprise uses. It is not a physical thing, but rather a categorical listing. On the other hand, MATERIAL_LOT refers to a physical manifestation of a material. It is a tangible thing. There are many types of entities of an enterprise that show this pairing of concept and a physical manifestation of the concept. We shall see further examples later.

Note that the total quantity received by the enterprise (1400) has been divided into three groupings. The first two groupings (1000, 300) are differentiated by the *manufacturer's lot number*, which is one of the differentiating factors in the *material lot number*. Thus, the delivered material was separated into two material lots because it was composed of two different manufacturer's lot numbers. It is always possible that material will be delivered by the same vendor on the same purchase order on another date that carries the same manufacturer's lot number as that of material previously received. This delivery would be assigned its own material lot number because it is distinguished from the previous delivery by the DATE_RECEIVED data field. This implies that it was delivered as part of a different shipment having different shipping conditions associated with it.

The *description* data field comes off the bill of laden. The *storage location* is the location in the warehouse to which the material is moved. It is necessary to have this information in the materials management system because, when the material is to be delivered into production, a fork truck operator should be given information concerning the location from which to retrieve the material. Also, materials management will prefer to use the oldest materials first in order to avoid spoilage. This can be accomplished with a good system of tracking material by lot number and storage location.

The warehouse location system used by the University Food Company defines four concepts: *area*, *aisle*, *tier*, and *bin*. These are shown in Figure 5.9. The area refers to major groupings of materials based on storage requirements. For example, fresh meat and poultry materials must be refrigerated, while flour and canned goods can be stored at ambient temperature in the dry goods part of the facility. For simplicity, we have defined three storage areas: refrigerator, freezer, and dry goods. Within each area, the next major subdivision is aisle, followed by tier, which is an upper or lower (two-tier) location within the aisle, followed by bin, which is a specific cube of space within the area/aisle/tier configuration. Using this system definition, the location of a material lot is defined to approximately a 5 cubic foot area (full pallet size).

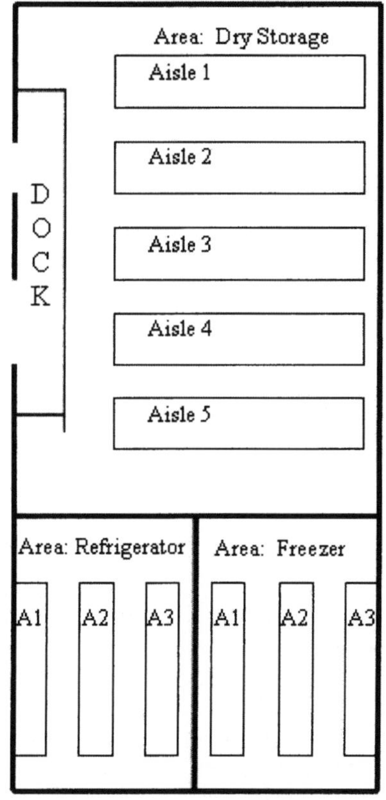

Figure 5.9 Warehouse layout.

The preceding narrative has given us new information on this case. Given this scenario, we shall address each new data element, taking each item in its turn.

5.3.2.1 Supplier

It has been pointed out in the narrative that *supplier* is synonymous with *vendor.* The only place a distinction is made is in the terminology being used on the purchase order and the receiving report. Since the VENDOR entity already exists, it is not necessary for us to do anything with this data field in the data model.

5.3.2.2 Date Received

This is the actual delivery date, which would appear to be an attribute. It is clearly not an attribute of PURCHASE_ORDER because it is not true that the entire contents of a purchase order are necessarily delivered on the same date. As mentioned earlier, it is related to the PROMISED_DEL_DATE found in the entity PO_ITEM. Therefore, it would seem reasonable to attach it to that entity. However, consider the case in which a vendor splits the delivery of a line item and delivers it on two or more different dates. This, of course, is a possible case since partial deliveries will occur if a vendor's inventory of a material is too low to completely fill the order when it is due to be shipped. Therefore, adding DATE_RECEIVED to PO_ITEM will not do.

The problem here is that we lack an appropriate entity, which should be our first clue to look for a new entity. Suppose we consider defining an entity called DELIVERY_DATE, with a composite primary key given by the tuple PO_NUMBER, PO_LINE_ITEM, and DATE_RECEIVED. There would be a 1:M cardinality relationship between PO_DETAIL and DELIVERY_DATE, retaining the desired relational architecture. Such a relationship is shown in Figure 5.10. Let us tentatively accept this solution. However, we will return to this decision shortly as we continue to develop the model.

5.3.2.3 Quantity, Manufacturing Lot No, Item Code, and Material Lot No

We will take these four data fields together because they are clearly related. The first two entries are distinguished by the need to create two material lots out of the delivery. We know from the narrative that the material lot number is a unique identifier for the most specific unit of material inventory kept track of by the enterprise. This implies that there is a new entity, the MATERIAL_LOT, about which we may wish to retain information. Hence, we define this entity in Figure 5.11. The attributes defining MATERIAL_LOT are those that were described in the narrative.

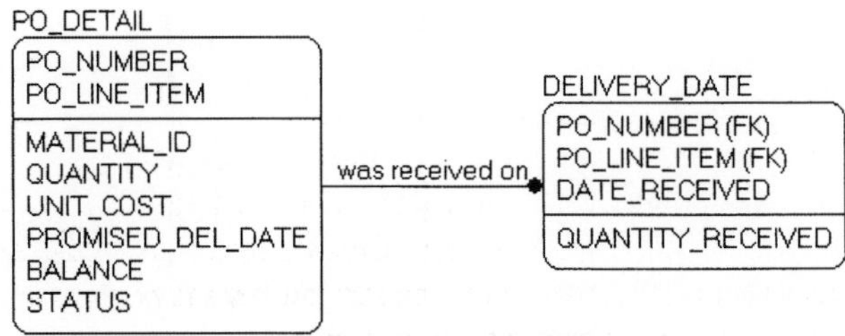

Figure 5.10 Tentative solution to the data field DATE_RECEIVED.

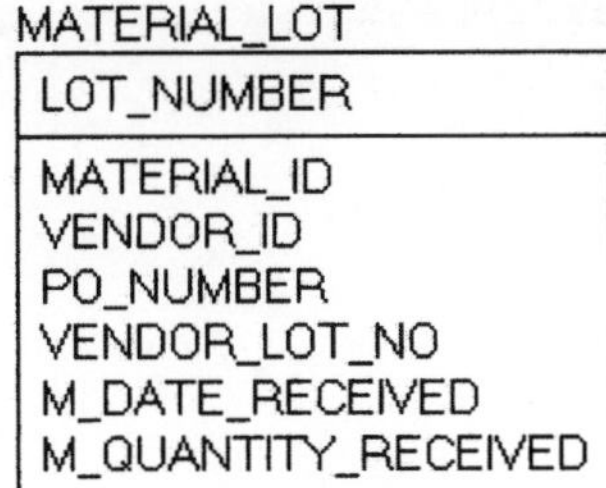

Figure 5.11 Tentative definition of material lot.

At this point, it is well to reconsider the decision made in Section 5.3.2.2 concerning the disposition of the attribute DATE_RECEIVED. It is clear now that DATE_RECEIVED is an attribute of MATERIAL_LOT. Furthermore, it is the MATERIAL_LOT, not the PO_DETAIL, that is actually being received. Therefore, the attribute DATE_RECEIVED logically belongs to the entity MATERIAL_LOT.

As a general rule, defining a date as an entity is not good practice. It may be true that the date is an important marker for an event that is necessary to keep track of. However, an entity can usually be defined with the name of the event or, as in this case, the physical entity (MATERIAL_LOT) associated with the event.

In comparing Figures 5.10 and 5.11, we note one inconsistency. The PO_LINE_ITEM is not accounted for in Figure 5.11. Upon some reflection, it is clear that the material lot is the actualization of a line item and is, therefore, related to the line item that authorized it. Therefore, Figure 5.11 should be augmented to include the foreign key PO_LINE_ITEM.

5.3.2.4 Storage Location

The MATERIAL_LOT is stored in one or more *storage locations*. Therefore, storage location is an entity of interest for keeping track of the material lot. Based on the narrative, the entity storage location (call it WAREHOUSE_LOCATION) is defined by the key attributes area, aisle, tier, and bin, as shown in Figure 5.9.

5.3.2.5 An Extended Data Model

Figure 5.12 shows the extended data model based on the newly defined entities and their attributes. Note that two quantity attributes exist in the entity MATERIAL_LOT. They are M_QUANTITY_RECEIVED and M_QTY_ON_HAND. The first quantity is the quantity received, as discussed in Section 5.3.2.3. The second quantity is the quantity on hand. As discussed in the narrative, one purpose of defining the material lot is to keep track of the actual material in inventory within the enterprise. The material on hand is the difference between the material of a particular lot that was delivered and the amount used in production. Therefore, these important attributes may as well be added to the MATERIAL_LOT entity at this point.

In Section 5.2, we noted that a relationship is optional to the parent in a 1 : M relationship if the M side allows the cardinality "0." So, for example, in Figure 5.12, EMPLOYEE is optional in the relationship with PURCHASE_ORDER since not all employees will appear on a purchase order.

In Figure 5.12, another symbol has been introduced on the 1 side of a 1 : M relationship. In the relationship between MATERIAL_LOT and WAREHOUSE_LOCATION, a diamond appears on the 1 side. This indicates that NULL values for the foreign key may exist on the M side. This may occur, for example, because a bin is empty at the time and the LOT_NUMBER attribute is NULL. The diamond indicates that the M side entity is optional to the relationship.

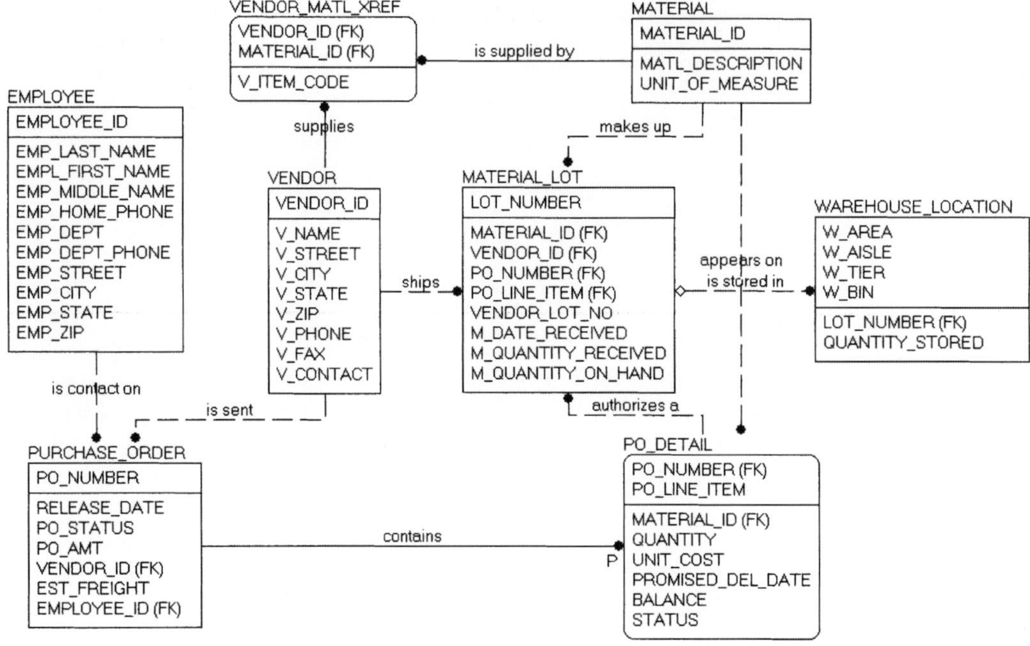

Figure 5.12 Data model after completed evaluation of nodes A311 and A313.

5.3.3 Analysis of Information Requirements for the Control of Stored Materials

The processes for the control of stored material were documented in Figure 4.17 using a data flow diagram. For convenience, we have reproduced the figure as Figure 5.13.

The data flow diagram in Figure 5.13 introduces three data stores, which are suggestive of database tables. These data stores are Inventory Log, Warehouse File, and Inventory File. Two of these data stores appear to have already been adequately captured in Figure 5.12. In particular, "Warehouse File" serves the same purpose as the entity WAREHOUSE_LOCATION. Similarly, the "Inventory File," which keeps track of material lots, is captured by the entity MATERIAL_LOT in Figure 5.12. Therefore, we shall try to build the remaining information requirements around these existing entities.

The data source titled "Production Scheduling" initiates the movement of materials by issuing a schedule. Production scheduling brings in a whole new set of production processes that have not yet been described here. From the point of view of the process *control stored materials*, it is just an exogenous activity that is important only insofar as it is a trigger to the process. Here we will only account for data flows that are initiated by the process. These data flows are those that are captured in the inventory log.

Based on the narrative in Section 4.6.3, the purpose of an entry made in the inventory log is to document the occurrence of a ***transaction event***. That is to say, the inventory log is a time-stamped event list of the movements of material lots into and out of storage. The concept of a transaction event list is important in database design. Some entities of a database exist to show the current status of the enterprise. For example, WAREHOUSE_LOCATION shows the current state of usage

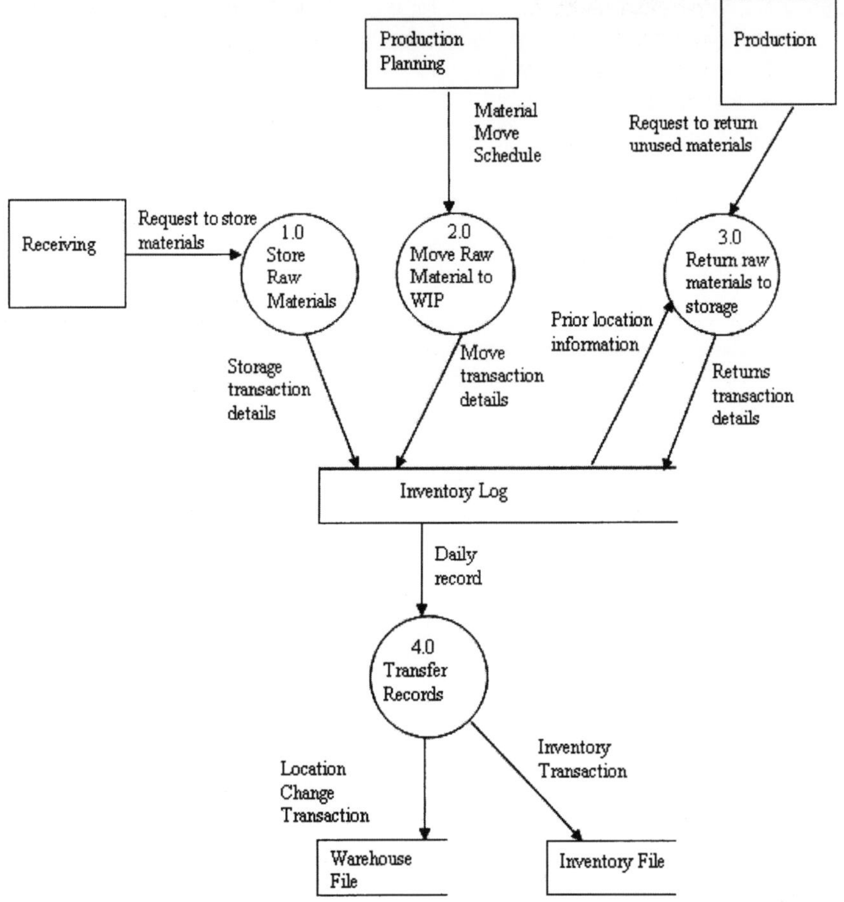

Figure 5.13 Decomposition of control stored materials.

of the warehouse, and MATERIAL_LOT carries the current on-hand inventory status of each material lot. Sometimes it is desirable to have a record of the events that have occurred over time that change the status of certain entities. The classic example of this occurs in accounting information systems where it is necessary to have a record over time of the debits and credits made to each account in order to audit the accounting records.

There are many practical reasons for wanting to record inventory transactions. For example, it is desirable to know how much material is used for a given day's production of a product. This information can be used to compute product yields and the actual material cost of the product. A *transaction entity* is analogous to a journal entry in accounting. In any data model design, it is useful to ask whether or not one wants to keep an "audit trail" of the reasons why the value of the attribute changes over time.

In Figure 5.13, forklift truck operators are withdrawing lots of material from the warehouse and returning lots of material to the warehouse after production. Each of these transactions are recorded in the inventory log used as the basis to update the inventory records at the end of the day. In effect, current practices of the enterprise provide an audit trail for each inventory transaction, albeit in paper form.

The equivalent data model entity will be a lot transaction record (LOT_TRANSACTION). It will be used to record withdrawals from, additions to, and adjustments to inventory lots. These transactions will affect the amount of inventory on hand and the amount of inventory in a particular warehouse location. Each transaction will only be allowed to affect one lot and one warehouse location. The following business rules apply.

Business rule 1. A lot is moved by one or more transactions. A transaction moves only one lot.

Business rule 2. A warehouse location has its contents changed by one or more transactions. A lot transaction changes the contents of only one warehouse location.

The appropriate model that reflects these rules is shown in Figure 5.14. The description of a LOT_TRANSACTION table follows:

Entity: LOT_TRANSACTION

Attribute Name	Description	PK or FK
TRANSACTION_NO	Unique identifier of a transaction	PK
TRANSACTION_DATE	Date on which transaction occurred	
TRANSACTION_TIME	Time at which transaction occurred	
MATERIAL_LOT_NO	Unique identifier of a lot	FK
LOCATION_ID	Unique identifier of a warehouse location	FK
TRANSACTION_TYPE	IN=1; OUT=2; Adjustment=3	
TRANSACTION_QTY	The amount of material moved	

The LOT_TRANSACTION entity is added to the existing database design in Figure 5.15. It should be noted that this final data model is not in 3rd normal form. In particular, in the entity MATERIAL_LOT, there is a transitive dependency between VENDOR_ID and PO_NUMBER, as well as between MATERIAL_ID and the composite key PO_NUMBER, PO_LINE_ITEM. If MATERIAL_ID and VENDOR_ID were eliminated from the MATERIAL_LOT table, it would still be possible to retrieve the vendor ID and material ID using PO_NUMBER and PO_LINE_ITEM, and the resulting data model would be in 3rd normal form. Here we have made a design

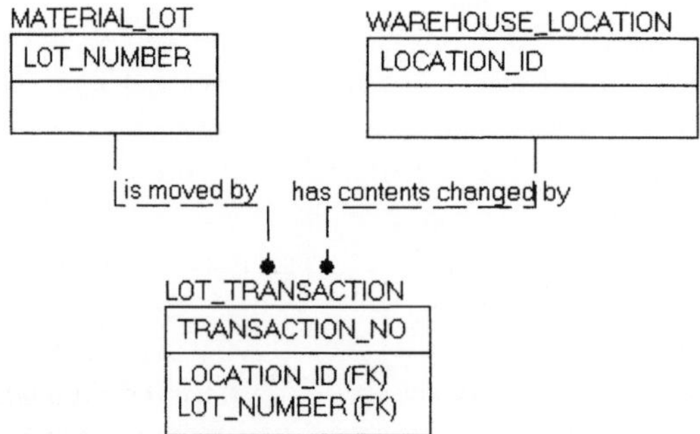

Figure 5.14 Transaction entity for movement of inventory.

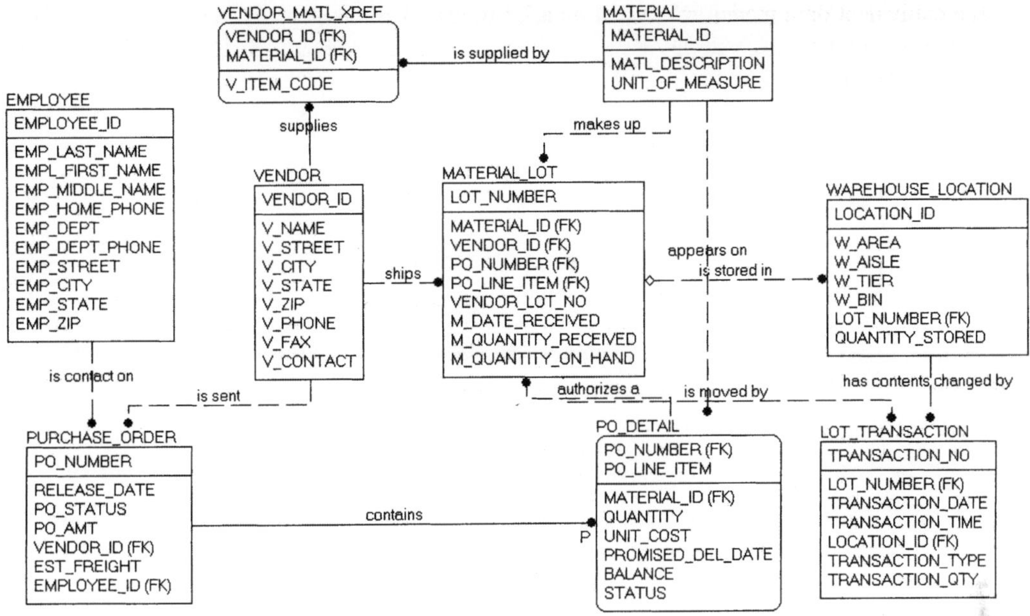

Figure 5.15 Final logical data model for activities A111, A113, and A32.

decision to allow some minor denormalization in order to enable more direct queries for vendor and material information related to lot numbers.

5.4 SUMMARY

In this chapter we have described the process of developing an informational architecture or logical database design. We introduced the entity-relationship modeling approach known as IDEF1X. Through a case study example, we have shown the process an analyst would go through in building such a model. This process relies on functions identified in the functional architecture design. Documents that are used by the individuals performing those functions are the sources of data for the informational architecture. This chapter has shown the way in which functional and informational architectures are coordinated in the design process. A more detailed description of the IDEF1X standard, as published by the National Institute of Standards and Technology (NIST), can be obtained from http://www.itl.nist.gov/fipspubs/idef1x.doe.

The IDEF1X formalism is not the only available technique widely used for documenting the information model. Other methods popularly used are information engineering and Oracle CASE*Method. These techniques are reviewed and compared to IDEF1X in the article by Hay (1999), to which the reader is referred.

Professional IDEF software is available from several companies and can be downloaded in trial version, which may be useful in completing the end-of-chapter exercises. We refer the reader to selected Web sites listed at the end of the bibliography section at the back of this book. A Visio tutorial for drawing IDEF1X diagrams is also available from the Web site supporting this book.

It can be argued that the design of the informational architecture is the most important step in the process of developing a database application. A poorly designed informational architecture will create problems when it comes to the implementation and development of user views, called *forms* and *reports*. These implementation issues will be discussed in the next chapter, which covers the design of forms and reports.

REVIEW EXERCISES

5.1 The data model shown in Figure 5.15 does not quite account for all of the information contained on the bill of laden and the receiving report shown in Figures 5.5 and 5.6. In particular, the following items on the bill of laden and receiving report were not discussed in the text:

- Comments
- Received by
- Quantity not accepted
- Bill of laden number
- Carrier
- Truck number
- Driver name
- License number

Augment Figure 5.15 to show how you would handle these items in the IDEF1X model.

5.2 Warehouse location can be further characterized by adding the concept of "location type." This concept refers to the freezer area, dry storage, and so on. Augment Figure 5.15 to incorporate the entity LOCATION TYPE.

5.3 Refer to Review Exercise 2.3 at the end of Chapter 2. Construct the IDEF1X model that is used to design the tables shown there.

5.4 Refer to Review Exercise 2.5 at the end of Chapter 2. Construct the IDEF1X model that is used to design the tables shown there.

5.5 Refer to Review Exercise 2.6 at the end of Chapter 2. Construct the IDEF1X model that is used to design the tables shown there.

5.6 Refer to Review Exercise 3.6 at the end of Chapter 3. Construct the IDEF1X model for that situation.

CASE STUDIES

5.7 ACME Maintenance Case (C)

Refer to ACME Maintenance Case (B) at the end of Chapter 4. Given the information in the case study, answer the following questions:

(a) Define an informational model using the IDEF1X methodology, indicating entities, attributes, and their relationships.
(b) Show the tables that will exist in the database, indicating column names and data types.
(c) Write structured query language (SQL) statements that will query the tables and present the data to the user as required in Exhibit CS4.5.

5.8 University Food Company Case (C)

Refer to University Food Company Case (B) at the end of Chapter 4. Answer the following related questions:

(a) Using the information shown in the exhibits, and your judgment, develop the IDEF1X information architecture. Since some of the information comes from tables in the databases of sales and production planning, you will have to include these entities. When you are finished, each data element on the exhibits should be an attribute in one of the entities of the IDEF1X model.
(b) Using the data model developed in part (a), write the SQL queries that would be used to generate the reports of Exhibits CS4.6 and CS4.7 of University Food Company Case (B).

5.9 Regal Foreign Car Repairs Case (C)

Refer to Case Study 3.10 at the end of Chapter 3. Construct an IDEF1X model based on this case study. Include in your model the capability to be consistent between the cost of parts at different points in time and the cost summary for parts on the repair order. Inconsistency can arise when the cost of a part changes over time and there is no history that relates the cost summary of parts to the cost of the part at the time the repair order was processed. We want the data model to include traceability between the cost summary and the cost of individual components at the time that the repair order was documented.

Chapter 6

Design of a User Interface

6.1 INTRODUCTION

In Section 3.1, there was a discussion of the ANSI/SPARC three-level architecture for designing a database application. The ***external level*** resided at the top of the architecture. The external level deals with the way different users view the database. Once a database model has been designed and implemented in a DBMS, the analyst must consider the design of screens that will allow users to work with the database. This is often referred to as the design of the "user view."

Designing the user view begins by identifying a function or set of related functions that the screens will service. Following this, the screen layout is designed and reviewed with the user to determine whether or not it meets its intended purpose. Finally, the screens are implemented. In this chapter, we will discuss this process and implement some typical screen designs using the Access Database Management System.

6.2 THE FUNCTIONAL/ENTITY INTERACTION MATRIX

The analyst has defined the functional and informational architecture at the conceptual level, as discussed in Chapters 4 and 5. The functional architecture is related to the informational architecture by the manner in which each function uses data from the entities that were defined in the informational model. The way in which each function uses data when interacting with the informational model, as implemented in database tables, can be specified in four categories:

1. *Insert.* If a user is allowed to create new records in the database tables. For example, a purchasing agent will create new records each time he or she creates a purchase order.
2. *Delete.* If a user is allowed to remove a record from a table. For example, if a purchase order time item is concealed, the purchasing agent may remove the record in the PO_DETAIL table.
3. *Update.* If the user is allowed to change the value of an attribute in the table. For example, when an open purchase order line item is closed out, the purchasing agent must be allowed to change the STATUS attribute value.
4. *Read only.* If the user is allowed to access the information in a table but is not allowed to change it. For example, the receiving clerk needs to access the information on purchase orders

ENTITY SET

I - Insert
D - Delete
U - Update
R - Read Only

ELEMENTAL FUNCTION	VENDOR/MATERIAL	PURCHASE ORDER	PO DETAIL	MATERIAL LOT	WAREHOUSE
Confirm Validity of Shipment		R	R		
Receive Materials	R			I	I

Figure 6.1 Functional/entity interaction matrix.

and their line items in order to decide whether or not to accept a delivery. However, the clerk would not be allowed to change the contents of those records.

The ***functional/entity interaction matrix*** is a tool that the analyst uses to summarize the allowed actions that a particular function can have on the entities it uses. An example is shown in Figure 6.1.

The explanation of the entries in Figure 6.1 is derived from the narrative for each elemental function. For example, in the *Confirm Validity of Shipment* function, the receiving clerk needs access to the purchase order information in order to confirm that the material being received is part of an open purchase order. Therefore, this function has read-only access to the PURCHASE_ORDER and PO_DETAIL tables.

During the function *Receive Materials*, the receiving clerk is inserting a new record into the tables for MATERIAL_LOT and WAREHOUSE, indicating the location at which the lot is stored. To do so, the clerk must also view the VENDOR_MATL_XREF, which relates the vendor's material ID to the material ID used by the enterprise for the same material.

The importance of Figure 6.1 is that it provides a map that tells the analyst what actions the users will be allowed to perform from the user screen after they log in. It is important to control the user actions by user function in order to maintain data integrity. Thus, Figure 6.1 defines the restrictions on data access associated with each function.

6.3 SCREEN DESIGN

There are basically two kinds of electronic documents for which the analyst develops screen designs. These are called forms and reports. A ***form*** is a computer screen that allows a user to view data and, possibly, change or add data. It allows the functions of insert, delete, update, and read only, as described in the previous section. A ***report*** is any document that retrieves information from the database and formats it for presentation. A report, as opposed to a form, does not allow the user to manipulate or change data.

6.3.1 Form Specification

A properly designed form is one that will allow the users to perform their functions with the easiest possible access to information. It is natural to look at the paper forms being used by the enterprise in order to get some idea of what the screen design should look like. In many cases, presenting the users with something that looks like what they are already using facilitates the users' transition from paper form to computer screen. On the other hand, the analyst should try to simplify forms as much as possible, removing extraneous items that are confusing or unnecessary.

The ***form specification*** needs to address several issues. The first is the ***layout*** of the data on the screen. Once again, this can be facilitated by reference to existing documents. However, given the size limitations of the computer screen, the analyst may need to consider multiple pages.

A second issue is the ***relationship of different blocks on the screen***. In many cases, forms are organized as master/detail records on a screen, usually referred to as a master/detail form. The master record is usually based on one table. The detail record is based on another table with a foreign key relationship to the table on which the master record is based. Such is the case, for example, with a purchase order. The master information from the table PURCHASE_ORDER is laid out on the screen as a contiguous block of data. The detailed line items from the table PO_DETAIL are shown in a detail block. Whenever a particular master block record is retrieved, there must be coordination between these two blocks to ensure that the detail data retrieved are related to the master record.

A third issue is ***data integrity requirements***. When an individual is using the form for the purpose of entering data, all efforts should be made to ensure that the data being entered are correct. For example, when the receiving clerk is filling out a receiving report, she or he must enter the material ID. It is always possible that the user could enter an incorrect ID. Since the allowable material IDs are known to the database, any material ID that does not appear in the table MATERIAL should not be allowed. An integrity check on fields inserted or updated by a user can be designed into a form.

Another issue in the design of forms is ***form navigation***. In many cases, form elements are related to each other in some manner. Also, some forms have multiple pages. It is important to design into each form the navigation keys or controls that will allow a user to move from one form element or one page to another in the easiest fashion.

The final important consideration is ***user interaction*** (i.e., the operations that will be made available to the user on the form). Recall that different levels of data access are allowed: insert, update, delete, and read only. The form should provide an authorized user with a way of selecting and executing the appropriate operation and should restrict those who are not allowed.

We will use these factors as a checklist when we design a series of forms later in this chapter.

Summary Considerations on Form Specification:

- **Layout of the form**
- **Relationship of blocks on the screen**
- **Integrity requirements**
- **Form navigation**
- **Providing user interaction**

6.3.2 Report Specification

A report is for information purposes only. Thus, it does not require the data integrity or user interaction considerations that a form requires. As in the case of forms, it is natural to look at the reports already being used by the enterprise as a starting point.

The ***report specification*** needs to consider several issues. The first is the ***layout of the report***. A key element of the layout is the size in terms of rows and columns. Remember, reports are usually printed out, and some consideration has to be given to the printer and paper size alternatives.

Another consideration concerns ***interaction with tables*** (i.e., the tables used in the data retrieval). One option in designing a report is to create a single table that contains all the records to be printed. As we have discussed, this is not necessarily the best solution from a database design point of view. However, it saves retrieval time, especially if the report is long. When the data are to be retrieved from many tables and the data fields have to be coordinated, extensive data retrieval time may be encountered.

Finally, some consideration must be given to ***derived data***. Reports usually have subtotals, totals, and other summary information. How this is presented and how the reports are coded to derive these summaries must be considered.

We will revisit these points in a later section when we design a typical report.

Summary Considerations on Report Specification:

- **Layout of the Report**
- **Interaction with tables**
- **Derived data**

6.4 A SINGLE TABLE FORM

A form may be based on one or more tables or queries. In the simplest case, only one table is involved. In this section we will implement a form based on a single table.

To implement a form, a specific DBMS must be used. Different DBMSs have different facilities for designing a form. Therefore, what one learns by using Microsoft Access is only in part transferable to other DBMSs. On the other hand, there are principles of form design that, when illustrated with a particular DBMS, will be applicable to others.

The first issue in form specification cited in Section 6.3.1 is the layout of the screen. We illustrate the single table form using the table VENDOR. The purpose of this form is the following:

1. Provide a screen that can be used by the purchasing agent for table maintenance. That is to say, the purchasing agent must add new vendors and update vendor information as it changes.
2. Provide a screen that others can use for viewing (read-only) vendor information.

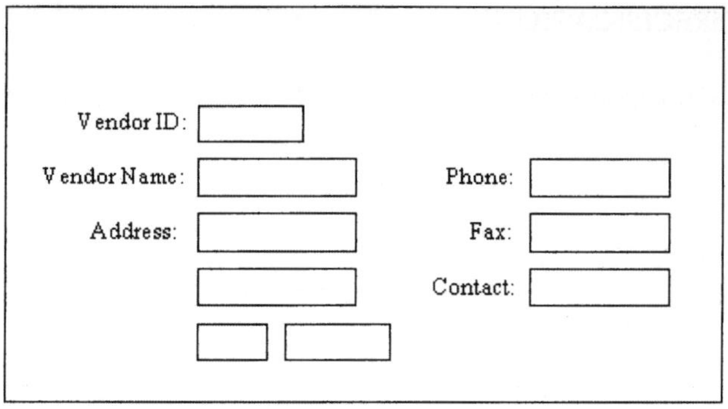

Vendor ID:

Vendor Name: Phone:

Address: Fax:

Contact:

Figure 6.2 Vendor form screen design.

The analyst defines the layout of information on the screen with the help of the users. We will propose to implement the simple layout shown in Figure 6.2.

6.4.1 Implementing a Single Table Form in Microsoft Access

The forms and reports that will be designed in this chapter are based on the data model that was shown in Figure 5.12. Before beginning the exercises offered in this chapter, the reader should load the database titled "Material_Manager_CH6_stu" from the Web site supporting this book. It contains a set of updated tables, populated with data, and with the attributes as they are shown in Figure 5.12. Alternatively, the populated tables are shown in Appendix 6A of this chapter. The user may create the database and enter the data manually. After loading the database, continue with the exercises in this chapter.

6.4.1.1 Creating a Default Form

There are two different ways in which Access can be used to establish a form. The first is the use of the ***form wizard***. The form wizard automatically generates a default form for the user by providing data fields on the form for the attributes chosen by the form designer. This is also a common option in other DBMSs. For example, the Oracle DBMS has a default option for creating a form that produces the form automatically. It is usually necessary to tailor the default form afterward in order to get what is eventually desired. In this section, we will use the form wizard to illustrate this point.

Another way to establish the form is to custom design it. For this case, the user starts with a blank form and manually adds each data field to the form. This gives the user more control over the design process. We will illustrate manually adding elements to a form in a later section when more complex forms are designed.

Access Exercise 6.1:

We want to establish a form named "VENDOR". This will be a form based on the **VENDOR** table.

1. Go to MS Access Main Window.
2. Open the database **Material_Manager_CH6_stu**.
3. From the database window, click on the "Form" tab and then double click "Create form by using wizard". This will bring you to the "Form Wizard" window ***(See Figure 6.3)***.

Forms can be based on either a table or a query. This form will be based on the table ***VENDOR***.

4. In the "Tables/Queries" pull down menu, select the table name ***VENDOR.***
5. In the "Available Fields" window, highlight the attributes you want on the form and click [>]. The chosen attribute will move to the "Selected Fields" selection box. Do this for all the attributes in the table. When this is done, click on the "Next" command button.
6. The Form Wizard will ask you to choose a form layout. Choose the "columnar" layout. Click the "Next" command button.
7. The Form Wizard will ask you to choose the style that will be used for the background display. Choose "Standard" and click the "Next" button.
8. The form wizard will ask for the form name. Use the name ***VENDOR***. Click the "Finish" command button.
9. After execution, the form should appear as shown in Figure 6.4. You have now created a single table default form for the ***VENDOR*** table.

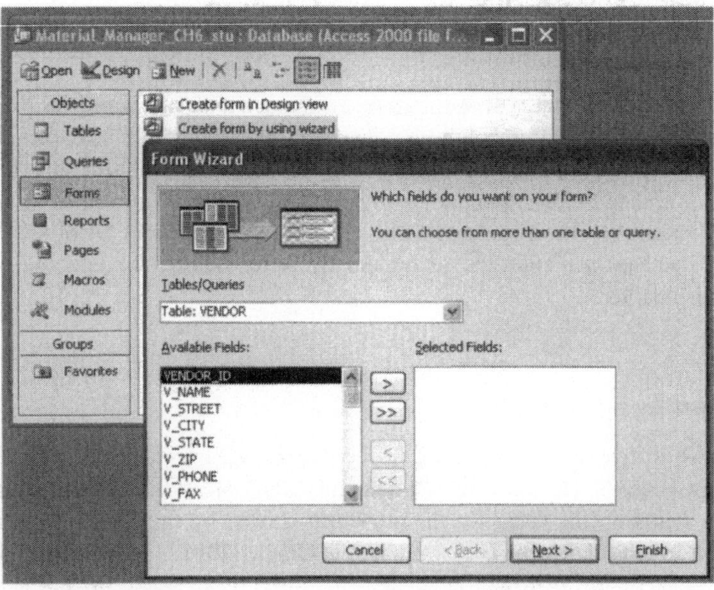

Figure 6.3 Step 3 in Access Exercise 6.1.

A default form consists of labels and text boxes. The ***labels***, shown on the left side of the form, default to the attribute name. The ***text boxes*** are "windows" for viewing each record of the attributes in the table. The navigation button, shown at the bottom, allows the user to move through each record of the table.

Note that the form does not look like the desired form layout shown in Figure 6.2. The labels on the default form of Figure 6.4 use the names of the attributes as they are defined in the table VENDOR. Also, the layout of the fields is not the same as it appears in Figure 6.2. At this point, we must tailor the default form to our design.

6.4.1.2 Tailoring a Default Form

There are different form "Views" available in Access. At design time, the user works in the ***design view***. When running a form, the user works in the ***form view***. When the form was generated in Figure 6.4, the DBMS left the user in the form view, with the form running. By clicking on the navigation button at the bottom of the form, the user will be able to move through the records that exist in the table VENDOR. Try it!

To change the form to look like Figure 6.2, the user must go to the design view. This can be done by clicking on the design icon on the toolbar, as shown in Figure 6.5. Using the design icon on the main menu, put the Vendor form into the design view. It should appear as shown in Figure 6.5.

The form appears in its various ***sections*** in the design view of Figure 6.5. The inner section is the ***form template***, and the outer section is the ***form border***. The form template is the region on which the form is designed. The form border is a boundary region that contains controls such as the scroll bars shown in Figure 6.5.

There are three sections of the form template. They are the form header, detail, and form footer. The ***form header*** is the region where the title information will appear. The designer adds title information in this region during the design process. We will use the form header region subsequently. The ***form detail*** is the area of the form template where the data elements are placed. The form wizard has populated this area with text boxes and labels. Interaction with the database is coded in the detail area. The ***form footer*** can be used to display instructions on how to use the form and can be used for certain Visual Basic for Applications (VBA) control elements, some of which will be described in a later section. The form header, detail, and form footer sections can be enlarged by placing the cursor on the section boundary and dragging the boundary to the desired height.

The form border exists around the form template. The border has scroll bars for moving the form template horizontally and vertically. The form border can be enlarged horizontally or vertically independent of the form template using the mouse by selecting the border at the boundary and dragging the boundary to the desired width or height.

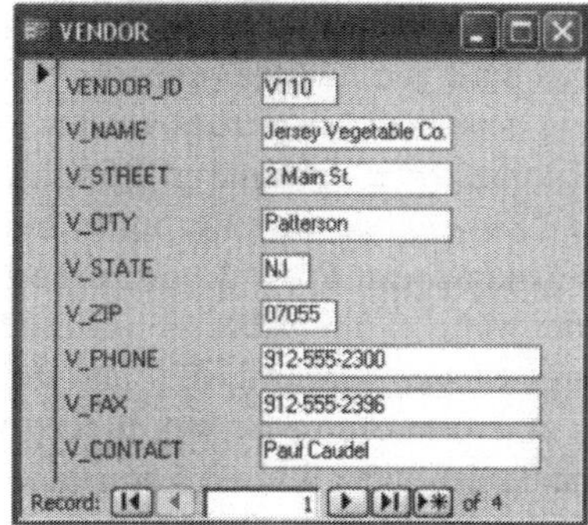

Figure 6.4 Default form for the VENDOR table.

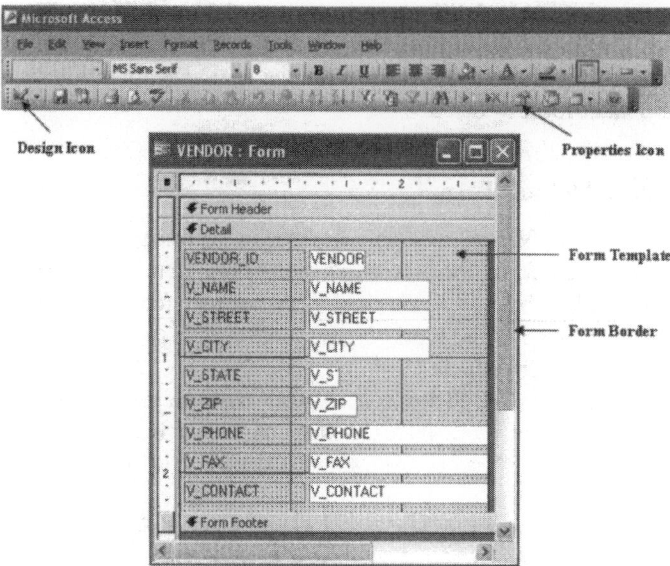

Figure 6.5 Vendor form in the design view.

To rearrange the fields on the form in the format of Figure 6.2, the width of the form must be enlarged. This brings us to the subject of ***form properties***. In the Access DBMS world, forms are thought of as objects. An ***object*** is an abstract representation of a real-world entity that has a unique identity, embedded properties, and the ability to interact with other objects. Access defines four classes of properties in relation to a form object. They are the ***layout properties***, ***data properties***, ***event properties***, and ***other properties***. We will be using a select subset of these properties as we go along. For a complete understanding of properties, the reader is referred to the Access manuals or Help menu item.

The layout properties have to do with the size and appearance of the form. One of the layout properties is "width," which gives the maximum width of the form. The default form of Figure 6.5 has insufficient width to view the data in the format of Figure 6.2, so we shall begin by enlarging the form. Observe Figure 6.6. Click on the form in the square at the upper left in order to select it and go to the "Properties" icon on the Menu bar as shown to the right and in Figure 6.5. Click on the "Properties" icon, and the properties window will appear as in Figure 6.6. Choose the "Format" tab. Go down the list of properties until you find the property "Width." Change its current value to "5.5 in" and left click anywhere outside the properties window.

At this point, the form template has been enlarged to 5.5 inches. However, you cannot see it until the form border is enlarged. Now enlarge the form border by placing the cursor on the right edge and the double arrow ⇔ should appear. Press down on the left mouse button and drag the right edge to open the form border width to 5.5 inches as indicated by the ruler along the top. At this point, the form should be large enough to rearrange it into the layout of Figure 6.2.

Some of the layout properties of a form can be altered directly by manipulating the form template. For example, open the form border width beyond 5.5 inches. You will observe that the form template remains at 5.5 inches. Using the cursor, left-click on the right edge of the form template

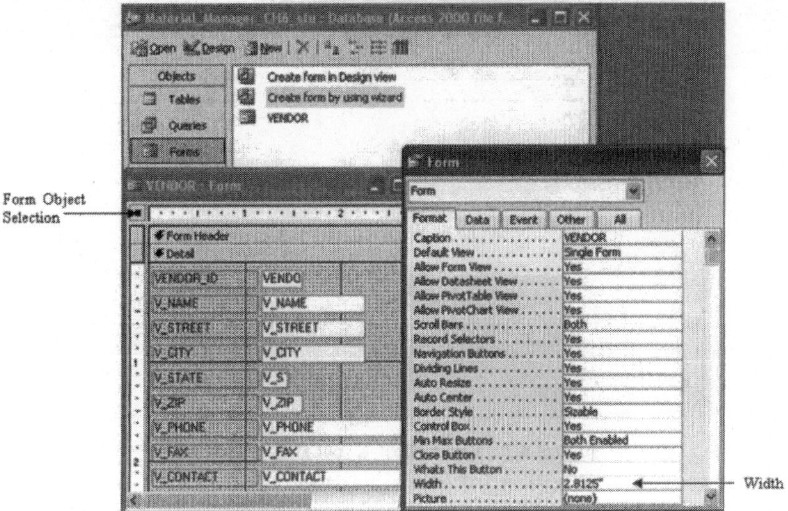

Figure 6.6 Form properties window.

until the double arrow appears. Drag the template to the right beyond 5.5 inches. Observe that the value of the Width property in the properties window changes to reflect the enlarged form template width.

We want to change the form of Figure 6.5 to look like that of Figure 6.7. The easiest thing to do first is to delete or change labels. There are two ways to do this. The first is to select a label by clicking on it. While it is selected, click on the properties button on the toolbar and the Properties window associated with the selected label object will open. If the Properties window is already open, it will automatically switch to the properties of the currently selected object.

Just as a form is treated as an object, a label is also treated as an object with its own embedded properties. One of the properties is the caption property. By changing the caption property, you will change the caption on the label. Alternatively, you can select a label by clicking on it, placing the cursor in the label box, deleting the existing caption, and typing in a new caption. This will automatically change the caption property of the label in the Properties window. To remove a label entirely, select it and hit the delete key. The label will cease to exist. The reader should refer to Figure 6.7 and make the appropriate changes to the labels.

To move labels and text boxes on the screen, the user selects the object by clicking on it and holding the left mouse button down. A hand will appear at the cursor position. While holding the mouse button down, drag the object to the new location and release the button. When these actions are completed, the form should look like Figure 6.7.

Figure 6.7 shows the Vendor form in design view. The user should now switch to form view by selecting the form view icon on the toolbar, which is found on the same pull-down menu list as the design view icon. Test the performance of the form by moving through the records using the navigation buttons at the bottom of the form.

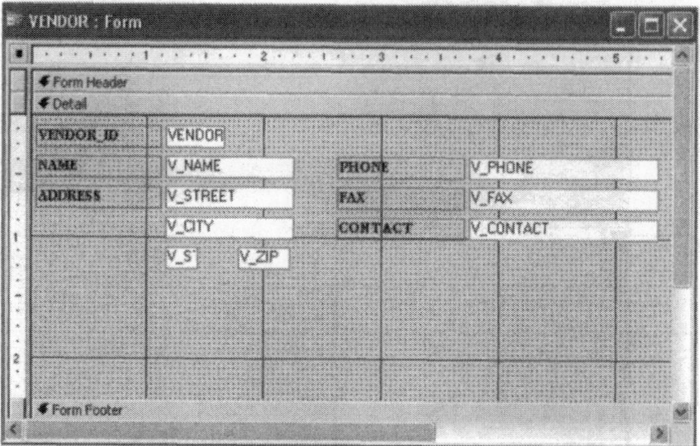

Figure 6.7 Vendor form layout after redesign.

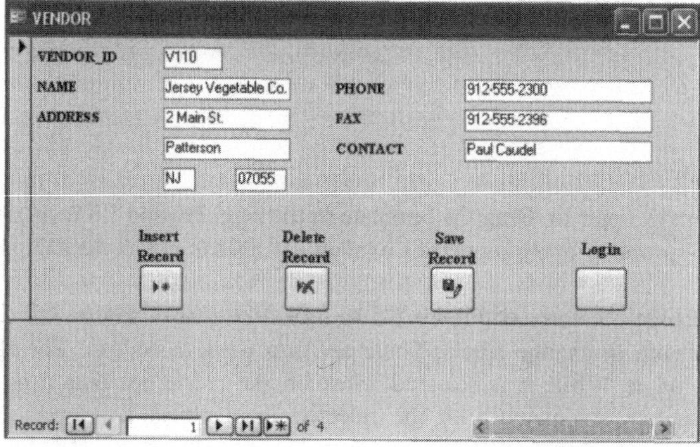

Figure 6.8 Vendor form with functionality added.

6.4.2 Providing User Interaction

Now that the form layout has been implemented, we shall turn our attention to other aspects of the form specification. Recall from Section 6.3.1 that defining and implementing the actions that the user will be allowed to use is part of the form specification. These actions include insert, delete, update, and read only. The design of the form should give the user an efficient way of performing these actions. This should be standardized as much as possible throughout the set of forms used in an application in order to present the user with a consistent interface.

Access provides the designer with a programmed set of command buttons that can be quickly implemented to give functionality to a form. These will be used in implementing the operations of insert, delete, and update. But first refer to Figure 6.8, which shows the Vendor form with this functionality added.

In Figure 6.8, we have added functionality through three command buttons: insert record, delete record, and save record. However, we wish to control the functionality allowed for any particular user. This will be the purpose of the Login button. The protocol that will be enforced on the user is as follows:

1. When a form is displayed, it will be available only for reading. The form will appear without the insert, delete, and save buttons displayed.
2. If a user wishes to insert, delete, or update a record, she or he will click on the Login command button. Depending on the functions that are allowed for that particular user, either the full function set will appear, a subset of the function buttons, or a message indicating that "read only" is available to the user.
3. The user can then utilize the operations that are allowed.

The important consideration in this protocol is that functions should be made available only to individuals authorized to use them based on the Functional/Entity interaction matrix, such as the one shown in Figure 6.1. Having said that, we will proceed to implement the form shown in Figure 6.8, but we will leave out the code that screens individuals for the authority to access a particular function. That will be considered in a later section.

Before continuing, some discussion of the design objects available in Access is necessary. Opening the ***design toolbox*** from the Access menu will expose these design objects. Put the Vendor form in design view. From the toolbar at the top of the screen, select "View," which will open a drop-down menu. Click on "Toolbox." This will open the design toolbox as shown at the bottom left in Figure 6.9.

The toolbox contains a number of useful objects that can be used to provide functionality to a form. The interested reader is encouraged to refer to the Access manuals or Help menu for a complete description. Here we will focus only on those toolbox objects that will be used in illustrating the design of the Vendor form.

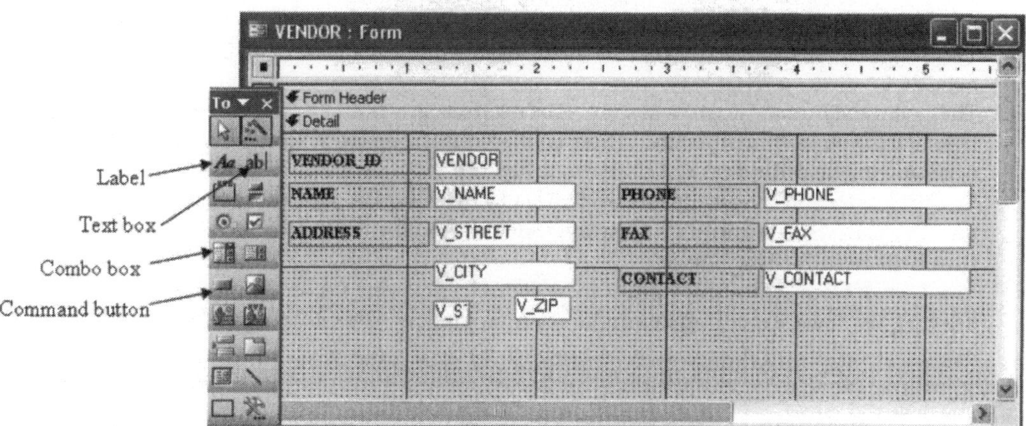

Figure 6.9 Access design toolbox.

6.4.2.1 Adding Command Buttons

The two objects that will be used in this illustration are the ***label*** and the ***command button***. When a user wants to display text on a form there are two choices: a text box or a label. The text box is appropriate when the user wishes to display data from the database. When the form wizard generated the default Vendor form, it automatically generated the text boxes that are used to display the values of attributes from the table VENDOR. The user will interact with these text boxes to insert, delete, and update records. The label is an appropriate way to display titles that the user will not be allowed to modify at run time. When the form wizard generated the default Vendor form, it automatically generated the labels that are used to provide titles to the text boxes. These were the labels whose captions we changed when we reorganized the layout of the default form. Labels will be used to place titles over the command buttons, as shown in Figure 6.8.

The command button is an object that carries out a command when the user clicks on it with the mouse. The command button may be user programmed or it may contain default code that is already programmed by Access. Just as in the case of text boxes and labels, command buttons have properties that can be set to give some desired performance features. We will continue to develop the Vendor form by adding command buttons and labels as per Figure 6.8.

Access Exercise 6.2:

We want to implement the command buttons for the operations ***Insert Record, Delete Record, and Save Record*** as shown in Figure 6.8.

1. Go to the Design View of the Vendor Form and open the Toolbox. From the toolbox, click on the command button icon (shown in Figure 6.9). This enables the icon. Move the cursor to the approximate position of the Insert Record button and click the left mouse button again. This will result in a rectangular button appearing on the screen and a menu window entitled "Command Button Wizard" with a default caption will appear ***(see Figure 6.10)***.

The Command Button Wizard has a series of buttons already coded to perform certain operations.

2. Under the list of "Categories", select "Record Operations". This will provide you with a list of the operations on database records that are already implemented by Microsoft as command buttons. One of the operations that will appear in the right window is "Add New Record". Select this option. Click on "Next".
3. The Command Button Wizard will display the picture that will appear on the Insert Button in Figure 6.8. Click on "Next".
4. The Command Button Wizard will show a default name in a text box. Type a meaningful name to refer to this button: "Insert_button". Click on the "Finish" command button and the Insert command button will appear as it is shown in Figure 6.8.
5. The "Delete Record" command button is added by following the same procedure, except selecting the "Delete Record" option in the selection window. Add the "Delete Record" command button. Give it the name "Delete_button".
6. The "Save Record" command button can be added by following the same procedure, except selecting the "Save Record" option in the selection window. Give it the name "Save_button".

Labels will now be added to the form.

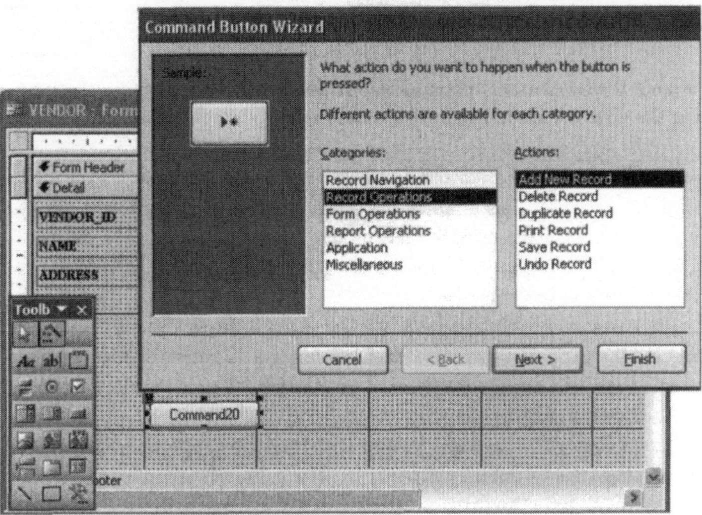

Figure 6.10 Adding a command button.

Access Exercise 6.3:

We want to add the labels for the operations ***Insert Record, Delete Record, and Save Record*** as shown in Figure 6.8.

1. Go to the Design View of the ***VENDOR*** Form and open the Toolbox. From the toolbox, click on the Label icon (shown in Figure 6.9). This enables the icon. Move the cursor to the approximate position of the Insert Record label and left click the mouse button again. While holding the left mouse button down, drag the mouse pointer until you create a label rectangle of the appropriate size.
2. Open the properties window of the label by clicking on the properties icon on the toolbar. Click on "Format" tab. Change the "Caption" property to "Insert Record" and change the "Text Align" property to "Center". This will center the caption within the Label. Click on "Other" tab. Change the "Name" property to "Insert Record."
3. Repeat this procedure to establish labels for "Delete Record" and "Save Record".
4. When these three labels are added, go into the Form View and confirm the appearance of these labels on the form.

In Section 6.4.2 we specified that when a form is opened, it should default to the read-only mode. From the read-only mode, a user will be able to gain access to the function only if he or she is allowed access. This will be done by clicking on the Login command button. In the next two exercises, we will implement this protocol.

Access Exercise 6.4:

We want to implement the label and command button for ***Login*** as well as partially implementing the protocol for giving the user access to the insert, delete and update function.

1. Go to the Design View of the ***VENDOR*** Form and open the Toolbox. From the toolbox, click on The Command Button icon. This enables the icon. Move the cursor to the approximate position of the Login command button and click the left mouse button again. The "Command Button Wizard" window will open. However, in this case we are going to create our own code, so choose the option "Cancel". This will close the window.
2. Resize the Login command button to look like the other buttons. Remove the caption on the button by opening the properties window and making the "Caption" property blank. Go to the "Other" properties tab and change the"Name" property to "Login_button". Add a label with the title "Login", as shown in Figure 6.8.
3. Click on the "Insert Record" command button and open the properties window. Click on the "Format" tab. Change the "Visible" property to "No". This establishes the default mode for the command button. Each time the form is opened, it will appear with the "Insert Record" command button invisible to the user.
4. Click on the "Insert Record" Label and open the properties window. Change the visible property to "No".
5. Follow the same procedure for the "Delete Record" and "Save Record" command buttons and labels.
6. Go to the File menu and save your work. Go into the Form View and observe that the form opens up with the three command buttons and labels invisible to the user.

At this point the mechanism for hiding a function from the user until he or she is cleared to use it has been established. However, the reader should note that the user still has access to the text fields of the attributes on the form and can make changes by inserting the cursor into the field. As previously specified, the Vendor form should open in a read-only mode. This is achieved by setting a default "Locked" property for the text boxes on the form.

Access Exercise 6.5:

We want to open the ***Vendor*** form in the read only mode by setting the default ***Locked*** property of the text boxes to "Yes".

1. Go to the Design View of the Vendor Form. Click on the text box for ***Vendor_ID***. Open the properties window. One of the "Data" properties is the Locked property. Set this property to "Yes". The locked property prevents the user from changing the data in this field. Go to the file menu and save the form. Open the form in Form View and try to change the current contents of the ***Vendor_ID*** text box. This field should be in the locked condition.
2. Go through the remaining text boxes and change the Locked property to "Yes" and save your work. This will result in a read only condition for each record when the form is opened.

6.4.2.2 Addressing Form Objects

To implement the requirement that a user must log in before being able to change or add data to the database, we must first describe how Access code can be written to address various objects on a form. To refer to a particular form, the following expression is used:

```
Forms![Form name]
```

The **Forms!** command indicates to Access that the argument that follows is the name of the form the user wishes to address. Thus, Forms![Vendor] is the expression used to address the Vendor form.

In writing code, the designer usually wants to address some specific object or property of the form. The syntax for addressing an object on the form is as follows:

```
Forms![Form Name]![Form Object]
```

Thus, in order to address the Vendor_ID text box on the form Vendor, the expression would be

```
Forms![Vendor]![Vendor_ID]
```

Finally, it is often the case that the designer wants to address some property of the object. This involves an expression with the following syntax:

```
Forms![Form Name]![Form Object].Property = Desired Property
```

Thus, in order to address the Visible property of the Insert Record button with the objective of making the object visible, the following expression would be used:

```
Forms![Vendor]![Insert_button].Visible = True
```

The name assigned to the Insert Record button when it was established in Access Exercise 6.2 was Insert_button. Access requires that an object is referred to by the exact name as it appears in the Name property. If you are following with your own Access implementation, you should use the names given to the command buttons used in your application. These can be found by opening the Property window for the button and clicking on the Other tab where the Name property is shown.

With this brief background, the protocol for opening the form in the read-only mode and allowing the user to access functions by clicking on the Login button shall now be programmed.

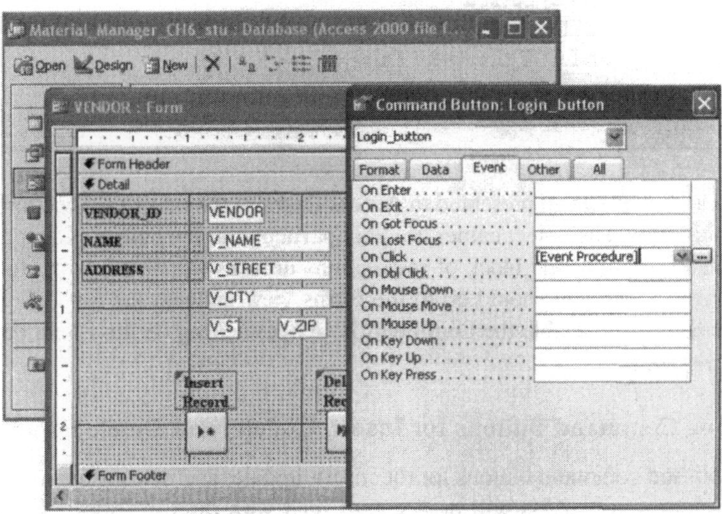

Figure 6.11 Adding an on-click event procedure.

Access Exercise 6.6:

We want to open the ***Vendor form*** in the read only mode and provide functionality to the user after the user clicks on the ***Login button***.

1. Go to the Design View of the ***Vendor Form***. Click on the "Login" command button. Open the properties window and choose the "Event" Properties. Scroll down to the "On Click" property. The "On Click" property is an event property. It defines what happens if a user clicks on the button. The "Insert Record", "Delete Record" and "Save Record" buttons have their "On Click" properties defined by Access when they were created. Here we will define the events that take place when the "Login" button is clicked.
2. Place the cursor in the "On Click" panel and click the left mouse button. Click on the down arrow and the option "Event Procedure" will appear. Select "Event Procedure". Click on the triple dots (Figure 6.11) and a window will appear to write the procedure.
3. Enter the following code in the "On Click" function. Then close the window and save your work.

```
  Private Sub Login_button_Click()
' Make labels and command buttons visible
    Forms![Vendor]![Insert_button].Visible = True
    Forms![Vendor]![Delete_button].Visible = True
    Forms![Vendor]![Save_button].Visible = True
    Forms![Vendor]![Insert Record].Visible = True
    Forms![Vendor]![Delete Record].Visible = True
    Forms![Vendor]![Save Record].Visible = True

' Unlock attribute fields for editing
    Forms![Vendor]![VENDOR_ID].Locked = False
    Forms![Vendor]![V_NAME].Locked = False
    Forms![Vendor]![V_STREET].Locked = False
    Forms![Vendor]![V_CITY].Locked = False
    Forms![Vendor]![V_STATE].Locked = False
    Forms![Vendor]![V_ZIP].Locked = False
    Forms![Vendor]![V_PHONE].Locked = False
    Forms![Vendor]![V_FAX].Locked = False
    Forms![Vendor]![V_CONTACT].Locked = False

End Sub
```

The first block of instructions presented in Access Exercise 6.6 make the function keys and their labels visible to the user. The object names used are particular to the names assigned to each object when it was created. The second block of instructions unlocks the fields on the form for editing. With these changes added you should go to the forms view and run the form. Note that no field can be changed before you enable the Login button. After clicking on the Login button, the fields are available for editing.

6.4.2.3 Using Command Buttons for Insert, Update, and Delete Functions

Having established command buttons for the insert, update, and delete functions, in this section we explain their proper use. This will be accomplished with three Access exercises. Follow the exercises exactly and your knowledge of the use of these command buttons will be complete.

Access Exercise 6.7:

We want to illustrate the use of the command buttons for inserting a new record into the VENDOR table.

1. With the ***VENDOR form*** in Form View, click the Login button to expose the other functions and unlock the data fields (textboxes).
2. Click on the ***Insert Record*** button. This action will navigate the form to a new record position, leaving blank fields in the text boxes.
3. Insert the following new record into the appropriate fields.
   VENDOR_ID V300
   V_NAME Finest Poultry
   V_STREET 20 Farm Lane
   V_CITY Farmington
   V_STATE NJ
   V_ZIP 08828
   V_PHONE 732-321-3000
   V_FAX 732-321-3111
   V_CONTACT Robert Tyson
4. After inserting the record on the form, click on the ***Save Record*** *button*. This will insure that the record is saved to the database.
5. Without closing the VENDOR form, go to the database window and click on the Tables Tab. Open the ***VENDOR table***. Check to see that the new record has indeed been inserted. Close the table.

You will note that when you pressed the Insert Record button, the navigation bar at the bottom of the form moved to the new record position. The Insert Record button essentially serves the same purpose as the navigation button in this instance. It navigates the user to a blank record. After inserting the new record on the form, clicking the Save Record button commits the data to the database.

Access Exercise 6.8:

We want to illustrate the use of the command buttons for updating a record in the ***VENDOR*** table.

1. With the ***VENDOR form*** in Form View, move to the record that was inserted in Access Exercise 6.7.
2. Insert the cursor into the text box labeled FAX.
3. Change the Fax number to 732-321-3001. At this point the fax number is updated on the form.
4. Click on the ***Save Record*** button to commit the update. At this point the updated record is saved *to* the database.
5. Without closing the VENDOR form, go to the database window and click on the Tables Tab. Open the ***VENDOR table***. Check to see that the updated record has indeed been saved. Close the table.

The only command button function required to update a record is the Save Record button. The user just has to navigate to the desired record using the navigation buttons and perform the update on the form. Clicking the Save Record button commits the update to the database.

Access Exercise 6.9:

We want to illustrate the use of the command buttons for deleting a record in the ***VENDOR*** table.

1. With the ***VENDORform*** in Form View, move to the record that was inserted in Access Exercise 6.7.
2. Click on the ***Delete Record*** button.
3. A prompt will appear informing you that the delete operation is irreversible. Select "yes". At this point the record should be erased from the form.
4. Click on the ***Save Record*** button and the deletion will be committed to the database.
5. Without closing the VENDOR form, go to the database window and click on the Tables Tab. Open the VENDOR table. Check to see that the deleted record has indeed been removed. Close the table.

The reader should note that the Delete Record button removes the record on the form, but should be followed by the Save Record button operation in order to commit this deletion to the database.

6.4.3 Implementing Data Integrity Requirements

Data integrity checks as part of the forms specification was discussed in Section 6.3.1. All efforts should be made to ensure that a data entry made by a user is correct. This is referred to as ***data validation***. Practically speaking, it is usually impossible to ensure the integrity of all data entered by a user. However, some validation techniques will become obvious to the form designer during the design process. Here we shall illustrate a couple of cases.

In our hypothetical enterprise, the VENDOR_ID attribute always begins with "V" followed by a number code. Hence, when a new vendor ID is entered, a simple check can be made on the existence of the correct letter before the number. If the number is limited to a specific number of fields, this can be checked also. The V_STATE is limited to a specific set of alternatives, which comprise the abbreviations of the various 50 states. So, for example, we can ensure that the state is represented by a text field that appears within that legal set. Finally, the telephone and fax numbers are 10-digit numbers with hyphens.

Access has provided a Validation Rule field as part of the Text box properties. This is a facility that makes it easy to implement the integrity checks mentioned earlier. In addition, Access has a Validation Text property that allows the user to enter an error message that will be given to the user when a data entry violates the validation rule. We will illustrate this using the Vendor form.

When the validation rule specifies a particular text character and text string length, the "Like" operator is a good choice to use in the validation rule. For example, Like "???-???-????" indicates a text string of length 12 with hyphens. The specified string length must be identical to the character length assigned to the attribute when the table was created. For a string of unspecified length beginning with "V," use Like "V*," where the asterisk is a wildcard. This validation rule would be appropriate for validating the VENDOR_ID.

When the validation rule specifies an expression from a set of possible entries, the "In" operator is appropriate. For example, to specify that V_STATE should be ***In*** the set of legitimate abbreviations, the validation rule would be as follows:

```
    In ("AL", "AK", "AZ", "AR", "CA", "CO", "CT", "DE", "FL", "GA", "HI", "ID",
"IL", "IN", "IA", "KS", "KY", "LA", "ME", "MD", "MA", "MI", "MN", "MS", "MO",
"MT", "NE", "NV", "NH", "NJ", "NM", "NY", "NC", "ND", "OH", "OK", "OR", "PA",
"RI", "SC", "SD", "TN", "TX", "UT", "VT", "VA", "WA", "WV", "WI", "WY")
```

Each time you exit the V_STATE field after making an entry, Access checks the validation rule to make sure it is true.

Access Exercise 6.10:

We want to enter validation rules and validation text for the ***Vendor form***.

1. Go to the Design View of the ***VENDOR Form***. Click on the text box for VENDOR_ID. Open the properties window. One of the "Data" properties is the "Validation Rule" property ***(see Figure 6.12)***. Click on the panel and enter the rule ***Like "V*"***. Click on the "Validation Text" panel and enter ***This is not a legitimate Vendor_ID. The Vendor_ID must begin with "V"***. Close the property window.
2. Click on the text box for V_STATE. Place the cursor in the "Validation Rule" property panel. Enter the ***"In"*** expression containing the legal state abbreviations defined previously. Click on the Validation Text panel and enter ***This is not a legitimate state code. Use the capital letter abbreviation for the state.*** Close the property window.
3. Complete the validation rules and Validation Text properties for V_ Phone and V_Fax as follows. Validation Rule: ***Like "???-???-????"***. Validation Text: ***This is not a valid entry. A valid entry is of the form*** xxx-xxx-xxxx. Close the properties window.
4. Go to the Forms View and test the validation rules by trying to enter a new record that does not conform to the validation rules.

Embedding integrity checks into the design of a form cannot be overemphasized. One of the major problems with any information system application is the corruption of data during the data acquisition and entry process. When an operator keys data in, there is always the chance of human error. Even when data are automatically collected, as in the case of bar code readers, a finite error rate will occur. A great deal of effort is usually put into the validation problem at design time to avoid the costly errors that can occur due to incorrect data in the database. For example, if an incorrect count of on-hand raw materials is in the information system when production is being planned, it can result in chaos on the day of production when it is discovered that there is insufficient raw material to carry out the production run. In this section, we have illustrated some fairly simple integrity checks; the reader will be given an opportunity to consider other possibilities in the exercises at the end of this chapter.

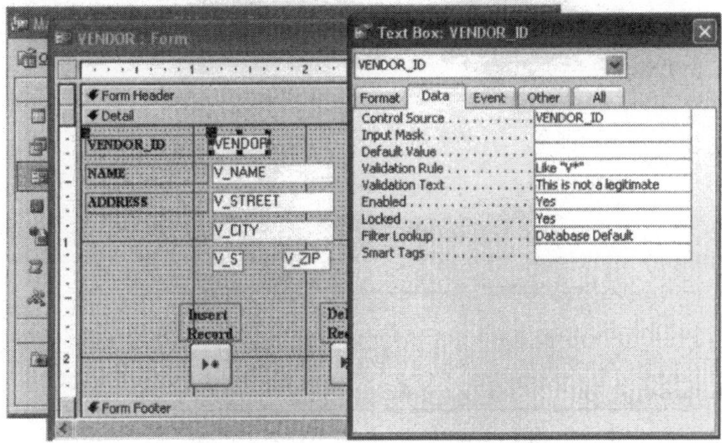

Figure 6.12 Entering data validation rules.

6.4.4 Form Navigation

Another form specification item discussed in Section 6.3.1 is ***form navigation***. Form navigation can include anything from providing a user with a method of moving between multiple pages on a multiple page form to allowing the user to move from text box to text box on a form in a logical manner consistent with the function being performed. Here we have been illustrating a rather simple single table form, and there are many issues of form navigation that do not apply. However, one area of application that is useful to consider even in a single table form is how to provide the user with a facility for moving quickly among the records in the table on which the form is based.

Consider the case of the Vendor form. One can imagine the user trying to locate a particular vendor record. As presently implemented, the user would have to index through each record using the navigation button until she finds the one she wants. This will be tedious when there are hundreds or thousands of records. The answer is to provide the user with a simple way of navigating through the records using some sort of help facility.

One way of implementing such a help facility is with a pull-down menu that lists key information from which the user may select the record for which she is looking. Such a facility is implemented on the Vendor form as a combo box and is shown in Figure 6.13.

A "Combo Box" combines a text box with a list box. It allows the user to select an item from a drop-down list. The form designer determines what fields will be shown in the combo box. The Vendor form of Figure 6.13 shows the vendor name (V_NAME). If the list of vendor names is long, scroll bars will be placed alongside of the combo box, allowing the user to quickly scroll through the list until she finds the correct vendor name. By clicking on the vendor name, a selection is made for the record to be shown on the form. The efficiency of this form navigation tool over searching through each record is obvious when several hundred vendors are in the database.

In this section, we create the navigation tool shown in Figure 6.13. This is done in Access Exercise 6.11.

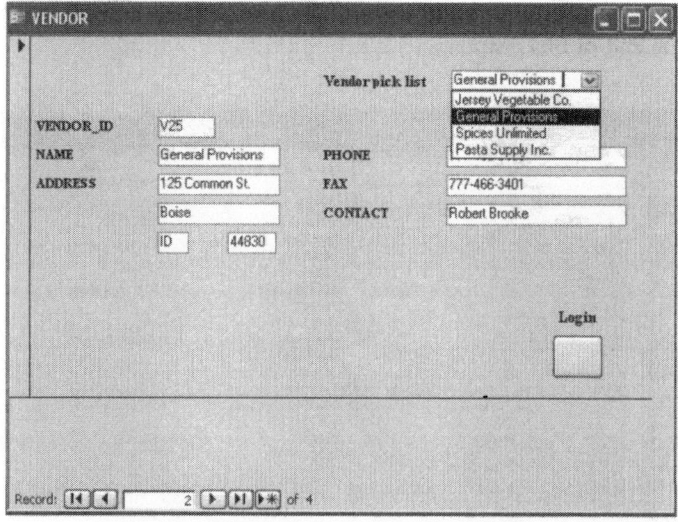

Figure 6.13 Record navigation using a combo box.

Access Exercise 6.11:

We wish to attach a combo box as a navigation aid to the ***VENDOR form***.

1. Go to the Design View of the Vendor Form. Allow some room on the "Detail" portion of the form as shown in Figure 6.13.
2. Open the Toolbox. Select a "Combo Box" and put it in the upper right of the Detail portion of the form (***see Figure 6.14***). The "Combo Box Wizard" will then appear.
3. In the Combo Box Wizard, choose the option: ***"Find a record on my form based on the value I selected in my Combo Box"***. Then click on "Next".
4. The Wizard will then allow you to select the fields for the Combo Box. Choose **VENDOR_ID** and **V_NAME**. Then click on the "Next" button.
5. The Wizard will display how the Combo Box will appear. Since the user is interested in the V_NAME to make the selection, the primary key (VENDOR_ID) should be hidden from the user. It will be used in the search to return the appropriate record. Click on "Next".
6. The Wizard asks you to give a name to the caption next to the Combo Box. Type: ***Vendor pick list.*** Click on "Finish".
7. Go to the forms view and run the form. Experiment with the Combo Box. If the label or text box is not large enough, return to the design view, select the label or text box, and enlarge it.

6.4.5 Implementing the Password

Database management systems have methods for enforcing password protection when a user logs on to the system. Access DBMS has such services, which are reviewed in Section 6.9. In this section, we will create our own password protection. This section has multiple purposes. One

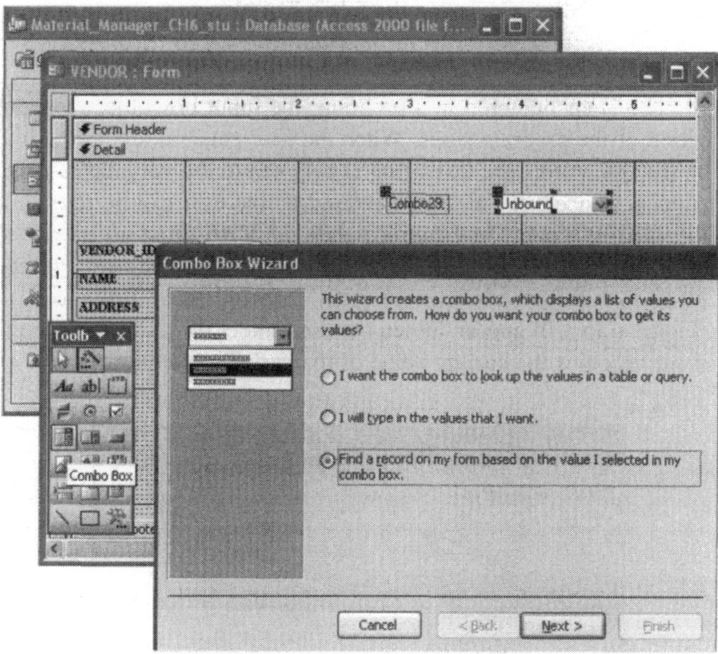

Figure 6.14 Combo box design window.

purpose is to illustrate a method for filtering user access at the form level. Another purpose is to introduce the reader to some useful Visual Basic for Applications (VBA) objects and methods that can be used in a variety of situations. A third purpose is to show how VBA can be used to write programs that retrieve data from the database for evaluation.

To provide users with some facility to write code for Access forms, Microsoft provides a subset of the Visual Basic Programming Language called Visual Basic for Applications (VBA). VBA gives the user the tools to write program control statements, such as "If . . . then . . . else," "Do while . . . ," and so on. It also provides a set of built-in objects and methods that can be used to achieve a specific result. In this book, we assume that the reader is familiar with some high-level programming language and that it is unnecessary to review the standard programming constructs, such as control statements. Objects and methods will be introduced as needed, and the user is referred to the MS Access Help menu to find the complete set of available objects. The reader is also referred to Appendix 6C for a refresher tutorial on programming fundamentals. In this section, we introduce two new useful VBA objects, the input box and the message box.

6.4.5.1 Input Box

An ***input box*** is a VBA object that accepts user input in a text string format and assigns it to a string variable. A typical input box is shown in Figure 6.15. It consists of the following elements:

- Title bar
- User prompt string
- Text input area
- OK button
- Cancel button

An input box is created in VBA code using the following syntax:

```
VariableName = InputBox("User prompt string", "Title bar")
```

When this line of code is encountered in a program, the input box object appears on the screen and further execution of the program stops. When the user enters a text string into the text input area and clicks on the OK button, the text string is assigned to the variable name and execution of the program continues. Henceforth, the value entered by the user can be referred to and recalled by reference to the variable name. For example, when the VBA program encounters the line:

```
Request = InputBox("Enter your password", "PASSWORD REQUEST"),
```

the input box in Figure 6.15 will appear. When the user enters the password and clicks on the OK button, the password becomes the current value of the variable "Request."

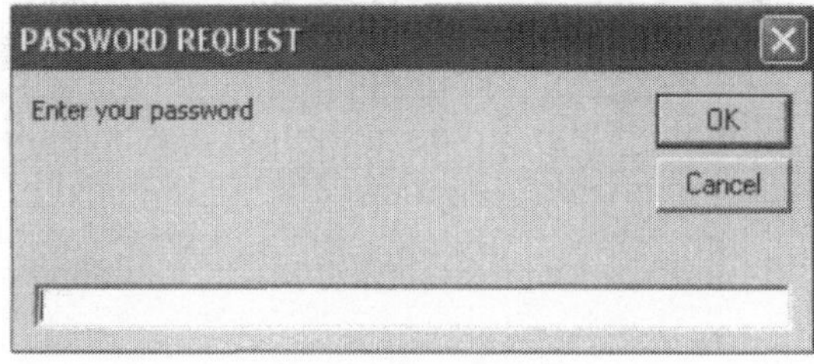

Figure 6.15 Input box.

The input box is a simple way to gather user information interactivity. However, it does not support the security of hiding the password by substituting an asterisk for the typed letter.

6.4.5.2 Message Box

Whereas the input box is used to assign user input to a variable, a ***message box*** is used to display a programmed message (output) to a user. A typical message box for the current application appears in Figure 6.16. A message box has four components:

- Title bar
- Message
- ICON
- OK button

A message box is created in VBA code using the following syntax:

```
VariableName = MsgBox("Message", ICON number, "Title bar")
```

When the VBA program encounters the preceding line of code, the message box appears on the screen. The user acknowledges (clears) the message by clicking on the OK button.

There are standard icons in VBA that can be used to add emphasis to the message. The icon number in the message box statement is a number that refers to the icon. Some examples are as follows:

The message box of Figure 6.16 is created with the following code.

```
Response = MsgBox("You have Read Only permission", 48, "PERMISSION DENIED")
```

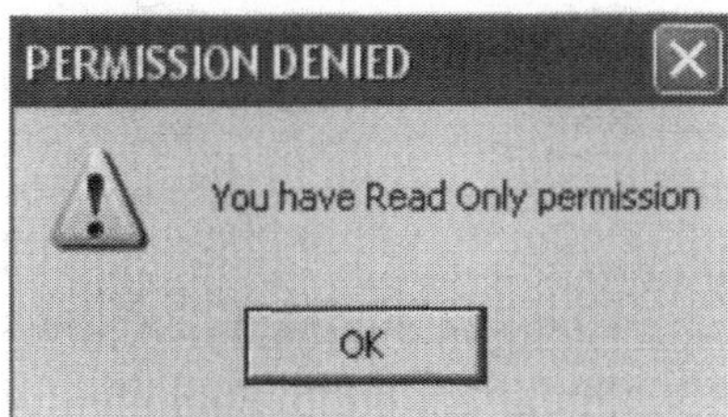

Figure 6.16 Message box.

6.4.5.3 Retrieving a Database Record

The input box and the message box will be used to confirm that the user has the permission level to use the insert, delete, and update functions. Recall that the Login command button of Figure 6.13 will be the control for enabling this action. Therefore, the new code for checking the password will be part of the "on_click" procedure of the Login command button.

One of the tables for the applications in this chapter, shown in the Appendix 6.A, is the PASSWORD table. The table contains a record for each user who will be allowed full functionality of the Vendor form. Any user whose password does not appear in this table has read only permission. Since each potential user will enter his or her password via the input box, a subroutine must be written that compares the variable "Request" with each password entry in the table. If a match is found, permission will be granted. If a match does not exist, the message box will be returned to the user indicating that the permission is denied.

In order to accomplish these ends, the VBA code must first open the PASSWORD table. Following that, a subroutine must search the entries for a match. A standard method for doing this will be described in this section.

Microsoft has provided more than one way to connect to a database using VBA objects. One of the first and simplest methods is the **Data Access Object (DAO)** model. DAO provides a set of methods for connecting to a database and retrieving records from tables. DAO was designed primarily for use with the Microsoft Jet database engine connection service. More recently, Microsoft introduced the **Active X Data Object (ADO)**, which facilitates connecting to a number of databases. ADO is discussed in Appendix 6C and will be described more thoroughly in Chapter 8. In this chapter, we illustrate database connectivity using DAO.

Before using DAO, make sure that the library components are available in your database application. This can be checked from the Visual Basic Editor window of Access. From the Access main menu, choose **Tools⇒Macro⇒Visual Basic Editor**. From the Visual Basic Editor main menu, select **Tools⇒References**. A pop-up window will present a list of currently selected Reference Libraries. A checkmark should be placed in the check box for **Microsoft DAO Object Library** (Figure 6.17). Click OK to close the window, and then close the VBA Editor window and return to Microsoft Access.

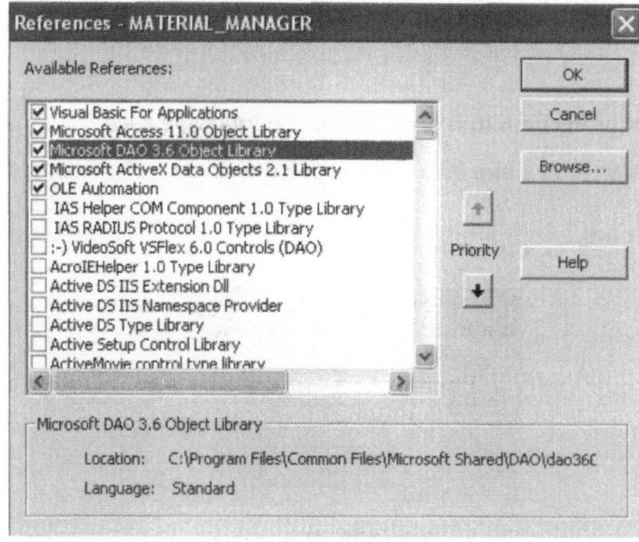

Figure 6.17 Including DAO object library.

The relevant DAO objects and methods will now be described. When a user is working with an open database, the database is said to be current and the workspace in which it is running is the current workspace. The user can connect to this database from VBA by using the Set keyword as follows:

```
Set MyDB = DBEngine.Workspace0.Database0
```

This command creates an object, MyDB, that has the following components:

- DBEngine—This refers to the Microsoft Access jet engine, which is a native set of drivers that communicate with an Access database.
- Workspaces0—This is a collection of active database sessions. Element 0 refers to the active workspace.
- Database0—The current database is element 0 in the workspace.

The previous line of code will always return the active database with a Microsoft jet connection. There are other ways to connect to a database that will be described in a later chapter. For this application, DAO and the Microsoft Access jet connection are the easiest to use.

With the database open and an Access jet connection established, it is necessary to retrieve the contents of a specific table of interest. In Visual Basic, the term ***recordset*** is used to describe the dataset selected from the table. The syntax for retrieving a recordset is as follows:

```
Set RecordSetName = DB.VariableName.OpenRecordSet("Table name",
dbOpenDynaset)
```

So, for retrieving the entire contents of the PASSWORD table, the VBA statement is written as follows:

```
Set MyRec = MyDB.OpenRecordSet("PASSWORD", dbOpenDynaset)
```

MyRec is defined as a recordset object using the OpenRecordSet method of the database object previously created with the name MyDB. The arguments of the OpenRecordSet method are the table name and the Open command. Upon execution, a recordset will be created in the current workspace.

To address a specific field in the recordset, the following syntax is used:

```
Set Field Name = RecordSetName.Fields("Attribute Name")
```

So, for addressing the PASSWD field of the PASSWORD table, the following command is used:

```
Set MyFld = MyRec.Fields("PASSWD")
```

Once the recordset is open, a pointer is set to the first entry in the recordset. The VBA code can be written to move the pointer through each record and compare the password entered by the user to the entry in the recordset. The following commands can be used to control the pointer:

MoveNext—Moves the pointer to the next sequential record
MoveFirst—Moves the pointer to the first record in the recordset
MoveLast—Moves the pointer to the last record in the recordset

6.4.5.4 A Completed Application

Figure 6.18 is a listing of the code that should be placed in the on-click event procedure of the Login command button. It contains the elements that were just discussed as well as the code for

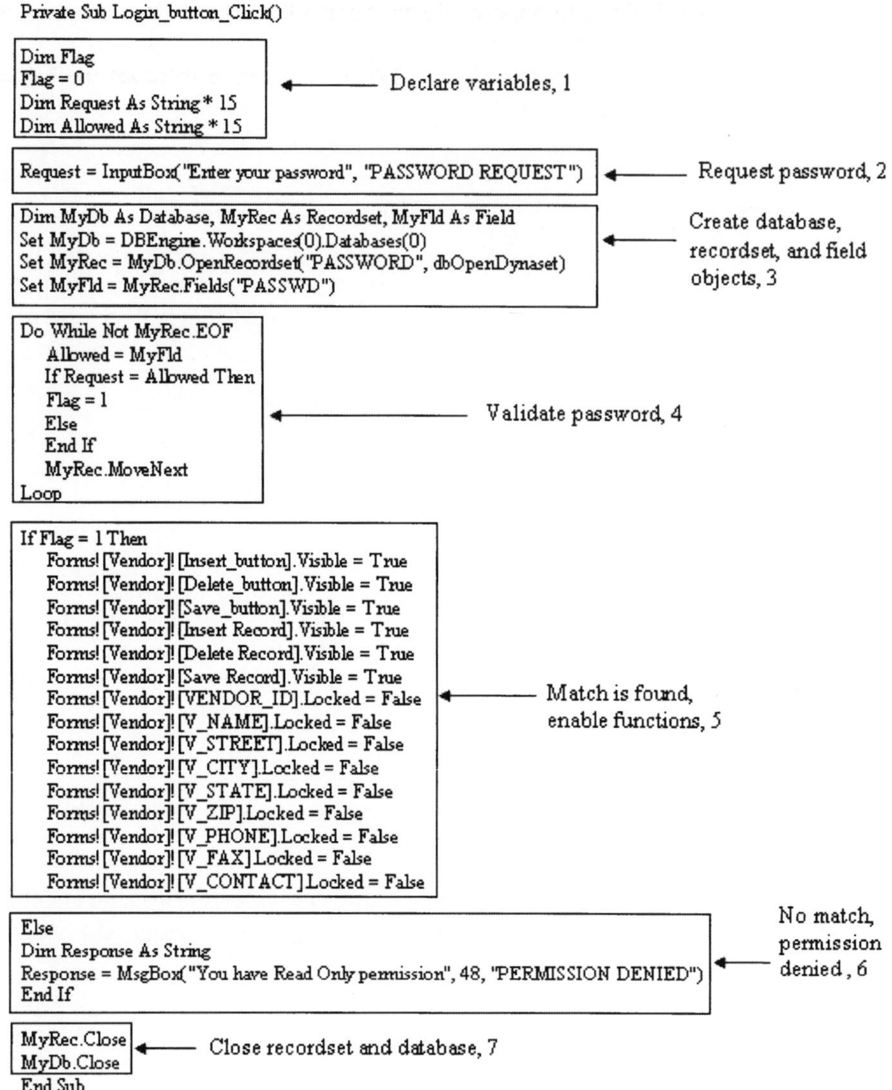

Private Sub Login_button_Click()
Dim Flag
Flag = 0
Dim Request As String * 15
Dim Allowed As String * 15
Declare variables, 1
Request = InputBox("Enter your password", "PASSWORD REQUEST")
Request password, 2
Dim MyDb As Database, MyRec As Recordset, MyFld As Field
Set MyDb = DBEngine.Workspaces(0).Databases(0)
Set MyRec = MyDb.OpenRecordset("PASSWORD", dbOpenDynaset)
Set MyFld = MyRec.Fields("PASSWD")
Create database, recordset, and field objects, 3
Do While Not MyRec.EOF
Allowed = MyFld
If Request = Allowed Then
Flag = 1
Else
End If
MyRec.MoveNext
Loop
Validate password, 4
If Flag = 1 Then
Forms! [Vendor]! [Insert_button].Visible = True
Forms! [Vendor]! [Delete_button].Visible = True
Forms! [Vendor]! [Save_button].Visible = True
Forms! [Vendor]! [Insert Record].Visible = True
Forms! [Vendor]! [Delete Record].Visible = True
Forms! [Vendor]! [Save Record].Visible = True
Forms! [Vendor]! [VENDOR_ID].Locked = False
Forms! [Vendor]! [V_NAME].Locked = False
Forms! [Vendor]! [V_STREET].Locked = False
Forms! [Vendor]! [V_CITY].Locked = False
Forms! [Vendor]! [V_STATE].Locked = False
Forms! [Vendor]! [V_ZIP].Locked = False
Forms! [Vendor]! [V_PHONE].Locked = False
Forms! [Vendor]! [V_FAX].Locked = False
Forms! [Vendor]! [V_CONTACT].Locked = False
Match is found, enable functions, 5
Else
Dim Response As String
Response = MsgBox("You have Read Only permission", 48, "PERMISSION DENIED")
End If
No match, permission denied, 6
MyRec.Close
MyDb.Close
Close recordset and database, 7
End Sub

Figure 6.18 VBA code for password using input box.

making the hidden command buttons visible and unlocking the text boxes. It also includes standard programming protocols and procedures with which the reader should already be familiar.

Note the way the program accesses the information in the database. After opening the database, a recordset is retrieved. All search operations are performed on the recordset, not the database table. In this case, the only operation is to read the entries in the PASSWD field from the beginning to the end of the recordset, assigning the values to the string variable "Allowed." The variable "Request" is compared to "Allowed" and a "true" comparison results in setting a variable called "Flag." If Flag has been set, the insert, delete, and update functions are enabled. If Flag is "false," permission is denied. Finally, the recordset is closed and the database is closed. The recordset must be closed first while the database is still open.

Note the manner in which the data types of Request and Allowed are dimensioned. They are dimensioned as strings of length 15, where 15 is the field width specified for the attribute PASSWD when the PASSWORD table was designed. When the VBA program retrieves the contents of PASSWORD for the recordset, the trailing white spaces between the end of the data and the field length of 15 is also retrieved. Therefore, when a comparison is made between Request and Allowed in the Do While loop, the two variables must match exactly, including white spaces. By setting both variable strings to length 15, the match will be exact if a match exists.

The flowchart shown in Figure 6.19 describes the logical structure of the subroutine. The reader should compare the flowchart to the code. It is left to the reader to enter the subroutine into the on-click event procedure of the Login command button and to test it for correct operation. The existing passwords are listed in the PASSWORD table of Appendix 6A. For illustrative purposes, we have simply used last names. Additional passwords can be added directly to the table.

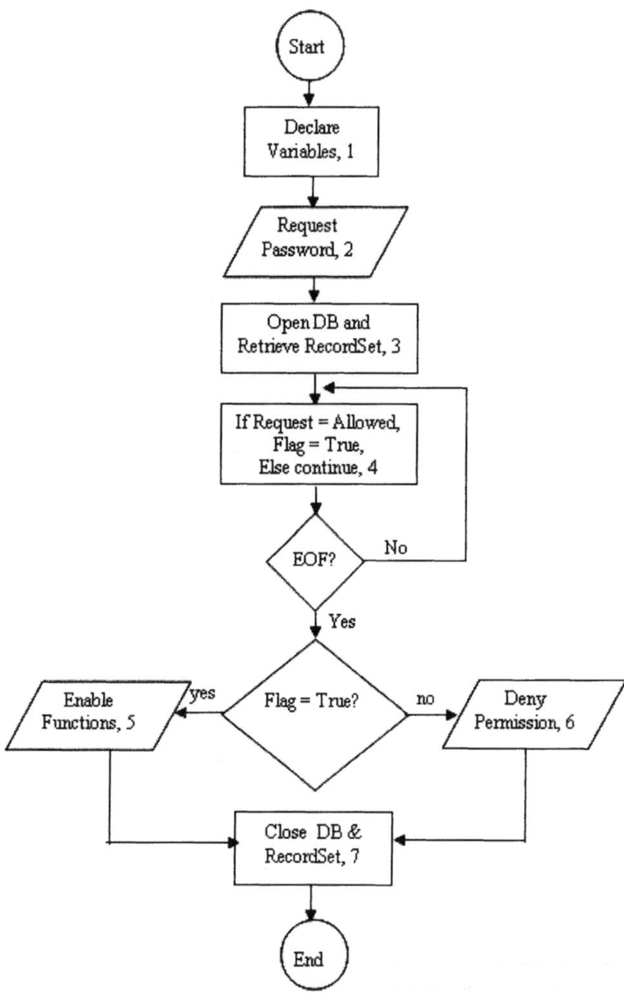

Figure 6.19 Flowchart of major sections of Figure 6.18.

6.4.6 Adding Titles in the Header Area

To complete the form design, the designer should add form title information in the Header section. This is a relatively simple task. These titles are Access labels. Access labels can be relocated on a form, its properties can be changed, and additional labels can be added at design time. In this section, we enter the "VENDORS" title and add labels for the company title, street address, and city/state/zip code. The procedure is given in Access Exercise 6.12, and the completed form is shown in Figure 6.20.

Access Exercise 6.12:

We want to add the header shown in Figure 6.20.

1. Open ***VENDOR Form*** in design view.
2. Using the mouse, pull down the "Detail" area of the form in order to make room for the header information.
3. Open the Toolbox. Click on the label button and add a new label for the "VENDORS" title. Put the cursor in the label and type ***VENDORS***. This will automatically change the label caption property to this title. Note that the typed caption is automatically right justified. This is the default text alignment. Change this by placing the focus on the label and opening its properties window. Change the "Text Align" property to "Center". Close the property window and note that the title is now centered.
4. Repeat steps 4 and 5 for the labels "The University Food Company", "1776 New England Avenue" and "Piscataway, NJ 08854".
5. Go to the "Form View" and observe that the header title now conforms to that of Figure 6.20.
6. Go to the File menu and save your work.

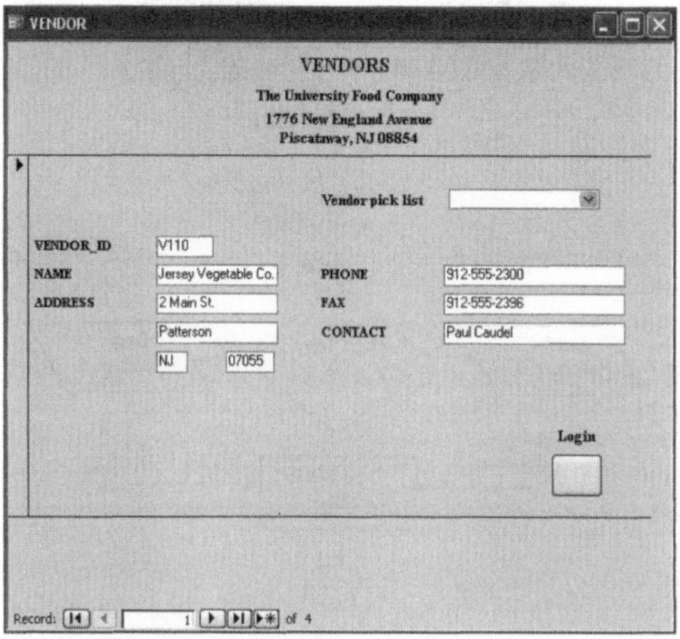

Figure 6.20 Vendor form with heading.

6.4.7 Summary of a Single-Table Form

It was the purpose of this section to introduce the reader to the design of single-table forms. Several principles of form design were illustrated. These principles included the initial specification of the layout of the form, the implementation of functionality on the form so that it is easy and intuitive to the user, the implementation of integrity checks on data being entered, and the application of navigation aids, such as a combo box that allows the user to move among records. We will expand on these themes in later sections as we discuss forms based on more than one table.

6.5 FORMS BASED ON MORE THAN ONE TABLE

In the previous section, the procedure for designing a form based on a single table was described. One of the issues of *form specification*, as discussed in Section 6.3.1, is the relationship of different blocks on the user screen. This specification becomes relevant when designing a form based on more than one table. In this section, we address the design of a form based on more than one table.

The term ***master/detail form*** is used to describe a form based on more than one table that shows one record of a given table (the master) together with associated records in one or more other tables (the details). The master/detail form is used to show related records in more than one table on a user screen. A good example of this is the purchase order, which naturally recommends itself to a master/detail format. If the reader looks at Figure 5.7, it is clear that the principal tables involved are PURCHASE_ORDER and PO_DETAIL. These entities have a 1:M relationship based on the foreign key PO_NUMBER. Note that the purchase order also contains information from the VENDOR table and the EMPLOYEE table. PURCHASE_ORDER is related to these tables by the foreign keys VENDOR_ID and EMPLOYEE_ID, and PURCHASE_ORDER is the M side of these relationships. Therefore, to construct a purchase order form there are several tables involved.

The term ***block*** is used to refer to the data and text on a form that corresponds to one table or query of a database. In Section 6.4 we developed a *single-block form* because all of the data elements were from the VENDOR table. The term ***page*** is used to define the part of a form that is displayed on a screen at one time. A single form may consist of one or more pages, and a single page may consist of one or more blocks. In this section we will be developing a single page form consisting of multiple blocks.

To illustrate the concept of a block, the reader is referred to Figure 6.21. This figure shows a four-block master/detail form that is based on the following tables: PURCHASE_ORDER, VENDOR, EMPLOYEE, and PO_DETAIL. Taken together, these tables are used when a new purchase order is created. When the purchasing agent creates a new purchase order, she or he inserts new information into the PURCHASE_ORDER and PO_DETAIL tables and uses the VENDOR and EMPLOYEE tables for read-only purposes.

Block 1 is based on the PURCHASE_ORDER table and is the M side of the 1:M relationship with the tables VENDOR and EMPLOYEE. Block 1 is the master block and is created as the master form, which is also referred to as the **main form**. Blocks 2, 3, and 4 are linked to the master block through their foreign keys. They are created from detail forms, also referred to as **subforms**.

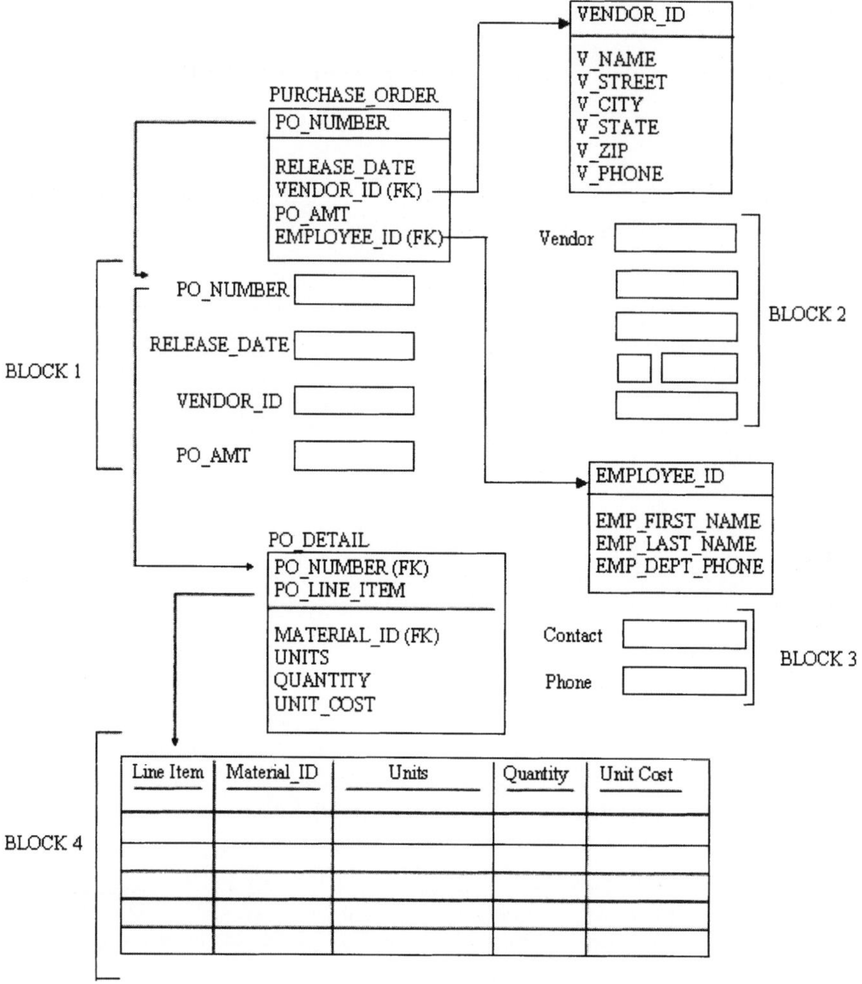

Figure 6.21 Template of a Master/Detail form with four blocks.

6.5.1 Creating a Master/Detail Form

In this section Microsoft Access will be used to create a master/detail form. The procedure is to create the master form and the detail form separately and then to bind the detail form to the master form. Before designing a master/detail form, it is necessary to establish the relationship between tables that will be used in designing the form.

6.5.2 Establishing Relationships in Access

Since there is more than one table involved in a master/detail form, it is first necessary to specify the relationship between tables using the Access ***Relationship Builder***. This is a facility that allows the user to graphically define the data model of the database. In this exercise, the Relationship Builder will be used to build the data model of Figure 5.12 in Access. Figure 5.12 is the data model showing the relationship between entities in IDEF1X format. The following exercise will establish that data model in the database Material_Manager_CH6_stu.

Access Exercise 6.13:

We want to establish the relationships between tables as shown in Figure 5.12.

1. Go to MS Access Main Window and open the database **Material_Manager_CH6_stu.**
2. Select the menu item ***Tools => Relationships*** or click on the "Relationships" button on the toolbar. The "Relationships" window will open as shown in Figure 6.22. If the relationships have already been entered, the background will appear as shown in Figure 6.23.
3. Assuming that the relationships have not been entered, select the menu item ***Relationships => Show Tables***. Add tables in the "Show Tables" window. As you select a table, it will appear in the Relationships window. After all the tables shown in Figure 6.23 have been added, close the Show Tables window.
4. The relationship between PURCHASE_ORDER and PO_DETAIL is based on the common attribute PO_NUMBER. To show this relationship, click on PURCHASE_ORDER.PO_NUMBER (the 1 side of the relationship) and hold down the button. Drag the cursor and place it over PO_DETAIL.PO_NUMBER (the M side of the relationship). Release the mouse button. A relationship dialog box will appear.
5. In the dialog box, check the item "Enforce relational integrity" and click on the "Create" button. A 1:M relationship is now established between the two entities.
6. Repeat steps 4 and 5 for all of the other relationships of Figure 5.12. When this is completed, the Relationships window should look like Figure 6.23. You can rearrange the tables by dragging them across the screen.
7. Save and close the relationships window.

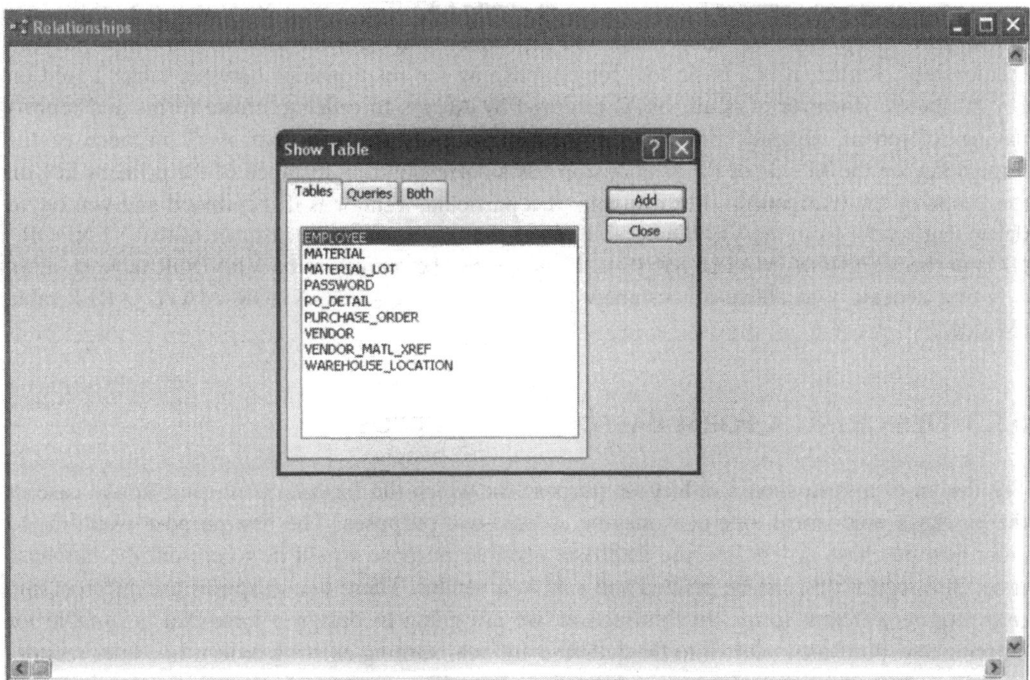

Figure 6.22 Relationships window and Show Table window.

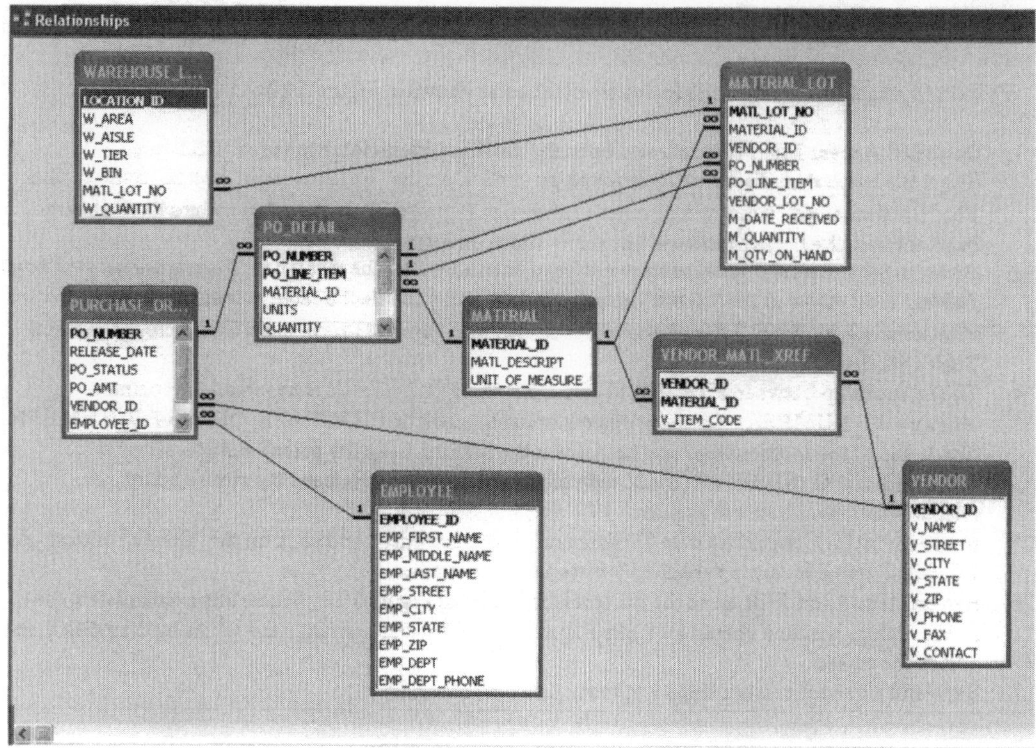

Figure 6.23 Relationships Window when E-R diagram is complete.

The reader should compare Figure 6.23 with Figure 5.12. The Access Relationship Builder does not support the concepts of independent and dependent entities or identifying and nonidentifying relationships. Rather, it is a basic tool for establishing the relationships between tables based on key attributes. However, it is all that is required by Access in order to create forms and reports having referential integrity. Enforcing referential integrity requires that every instance of the foreign key on the M side of the relationship has a corresponding instance of the primary key on the 1 side of the relationship. For example, if a particular vendor is discontinued and you try to delete that vendor from the VENDOR table, it will not be allowed if an instance of that VENDOR_ID is in the VENDOR_MATL_XREF table. To delete the record in the VENDOR table (1 side), it is first necessary to delete all instances of the vendor in the VENDOR_MATL_XREF table (M side).

6.5.3 Designing a Form Based on Its Purpose

The design of a form should reflect the purpose for which the form is to be used. In the case of the purchase order form, one can imagine at least two purposes. The first purpose would be to enter new purchase orders into the database. Another purpose would be to present the purchase order in a format that can be printed and sent to a vendor. These two purposes are different and may require different forms. In this section, we are going to design a form that is suitable for entering new purchase orders into the database and maintaining existing orders. In a later section, we will design a form based on the requirements for ordering materials from a vendor.

6.5.4 Designing a Master/Detail Form for Data Entry

When a form is used for data entry, it is important to distinguish between data that are being newly committed to the database and data that are on the form for viewing only. Figure 6.24 illustrates the issue for the case of the Purchase Order form.

Recall that a *block* is a set of data on a form that is associated with one table. Some blocks in Figure 6.24 correspond to new data that must be added to tables of the database. Other blocks are there to show existing records from other tables. For example, in Figures 6.21 and 6.24, the fields of the PURCHASE_ORDER table (block 1) and the PO_DETAIL table (block 4) must be entered in order to establish a new purchase order. This includes the VENDOR_ID and EMPLOYEE_ID fields in block 1. The data from the VENDOR (Block 2) and EMPLOYEE (Block 3) tables are related to block 1 by the foreign keys of VENDOR_ID and EMPLOYEE_ID and do not represent new data. By using the foreign keys of VENDOR_ID and EMPLOYEE_ID in block 1 to retrieve the existing records in block 2 and block 3, the only new records that will be created when the form is saved are those of the purchase order and its detail. In general, when a foreign key attribute exists in the table on which the form is based, that foreign key points to the primary key in another table that can be used for display on the form.

6.5.4.1 Designing the Master Form

The layout of Figure 6.21 is implemented in Figure 6.24 by establishing a main form based on the table PURCHASE_ORDER. This will also be block 1. The other blocks are added as subforms.

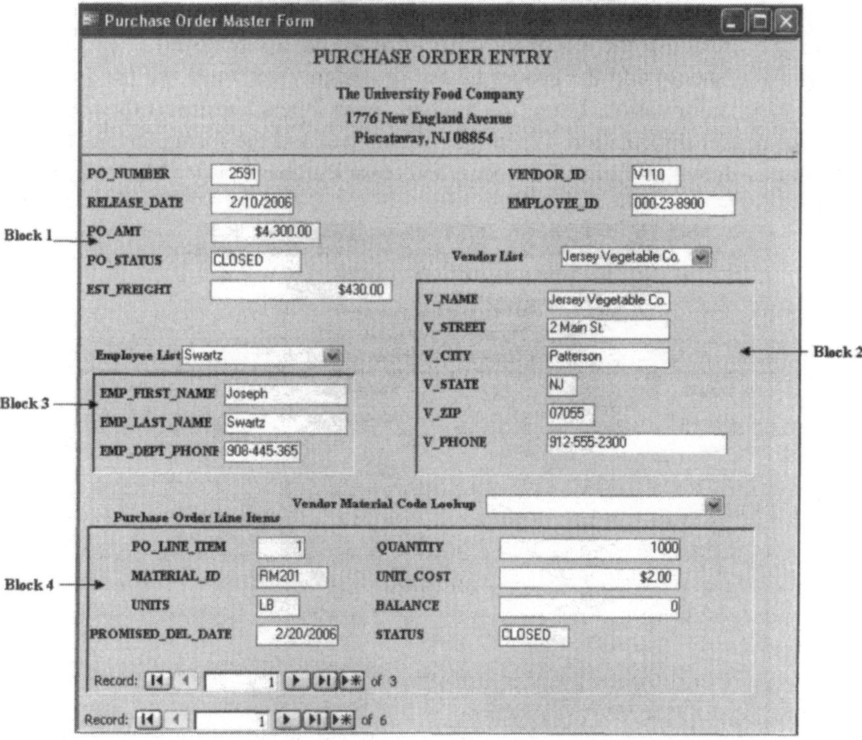

Figure 6.24 Purchase order form for data entry in form view.

This will be explained in the Access exercises that follow. The final objective is a form that looks like that shown in Figure 6.24. In Figure 6.24, some additional elements have been added as combo boxes to enable the user to quickly reference vendor, employee, and material information. They will be explained in due course.

> ***Access Exercise 6.14:***
>
> We want to establish the main form based on the ***PURCHASE_ORDER table***.
>
> 1. Open the database **Material_Manager_CH6_stu** and left click on the "Forms" tab.
> 2. Select "***Create form by using wizard***" and, from the pull down menu, choose the table that will serve as the master table. This is the PURCHASE_ORDER table.
> 3. Choose all the fields. Click on Next.
> 4. Choose Columnar Layout. Click on Next.
> 5. Choose Standard style. Click on Next.
> 6. Use the title: Purchase Order Master Form. Click on Finish.

Open Purchase Order Master Form in design view, enlarge the form, and rearrange the text boxes and labels so that they appear as shown in Figure 6.25. Put the form in form view and note the existence of the "record selector" arrow on the left-hand side, which does not appear in Figure 6.25, and the scroll bar on the right-hand side that does not appear in Figure 6.25. These are unnecessary in our form design. They can be removed by first putting the form in design view, selecting the form (click on the upper left hand box), then opening the properties window and selecting "No" for the Format property "Record Selectors" and "Neither" for the Format property "Scroll Bars." Return to the form view to confirm that the record selector and scroll bar are removed.

At this point you should add the header labels. In design view, open the header area to make room for the header information. Using the toolbar, select labels and insert them into the header area, adding the header information. When this is complete and the form is opened in form view, it should appear as shown in Figure 6.25. Save and close Purchase Order Master Form.

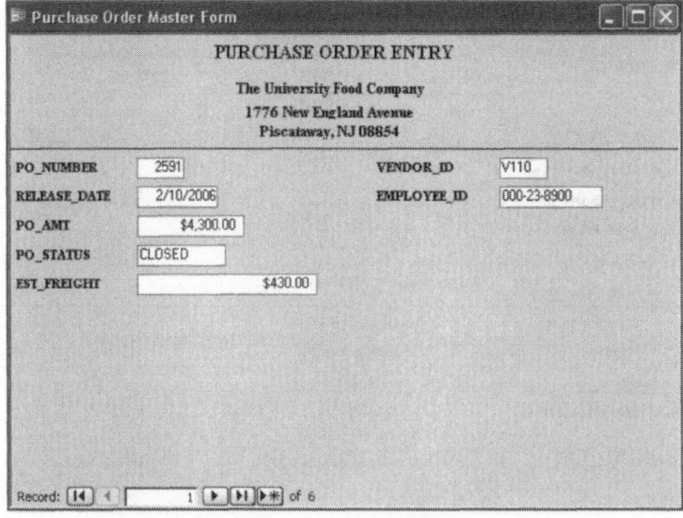

Figure 6.25 Main form for purchase order data entry.

6.5.4.2 Designing Subforms

As previously stated, the VENDOR table and EMPLOYEE table information are for display only. Their corresponding blocks are created as subforms and linked to the main form using the appropriate master and child fields. The Vendor Subform will be created in the next Access Exercise.

Access Exercise 6.15:

*We want to establish the **VENDOR subform**.*

1. Open the database **Material_Manager_CH6_stu** and left click on the "Forms" tab.
2. Select "***Create form by using wizard***" and, from the pull down menu, choose the table that will serve as the form table. This is the VENDOR table.
3. Choose the fields shown in Figure 6.24: V_NAME, V_STREET, V_CITY, V_STATE, V-ZIP, V_PHONE. Click on Next.
4. Choose Columnar Layout. Click on Next.
5. Choose Standard style. Click on Next.
6. Use the title: Vendor Subform. Click on Finish.

Open the form in form view in order to confirm that it is functional. Leave the Vendor Subform in the default layout. Note the existence of the record selector on the left, the navigation button on the bottom, and the dividing line at the bottom of the display area. As shown in Figure 6.24, these do not appear in the Vendor block. Put the Vendor Subform in design view and use the appropriate Format properties to remove these elements. When this is completed, the Vendor Subform should appear as shown in Figure 6.26 when placed in form view.

Now the Vendor Subform will be bound to the Purchase Order Master Form to make a master/detail form. Close Vendor Subform and open Purchase Order Master Form. There are two steps involved in binding the subform to the main form. First, the subform is placed on the main form. Next the referential integrity is established between the two forms so that their records are coordinated. This is all explained in Access Exercise 6.16.

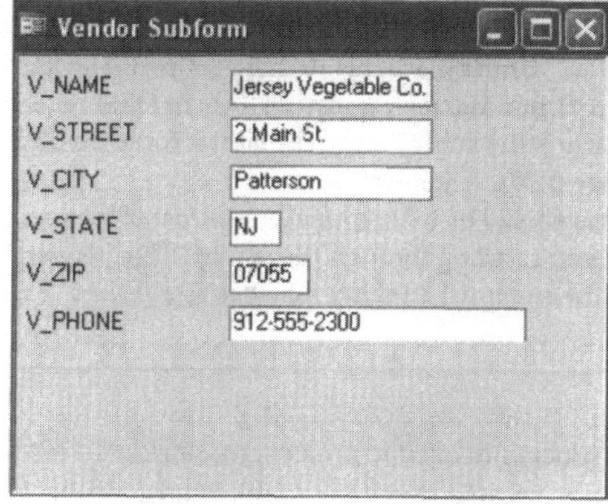

Figure 6.26 Vendor Subform in form view.

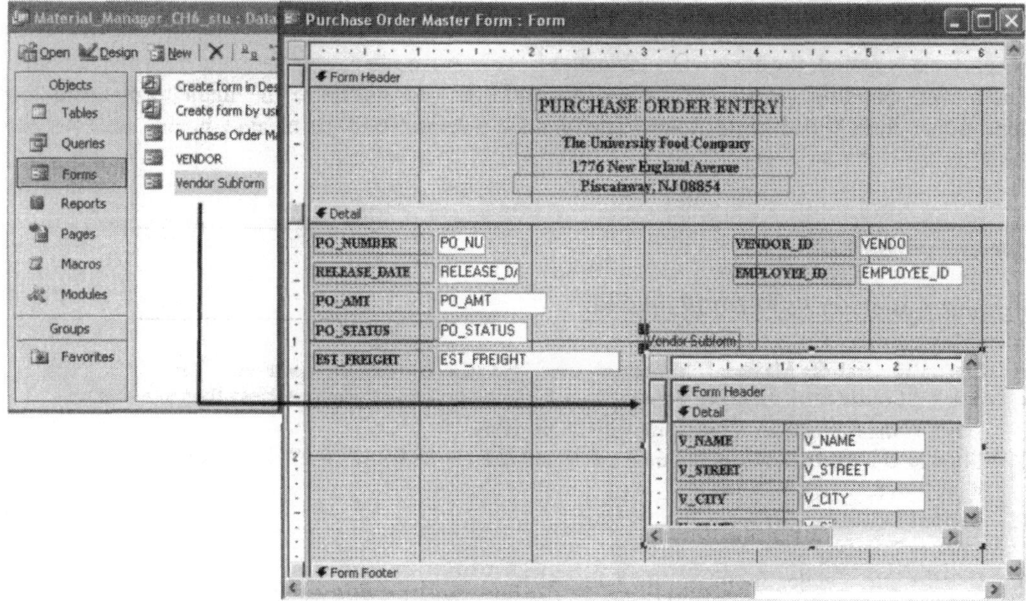

Figure 6.27 Binding the subform to the main form.

Access Exercise 6.16:

We want to bind the master/detail form.

1. With the Vendor Subform closed, open the Purchase Order Master Form in Design View. Enlarge the form to make some room for the Vendor Subform as shown in Figure 6.27.
2. Move the Purchase Order form away from the **Material_Manager_CH6_stu** database window so that the Vendor Subform name is visible.
3. Left click the mouse button on the Vendor Subform name and hold the button down. Drag the cursor on to Purchase Order Master Form and place it on the Detail section where you would like to place the subform. Release the mouse button and the subform will appear on the main form. Position it below the main form information (**see Figure 6.27**).
4. Delete the Vendor Subform label.
5. Relational integrity must now be established. Select the Vendor Subform by clicking on it. Then open the Properties window. Under the Data tab there are two relevant properties of the Subform: **"Link Child Fields"** and **"Link Master Fields"**. These fields must contain the names of the foreign key attributes that link the subform and the master form. In both fields enter the following: **VENDOR_ID (see Figure 6.28)**. Close the window.
6. At this point the two forms should be a functional master/detail form. Check the functionality by going to the Forms view and scrolling through the records. The records should be coordinated with each other through the relational integrity that was established. Figure 6.29 shows the form in form view at this point.

Let us briefly review the procedure used to create a master/detail form. A master form is created first. Separately, the subform is created based on a table whose primary key is a foreign key of the table on which the master form was based. Therefore, in Figure 6.29, each record of Purchase Order Master Form is related to a unique record in Vendor Subform. The records are coordinated with each other by binding the forms together based on the appropriate Foreign key. This is done using the "Link Child Fields" and "Link Master Fields" data properties of the subform.

Using the same procedure as Access Exercise 6.15, implement the EMPLOYEE subform. Select the fields: EMP_FIRST_NAME, EMP_LAST_NAME, and EMP_DEPT_PHONE. Name the form

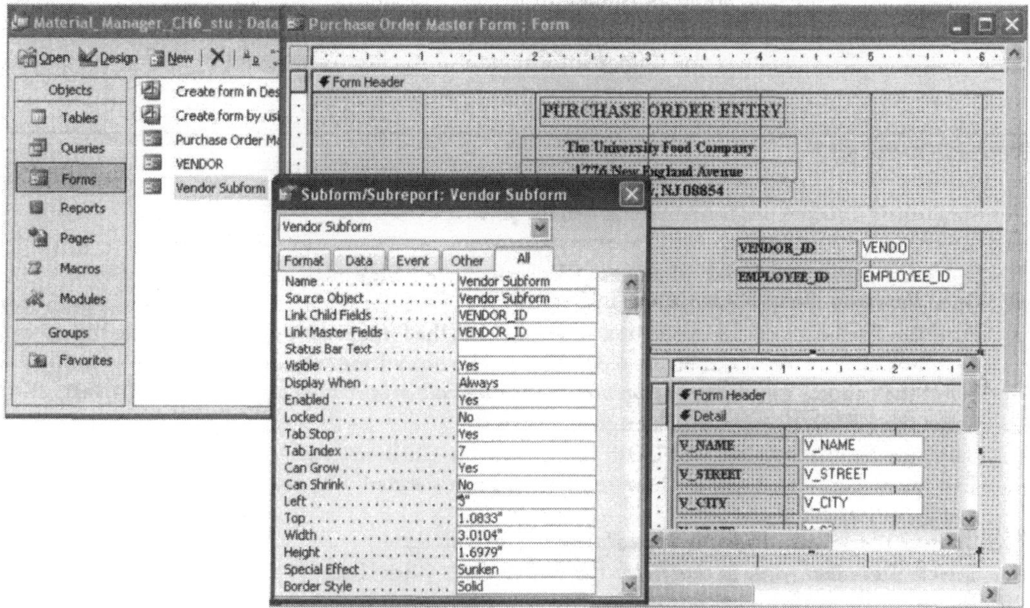

Figure 6.28 Linking child and master fields.

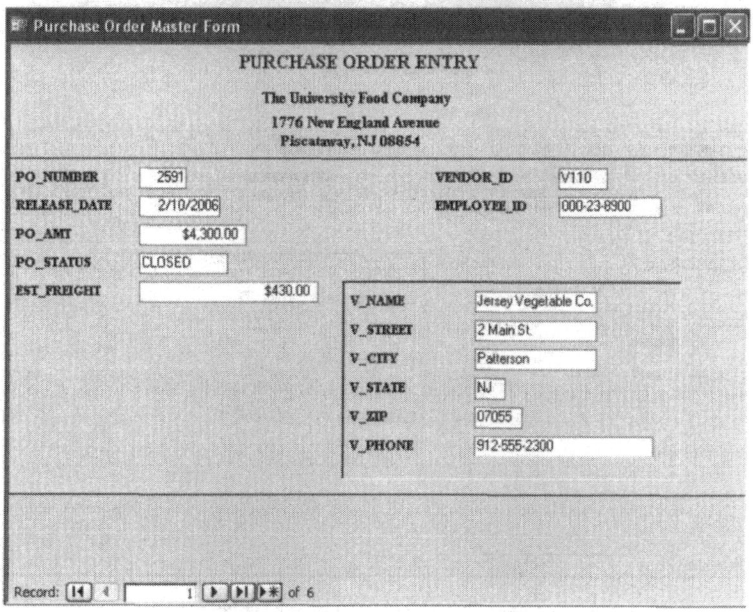

Purchase Order Master Form
PURCHASE ORDER ENTRY
The University Food Company
1776 New England Avenue
Piscataway, NJ 08854
PO_NUMBER 2591
RELEASE_DATE 2/10/2006
PO_AMT $4,300.00
PO_STATUS CLOSED
EST_FREIGHT $430.00
VENDOR_ID V110
EMPLOYEE_ID 000-23-8900
V_NAME Jersey Vegetable Co.
V_STREET 2 Main St.
V_CITY Patterson
V_STATE NJ
V_ZIP 07055
V_PHONE 912-555-2300
Record: 1 of 6

Figure 6.29 Main form and subform in form view.

"Employee Subform." Bind it to Purchase Order Master Form using the procedure of Access Exercise 6.16. Use EMPLOYEE_ID to link the subform to the main form. Eliminate the record selector, navigation button, and dividing line. When this is completed, the Purchase Order Master Form, when placed in form view, should appear as shown in Figure 6.30.

The purpose for which a form is to be used should inform the designer about additional features to add to assist the user. It is natural to imagine that the user of Purchase Order Master Form for data entry would appreciate some assistance in locating vendor and employee IDs from existing information in the database. For this reason, combo boxes are added as shown in Figure 6.24. The reader will add the combo box for vendor information in Access Exercise 6.17.

Access Exercise 6.17:

We want to add a combo box for retrieving vendor ID's.

1. Open the database **Material_Manager_CH6_stu** and left click on the "Forms" tab.
2. Open ***Purchase Order Master Form*** in Design View.
3. From the Toolbox, select a combo box and place it on the form in the position shown for the "Vendor List" combo box in Figure 6.31. The combo box wizard window will appear.
4. Select the option: "***I want the combo box to look up values in a table or query***." Click Next.
5. Select the ***VENDOR table***. Click on Next.
6. Select the following fields: VENDOR_ID, V_NAME. Click on Next.
7. The combo box wizard will ask how you want the fields sorted. Choose the field V_NAME and sort in the Ascending order.
8. Choose the default "***Hide the key column***". Click on Next.
9. Select "***Store that value in this field: VENDOR_ID***". Click Next.
10. Name the combo box: ***Vendor List***.

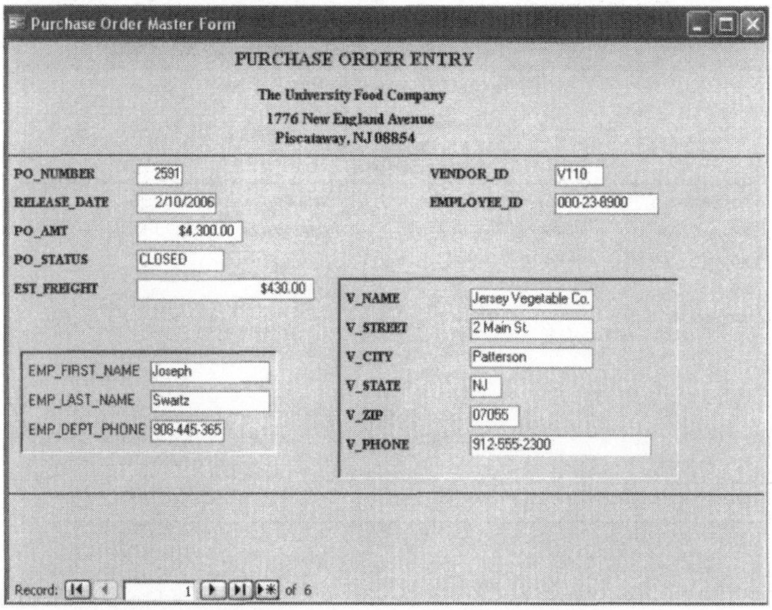

Figure 6.30 Purchase Order Master Form with subform blocks 2 and 3.

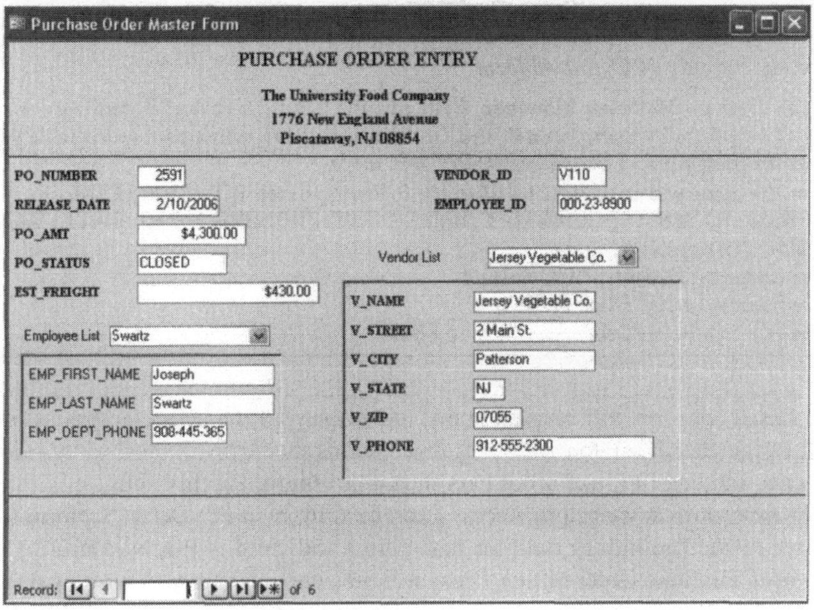

Figure 6.31 New purchase order form with combo boxes.

Repeat the steps shown in Access Exercise 6.17 for the Employee List combo box. Choose the fields EMPLOYEE_ID and EMP_LAST_NAME. When this is complete, Purchase Order Master Form, when opened in form view, should appear as shown in Figure 6.31.

At this point, Purchase Order Master Form can be used for data entry into the PURCHASE_ORDER table. The form does not yet accept the corresponding data for PO_DETAIL. To experiment with the partially completed form, the reader should put Purchase Order Master Form into the form view. Move to a new record using the navigation button at the bottom of the form. Add the following data in this order:

PO_NUMBER	9999
RELEASE_DATE	8/1/06
PO_AMT	0
PO_STATUS	OPEN

From the combo boxes, select:

Employee list	Swartz
Vendor list	Spices Unlimited

Note that the combo boxes have eased the completion of the form by adding VENDOR_ID and EMPLOYEE_ID in block 1 and displaying detail information for blocks 2 and 3 when the user chooses the vendor name and employee name from the combo boxes. Referential integrity enforces this through the "relationships" model established in Access Exercise 6.13. On closing the form, the data is saved. Once the form is closed, the reader should open the PURCHASE_ORDER table and confirm that the new record has been added.

The purchase order detail block should now be created and added to Purchase Order Master Form.

Access Exercise 6.18:

We want to establish the ***PO Detail subform***.

1. Open the database **Material_Manager_CH6_stu** and left click on the "Forms" tab.
2. Select "Create form by using wizard" and, from the pull down menu, choose the table that will serve as the form table. This is the ***PO_DETAIL table***.
3. Choose the fields shown in Figure 6.24 in the following order: PO_LINE_ITEM, MATERIAL_ID, UNITS, QUANTITY, UNIT_COST, PROMISED_DEL_DATE, BALANCE, STATUS. Click on Next.
4. Choose Columnar Layout. Click on Next.
5. Choose Standard style. Click on Next.
6. Use the title: PO Detail Subform. Click on Finish.

The PO Detail Subform will be used to add and display all the detail records associated with a particular PO number. Therefore, the navigation buttons should be retained on this subform. The record selector, dividing line, and scroll bars should be removed at this time.

Using the procedure described in Access Exercise 6.16, bind PO Detail Subform to Purchase Order Master Form. The linking field for both parent and child is PO_NUMBER. Once this is completed, open Purchase Order Master Form in form view to confirm its functionality. Note that coordination exists among records of the master form and subforms. Change the label of PO_Detail Subform to "Purchase Order Line Items," as shown in Figure 6.24.

The last remaining item on Figure 6.24 is the combo box titled "Vendor Material Code Lookup." Again, one has to imagine how the form will be used. The process of ordering materials from vendors in most firms begins with a requisition. A *requisition* is a request sent from a department of the enterprise to the purchasing department to buy a certain amount of a material on its behalf. From the requisition, the purchasing agent completes a purchase order for the materials. The material requisition will include the material description and may include the material ID. The purchasing agent will have to correlate that information with the vendor list to see which vendors can provide the material. As we previously pointed out, the vendor item code, not the enterprise Material ID, will appear on the purchase order sent to the vendor. Therefore, a Material List combo box will be used to ensure that the enterprise material ID has a corresponding vendor item code for the vendor to whom the purchase order will be sent.

Access Exercise 6.19:

We want to add a ***combo box*** for retrieving material ID and vendor material code by vendor.

1. Open the database **Material_Manager_CH6_stu** and left click on the "Forms" tab.
2. Open Purchase Order Master Form in design view.
3. From the toolbox, select a combo box and place it on the form in the position shown for the ***"Vendor Material Code Lookup"*** combo box in Figure 6.24. The combo box wizard window will appear.
4. Select the option: ***"I want the combo box to look up values in a table or query"***. Click Next.
5. Select the table: VENDOR_MATL_XREF. Click on Next.
6. Select the following fields: VENDOR_ID, MATERIAL_ID, V_ITEM_CODE. Click on Next.
7. Adjust column widths so that all attributes are visible. Click Next.
8. Choose the fields VENDOR_ID, MATERIAL_ID ascending. Click on Next.
9. Name the combo box: Vendor Material Code Lookup. Click Finish.
10. Put Purchase Order Master Form in Form View and observe that you can verify that a VENDOR_ID is able to provide a particular MATERIAL_ID and that there exists an associated V_ITEM_CODE. This is illustrated in Figure 6.32.

The data entry form for purchase orders is now complete, and it should appear as shown in Figure 6.24 when opened in form view. The form can be tested by using the navigation button to move to the last record, PO_NUMBER 9999. Make the following entries in the detail section:

	1st Record	2nd Record
PO_LINE_ITEM	1	2
MATERIAL_ID	RM305	RM311
UNITS	LB	LB
QUANTITY	1000	500
UNIT_COST	0.5	0.25
PROMISED_DEL_DATE	8/20/06	8/20/06
BALANCE	1000	500
STATUS	OPEN	OPEN

Using the Vendor Material Code Lookup combo box, confirm that RM305 and RM311 are sold by vendor V250, the vendor selected in block 2. If RM305 or RM311 are not provided by vendor V250, it will not be possible to retrieve a vendor product code when it is time to print the purchase order and send it to the vendor. After closing Purchase Order Master Form, open PO_DETAIL table and confirm that the records have been entered.

In this section, we discussed some aspects of designing a form for data entry. The use of blocks as an organizing principle was emphasized. There are many other enhancements that can be added to ease the use of the form. The reader will be asked to make further enhancements in the problem set at the end of the chapter.

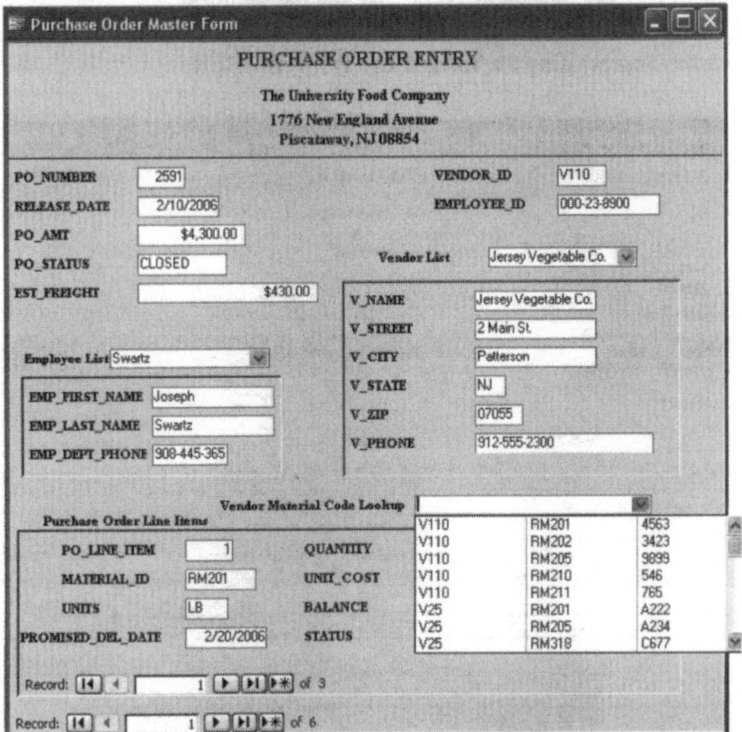

Figure 6.32 Using a combo box to confirm vendor codes.

6.5.5 Designing a Subform Based on a Query

As previously discussed, the design of a form should reflect the purpose for which the form is to be used. The design of a master/detail form for data entry based on the principle of blocks was illustrated in Section 6.5.4. This is a safe way to proceed when the form is used for data entry and each block is associated with one table. However, one can imagine a form designed for output or "display only" for which this principle may not apply. Such is the case for a printed purchase order form of the type shown in Figure 5.7. Here the PO Detail Subform shows information from more than one table. For example, the material description is in the MATERIAL table and the vendor item code is the the table VENDOR_MATL_XREF. Thus, the PO Detail Subform block will have to retrieve data from more than one table. When encountering a situation such as this, it is often better to base the form on a query.

6.5.5.1 Establishing the Query

Figure 6.33 shows a form proposed for the purpose of printing a purchase order to send to a vendor. Note how it differs from Figure 6.24. Since the data on which the printable purchase order is based have already been entered into the database, there is no need to have combo boxes. They have been deleted. There is also no need to have visible text boxes, such as vendor ID and employee ID, information that would not appear on a printed form for ordering material. Also, instead of a columnar form layout, which shows only one PO line item at a time, the printable purchase order shows all the line items. This layout is called a datasheet layout.

The datasheet subform in Figure 6.33 will be referred to as "Printable PO_Detail Subform." The data items that come from tables other than PO_DETAIL are the material description and the vendor item code. Therefore, the SQL command for the subform must retrieve data from PO_DETAIL, MATERIAL, and VENDOR_MATL_XREF tables. In addition the records must be coordinated with the master using the foreign key PO_NUMBER.

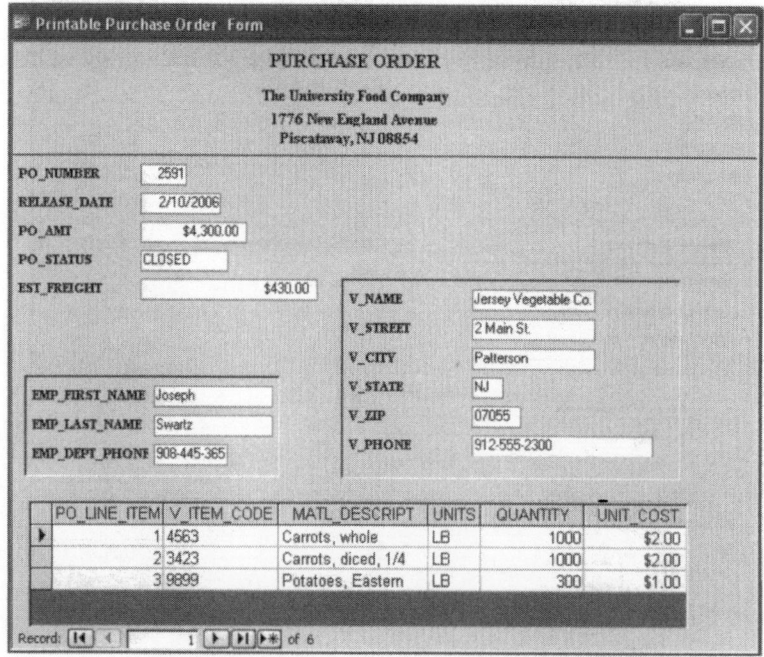

Printable Purchase Order Form

PURCHASE ORDER

The University Food Company
1776 New England Avenue
Piscataway, NJ 08854

PO_NUMBER 2591
RELEASE_DATE 2/10/2006
PO_AMT $4,300.00
PO_STATUS CLOSED
EST_FREIGHT $430.00

V_NAME Jersey Vegetable Co.
V_STREET 2 Main St.
V_CITY Patterson
V_STATE NJ
V_ZIP 07055
V_PHONE 912-555-2300

EMP_FIRST_NAME Joseph
EMP_LAST_NAME Swartz
EMP_DEPT_PHONE 908-445-365

PO_LINE_ITEM	V_ITEM_CODE	MATL_DESCRIPT	UNITS	QUANTITY	UNIT_COST
1	4563	Carrots, whole	LB	1000	$2.00
2	3423	Carrots, diced, 1/4	LB	1000	$2.00
3	9899	Potatoes, Eastern	LB	300	$1.00

Record: 1 of 6

Figure 6.33 Purchase Order Form for printing.

Figure 6.34 shows an appropriate SQL statement for retrieving the recordset. Note the conditions are enforced in the WHERE clause to ensure that the records are coordinated. At this point, the reader should create the query in Access. Name it PO_DETAILQuery.

6.5.5.2 Creating the Subform from the Query

Access Exercise 6.20 shows the steps required to create Printable PO_Detail Subform from the PO_DetailQuery.

Access Exercise 6.20:

We want to design a subform based on the Query: ***PO_DetailQuery***.

1. Open the database **Material_Manager_CH6_stu** and left click on the "Forms" tab.
2. Select ***"Create form by using the Wizard."***
3. From the Form Wizard window select ***"Query: PO_DetailQuery"*** from the Tables/Queries pull down menu.
4. Under ***"Available Fields"*** select only those fields that will appear on the subform and in the order they will appear as follows: PO_LINE_ITEM, V_ITEM_CODE, MATL_DESCRIPT, UNITS, QUANTITY, UNIT_COST. Click Next.
5. Choose Datasheet Layout. Click on Next.
6. Choose Standard Style. Click on Next.
7. Name the Form: Printable PO_Detail Subform. Click Finish.
8. On putting the form into Form View, it should appear as Figure 6.35.

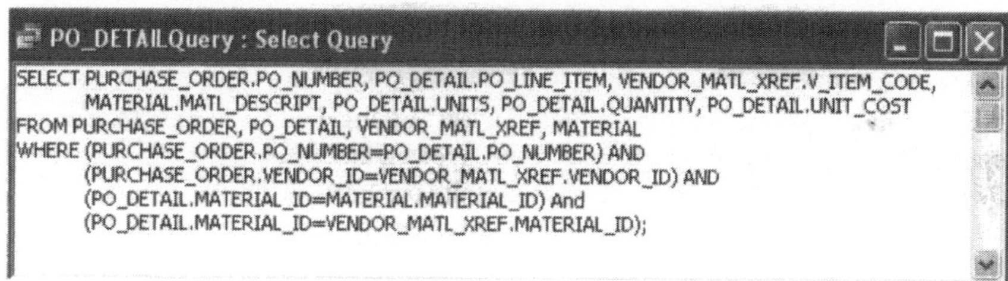

PO_DETAILQuery : Select Query

```
SELECT PURCHASE_ORDER.PO_NUMBER, PO_DETAIL.PO_LINE_ITEM, VENDOR_MATL_XREF.V_ITEM_CODE,
        MATERIAL.MATL_DESCRIPT, PO_DETAIL.UNITS, PO_DETAIL.QUANTITY, PO_DETAIL.UNIT_COST
FROM PURCHASE_ORDER, PO_DETAIL, VENDOR_MATL_XREF, MATERIAL
WHERE (PURCHASE_ORDER.PO_NUMBER=PO_DETAIL.PO_NUMBER) AND
       (PURCHASE_ORDER.VENDOR_ID=VENDOR_MATL_XREF.VENDOR_ID) AND
       (PO_DETAIL.MATERIAL_ID=MATERIAL.MATERIAL_ID) And
       (PO_DETAIL.MATERIAL_ID=VENDOR_MATL_XREF.MATERIAL_ID);
```

Figure 6.34 SQL code for PO_DETAILQuery.

Printable PO_Detail Subform

PO_LINE_ITEM	V_ITEM_CODE	MATL_DESCRIPT	UNITS	QUANTITY	UNIT_COST
1	4563	Carrots, whole	LB	1000	$2.00
2	3423	Carrots, diced, 1/4	LB	1000	$2.00
3	9899	Potatoes, Eastern	LB	300	$1.00
1	F444	Olive oil	GAL	800.5	$0.50
2	F456	Vinegar, white	GAL	210.5	$0.50
1	546	Peas, shelled	LB	1000	$2.00
2	765	Tomatoes, whole	LB	2000	$1.00
1	2-414	Garlic, whole	LB	4000	$0.50
2	2-564	Garlic powder	LB	2000	$0.25
3	3-212	Salt, iodized	LB	2000	$0.25
4	3-675	Sugar, bulk	LB	560	$0.50
1	2-312	Paprika	LB	400	$0.50
2	2-897	Onion Salt	LB	1200	$0.25
1	666	Sugar, brown	LB	5000	$0.20

Figure 6.35 Printable PO_Detail Subform in form view.

Note that Figure 6.35 lists all the appropriate records of the database with coordination among tables. Once this subform is appended to the main form and bound to the main form using the foreign key PO_NUMBER, the records displayed on the subform will be limited to those associated with the purchase order number on the master form.

6.5.5.3 Binding the Subform to the Master Form

If the reader is following along with the database, close Printable PO_Detail Subform. Open Purchase Order Master Form in design view. Delete the PO_Detail Subform that is currently on the master form. Also, delete the three combo boxes as they are not needed. Hide the text boxes for vendor ID and employee ID by changing their Visible properties to "No." Change the Header label to "Purchase Order" as shown in Figure 6.33. Save the resulting form under the name "Printable Purchase Order Form."

With Printable Purchase Order Form in design view and following the steps of Access Exercise 6.16, drag Printable PO_Detail Subform onto the master form and bind it by linking the child and master with the foreign key attribute PO_NUMBER. Place the form in form view and arrange fields until it looks like Figure 6.33.

6.5.5.4 Adding Derived Attributes to a Form

A derived attribute is one that is computed from other attributes. There are no derived attributes in Figure 6.33. However, on an actual purchase order there is usually a "Total Cost" column in the detail section that is computed as QUANTITY times UNIT_COST. Access provides user-friendly facilities for adding derived attributes. This is illustrated in Access Exercise 6.21.

Access Exercise 6.21:

We want to add the attribute "Total Cost" to ***Printable PO_Detail Subform***.

1. Open Printable PO_Detail Subform in Design View. Changes made on this subform will automatically be reflected in the subform as displayed on the master form.
2. Make room in the Detail section for another attribute text box and label.
3. Open the Toolbox. Take a text box from the toolbox and position it under the UNIT_COST text box. A label should also appear next to the text box.
4. Change the caption of the label to "Total Cost".
5. Place the cursor in the text box and type "=[QUANTITY]*[UNIT_COST]". This equation defines the value of this field and the computation is automatic each time the quantity and unit cost fields change.
6. Open the Properties window for the text box and change the Format property to "Currency". Change the Text Align property to "Right".
7. Save the form and confirm the results by opening Printable PO_Detail Subform in Form View. Adjust the column widths as required. Save and close the subform.
8. Open Printable Purchase Order Form in Form View. Note that the changes made to Printable PO_Detail Subform now appear on the subform section of the main form. Go to the design view and adjust the boarders of the form until it appears as shown in Figure 6.36.

To print the form shown in Figure 6.36, the user should go to the form view and select File ⇒ Print from the main menu. Choose "Selected Records" and Landscape layout from Properties. Click "OK." The printed document will appear as Figure 6.36, but without the navigation button.

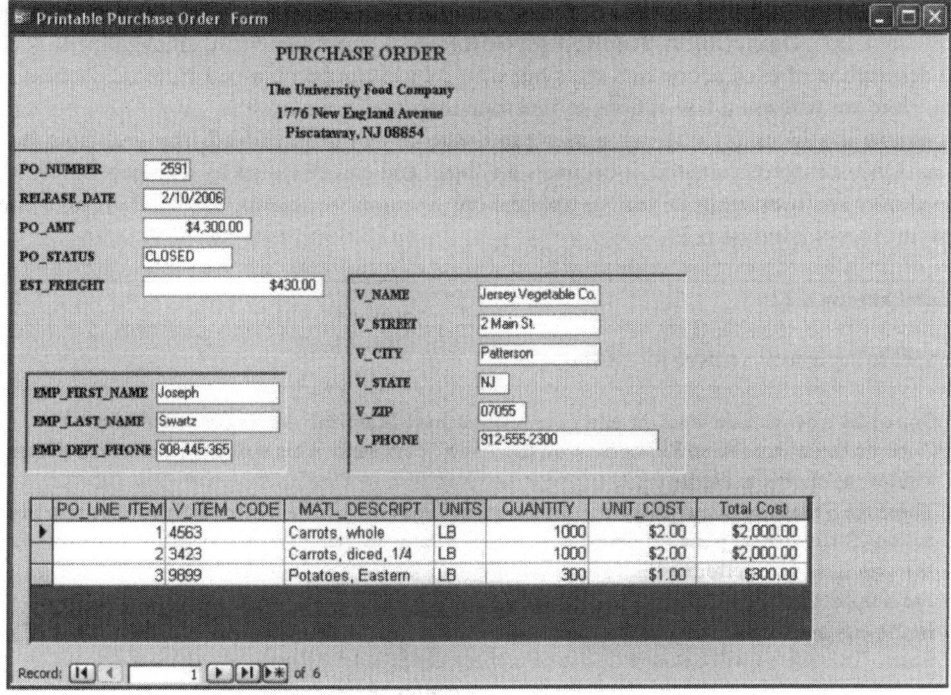

Figure 6.36 Printable Purchase Order Form with total cost column.

6.5.6 Summary of a Master/Detail Form

It was the purpose of this section to introduce the reader to the design of a master/detail form. Creating forms based on tables and based on queries was illustrated. Several principles were involved. These included the requirement to establish relationships between tables that will be queried when using the form, and the linking of master and detail forms using the Link Master Fields and Link Child Fields properties. We also make the distinction between a form used to enter data as opposed to one used to view data. Data entry is most easily accomplished when the form is based on the principle of independent blocks for data entry.

6.6 SOME ADDITIONAL ACCESS TOOLS

Access has many capabilities that can be used to add functionality to a form without having to develop much code. Some of these capabilities will be briefly discussed in this section. This will include macros and actions, unbounded text boxes, and form navigation tools.

6.6.1 Macros and Actions

If a user wishes to have functionality in a form beyond just displaying data, it is necessary to have code behind a command button to make that functionality occur. Access has some predefined VBA code that can be used directly in an application by just selecting it. We observed some of these functions in Section 6.4.2 when preprogrammed command buttons for the insert, delete, and save operations were used on the Vendor form.

Access 2003 provides 49 functions, called *actions*. Each action is a VBA subroutine. Typical actions are Close, DeleteObject, FindRecord, GoToRecord, and OpenForm, among others. A complete description of each action and the syntax for its use can be obtained from the Access Help menu. Here we will use a few actions to illustrate their use in a macro.

A ***macro*** is a block of code that a user can create or that is assembled from available Access actions. Once a macro is created, it becomes an object and can be called by its object name. It can be used over and over again in various applications to automate actions. We shall create a simple macro in Access Exercise 6.22.

Access Exercise 6.22:

We want to implement a ***macro*** for closing a form.

1. Go to MS Access main window and open the database **Material_Manager_CH6_stu**.
2. Click on the macro tab and then click on the "New" command. This will open the macro design window as shown in Figure 6.37.
3. There are 3 main sections on the screen shown in Figure 6.37: 1) Action pane for choosing the action, 2) Comments pane for documentation, 3) Argument pane for defining the object on which the action is to be performed.
4. From the action pane, choose the Close action.
5. In the arguments pane, choose the object type " form" and the object name Purchase Order Master Form. This macro will be used to close Purchase Order Master Form.
6. In the argument pane, choose "yes" for the save option. Any new contents of the form will be saved automatically when the form is closed.
7. As the reader will see in the title bar, the macro is automatically assigned a default name. The name can be changed to reflect the application and make it easier to remember when the macro is saved. For this exercise, we will use the name "**Macro Close**".
8. Close the macro design window and save the macro under the name MacroClose.

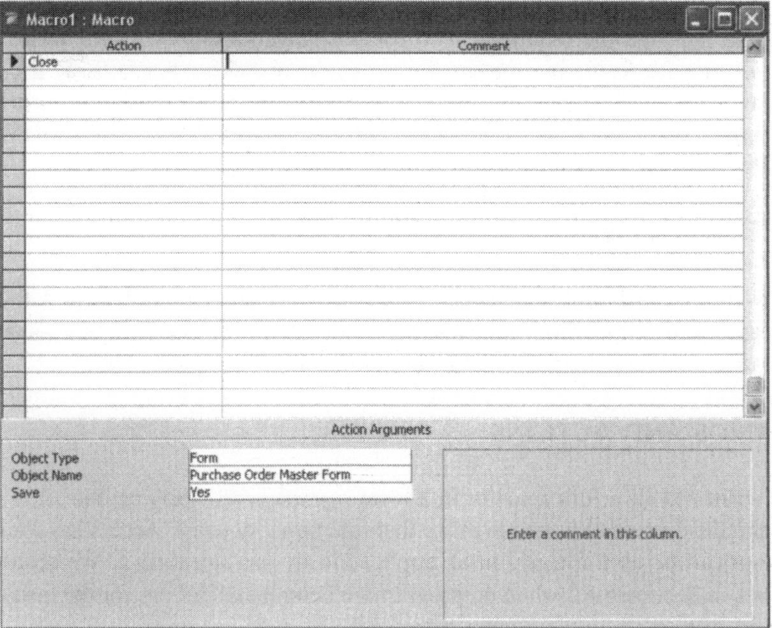

Figure 6.37 Macro design window.

Macros can be used at all levels of a form. For example, at the form level, a macro can be inserted into the On_Open or On_Load event property. When the form is opened or loaded, the macro will automatically run. At the form element level, it can be inserted into one of the event properties of an object, such as the On_Click property of a command button. To illustrate, we shall add a command button to close Purchase Order Master Form using MacroClose in Access Exercise 6.23.

Access Exercise 6.23:

We want to add the MacroClose button to ***Purchase Order Master Form***.

1. Go to MS Access Main Window and open the database **Material_Manager_CH6_stu.**
2. Click on the forms tab and then open Purchase Order Master Form in design view.
3. Select a command button from the design tools and place it on the form in roughly the position shown in Figure 6.38. The command button wizard will appear.
4. Choose the category "Miscellaneous" and choose the action "**Run Macro**." Click Next.
5. When asked to select the macro to run, choose "MacroClose." Click Next.
6. Choose to display text on the button and display the word "Close". Click Next.
7. Name the button "**Close_button**." Click Finished.
8. At this point Purchase OrderMaster Form, when in form view, should look like Figure 6.38.

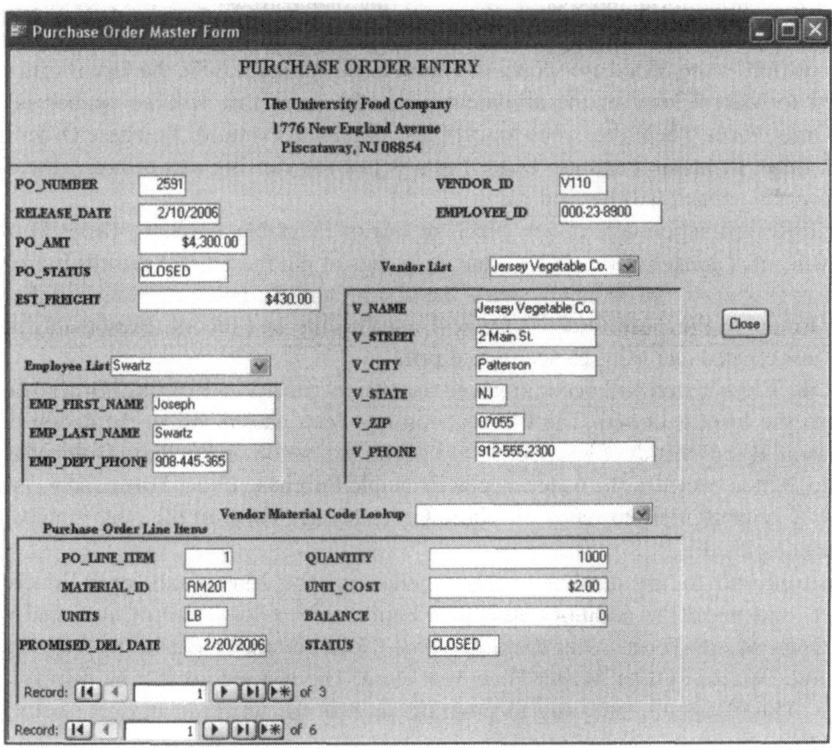

Figure 6.38 Purchase Order Form with close button.

Test the functionality of the button by putting the form into form view. Click on the Close command button and note that the form closes. Reopen the form, and put it in design view. Click on the Close command button and open the properties window. From the event properties, choose the On_Click property and open the event procedure. Here you see the VBA code that Access generated automatically when you created the command button. Note the following components:

```
'A string variable is defined called stDocName.
  Dim stDocName As String

'The macro object name is assigned to the string variable.
  stDocName = "MacroClose"

'The action "RunMacro" is commanded for the macro assigned to the string
variable.
  DoCmd.RunMacro stDocName
```

The "DoCmd" statement is a VBA command that tells Access to run the action that follows on the argument given by the string variable. In general, the syntax is DoCmd.ActionName [argument].

The use of a macro is a convenient way to execute a sequence of actions. In the preceding example, a single action, Close, was used. It is useful to illustrate a macro with more than one action step explicit in the macro design. One can imagine the user of Purchase Order Master Form entering a new record for a purchase. Once the record is entered, the next logical step would be to review the format of the PO to be sent to the vendor (i.e., the Printable Purchase Order Form). The user can then print the form and send it. Based on this scenario, we will add two more actions to MacroClose.

The two actions to be added are shown in Figure 6.39. In Figure 6.39, the OpenForm action has been added to MacroClose. In the argument pane, the form that will be opened is Printable Purchase Order Form. Therefore, when MacroClose executes, Printable Purchase Order Form will be opened. Since Printable Purchase Order Form is just for viewing and printing, the data mode when the form is opened will be read only.

The GoToRecord action determines which record of Printable Purchase Order Form will be displayed when it is opened. One can imagine thousands of purchase order records in the database. When a form is opened, the default record is the first record. By choosing "Last" in the "Record" argument, Printable Purchase Order Form will open to the last record. Presumably, this is the record we just created and want to review and print.

Finally, the Close action will occur in which the current record on Purchase Order Master Form is saved and the form is closed. The Close action has been placed last in the action order for a specific reason. Recall that the Close_button is a command button on Purchase Order Master Form. Since Close_button executes the code to open Printable Purchase Order Form and to move to the last record, it is appropriate to retain Purchase Order Master Form in the open state while these actions are completed.

At this time, go to the database window, click on the Macro tab, open MacroClose in design view and make the additions shown in Figure 6.39. Close MacroClose, and then open Purchase Order Master Form. After clicking on the Close button, Printable Purchase Order Form will open and Purchase Order Master Form will close. The last record will be displayed in read-only mode. This illustrates one way to program so that the user can navigate between forms automatically.

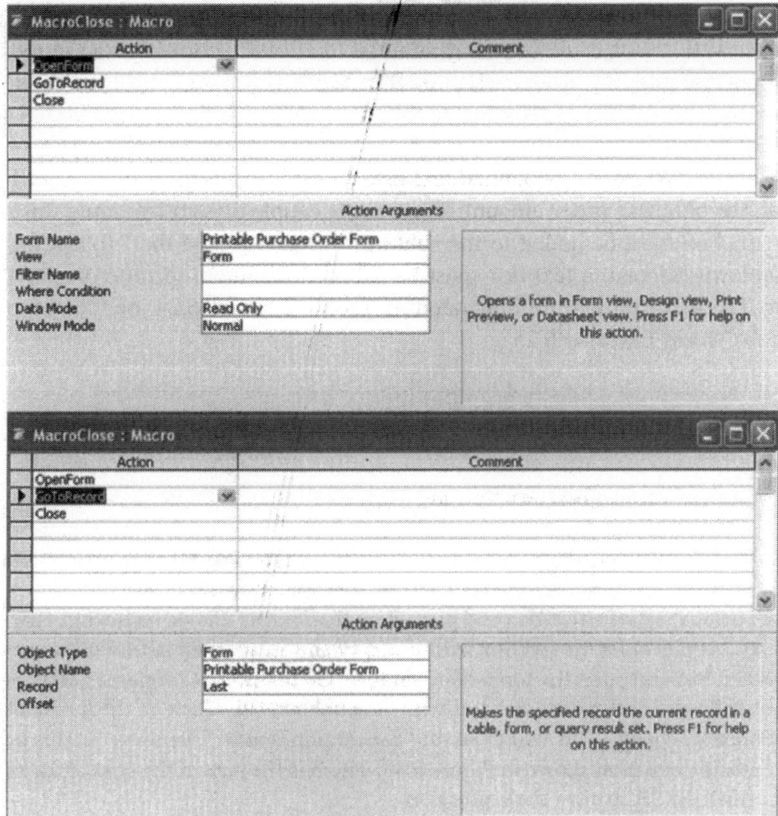

Figure 6.39 MacroClose with two additional actions.

6.6.2 UNBOUNDED TEXT BOXES

Thus far, when text boxes have been used on a form, they have been a kind of "window" to an attribute value in the table(s) on which the form is based. When the user enters a value into the text box and saves the form, the entered value is saved as a record in the table to which that text box applies. This is because the text box is "bound" to a particular "data source" in the database at the time it is created.

Text boxes can be used more generally to display any information on a form, even when it is not related to a specific table. This is done by using an "unbounded" text box. An ***unbounded text box*** is not bound to a particular data source when it is created. Programming it to display or handle a particular entry can control its contents. In this section, we illustrate some uses for unbounded text boxes that will enhance the readers' ability to create interesting Access applications.

For example, consider Purchase Order Master Form shown in Figure 6.38. The detail, or subform, lists the items to be ordered and their unit costs. In Figure 6.38, the total value is $4300. Note that this is the total amount on the master, or main form. To enter a new purchase order, the user enters each line item on the form detail and then must compute the total cost and enter that amount in the PO_AMT text box on the main form. This is necessary because the PO_AMT text box on the main form is bounded to the PO_AMT record in the PURCHASE_ORDER table and it must be entered explicitly by the user. This raises two observations. The first observation is that the user should not have to total up the cost on the detail, since this can be done automatically. The second observation is that the user may sum the total cost incorrectly and introduce an error into the table by entering the wrong value for PO_AMT. This is an excellent case for using an unbounded text box to automatically compute the PO_AMT from the data entered in the purchase order detail.

We will illustrate the use of unbounded text boxes for the purpose of automatically computing and displaying the purchase order amount. There are a couple of steps in doing this. First, a new (unbounded) text box must be added to the subform in order to sum the "Total Cost" column of PO_Detail Subform. Second, a text box must be added to the main form to display the purchase order amount. The first step will be illustrated in Access Exercise 6.24, and the second step will be illustrated in Access Exercise 6.25.

Access Exercise 6.24:

We want to add an ***unbounded text box*** for computing the total amount on the ***PO_Detail Subform***.

1. Go to MS Access Main Window and open the database **Material_Manager_CH6_stu.**
2. Click on the Forms tab and then open PO_Detail Subform in Design View.
3. From the Toolbox, select a text box and place it on the form as shown on the right hand side in Figure 6.40. At this point the text box will be empty and an arbitrary label will be assigned.
4. Select the text box and open the properties window. Go to the Data properties and note that there is no Control Source assigned to this text box. It is unbounded. Click on the three dots next to the Control Source property. This will open the "Expression Builder" window. At this point it will be empty. Enter the equation shown in Figure 6.40, which is the sum of the cost of the items on the subform. Click on OK to save the expression.
5. In the text box properties window, go to "Other" and change the ***Name*** property to ***"Total_Amount"***, a name that reflects its contents and purpose. Click on the Format tab and change the ***Format property*** to "***Currency***". Close the Properties window.
6. Put the form in the Forms View and note that a new column is added that shows the Total Amount. At this point it still has the default label. .
7. Close and save the changes to PO_Detail Subform

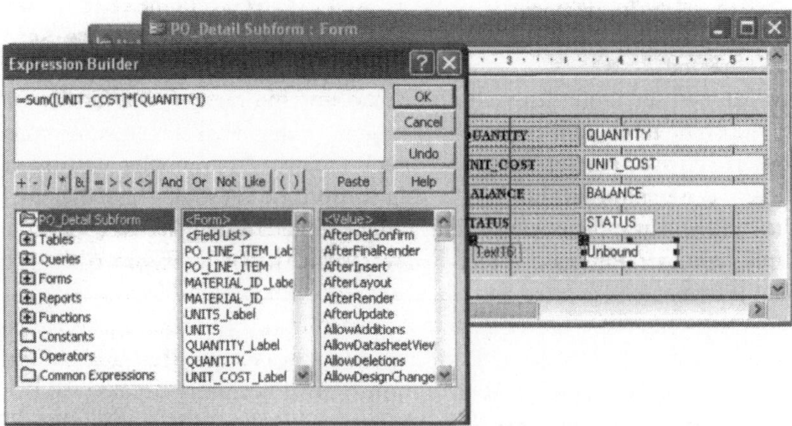

Figure 6.40 Adding an unbounded text box to a PO_Detail subform.

At this point, the automatic computation of the total amount on the purchase order detail has been implemented. Henceforth, there will be a running total of all the items that are added to the purchase order. However, it will not be reflected on the main form. To do this, an unbounded text box must be added to the main form, and it must be linked it to the Total_Amount text box on the subform.

Access allows an object on a main form to refer to an object on a subform. The syntax is as follows:

```
[Subform Name].Form![Subform Object Name]
```

The first element is the name of the subform on which the object is located. This is followed by the command "Form!" which indicates to Access that the argument that follows is the name of an object on the form. This is not to be confused with the command "Forms!" that we used earlier in which the argument was the name of a form. Finally, the name of the form object being referenced is given. So, for example, to refer to the Total_Amount text box on the PO_Detail Subform, the syntax is as follows:

```
[PO_Detail Subform].Form![Total_Amount].
```

This syntax will be implemented in Access Exercise 6.25.

Access Exercise 6.25:

We want to add an unbounded text box on the main form for displaying the Total Amount from the subform.

1. Go to MS Access Main Window and open the database **Material_Manager_CH6_stu.**
2. Click on the forms tab and then open Purchase Order Master Form in design view.
3. From the tools menu, select a text box and place it on the main form as shown on the left hand side in Figure 6.41. At this point the text box will be empty and an arbitrary label will be assigned, such as Text41 in Figure 6.41.
4. Select the text box and open the properties window. Go to the Data properties and note that there is no Control Source assigned to this text box. It is unbounded. Click on the three dots next to the Control Source property. This will open the "Expression Builder" window. At this point it will be empty. Enter the equation shown in Figure 6.41, which is the Total_Amount text box of the subform. Click on OK to save the expression.
5. In the text box properties window, go to "Other" and change the ***Name*** property to ***"Total_PO_Amount"***, a name that reflects its contents and purpose. Click on the Format tab and change the ***Format*** property to ***"Currency"***. Close the Properties window.
6. Put the form in the Forms View and note that a text box is added that shows the Total_PO_Amount. At this point it still has the default label.
7. You can experiment with the form by using the scroll bar to go to a new record. As you add line items to the purchase order, you will note that the total purchase order amount is displayed in the new text box.

At this point, two important uses of the unbounded text box have been illustrated. The first use is to compute derived data on a form. The second use is to retrieve data for the current form (Purchase Order Master Form) from another form (PO_Detail Subform). However, we are not quite finished. The PO_AMT text box on the main form, which is bounded to the PO_AMT field in the PURCHASE_ORDER table, still must be filled in by the user because this is the value that will

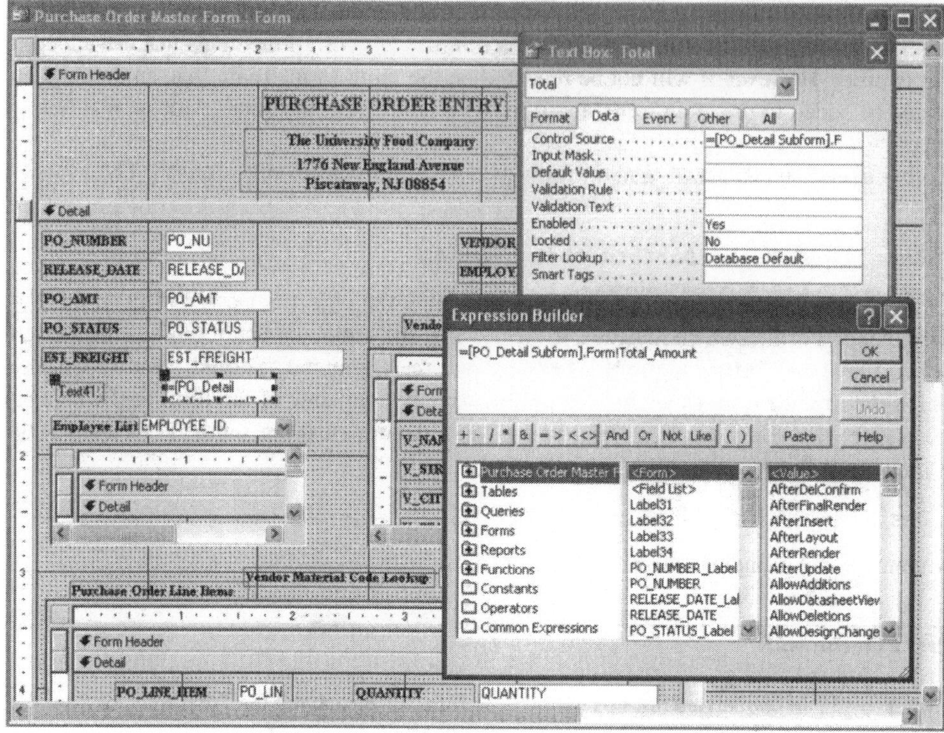

Figure 6.41 Adding an unbounded text box to the Purchase Order Master Form.

be saved in the database table. It would be preferable to have the already computed Total_PO_Amount automatically entered into the PO_AMT field. This can be done in two steps. First, the value in Total_PO_Amount can be copied into the PO_AMT text box. Then the contents of the form can be saved, which will insert the entire record into the table. This is the subject of Access Exercise 6.26.

Access Exercise 6.26:

We want to transfer the Total_PO_Amount from the unbounded textbox on the Purchase Order Master form to the bounded text box for PO_AMT when closing and saving New Purchase Order Form.

1. Go to MS Access Main Window and open the database **Material_Manager_CH6_stu.**
2. Click on the forms tab and then open Purchase Order Master Form in design view.
3. Click on the Close command button and open the properties window.
4. Select the Event property On_Click and open the On_Click Event Procedure.
5. Add the line of code: **Me.PO_AMT.Value=Me.Total_PO_Amount.Value,** as shown in Figure 6.42.
6. Close the window and save the database.

```
Private Sub Close_button_Click()
On Error GoTo Err_Close_button_Click

    Me.PO_AMT.Value=Me.Total_PO_Amount.Value

    Dim stDocName As String

    stDocName = "MacroClose"
    DoCmd.RunMacro stDocName

Exit_Close_button_Click:
    Exit Sub

Err_Close_button_Click:
    MsgBox Err.Description
    Resume Exit_Close_button_Click

End Sub
```

Figure 6.42 Code for the Close_button event procedure.

At this point, we have added one additional step when Purchase Order Master Form is closed. Before it is saved and closed, the value in the Total_PO_Amount text box is transferred to the PO_AMT text box. This transfer must occur while the form is open. Thus, the line of code must appear before MacroClose.

The form designation "Me." should be used whenever the action being performed is related to the current form. "Me." is the proper form level object designation when referring to the active form.

The reader should now experiment with the Purchase Order Master Form by entering new data. Note that the total amount is accumulated as the line items are entered. On closing the form, the total amount is transferred to the PO_AMT text box and saved into the database. Trial data sets are given in the appendix to this chapter so that the user can experiment with the form.

To improve the appearance of the form, the reader should make Total_Amount text box on the PO_Detail Subform invisible. The value in that text box now appears on the main form. Also, the PO_AMT text box on the main form can be rendered invisible since it is only there to receive the value in Total_PO_Amount text box when the form is closed.

6.7 IMPLEMENTING A REPORT

A ***report*** is designed to provide a summary view of data as a read-only document, most often on paper. Unlike a form, it is not used for updating or inserting information. Although there are different formats for creating a report, a typical report shows rows and columns of grouped data and has summary totals where relevant. In this section, we design a typical *tabular* report using the Access report writing facility.

Section 6.3.2 described three major steps for designing a report: (1) lay out the report, (2) define the interaction with tables, and (3) specify the derived data. In the following sections, we discuss these three issues and implement the resulting design.

Inventory Report

Material	Code	lot number	location	units	quantity	unit price	total value
		XXXXX	XXXXX	XXX	XXXXX	XXXXX	XXXXX
		XXXXX	XXXXX	XXX	XXXX	XXXXX	XXXXX
		XXXX	XXXXX	XXX	XXXXX	XXXXX	XXXXX
					Subtotal		XXXXXX
		XXXXX	XXXXX	XXX	XXXXX	XXXXX	XXXXX
		XXXXX	XXXXX	XXX	XXXX	XXXXX	XXXXX
		XXXX	XXXXX	XXX	XXXXX	XXXXX	XXXXX
					Subtotal		XXXXXX
				GRAND TOTAL			XXXXXXXX

Figure 6.43 Template for an inventory report.

6.7.1 Layout of the Report

The starting point is to have a good general idea of what you want to portray on the report and how it should be laid out. This is usually done by making a sketch of how the report will appear. Figure 6.43 is an example. Here we are interested in a report on the current inventory of the enterprise, and we want to view it by each material. The report is sketched out so that it shows the placement of the different fields. Recordsets are to be grouped by the material name and code (MATERIAL_ID). The group of records will be listed by lot number. Thus, it will be possible for the user to review the status of each material by its lot number, the quantity remaining of that lot number, and where it is physically located in storage. Figure 6.43 provides the template for designing the report.

6.7.2 Interaction with Tables

Each of the attributes of the inventory report will be queried from tables. In fact, the report is constructed by first assembling the data from tables into a recordset and then sorting the data into the report format. Therefore, the sources of the data for each attribute must be specified by table.

Table 6.1 specifies the source of each column of data. The column name is as it appears in the template shown in Figure 6.43. The source is the attribute and table name in the database. Note that the last column is a derived attribute.

6.7.3 Derived Attributes

Reports usually present summaries of some aspect of the enterprise, and derived attributes, such as totals and averages, are typical in such reports. Figure 6.43 shows derived attributes across rows (total value) and down columns (subtotal, grand total). Other derived attributes could have been added, such as average unit price and subtotals at the bottom of each page. Defining derived attributes is a design decision based on the objectives of the report. Each derived attribute has a computation time associated with it. Each time the report is generated, the values of these attributes are computed. Therefore, the report designer should consider the minimum require-

Table 6.1
Data Sources by Table

Column	Source
Material	MATL_DESCRIPTION in MATERIAL
Code	MATERIAL_ID in MATERIAL
Lot number	MATL_LOT_NO in MATERIAL_LOT
location	LOCATION_ID in WAREHOUSE_LOCATION
units	UNITS in PO_DETAIL
quantity	W_QUANTITY in WAREHOUSE_LOCATION
unit price	UNIT_COST in PO_DETAIL
total value	Derived: = [quantity * unit price]

ments for achieving the objectives of the report. Reports with unnecessary totals, products, averages, and so on can create unnecessary clutter that detracts from the readability of the report.

6.7.4 Implementing a Report in Access

Access provides several utilities for implementing reports. At one extreme, the ***report wizard*** is a way of quickly building a report. However, the designer does not have much control over shaping the design of the report. At the other extreme, the designer can completely customize a report by starting with a blank report and adding labels, text boxes, and other controls to the report.

A report can be based on either a table or a query. Reports that use data from more than one table require a query to assemble the data into a recordset. The first step in creating a report is to create the underlying relationships among tables used in the query. This was previously done in Section 6.5.2 and in Access Exercise 6.13. Therefore, it is unnecessary to do it again.

6.7.4.1 Creating the Query

The underlying tables for the report of Figure 6.43 are listed in Table 6.1. They are MATERIAL, MATERIAL_LOT, WAREHOUSE_LOCATION, and PO_DETAIL. The attributes to be retrieved from these tables are also listed in Table 6.1. The query we wish to create will return a recordset consisting of the desired attributes as specified in Table 6.1. This can be accomplished by writing the query in SQL, as we did in Section 6.5.5.1, or using query-by-example, as was shown in Appendix 2A. Here we will use the Access simple query wizard to create the SQL code for us.

Access Exercise 6.27:

We wish to establish the query for the ***Inventory Report***.

1. Go to MS Access Main Window and open the database **Material_Manager_CH6_stu.**
2. Click on the "Queries" tab and select ***"Create query using the wizard"***. The "Simple Query Wizard" window will open.
3. Using the Source columns of Table 6.1, select the appropriate tables and available fields in the order given in Table 6.1 ***(see Figure 6.44)***. The derived attribute, total value, should be ignored at this time. Left click on Next.
4. You will be asked whether you want the detail or a summary. Select detail. Click Next.
5. Save your work under the name: ***Inventory Report Query***.

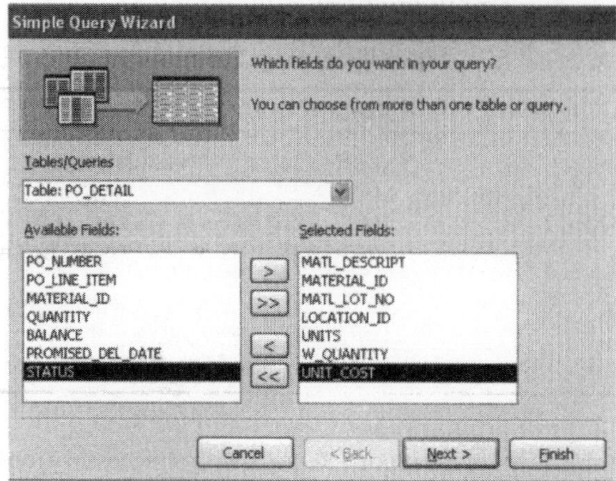

Figure 6.44 Creating a query using the simple query wizard.

The simple query wizard has created the query for us. To view the SQL code, open the inventory report query in design view. Then switch to the SQL view. The SQL query could also have been written directly.

6.7.4.2 Creating the Report Based on the Query

The report visualized in Figure 6.43 is a tabular report. It is most easily generated using the report wizard utility in Access.

Access Exercise 6.28:

We wish to create the report based on the query of Access Exercise 6.27.

1. Go to MS Access Main Window and open the database **Material_Manager_CH6_stu.**
2. Click on the "Reports" tab and select ***"Create report by using wizard"***. The "Report Wizard" window will open.
3. In the pull down menu, choose the query on which the report is based: ***Inventory Report Query*** and move all the available fields into the selected fields window. Click on Next.
4. At this point you will be asked to add grouping levels. Note that, in Figure 6.43, grouping is done on MATL_DESCRIPTION and MATERIAL_ID (Code). Move MATL_DESCRIPTION and MATERIAL_ID into the right panel. Click Next.
5. You will be asked if you want to sort fields. Sort the MATL_LOT_NO in the ascending order. Click Next.
6. You will be asked to choose a layout. Select Block and Landscape. Click on Next.
7. Choose Corporate Style and click Next.
8. Save your work under the report title: ***Inventory Report.***

Open the Inventory Report in design view and change the labels so that they agree with the column names in Table 6.1. When this is completed, the form should look like Figure 6.45. You may have to go to design view and rearrange columns to make the report more compact.

6.7.4.3 Report Format

In Access, reports are described by their sections. Figure 6.46, which is the Inventory Report in design view, shows six kinds of sections: Report header, Page header, Group header, Detail, Page footer, and Report footer. Each of these sections has a specific purpose as follows:

Report header. This section prints its contents once at the beginning of the report. Therefore, it is used for a title that references the entire report.

Page header. This section prints at the top of each page. It is used for titling columns that refer to data on the entire page.

Group header. This section prints the instances of group data.

Detail. This section prints each record of the recordset.

Page footer. This section prints at the bottom of each page. It usually contains summary information by page. In Figure 6.41, it contains the current date (= Now()) and the page number (= "page" . . .).

Report footer. This section prints once at the end of the report.

Inventory Report

MATL_DESCRIPT	*MATERIAL_ID*	*MATL_LOT_NO*	*LOCATION_ID*	*UNITS*	*W_QUANTITY*	*UNIT_COST*
Carrots, diced, 1/4	RM202	1009	RG010102	LB	250	$2.00
		1008	RG010104	LB	100	$2.00
		1001	FR010102	LB	300	$2.00
Carrots, whole	RM201	1007	RG010103	LB	100	$2.00
		1000	RG010101	LB	400	$2.00
Olive oil	RM805	1010	WH010105	GAL	200	$0.50
		1010	WH010104	GAL	300	$0.50
		1006	WH010103	GAL	200	$0.50
Peas, shelled	RM210	1005	FR010103	LB	200	$2.00
		1004	FR010101	LB	100	$2.00
Potatoes, Eastern	RM205	1002	WH010102	LB	200	$1.00

Figure 6.45 Inventory Report after Access Exercise 6.25.

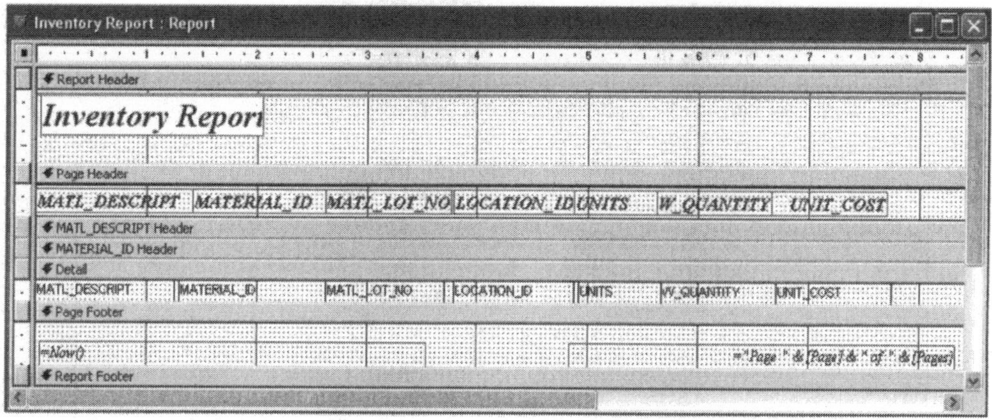

Figure 6.46 Default Inventory Report at the end of Access Exercise 6.28.

Figure 6.47 shows the components of a report in template format.

6.7.4.4 Adding Derived Attributes to the Report

Figure 6.45 does not include the derived attributes shown in the template of Figure 6.47. These derived attributes are added by placing text boxes in the appropriate sections. An expression is written in each text box to describe the computation required for deriving the attribute value. Access Exercise 6.29 takes the reader through the process of defining the "Total Value" derived attribute.

Access Exercise 6.29:

We wish to add a derived field for Total Value to the Inventory Report.

1. Open Inventory Report in Design view. Open the Toolbox.
2. Add the label "Total Value" in the Page header section.
3. Add a text box in the Detail section under the heading "Total Value". Delete the label of that text box.
4. Open the text box properties window. Under the "Data" tab, place the cursor in the ***"Control Source"*** properties field and enter the following; =*[W_QUANTITY]*[UNIT_COST]*
5. Under the "Format" tab, set the "**Format**" property to ***Currency***. Set the "**Decimal Place**" property to 2.

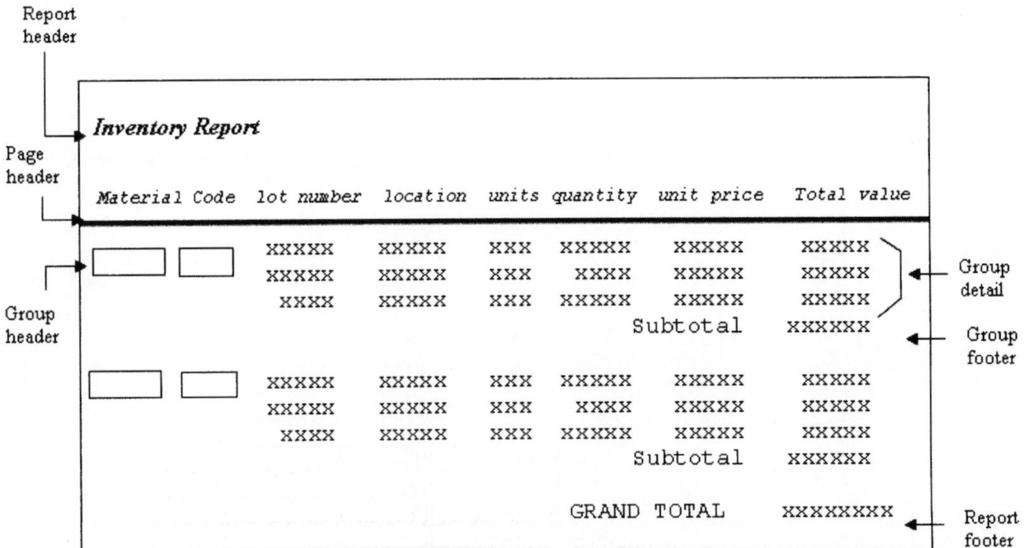

Figure 6.47 Template for inventory report by sections.

At this point, the reader should resize the labels in the page header and the text boxes in the details section so that the "Total Value" label and the text box can fit on the form. Figure 6.48 shows the resulting design view and form view. If you want borders around the total values, go to the design view, open the Format Properties window of the text box, and change the Borders property to Solid. If you wish to eliminate all borders of the text boxes, go to the Properties window of each text box and change the Borders property to Transparent.

The other derived attributes that appear on the template of Figure 6.41 are the column totals: ***subtotal*** and ***grand total***. The subtotal is taken on the grouped data of each material. The grand total is taken once in the report footer. Since the subtotal is taken on the grouped data, it is necessary to add group footer to the inventory report format. We illustrate the addition of group footer in Access Exercise 6.30.

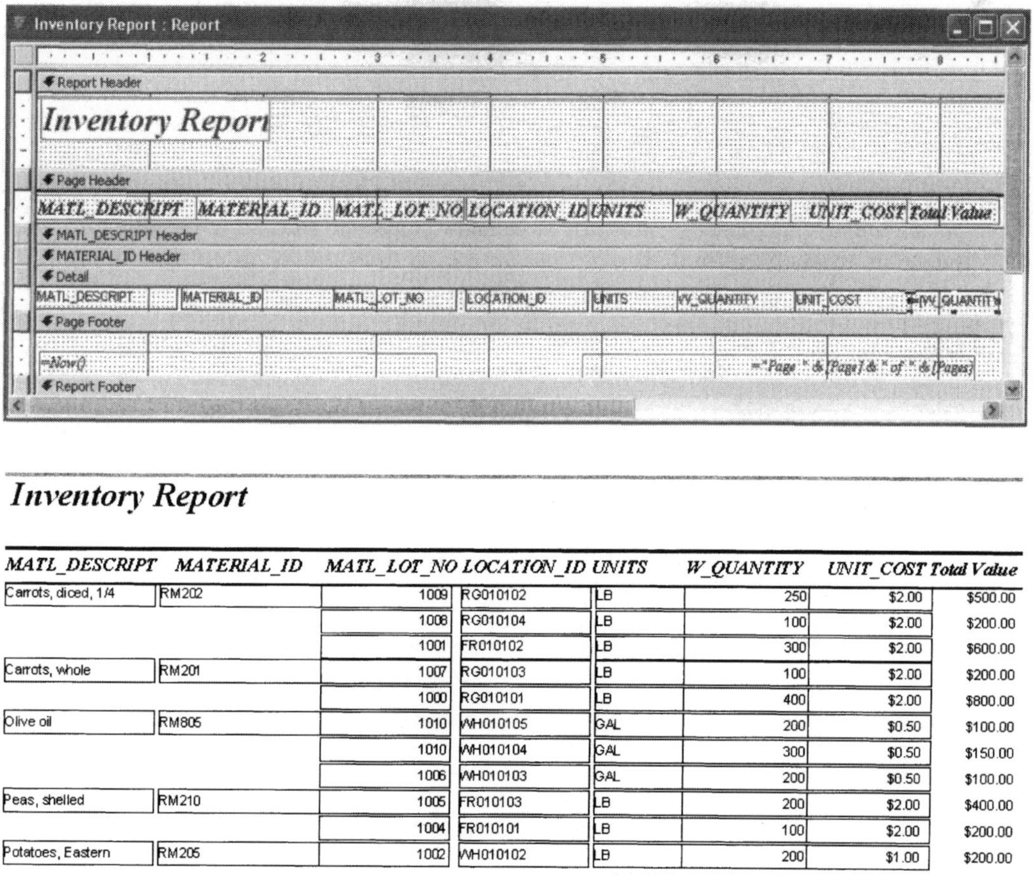

Inventory Report

MATL_DESCRIPT	MATERIAL_ID	MATL_LOT_NO	LOCATION_ID	UNITS	W_QUANTITY	UNIT_COST	Total Value
Carrots, diced, 1/4	RM202	1009	RG010102	LB	250	$2.00	$500.00
		1008	RG010104	LB	100	$2.00	$200.00
		1001	FR010102	LB	300	$2.00	$600.00
Carrots, whole	RM201	1007	RG010103	LB	100	$2.00	$200.00
		1000	RG010101	LB	400	$2.00	$800.00
Olive oil	RM805	1010	WH010105	GAL	200	$0.50	$100.00
		1010	WH010104	GAL	300	$0.50	$150.00
		1006	WH010103	GAL	200	$0.50	$100.00
Peas, shelled	RM210	1005	FR010103	LB	200	$2.00	$400.00
		1004	FR010101	LB	100	$2.00	$200.00
Potatoes, Eastern	RM205	1002	WH010102	LB	200	$1.00	$200.00

Figure 6.48 Inventory Report after Access Exercise 6.29.

Access exercise 6.30:

We wish to add ***group footer*** to the inventory report.

1. Open ***Inventory Report*** in design view.
2. Select menu item ***View=>Sorting and Grouping***.
3. Fill out the Sorting and Grouping window as shown in Figure 6.49. This instructs Access to group on the major group MATERIAL and, within MATERIAL, to sort by MATL_LOT_NO. The group footer is required by indicating "yes" in the appropriate box.
4. Once this is complete, close the sorting and grouping window. A group footer for MATERIAL should appear on the inventory report.

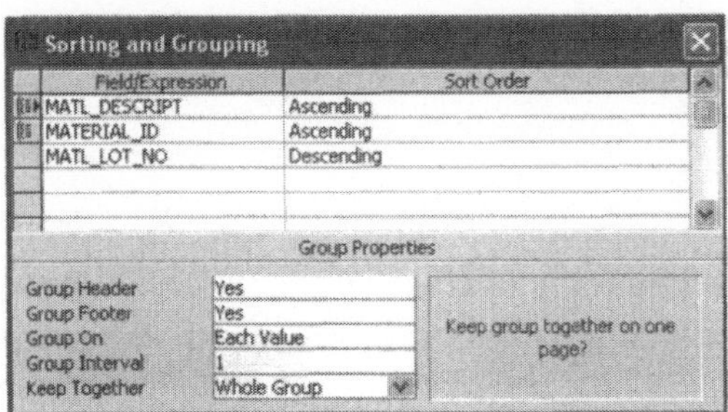

Figure 6.49 Sorting and grouping window.

Having established a group footer for summing grouped data (subtotals) and having a report footer for summing report data (grand total), we can now proceed to add text boxes for these totals. This is done in Access Exercise 6.31.

Access Exercise 6.31:

We wish to add ***derived fields*** to the Inventory Report for Subtotals and Grand Total.

1. Open ***Inventory Report*** in Design View. Open the Toolbox.
2. Add a text box in the MATERIAL_ID footer section where the group subtotal should appear. Enter the following caption on the text box label: ***Subtotal***.
3. Open the text box properties window and select the "Data" tab. Enter the "Control Source" property as follows: ***=Sum([W_QUANTITY]*[UNIT_COST])***.
4. Under the "Format" tab set the "Format" property to ***currency*** and the "Decimal Place" property to ***2***. This completes the addition of the derived attribute for group subtotal.
5. Add a text box in the Report footer section where the grand total should appear. Enter the following caption on the text box label: ***Grand Total***.
6. Open the text box properties window and select the "Data" tab. Enter the "Control Source" property as follows: ***=Sum([W_QUANTITY]*[UNIT_COST])***.
7. Under the "Format" tab set the "Format" property to ***currency*** and the "Decimal Place" property to ***2***. This completes the addition of the derived attribute for grand total.
8. Go to the report view and confirm that the Inventory Report has the desired format as per Figure 6.50.

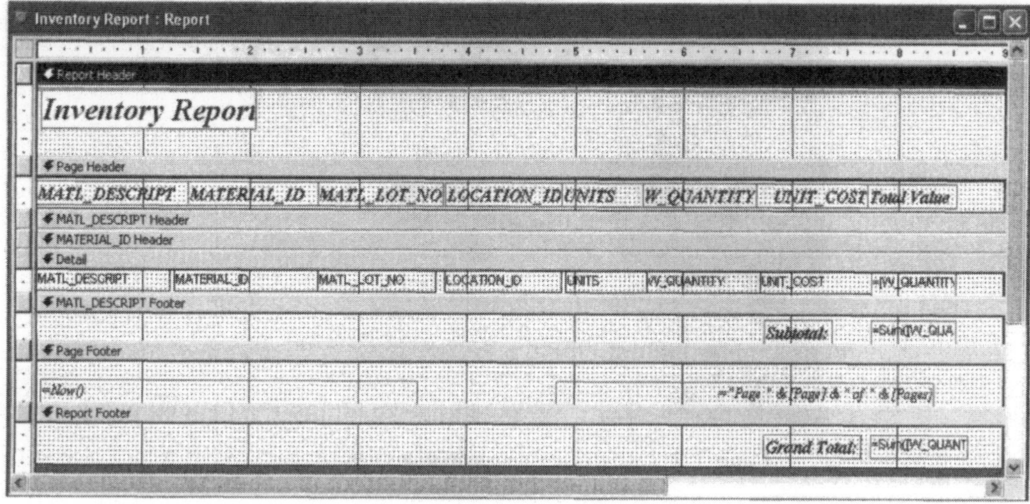

Inventory Report

MATL_DESCRIPT	*MATERIAL_ID*	*MATL_LOT_NO*	*LOCATION_ID*	*UNITS*	*W_QUANTITY*	*UNIT_COST*	*Total Value*
Carrots, diced, 1/4	RM202	1009	RG010102	LB	250	$2.00	$500.00
		1008	RG010104	LB	100	$2.00	$200.00
		1001	FR010102	LB	300	$2.00	$600.00
						Subtotal:	$1,300.00
Carrots, whole	RM201	1007	RG010103	LB	100	$2.00	$200.00
		1000	RG010101	LB	400	$2.00	$800.00
						Subtotal:	$1,000.00
Olive oil	RM805	1010	WH010105	GAL	200	$0.50	$100.00
		1010	WH010104	GAL	300	$0.50	$150.00
		1006	WH010103	GAL	200	$0.50	$100.00
						Subtotal:	$350.00
Peas, shelled	RM210	1005	FR010103	LB	200	$2.00	$400.00
		1004	FR010101	LB	100	$2.00	$200.00
						Subtotal:	$600.00
Potatoes, Eastern	RM205	1002	WH010102	LB	200	$1.00	$200.00
						Subtotal:	$200.00
						Grand Total:	$3,450.00

Figure 6.50 Completed Inventory Report in Design view and Report view.

6.8 ORGANIZING FORMS AND REPORTS INTO APPLICATIONS

Once the forms and reports of an application have been created, it is often desirable to customize the application for the user such that she or he interfaces with a simple menu instead of the database. The menu hides the details of the application and prevents the user from directly accessing the underlying tables. It presents the user with a simple point-and-click navigation to each of the forms and reports.

Such a menu is shown in Figure 6.51 for the forms and reports that were developed in this chapter. In the Microsoft Access world, this menu is referred to as a ***switchboard***. It is composed of a number of command buttons and associated labels that describe the functions of the buttons. The On_Click event property of each command button is programmed by the database designer. Typically, clicking a command button will open the selected document for reading, editing, or printing.

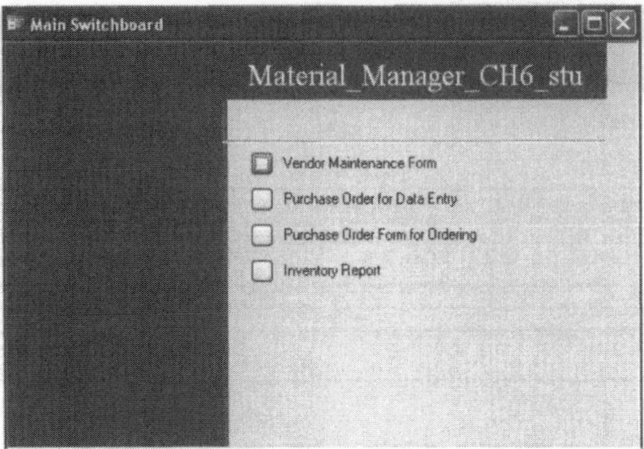

Figure 6.51 Menu design for purchase order application.

A switchboard is nothing more than another form composed of command buttons and labels. Instead of designing it using the form design methods previously introduced in this chapter, Access has automated the process by providing a facility called ***switchboard manager***. The switchboard manager will be used to implement the menu shown in Figure 6.51 in Access Exercise 6.32.

Access Exercise 6.32:

We wish to implement the ***switchboard*** of Figure 6.51.

1. Open **Material_Manager_CH6_stu** database.
2. From the main menu, select ***Tools => Database Utilities => Switchboard Manager***.
3. If Access asks you if you want to create a switchboard, click "Yes". This will take you to a default switchboard as shown in Figure 6.52 in the window "Switchboard Manager." From here you can either create a New Switchboard or use the default Main Switchboard. We will use the default switchboard.
4. Click on Edit. Access will respond with the Edit Switchboard Page window as shown in Figure 6.52.
5. The "Items on the Switchboard" panel is used to declare the names of the menu items. Click "New" and this will launch the "Edit Switchboard Item" window as shown in Figure 6.52. Here you add the name and function of each command button.
6. Fill in the first entry as follows:
   Text: Vendor Maintenance Form
   Command: Open Form in Edit Mode
   Form: VENDOR
   Click OK. The command button for Vendor Maintenance Form will appear in the panel.
7. Repeat step 6 for the other entries as follows:
   Text: Purchase Order for Data Entry, ***Command:*** Open Form in Edit Mode, ***Form:*** Purchase Order Master Form, OK.
   Text: Purchase Order for Ordering, ***Command:*** Open Form in Edit Mode, ***Form:*** Printable Purchase Order Form, OK.
   Text: Inventory Report, ***Command:*** Open Report, ***Report:*** Inventory Report, OK.
8. When you have completed the above entries, close the switchboard windows and the switchboard should have been created.

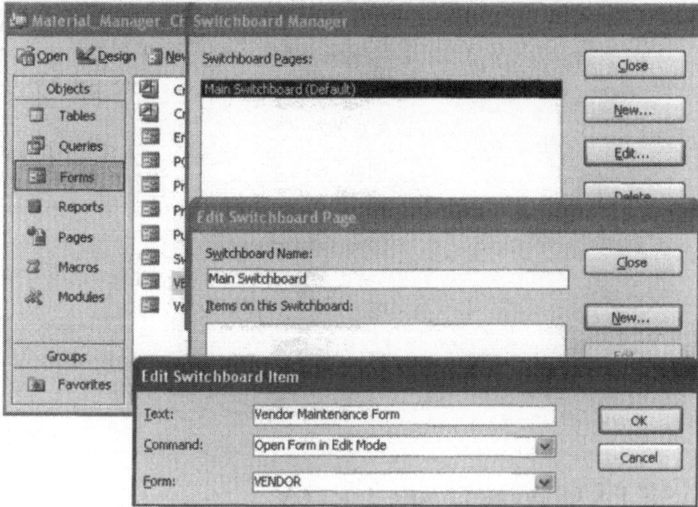

Figure 6.52 Implementing a switchboard.

The switchboard manager has created two objects in the database. The reader can view them in the database window. The first object is a form titled "Switchboard," which can be found under the Forms tab. This form is the design of Figure 6.51. The second object created by switchboard manager is a table titled "Switchboard Items." This table contains the organization of the switchboard menu. It can be found under the Tables tab of the database window.

A database application can be programmed to launch any form automatically when it opens. To launch the switchboard, it must be set as the default startup form when the database is opened. The procedure for doing this will be illustrated in Access Exercise 6.33.

Access Exercise 6.33:

We wish to launch the switchboard automatically when the application is opened by a user.

1. Open **Material_Manager_CH6_stu** database.
2. From the main menu, select ***Tools => Startup***. This will launch the Startup window shown in Figure 6.53.
3. Under the "Display Form/Page" panel, select ***"Switchboard"***.
4. Deselect ***"Display Database Window"*** by removing the check mark. This will result in a startup window that displays the switchboard and hides the database window.
5. Click OK. The switchboard will be launched each time the database is opened. Close the database and reopen it to confirm.

At this point, the switchboard is the interface to the application. It will remain so until the designer goes to the startup menu and removes the switchboard as the default startup screen. The Startup window shown in Figure 6.53 can also be used to remove other elements of the database window, for example, the toolbars and status bar can be eliminated at startup. If you wish to bypass the default screen and open the database window at startup, do this by holding down the **Shift key** on the keyboard while you are opening the database. Doing this will bypass the default switchboard and open the database window instead.

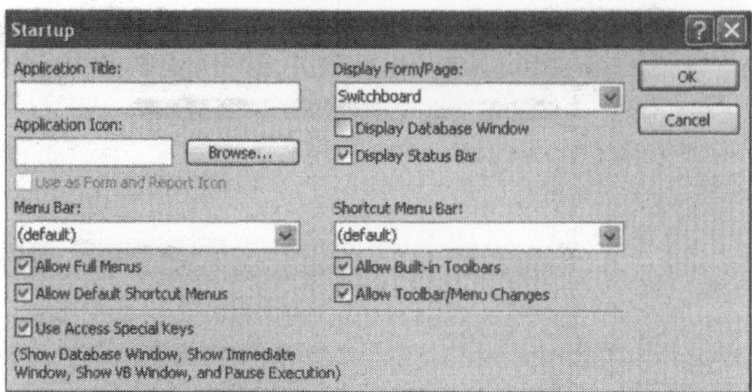

Figure 6.53 Startup window.

6.9 DATABASE PASSWORD SECURITY

In Section 6.4.5, the reader was introduced to some useful VBA objects and methods for dynamically connecting to and opening a database. An application was illustrated in which user password authorization was implemented by querying a database table. In general, password security is not implemented in this way. Database management systems usually provide a facility for declaring user privileges at several levels without the need to write programming code. In this section, we present an overview of this topic.

Data security and integrity are of great concern to the enterprise using a DBMS. The individual responsible for security, as well as other database administrative functions, is the database administrator (DBA). The DBA is a person who has both managerial and technical skills and is responsible for the overall strategy of how the enterprise evolves its information systems and how the organization interacts with the database. This is a crucial function in an organization that has an enterprise-wide information system. Among the ways the DBA tries to maintain security is to limit access only to authorized users. One approach is to define "user groups," which are classes of users who have common requirements for accessing data. That is to say, they have read, insert, update, and delete requirements for the same tables, forms, and reports. In this way, database privileges are defined as group rights.

Individual users are then assigned to be members of a user group, with all the privileges and rights that pertain to the group. Each user is given (or creates) a unique password. The password is required when accessing the database. The password identifies the user and, by association, the user group. The privileges granted to the group are made available to the individual.

In Microsoft Access, the process of defining groups, users, and privileges is done by setting **user-level security**. Access provides a **user-level security wizard** to assist in this process. In order to apply user-level security, you must have administrative privileges. A detailed description of assigning administrative privileges for a database and implementing user-level security is outside the main purposes of this chapter. However, the topic is covered in Appendix 6D for the interested reader.

Access is usually used in stand-alone applications. Sometimes an individual application is implemented on a department computer or a machine accessible by others. If the owner of a database application running on Access wishes to protect his or her database from being used by others, Access has provided a simple password security facility that prevents the database from being opened without the password. The owner can set his or her own password protection from the

Access main menu. Anyone trying to open the database must use the password. If the database owner forgets the password, there is no way to recover it and the database cannot be opened. Therefore, it is incumbent on the owner to maintain a record of the password. To show the process of setting password authorization, we introduce Access Exercise 6.34.

Access Exercise 6.34:

We wish to implement ***password protection*** for the **Material_Manager_CH6__stu** database.

1. Make a backup copy of **Material_Manager_CH6_stu** database and store it in a separate folder. Password protection will be demonstrated on this copy of the database.
2. Open Microsoft Access with the database closed.
3. From the main menu, select ***File*** => ***Open***. This will launch the "Open" window.
4. In the Open File window, browse to the folder in which the copy of the database is stored. Click on the database file name.
5. Instead of opening the database in normal mode, click on the arrow to the right of the Open button, then select "***Open Exclusive.***" The database will now open in the exclusive mode.
6. From the main menu, select ***Tools*** => ***Security*** => ***Set Database Password***. This will launch the window (input box) for setting your password.
7. Type your chosen password into the password box (must be 4-20 alphanumeric characters). Retype your password in the verify box and click OK. The password is now set.
8. Close the copy of **Material_Manager_CH6_stu** database.
9. Reopen the copy of **Material_Manager_CH6_stu** database using the "Open" button (normal mode) to confirm that it is password protected.

Henceforth, a perspective user will need the password to gain access to your database. If you wish to remove password protection, follow steps 2 to 5 of Access Exercise 6.34. In step 6, select ***Tools⇒Security⇒Unset Database Password.*** This will launch an "Unset Database Password" input box. After you type your password into the input box and click OK, password protection will be removed.

6.10 SUMMARY

Once an information system model has been developed, the analyst must consider the design of screens that will become the user views of the database. The major categories of user views are forms and reports. Although the design process is a creative process, there are certain considerations that are used to guide the designer through the design process. In this chapter we have demonstrated the application of these considerations as steps in creating functional forms and reports.

We have also introduced the reader to the use of the Access DBMS in the design of forms and reports. Our objective is to provide the reader with some hands-on experience, not to learn all the features of Access. However, enough key topics were covered to enable the reader to create basic forms and reports. This included an introduction to some features of Visual Basic for Applications. Interested readers can pursue their knowledge of the subjects covered in this chapter by reading Appendix 6C (Visual Basic for Applications Overview) and the following references, in the bibliography of this book (Dobson, 2003; Gosmall, 2004; Microsoft, 2003; Prague et al., 2003; and Viescas, 2005).

REVIEW EXERCISES

6.1 Construct a functional/entity interaction matrix like that of Figure 6.1 for the "Control Stored Materials" function in Figure 5.13. Assume that the functions (1) store raw material, (2) move raw material to work in process, and (3) return raw material to storage are now associated with direct access to the tables Material_Lot, Warehouse_Location, and Lot_Transaction. That is, the "to be" implementation of the functions eliminates the paper entries into the "Inventory Log." The information requirements and relationship between functions and entities is described in Section 5.3.3.

6.2 The figure that follows shows a single table form for the MATERIAL table, which is in the Material_Manager_CH6_stu database. It is used to maintain the table (i.e., to add new material descriptions and to delete obsolete materials).
(a) Implement this single table form in Access as shown.
(b) Add function buttons as in Access Exercise 6.2.

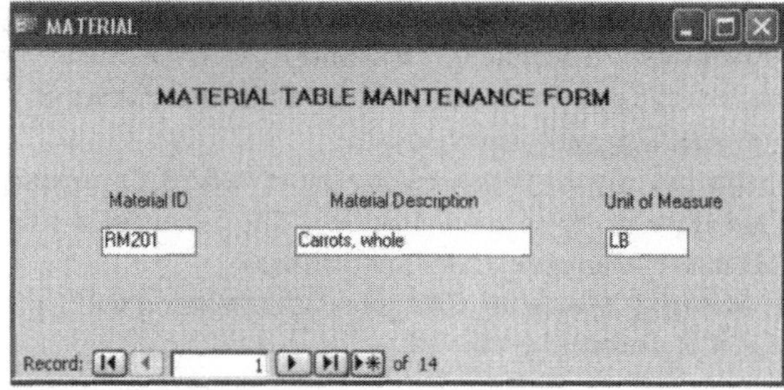

6.3 For the material form in Review Exercise 6.2, add the following data integrity requirements (see examples of table data in Appendix 6A).
(a) MATERIAL_ID must begin with RM.
(b) UNIT_OF_MEASURE is limited to four entries: LB, GAL, PCS (pieces) or CASES.

6.4 Shown here is a master/detail form based on the following tables: MATERIAL, VENDOR, and VENDOR_MATL_XREF. The form is used to view the vendor material IDs in the cross-reference table associated with each material of the enterprise. It is also used to add new materials and their cross reference with vendor materials.
(a) Implement this form in Access without command button functions.
(b) Add command buttons for the insert, delete, and save functions.

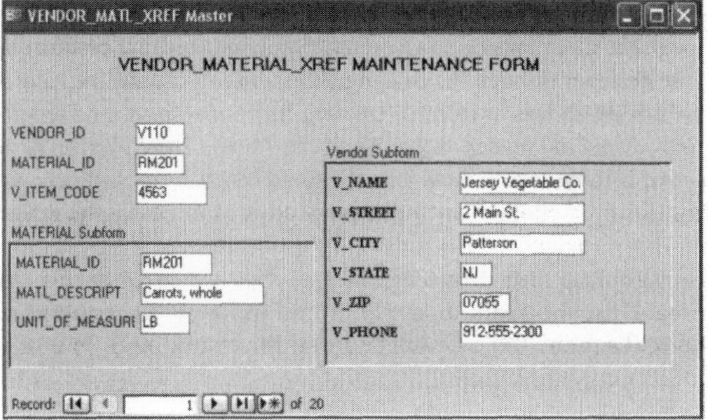

6.5 For the Vendor_Matl_Xref maintenance form in Review Exercise 6.4, add data integrity requirements for VENDOR_ID, MATERIAL_ID, and UNIT_OF_MEASURE.

6.6 Create a menu of forms (Switchboard) for the forms shown in Review Exercises 6.2 and 6.4. Note: only one switchboard can be active in a database. You must make this your Default switchboard to activate it.

6.7 For the material form of Review Exercise 6.2, add a combo box that will allow the user to pick a material from a list instead of having to enter it manually.

6.8 In Figure 6.34 a query was created as the basis for the PO_DETAIL subform shown in Figure 6.33. Create a report based on this query in which the details of all purchase orders are shown, grouped by purchase order number, as illustrated here.

PO_DETAIL_Report

PO_NUMBER	PO_LINE_ITEM	V_ITEM_CODE	MATL_DESCRIPT	UNITS	QUANTITY	UNIT_COST
2591	1	4563	Carrots, whole	LB	1000	$2.00
	2	3423	Carrots, diced, 1/4	LB	1000	$2.00
	3	9899	Potatoes, Eastern,	LB	300	$1.00
2592	1	F444	Olive oil	GAL	800.5	$0.50
	2	F456	Vinegar, white	GAL	210.5	$0.50
2593	1	546	Peas, shelled	LB	1000	$2.00
	2	765	Tomatoes, whole	LB	2000	$1.00
2594	1	2-414	Garlic, whole	LB	4000	$0.50
	2	2-564	Garlic powder	LB	2000	$0.25
	3	3-212	Salt, iodized	LB	2000	$0.25
	4	3-675	Sugar, bulk	LB	560	$0.50
2595	1	2-312	Paprika	LB	400	$0.50
	2	2-897	Onion Salt	LB	1200	$0.25
2596	1	666	Sugar, brown	LB	5000	$0.20

6.9 Add the following to the report created in Review Exercise 6.8:
(a) A derived attribute, "Total Cost," which is the unit cost times the quantity
(b) A subtotal of "Total Cost" by each purchase order
(c) A grand total of "Total Cost" for all purchase orders

6.10 In ACME Machine Shop (C), the plant supervisor requested two reports, which are reproduced next. Prepare and implement these reports. The second report should be grouped by machine.

1. A list of warranty expiration dates by machine. We regularly extend our warranty dates by purchasing new contracts. I would like to be able to review them when I please.

MACHINE NAME	MACHINE SERIAL NO.	FIRST DATE IN SERVICE	WARRANTY EXP. DATE
2120 CNC Mill	E41520	12/1/98	12/1/03
Cincinnati CNC Lathe	C5001	2/3/99	2/3/01
Cincinnati CNC Lathe	C4010	4/1/96	4/1/01
.	.	.	.
.	.	.	.

2. I would also like to be able to obtain a preventive maintenance history by machine and by task.

MACHINE NAME	PM TASK ID.	DATE DONE	MACHINE HOURS	MINUTES WORKED
Cincinnati CNC Lathe	T7	3/10/2002	1000	60
Cincinnati CNC Lathe	T8	3/10/2002	1000	20
Cincinnati CNC Lathe	T3	3/8/2002	1000	30
Cincinnati CNC Lathe	T4	3/8/2002	1000	20
2120 CNC mill	T1	3/1/2002	2100	120
.	.	.	.	.
.	.	.	.	.
.	.	.	.	.

CASE STUDIES

6.11 University Food Receiving Department Case (D)

Chapters 5 and 6 detailed the work of the receiving clerk using IDEF0 and IDEF1X. Among the tasks she is responsible for are accepting shipments from vendors, creating material lots for storage in inventory, and filling out the receiving report. The management of University Food would like to automate these processes by implementing the database of Figure 5.12. The populated tables of this database are shown in Appendix 6A, and the database is on the Web site supporting this book.

You will design two forms to assist the receiving clerk. The first form will be used for accepting the shipment and creating the material lot. The second form will be used to assign the material lot to a warehouse location.

(a) Develop a form for creating material lots.
When the receiving clerk gets the bill of laden from the truck driver, she performs the following steps:
1. Compare the PO number on the bill of laden with the purchase order in the database.
2. Ensure that the material ID and quantity correspond with a line item of the purchase order.

3. Group the items of the shipment into lots by assigning lot numbers and enter the lot assignments into the material lot table.

Figure CS6.1 is the layout of the proposed form. The form is based on the PURCHASE ORDER table as the master form because the purchase order table is the link to the information that the receiving clerk needs to compare to items on the bill of laden and needs to complete the required fields for the MATERIAL LOT table. For example, the PURCHASE ORDER table has the vendor ID that can be used to retrieve the vendor name and address, which appears on the bill of laden. The purchase order number and vendor name and address from the bill of laden should both agree with what is in the database. The vendor subform information that is linked to the purchase order through the relationship model is shown on the lower left of Figure CS6.1.

The purchase order is also related to the PO detail, which contains line item information on the materials ordered and quantities. This will be needed to compare with the bill of laden. This information is shown on the left side center of Figure CS6.1. Note that navigation buttons are available for navigating among the line items declared within the purchase order number. Finally, the bill of laden refers to the material of the line item by the vendor item code. However, material lots are classified by the material ID system used by University Food. Therefore, it is necessary to cross reference the vendor code with the material ID before entering the latter into the MATERIAL LOT table. The cross reference comes from the VENDOR MATERIAL XREF table. This is shown in the lower right of Figure CS6.1. All of the aforementioned blocks are "read only" and should not be allowed to be other than read only during use. The only block for data entry is the block based on the table MATERIAL_LOT.

The data entry block is on the upper right of Figure CS6.1. These are the data items that will be entered by the receiving clerk. Finally, there is a Close button that should automatically save the data entered on the form to the database, followed by closing the form.

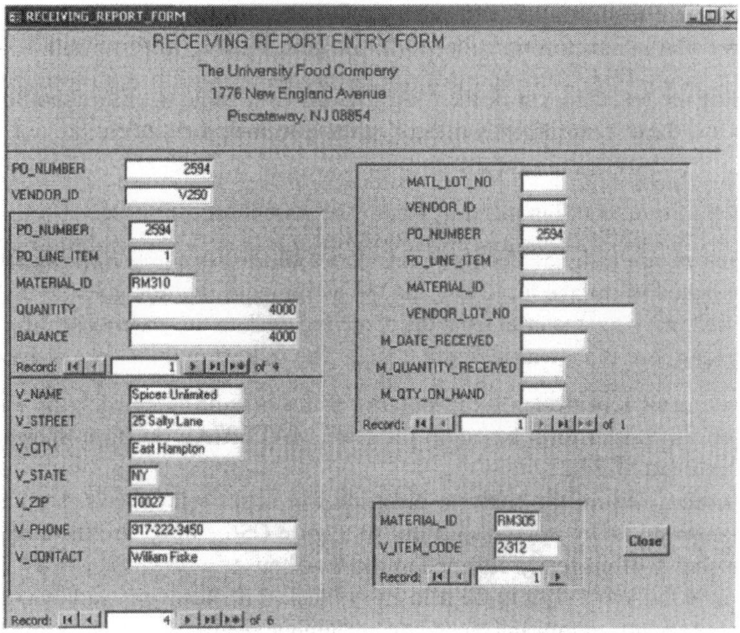

Figure CS6.1 Receiving report data entry form.

The following sequence outlines the way the receiving clerk uses the form:

1. The receiving clerk reads the PO number on the bill of laden and navigates to that number on the form. Then she checks to see that the vendor associated with that purchase order in the database is the one who appears on the bill of laden.
2. Using the vendor item code that appears on the bill of laden, she navigates through the vendor material cross reference block, which reveals the related material ID.
3. She navigates through the PO detail block to find the line item having that material ID and checks to see that an appropriate line item exists with a remaining balance to be delivered that is equal to or greater than the current shipment.
4. At this point, enough information is available to begin filling out the material lot block.
5. Other data are obtained as follows. The material lot number is the next available sequential number for material lots. The vendor lot number must be taken from the pallets as they come off the truck. The date received is the current date. The quantity received and the quantity on hand is the amount received in the shipment.

Create the receiving report entry form as shown in Figure CS6.1. Remember to link the subforms to the main form using the appropriate foreign keys.

The following data can be used as test data for executing the form. Assume a new shipment has arrived with the following information on the bill of laden.

Reference: PO2594
From: Spices Unlimited, 25 Salty Lane, East Hampton, NY 10027
Descriptions: Garlic, Item #2-414, 4000 lbs.
Salt, Item #3-212, 1000 lbs.

Other information needed:

Date received: 6/25/2005
Next two available sequential lot numbers: 1011, 1012
Manufacturer's lot numbers from pallets: 925 for Item #2-414, 1000 for Item #3-212

After entering the test data, check the MATERIAL LOT table to ensure that the information was saved. To use these material lot numbers again, you must first delete the test data from the table.

(b) Develop a form to assign material lots to the warehouse.

After lot numbers are assigned, the receiving clerk looks for available storage in the warehouse. Lots will be assigned to storage locations, and the assignments will be given to the forklift truck operator, who will move the material from the receiving dock to the warehouse. The form that will be used for this purpose is shown as Figure CS6.2. The following steps are involved.

1. The receiving clerk navigates to the lot number that is to be assigned to the warehouse. This is done on the master form based on the table MATERIAL_LOT, as shown on the upper left side of Figure CS6.2.
2. The datasheet view in the subform block on the upper right shows the quantities of the selected material lot by storage location. As Figure CS6.2 indicates, there are no quantities of lot number 1011 currently assigned to the warehouse.
3. The block in the lower right is the data entry block. The receiving clerk navigates to a location that is currently empty and enters the lot number and the amount to be stored. This block is based on the WAREHOUSE_LOCATION table, but is not bound to the master form.

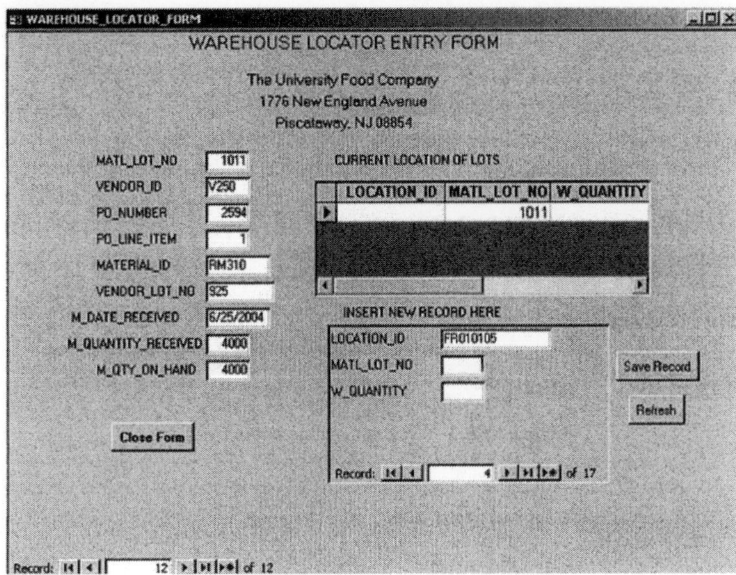

Figure CS6.2 Warehouse locator entry form.

4. After a new record is saved using the Save Record command button, the screen is refreshed using the Refresh command button. This will update the text boxes on the screen to the current values in the database, thus confirming that the data have been properly entered into the database.

The "Current Location of Lots" data sheet should be bound to the master form using the appropriate foreign key. This block and the master block should be read only. The "Refresh" command button is found under the command button category "Form Operations" and the associated action "Refresh Form Data." It will refresh the data on the screen from the data sources.

6.12 ACME Machine Shop Case (D)

A functional architecture was completed for the ACME machine shop maintenance function based on the case study at the end of Chapter 4. The corresponding information architecture was completed based on the exercise at the end of Chapter 5. Using that information and the scenario in ACME Machine Shop (B), do the following exercises. The populated database can be downloaded from the Web site supporting this book.

(a) Define the functional/entity interaction matrix.

(b) The plant supervisor wants to have a form for performing the function "Update Machine Operating Hours." He has suggested a layout for the form, which is shown in Figure CS6.3 in form view and Figure CS6.4 in design view. The form would function as follows:

   (1) When the form is loaded, it will show the basic information from the table MACHINE, such as the MACHINE_ID, MACHINE_NAME, MANUFACTURER_ID, and MACHINE_TOTAL_HOURS. It will also show the corresponding records from the

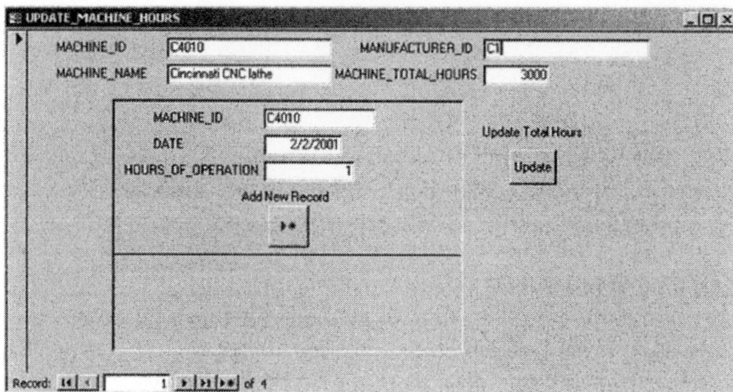

Figure CS6.3 Update machine hours form.

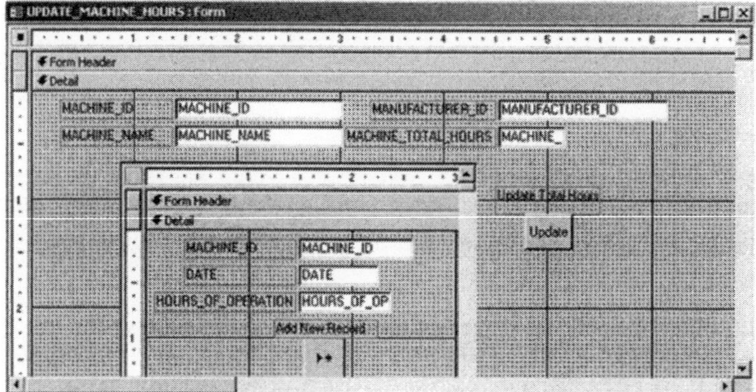

Figure CS6.4 Update machine hours form in design view.

table MACHINE_USE_RECORD as a subform. All text boxes will be locked when the form is loaded. The form appearance is shown in Figure CS6.3.

(2) A command button will appear on the subform that is used to insert a new record. When clicked, it should move to a new record (blank text box) and set the focus on the text box for DATE. It should also unlock the DATE and HOURS_OF_OPERATION text boxes so the user can enter a new record.

(3) After entering the DATE and HOURS_OF_OPERATION, clicking on the Update command button should result in adding the value in the HOURS_OF_OPERATION text box to the value in the MACHINE_TOTAL_HOURS text box. At this point, the record is complete for both the MACHINE and MACHINE_USE_RECORD tables.

(4) Clicking on the Update button should also lock the DATE and HOURS_OF_OPERATION text boxes so that the cycle can begin again with a new record to be added to the subform.

Suggestion: The following steps may help you to organize your work.

1. Create the relationships among tables.
2. Create the main form.
3. Create the subform.
4. Add the command button to the subform.
5. Bind the subform to the main form.
6. Add the Update command button to the main form.

(c) The plant supervisor wants to have a form for performing the function "Monitor Hours of Operation." It will show a comparison between the total hours elapsed since a maintenance task on a machine was performed and the required frequency of that task. The form is shown in Figure CS6.5. The Elapsed Hours are computed as [MACHINE_TOTAL_HOURS] – [MACH_HRS_LAST_DONE]. In this way, the plant supervisor can see which PM tasks are currently required to be performed. If the elapsed hours are greater than or equal to the PM_TASK_FREQUENCY, a work order should be released. Design and implement the form.

Suggestions: Base the "MONITOR_HOURS_OF_OPERATION_SUBFORM" on an SQL query. To bind the subform to the main form and to relate subform data from more than one table, you will require additional attributes in the query that are hidden on the form shown in Figure CS6.5.

(d) The plant supervisor has pointed out to you that most of the data required to create a preventive maintenance work order appear on the MONITOR_HOURS_OF_OPERATION form. He has suggested that he should be able to generate a PM work order by simply clicking on a command button on the MONITOR_HOURS_OF_OPERATION form screen. He has suggested the following:

(1) Add a new text box to the MONITOR_HOURS_OF_OPERATION main form. If, after comparing the elapsed hours with the required frequency, a work order should be issued, the user will enter the PM_TASK_ID into the text box (see Figure CS6.6).

(2) A new command button will be added to the MONITOR_HOURS_OF_OPERATION main form titled "Make WO." When clicked, a PM work order form will be opened and automatically filled in with the following information: MACHINE_ID, PM_TASK_ID, MACHINE_HOURS, and WO_DATE. The WO_DATE should be the current date. The other data are transferred from the MONITOR_HOURS_OF_OPERATION main form.

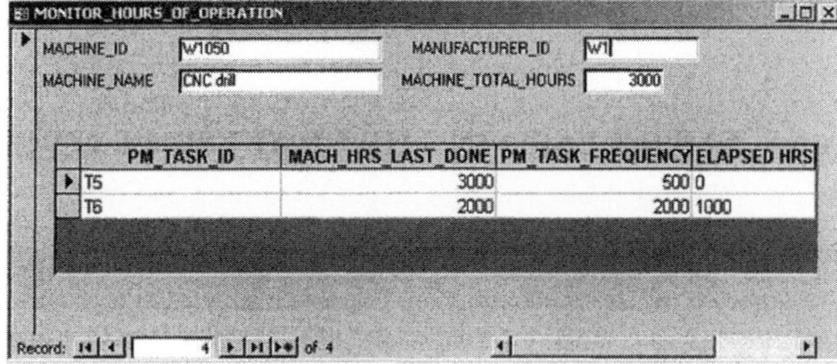

Figure CS6.5 Monitor hours of operation form.

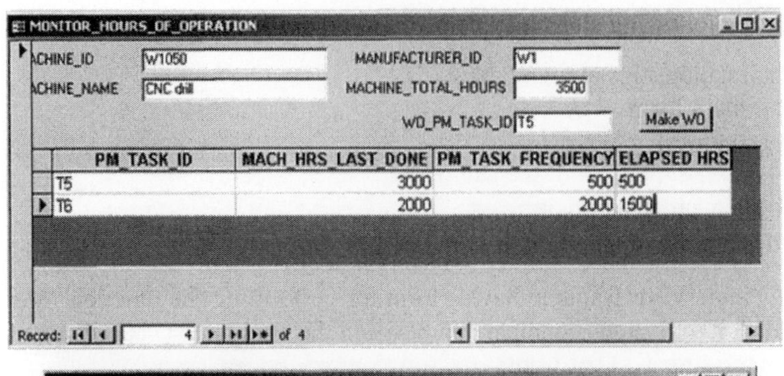

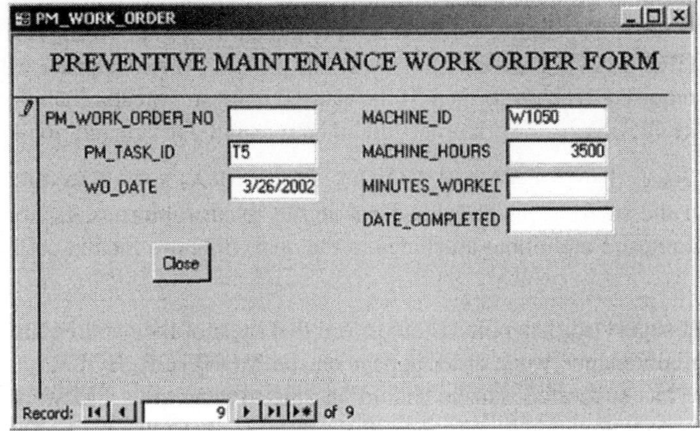

Figure CS6.6 Monitor hours of operation and PM work order forms.

(3) Once the work order form is opened and populated with the preceding data, the MONITOR_HOURS_OF_OPERATION form is automatically closed. This completes the action of the Make WO command button.

(4) The user then adds the PM_WORK_ORDER_NO to the PM_WORK_ORDER form and saves the record as an "open" work order when closing the form. After the work has been completed, the plant supervisor will be able to reopen the form and input the remaining data, thus closing the work order.

The layouts of the modified MONITOR_HOURS_OF_OPERATION form and the PM_WORK_ORDER form are shown in Figure CS6.6. Design and implement these forms.

APPENDIX 6A TABLES USED IN THE CHAPTER 6 EXERCISES

EMPLOYEE : Table

EMPLOYEE_ID	EMP_FIRST_N	EMP_MIDDLE_	EMP_LAST_NA	EMP_STREET	EMP_CITY	EMP_STA	EMP_ZIP	EMP_DEPT	EMP_DEPT_PH
000-23-8900	Joseph	John	Swartz	223 East	Fort Lee	NJ	06655	Purchasing	908-445-3654
002-07-6110	John	Thomas	Smith	24 Grove Street	Clifton	NJ	02215	Receiving	908-445-2100
042-38-6132	Randy	Randolf	Kim	23 Jay Ave.	Secaucus	NJ	06673	Purchasing	908-445-3657
098-67-9000	Thomas	Owen	Baker	34 Douglas	Montclair	NJ	07043	Purchasing	908-445-9999

MATERIAL : Table

MATERIAL_ID	MATL_DESCRIPT	UNIT_OF_MEASURE
RM201	Carrots, whole	LB
RM202	Carrots, diced, 1/4	LB
RM205	Potatoes, Eastern,	LB
RM210	Peas, shelled	LB
RM211	Tomatoes, whole	LB
RM305	Paprika	LB
RM308	Onion Salt	LB
RM310	Garlic, whole	LB
RM311	Garlic powder	LB
RM318	Salt, iodized	LB
RM340	Sugar, bulk	LB
RM502	Sugar, brown	LB
RM805	Olive oil	GAL
RM810	Vinegar, white	GAL

MATERIAL_LOT : Table

MATL_LOT_NO	MATERIAL_ID	VENDOR_ID	PO_NUMBER	PO_LINE_ITEM	VENDOR_LOT	M_DATE_RECEIV	M_QUANTITY	M_QTY_ON_HAN
1000	RM201	V110	2591	1	23596	2/10/2006	500	400
1001	RM202	V110	2591	2	24096	2/20/2006	500	300
1002	RM205	V110	2591	3	00597	2/20/2006	200	200
1003	RM205	V110	2591	3	00797	2/20/2006	100	0
1004	RM210	V110	2593	1	25096	2/21/2006	300	100
1005	RM210	V110	2593	1	25596	2/21/2006	200	200
1006	RM805	V25	2592	1	01097	2/22/2006	200	200
1007	RM201	V110	2591	1	23696	2/22/2006	500	100
1008	RM202	V110	2591	2	24096	2/22/2006	200	100
1009	RM202	V110	2591	2	24596	2/22/2006	300	250
1010	RM805	V25	2592	1	01097	2/25/2006	600	600

PURCHASE_ORDER : Table

PO_NUMBER	RELEASE_DATE	PO_STATUS	PO_AMT	VENDOR_ID	EMPLOYEE_ID	EST_FREIGHT
2591	2/10/2006	CLOSED	$4,300.00	V110	000-23-8900	$430.00
2592	2/10/2006	OPEN	$505.50	V25	042-38-6132	$50.55
2593	2/11/2006	OPEN	$4,000.00	V110	098-67-9000	$400.00
2594	2/12/2006	OPEN	$3,280.00	V250	000-23-8900	$328.00
2595	2/15/2006	OPEN	$500.00	V250	098-67-9000	$50.00
2596		HOLD	$1,000.00	V75	098-67-9000	$100.00

PO_DETAIL : Table

PO_NUMBER	PO_LINE_ITEM	MATERIAL_ID	UNITS	QUANTITY	BALANCE	PROMISED_DEL_DATE	UNIT_COST	STATUS
2591	1	RM201	LB	1000	0	2/20/2006	$2.00	CLOSED
2591	2	RM202	LB	1000	0	2/20/2006	$2.00	CLOSED
2591	3	RM205	LB	300	0	2/20/2006	$1.00	CLOSED
2592	1	RM805	GAL	800.5	0	2/25/2006	$0.50	CLOSED
2592	2	RM810	GAL	210.5	210	3/10/2006	$0.50	OPEN
2593	1	RM210	LB	1000	500	3/12/2006	$2.00	OPEN
2593	2	RM211	LB	2000	2000	3/12/2006	$1.00	OPEN
2594	1	RM310	LB	4000	4000	3/22/2006	$0.50	OPEN
2594	2	RM311	LB	2000	2000	3/22/2006	$0.25	OPEN
2594	3	RM318	LB	2000	2000	3/22/2006	$0.25	OPEN
2594	4	RM340	LB	560	560	3/22/2006	$0.50	OPEN
2595	1	RM305	LB	400	400	2/27/2006	$0.50	OPEN
2595	2	RM308	LB	1200	1200	2/27/2006	$0.25	OPEN
2596	1	RM502	LB	5000	5000		$0.20	OPEN

VENDOR : Table

VENDOR_ID	V_NAME	V_STREET	V_CITY	V_STATE	V_ZIP	V_PHONE	V_FAX	V_CONTACT
V110	Jersey Vegetable Co	2 Main St.	Patterson	NJ	07055	912-555-2300	912-555-2396	Paul Caudel
V25	General Provisions	125 Common St.	Boise	ID	44830	777-466-3400	777-466-3401	Robert Brooke
V250	Spices Unlimited	25 Salty Lane	East Hampton	NY	10027	917-222-3450	917-222-3455	William Fiske
V75	Pasta Supply Inc.	34 Henry St.	Philadelphia	PA	09098	402-450-5555	402-450-5566	Joseph Park

VENDOR_MATL_XREF : Table

VENDOR_ID	MATERIAL_ID	V_ITEM_CODE
V110	RM201	4563
V110	RM202	3423
V110	RM205	9899
V110	RM210	546
V110	RM211	765
V25	RM201	A222
V25	RM205	A234
V25	RM318	C677
V25	RM340	C546
V25	RM805	F444
V25	RM810	F456
V250	RM305	2-312
V250	RM308	2-897
V250	RM310	2-414
V250	RM311	2-564
V250	RM318	3-212
V250	RM340	3-675
V75	RM305	345
V75	RM308	555
V75	RM502	666

WAREHOUSE_LOCATION : Table

LOCATION_ID	W_AREA	W_AISLE	W_TIER	W_BIN	MATL_LOT_NO	W_QUANTITY
FR010101	FR	01	01	01	1004	100
FR010102	FR	01	01	02	1001	300
FR010103	FR	01	01	03	1005	200
FR010104	FR	01	01	04		
FR010105	FR	01	01	05		
FR010201	FR	01	02	01		
RG010101	RG	01	01	01	1000	400
RG010102	RG	01	01	02	1009	250
RG010103	RG	01	01	03	1007	100
RG010104	RG	01	01	04	1008	100
RG010105	RG	01	01	05		
RG010106	RG	01	01	06		
RG010201	RG	01	02	01		
WH010102	WH	01	01	02	1002	200
WH010103	WH	01	01	03	1006	200
WH010104	WH	01	01	04	1010	300
WH010105	WH	01	01	05	1010	200
WH010106	WH	01	01	06		
WH010107	WH	01	01	07		
WH010201	WH	01	02	01		

PASSWORD : Table

PASSWD
Boucher
Jones
Smith

APPENDIX 6B DATA FOR TRIAL EXERCISES

PURCHASE_ORDER : Table

	PO_NUMBER	RELEASE_DATE	PO_STATUS	PO_AMT	VENDOR_ID	EMPLOYEE_ID
+	9999	8/1/2003	OPEN	$1,000.00	V250	000-23-8900
+	9888	4/5/2003	OPEN	$2,000.00	V25	042-38-6132

PO_DETAIL : Table

PO_NUMBER	PO_LINE_ITEM	MATERIAL_ID	UNITS	QUANTITY	BALANCE	PROMISED_DEL_DATE	UNIT_COST	STATUS
9999	1	RM305	LB	1000	0	8/20/2003	$1.00	OPEN
9888	1	RM205	LB	2000	2000	5/2/2003	$1.00	OPEN

APPENDIX 6C VISUAL BASIC FOR APPLICATIONS OVERVIEW

Visual Basic is a high-level, event-driven programming language that evolved from the earlier DOS version called Basic, and it is usually used to design user interfaces. Visual Basic for Applications (VBA) is a subset of the more powerful Visual Basic, and the programming is done in a graphical environment. The users select certain *controls* (for example, a command button) and each control is programmed independently to be able to respond to user *actions* (for example, On_click). A VBA program is made up of many subprograms, each with their own program codes, and each can be executed independently or linked together.

VBA is included with several Microsoft applications including Microsoft Access. By extending a database to include VBA procedures, the user can customize the way the tables, forms, reports, and queries work together in the database. The user can create an *event procedure* by adding code to an event on a form or report or any of the controls within them.

THE VBA ENVIRONMENT IN MICROSOFT ACCESS 2003

To start up VBA, click on the Forms or Reports tab in the database wimdow. Then double-click on "*Create Form in Design View*" or "*Create Report in Design View*" in the Database window as shown in Figure 6C.1.

Double-clicking "*Create Form in Design View*" will display the screen shown in Figure 6C.2. If any of these windows do not appear on the screen, go to the "*View*" menu item and select these windows. The screen has three windows. From left to right they are as follows:

1. *Toolbox window*, which consists of all the controls essential for developing a form or a report. Controls are tools such as text boxes, buttons, labels, and other objects drawn on a form to get input or display output.
2. *Form window*, which will contain and display the user interface.
3. *Properties window*, which displays the properties associated with the controls added to the forms and reports.

The view displayed in Figure 6C.2 is what is referred to as the *Object View*, where the user can design the appearance of the form by dragging and dropping objects from the toolbox and editing their properties in the Properties window. The *Code View*, shown in Figure 6C.3, is where the user creates event procedures (software code) that are executed during run-time. The code view can be displayed by clicking on "*Code*" under the "*View*" menu item.

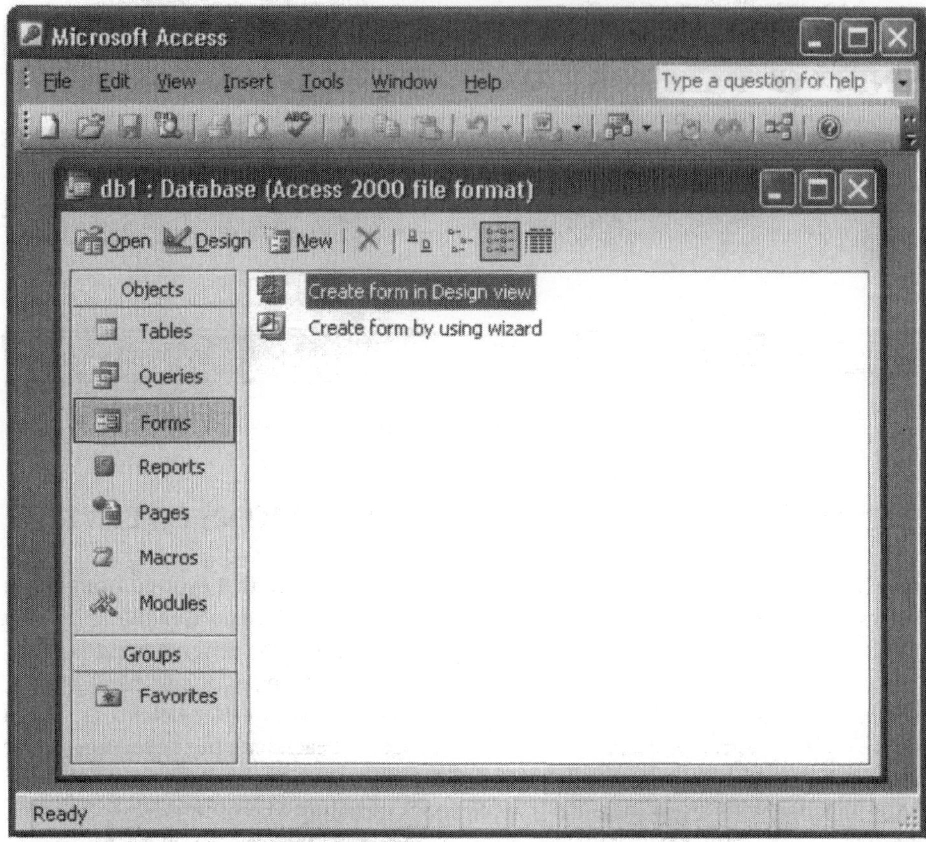

Figure 6C.1 The database window.

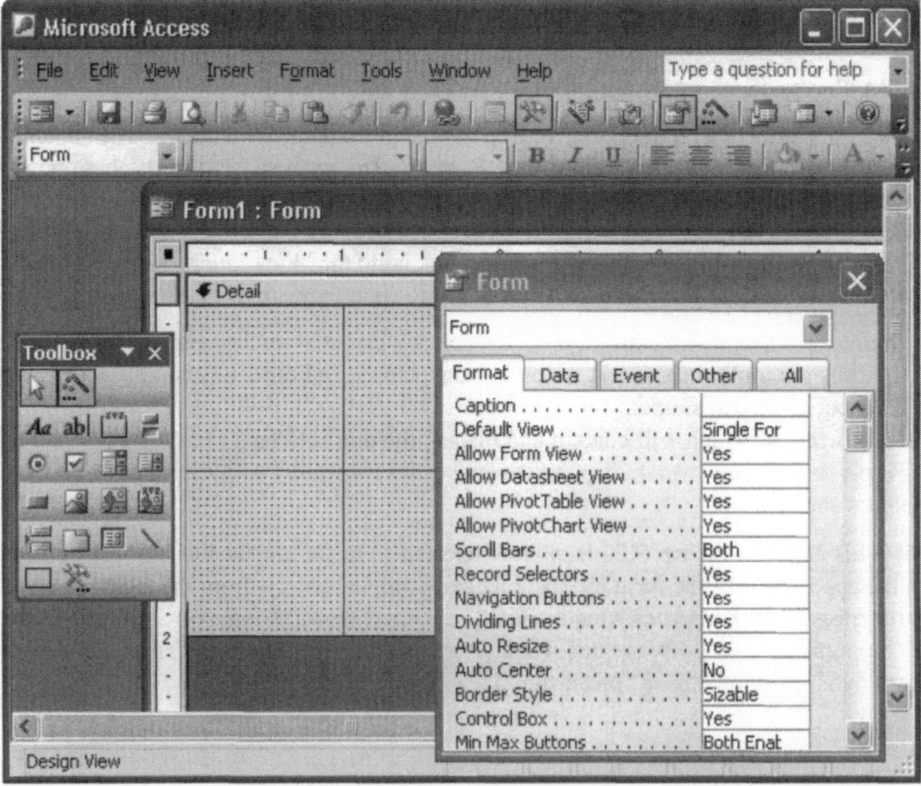

Figure 6C.2 Object view.

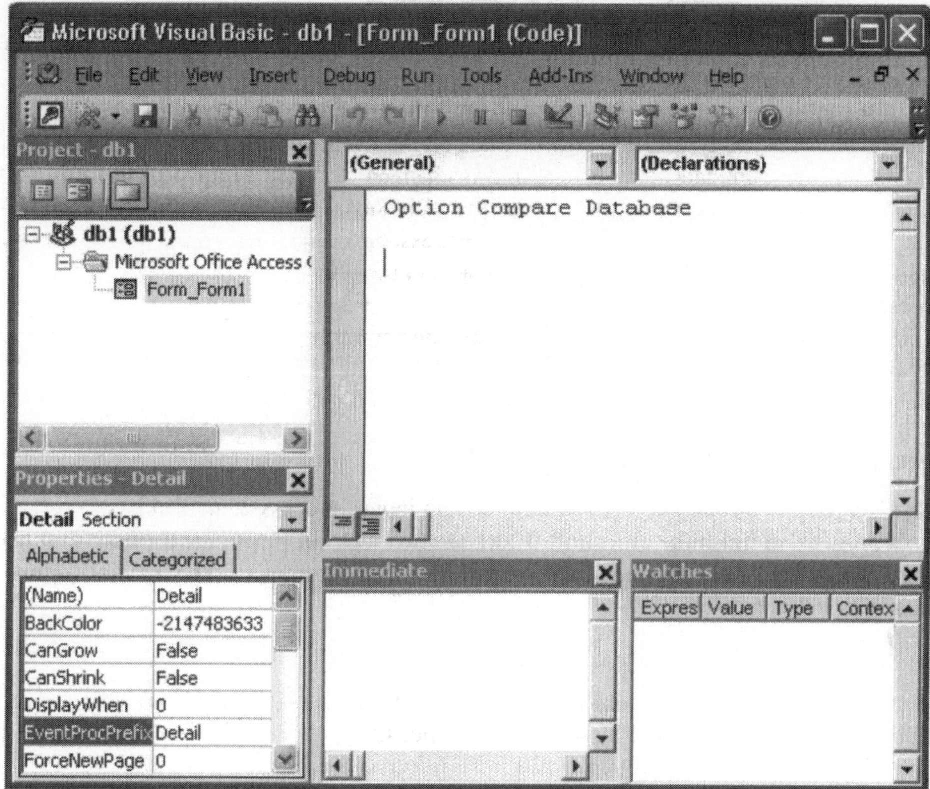

Figure 6C.3 Code view.

WORKING WITH CONTROLS

Before writing an event procedure for a control to respond to a user's action, certain properties for the control have to be set to determine its appearance and how it will work with the event procedure. The properties of the controls can be set in the Properties window shown in the lower left-hand corner of Figure 6C.3. While we will not describe all the properties, we will discuss some commonly used ones. For further information, the user should consult the Help menu item provided with Microsoft Access.

A meaningful name for the Name property should be set since this is the handle that is used in the event procedures to manipulate the properties and the functionality of the control.

The Caption property of a control is displayed in the form or report and should describe the functionality of the control. For example, a button control should display what would happen if the button is clicked by the user at run-time (e.g., "Close Window" or "Add Record").

Another property that is important is the Visible property, which determines whether the control is visible or not at startup. This property can only be set as true or false. The *Enabled* property is similar to the visible property and determines if the events of the associated control are executable or not.

As a quick exercise to demonstrate controls and events, adding a command button control to a form, open the design view of a form as shown in Figure 6C.2. From the toolbox, click on the "command button" control. This enables the icon. Move the cursor to the middle of the Form window, and click the left mouse button again. A rectangular button will appear on the screen and a window titled "*Command Button Wizard*" will appear. Close the wizard by clicking on the *Cancel* button. In the Properties window that opens up after the creation of the command button, click on the Caption property and type "Hide Form." Click on the Event tab in the Properties window. You will notice that the command button's caption is changed to "Hide Form" in the form, and the Properties window is displaying a list of all the events (actions) associated with a command button control. Writing code to be executed when an event associated with the command button control, such as the "on_click" event, will be discussed later in the tutorial.

Access Exercises in Section 6.4.1 through Section 6.4.3 demonstrate how these properties are set at design time using the Properties window and during runtime in the event procedures, using the *dot notation*.

Working with Built-in Functions

The main purpose of functions is to accept certain inputs, process them, and pass a value on to the main program to finish the execution. There are two types of functions, the built-in functions (or internal functions) and the functions created by the programmers. The general format of a function is as follows:

```
functionName(arguments)
```

where arguments are values that are passed on to the functions. Two of the most useful built-in functions are the MsgBox ( ) and InputBox ( ) functions.

MsgBox ( ) Function

The objective of MsgBox is to produce a popup message box and prompt the user to click on a command button before he or she can continue. The message box format is as follows:

```
myMsg=MsgBox(Prompt, Style Value, Title)
```

The first argument "Prompt" will display the message in the message box. The "Style Value" determines what types of Command buttons appear on the message box. Table 6C.1 lists the types of command buttons displayed. The named constant can be used in place of integers for the second argument to make the programs more readable. The "Title" argument will display the title of the message box.

Table 6C.1
Style Values

Style value	Named constant	Buttons displayed
0	vbOkOnly	OK button
1	vbOkCancel	OK and Cancel buttons
2	vbAbortRetryIgnore	Abort, Retry, and Ignore buttons
3	vbYesNoCancel	Yes, No, and Cancel buttons
4	vbYesNo	Yes and No buttons
5	vbRetryCancel	Retry and Cancel buttons

Table 6C.2

Return Values and Command Buttons

Value	Named constant	Button clicked
1	vbOk	Ok button
2	vbCancel	Cancel button
3	vbAbort	Abort button
4	vbRetry	Retry button
5	vbIgnore	Ignore button
6	vbYes	Yes button
7	vbNo	No button

Table 6C.3

VBA Icons

Value	Named constant	Icon
16	vbCritical	
32	vbQuestion	
48	vbExclamation	
64	vbInformation	

The variable "myMsg" is an integer variable that holds values that are returned by the MsgBox ( ) function, which are determined by the button clicked by the user. Table 6C.2 shows the values, the corresponding named constant, and buttons.

To make the message box appear more sophisticated, the user can add an icon next to the message. Four types of icons are available in VBA as shown in Table 6C.3.

The InputBox( ) Function

An InputBox( ) function displays a message box where the user can enter a value or a message in the form of text. The format is as follows:

```
myMessage=InputBox(Prompt, Title, default_text, x-position, y-position)
```

Note that "myMessage" is a variant data type. The Prompt and Title arguments are similar to the MsgBox ( ) function. "Default_text" appears in the input field for the user to accept or modify. x-position and y-position is the position of the input box on the screen. The default position is to center the message box on the screen. Exercises in Section 6.4.4 through Section 6.4.5 demonstrate the use of the built-in MsgBox ( ) and InputBox ( ) user functions.

User Defined Functions and Subroutines

In addition to built-in functions such as the MsgBox( ) and InputBox ( ), and the event procedures associated with controls such as _Click and _Load, the user can define customized functions and subroutines referred to as *general procedures.* The main difference between functions and subroutines is that, while subroutines only execute a piece of code, functions, in addition, return a value to the main program from which they are called. General procedures are declared and defined in the *General* section of Forms, Reports, or Modules. Modules, unlike forms and reports, do not have a graphical user interface and simply contain user-defined code. Modules can be added to an application by clicking on "*Module*" under the "*Insert*" menu item. To insert a general procedure, select the Code view (Figure 6C.3), and in the Project window (top left), double-click the form, report, or module, which will house the user defined function or subroutine. Click on "*Procedure*" under the Insert menu item. Figure 6C.4 shows the Add Procedure window, which contains the options to create the user-defined function or subroutine.

"Public" procedures in modules can be called anywhere in the project, and "Private" indicates that the procedure is only applicable to the module in which it is declared. If a procedure will only be called from a certain form or report, then it can be declared within that form's or report's *General* section. After setting the desired parameters, click the OK button. The function or subroutine will automatically appear in the Code view. The user can now enter Visual Basic statements.

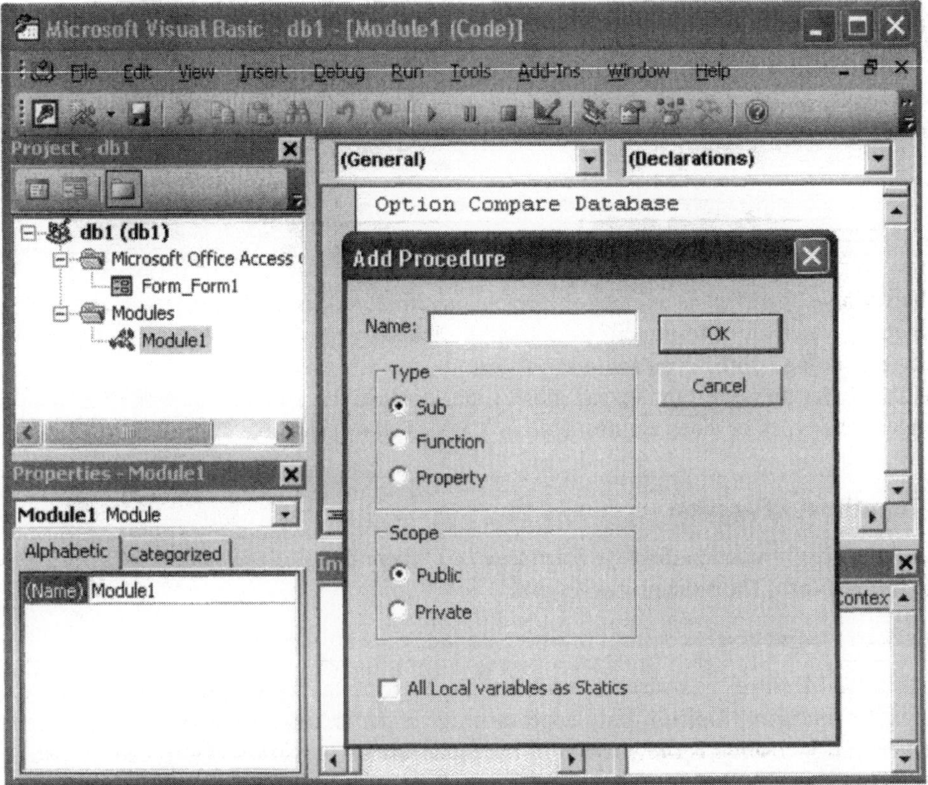

Figure 6C.4 Add Procedure window for user-defined procedure.

In order to enter Visual Basic code, add a module as described previously. Note that in VBA, modules are automatically named as Module 1, Module 2, and so on. To add a procedure to Module 1, double-click on "Module 1" in the Project window and click on "*Procedure*" under the Insert menu item. The Add Procedure window will appear as shown in Figure 6C.4. For the name of the subroutine, type "hide_form". Do not modify the other options so that the subroutine created is a Public procedure. Click the OK button. You should see the following code in the code window:

```
Public Sub hide_form()

End Sub
```

Between these two statements, type the following VBA statement:

```
Form_Form1.Visible=False
```

To execute this statement, which will make Form1 invisible, switch to the object view by clicking on "*Object*" under the View menu item. Click on the command button that was created with caption "Hide Form," and in the Properties window click on the Event tab. Now click on the "On_Click" event, and, from the pull-down menu, select "[Event Procedure] and click on the icon with three dots that appears to the right of the "On_click" event. This will bring up the *Code View* with a procedure for the on_click event of the command button control. In the procedure type "Call hide_form". In the code view, you should see the following:

```
Private Sub Command1_Click()
Call hide_form
End Sub
```

In the Database window, double-click on Form1 to execute the form, and click on the Hide Form button. This will call the hide_form ( ) procedure, which in turn will make Form1 invisible.

Data Types

The data types in VBA can be divided into two categories as numeric and non-numeric data types. Numeric data consist of numbers, which can be computed mathematically with various standard operators such as add, subtract, multiply, divide, and so on. The numeric data are divided into seven types as shown in Table 6C.4. The non-numeric data types are summarized in Table 6C.5.

Table 6C.4

Numeric Data Types

Type	Storage	Range of values
Byte	1 byte	0 to 255
Integer	2 bytes	–32,768 to 32,767
Long	4 bytes	–2,147,483,648 to 2,147,483,648
Single	4 bytes	–3.402823E+38 to –1.401298E–45 for negative values
		1.401298E–45 to 3.402823E+38 for positive values
Double	8 bytes	–1.79769313486232e+308 to –4.94065645841247E–324 for negative values
		4.94065645841247E–324 to 1.79769313486232e+308 for positive values
Currency	8 bytes	–922,337,203,685,477.5808 to 922,337,203,685,477.5807
Decimal	12 bytes	+/– 79,228,162,514,264,337,593,543,950,335 if no decimal is use
		+/– 7.9228162514264337593543950335 (28 decimal places)

Table 6C.5

Non-numeric Data Types

Data type	Storage	Range
String (fixed length)	Length of string	1 to 65,400 characters
String (variable length)	Length +10 bytes	0 to 2 billion characters
Date	8 bytes	January 1, 1,000, to December 31, 9999
Boolean	2 bytes	True or false
Object	4 bytes	Any embedded object
Variant (numeric)	16 bytes	Any value as large as double
Variant (text)	Length +22 bytes	Same as variable-length string

Table 6C.6

Conditional Operators

Operator	Meaning
=	Equal to
>	More than
<	Less than
>=	More than and equal
<=	Less than and equal
<>	Not equal to

MANAGING VARIABLES

In Visual Basic, variables are areas allocated in the computer memory to hold data. Each variable must be given a name. The name of a variable must be fewer than 255 characters and cannot begin with a number. No spaces or periods are allowed in variable names. The user needs to declare the variables before using them by assigning names and data types. They are normally declared in the General section of the Code window using the **Dim** statement. These variables are global variables and their values can be used or updated by any procedure within the program. Variables local to subroutines and functions can be declared at the beginning of each subroutine or procedure. The values of local variables can only be used by the subroutine or function within which they are declared. The declaration format is as follows:

```
Dim VariableName As DataType
```

If the data type is not specified, the variable is automatically declared as a variant (see Table 6C.5). Declarations of variables are demonstrated in the "Working with Data Files" section.

CONTROLLING PROGRAM FLOW

Conditional Operators

Conditional operators are one of the tools that can be used to control program flow. They resemble mathematical operators and let the programmer compare data values and decide what action to take from a set of actions associated with the outcomes of the comparisons. Conditional operators are shown in Table 6C.6.

Logical Operators

In addition to conditional operators, there are logical operators that can be used to control program flow. These are shown in Table 6C.7.

Table 6C.7

Logical Operators

Operator	Meaning
And	Both sides must be true
Or	One side and/or other must be true
Xor	One side or other must be true but not both
Not	Negates truth

Using If . . . Then . . . Else Statements

To control the program flow, an ***If . . . Then . . . Else*** statement can be used with the conditional and logical operators. The general format for the if . . . then . . . else statement is

```
If <conditions> Then
      <instruction>
Else
      <instruction>
End If
```

It may not always be necessary to use the Else clause, if the complementary conditions of the If clause does not require an action. Consider the following example:

```
'The following program evaluates your division skills
userinput = InputBox("what is 100 divided by 5?")
If userinput = "20" then
        Response = MsgBox("CORRECT!!", vbOkOnly)
Else
        Response = MsgBox("Sorry, the correct answers is 20.", vbOkOnly)
End If
```

Using Select Case Statements

If there are a number of conditions that lead to different outcomes, then the ***If . . . Then . . . Else*** structure may not be suitable. For multiple conditional statements, it may be more efficient to use the ***Select Case*** structure, whose format is as follows:

```
Select Case <expression>
Case <value1>
      <instruction>
Case <value2>
      <instruction>
.
.
Case Else
      <instruction>
End Select
```

Consider the following example:

```
'Course Grades
grade = InputBox("What is your grade for the Calculus I course?")
Select Case grade
Case "A"
       Response = MsgBox("Excellent!!", vbOkOnly)
Case "B"
       Response = MsgBox("Good!", vbOkOnly)
Case "C"
       Response = MsgBox("Satisfactory", vbOkOnly)
Case "D"
       Response = MsgBox("Please try a little harder.", vbOkOnly)
Case "F"
       Response = MsgBox("You will have to take the course again.
Sorry", vbOkOnly)
Case Else
       Response = MsgBox("That is not a valid grade for the course",
vbOkOnly)
End Select
```

Looping

Looping allows a procedure to be repeated as many times as the processor can support. Looping can be accomplished by the **For . . . Next** or the **Do . . . Loop** structure. The structures of the **For . . . Next, the top-test Do . . . Loop,** and **the bottom-test Do . . . Loop** are as follows:

```
For <counter> = <expression> to <expression> Step <step size>
       <instruction>
Next <counter>

Do [While <condition> or Until <condition>]
     <instruction>
Loop

Do
     <instruction>
Loop [While <condition> or Until <condition> ]
```

For . . . Next structures are used when the number of times the instructions must be executed are known. The **Do . . . Loop** can be in the form of a top-test loop or a bottom-test loop. The top-test Do . . . Loop is used when the condition must be tested before the instructions are executed, and the bottom-test Do . . . Loop is used if the instructions are to be executed once before the condition is tested. Both of these loops can be exited using the **Exit For** and the **Exit Do** commands during execution of the instructions within the loops. Examples that use **For . . . Next** structures are shown in the "Working with Files" section. Consider the following examples for the **Do . . . Loop** structures:

```
'The program below will loop until the string entered in the Inputbox
is "1234"
Password = "1234"
Do While Password <> Userinput
     Userinput = InputBox("Enter Password")
```

```
Loop

'The program below will loop until the user clicks the "No" button
Do
     Response = MsgBox("Do you want to continue looping?", vbYesNo)
Loop While Response = vbYes
```

ARRAYS

An array is a list of variables with the same data type and name. Arrays are useful when the programmer is dealing with a number of variables of the same data type that have a common use or function. For example, if the programmer requires the name of 100 students, instead of declaring 100 different variables, an array with dimension 100 (indexed 0 to 99) can be used. Each item in the array is differentiated by using the index value of each item; for example, Studname(0) refers to the name of the 1st student in the array, whereas Studname(52) refers to the name of the 53nd student. Arrays of multiple dimensions (matrixes) can also be used to store values. The Dim statement is used to declare an array just like variables. The general format to declare an array is as follows:

```
Dim ArrayName(arraysize) As DataType
```

The statement "**Dim** StudName(100) **As** String" would define an array of size 101 for student names starting from StudName(0) and ending with StudName(100). If the statement **Option Base 1** appears in the declaration area, StudName(0) is not defined and the array is defined for StudName(1) to StudName(100) for a total of 100 variables. Declarations and the use of arrays are demonstrated in the next section.

WORKING WITH DATA FILES

Variables and arrays are created and deleted during run-time. When the program terminates, all the values associated with variables and arrays are deleted from memory. Some applications may require that the information generated during program execution is not deleted after the program terminates. Other applications may require that data inputs to a program are retrieved from files (these may be information recorded by other programs) rather then users. One of the methods of capturing or retrieving information during program execution is to write or read the information in data files.

To write to a data file, the programmer must first create (open) a file for *output*. The format is

```
Open "filename" For Output As # filenumber
```

Each file created must have a filename and a file number for identification. The filename can include a path if different from the path of the program. For example, the statement

```
Open "C:\My Documents\studentname.txt" For Output As #1
```

will create a text file by the name of "studentname.txt" in the My Documents folder under the C directory. The file number is 1. The command that can be used to write to an already open data file is

```
Write #filenumber, <expression>
```

After a file is used for output, it must be closed using the

```
Close #filenumber
```

command. The following example opens a data file named "Studentgrades.txt," inputs 10 student names and grades from the user, writes these to the data file, and closes the data file. Note that the

Studentsname and Studentgrade variables are declared within the procedure that the code is executed in and therefore are local to the procedure. If it is necessary to access these variables from outside the procedure that follows, then they should be declared in the general section of a form, report, or module.

```
Dim Studentsname As String
Dim Studentgrade As Integer
Open "C:\My Documents\studentgrades.txt" For Output As #1
For i = 1 to 10
Studentname = InputBox("Enter the student name")
Studentgrade = InputBox("Enter the student grade")
Write #1, StudentName, Studentgrade
Next i
Close #1
```

To read a data file, the "**Input** #filenumber" statement is used after the data file is opened for input. The following program reads the information associated with the 10 students recorded using the preceding program and assigns these values to an array. Note that the conditions associated with the declarations of the following arrays are the same as the conditions associated with the declarations of the variables as discussed earlier.

```
Dim studentname(10) As String
Dim studentgrade(10) As Integer
Open "c:\My Documents\studentgrades.txt" For Input As #1
For i = 1 to 10
Input #1, studentname(i), studentgrade (i)
Next i
Close #1
```

Print # is similar to the **Write** # statement; however, there are important differences. If, at some future time, the data will be read back from a data file using the **Input** # statement, use the **Write** # statement instead of the **Print** # statement to write the data to the file. Using **Write** # ensures the integrity of each separate data field by properly delimiting it, so it can be read back in using **Input** #. **Print** # writes an image of the data to the file, therefore the programmer must delimit the data so they prints correctly.

Working with Databases

Databases may replace the use of data files and are a better and more organized alternative in storing and retrieving data. Microsoft has provided more than one way to connect to a database using VBA objects. One of the first and simplest methods is the Data Access Object (DAO) model, which is discussed in Section 6.4.5.3.

ActiveX Data Objects (ADO) is an upgrade to DAO and can be used for database manipulation in situations not covered by DAO. ADO is more in tune with database access on the Internet. A detailed description on how ADOs can be used to establish connectivity with databases is included in Section 8.3.5.1 in relation to the use of databases with web pages. If the VBA code that interacts with the database is housed within an Access Database project (e.g., in a form, report, or module), then connecting to the database is accomplished using the *CurrentProject Object*, which refers to the database project that contains the VBA code. The following set of statements connects to the database in the Access project that the VBA code is in and opens a recordset:

```
Dim myconnection As ADODB.Connection
Dim myrecordset As ADODB.Recordset
Set myconnection = CurrentProject.Connection
Set myrecordset = New ADODB.Recordset
myrecordset.Open "[table name]", myconnection, adOpenKeyset,
  adLockOptimistic, adCmdTable
```

The following example code shows how the student names and grades previously recorded in a data file can be recorded in a database. Before executing the following code, make sure that the database project is created containing a table named "students" and it has two fields named "name" and "grade." It is also good programming practice to close the ADO objects when they are no longer needed.

```
Dim myconnection As ADODB.Connection
Dim myrecordset As ADODB.Recordset
Dim Studentsnam As String
Dim Studentgrad As Integer
Dim studentname(10) As String
Dim studentgrade(10) As Integer

Set myconnection = CurrentProject.Connection
Set myrecordset = New ADODB.Recordset
myrecordset.Open "students", myconnection, adOpenKeyset,
  adLockOptimistic, adCmdTable
For i = 1 To 10
Studentnam = InputBox("Enter the student name")
Studentgrad = InputBox("Enter the student grade")
With myrecordset
     .AddNew
     .Fields("name") = Studentnam
     .Fields("grade") = Studentgrad
     .Update
End With
Next i

myrecordset.Close
myconnection.Close
```

After executing the code, check the students' table to verify that the entries are recorded. Similarly, the recorded data in the students' table can be retrieved by adding the following statements before the myrecordset. Close statement:

```
With myrecordset
     .MoveFirst
     For i = 1 To 10
      studentname(i) = .Fields("name").Value
      studentgrade(i) = .Fields("grade").Value
     .MoveNext
     Next i
End With
```

The *.movefirst* and *.movenext* methods are used to navigate within a recordset, and the *value* property returns the value in the specified field for the current record. Many other methods and properties are associated with ADO objects, which are covered in Section 8.3.5.1 and in the Help menu item. We recommend that VBA programmers dealing with ADO objects read the relevant sections to effectively use these methods and properties.

APPENDIX 6D GROUP LEVEL AND USER LEVEL SECURITY

Data security and integrity are of great concern to the enterprise using a DBMS. The individual responsible for security, as well as other database administrative functions, is the database administrator (DBA). The DBA is a person who has both managerial and technical skills and is responsible for the overall strategy of how the enterprise evolves its information systems and how the organization interacts with the database. This is a crucial function in an organization that has an enterprise-wide information system. Among the ways the DBA tries to maintain security is to limit access only to authorized users. One approach to doing this is to define "user groups," which are classes of users who have common requirements for accessing data. That is to say, they have read, insert, update, and delete requirements for the same tables, forms, and reports. In this way, database privileges are defined as group rights.

Individual users are then assigned to be members of a user group, with all the privileges and rights that pertain to the group. Each user is given (or creates) a unique password. The password is required when accessing the database. The password identifies the user and, by association, the user group. The privileges granted to the group are made available to the individual.

In Microsoft Access, the process of defining groups, users, and privileges is done by setting **user-level security**. Access provides a **user-level security wizard** to assist in this process. To apply user-level security, you must have administrative privileges on the machine. In this appendix, we review the steps necessary to use the wizard. Again, to set security levels, you must be logged on the machine as the administrator. The main steps in the process are as follows:

1. Establish a workgroup information file.
2. Specify the database objects that you wish to secure.
3. Specify the user groups that you want to include in the workgroup information file.
4. Grant permissions to users.
5. Assign individual users and their passwords to user groups.

Before beginning this process, make a copy of the database you want to secure in a separate folder. Rename the database. Use this database to run the example in this appendix. In the event you make an error, you may want to delete your work, so this will prevent you from losing your original database.

ESTABLISH A WORKGROUP INFORMATION FILE

Launch Access DBMS, and open the database you want to secure. Open the user-level security wizard by following the path **Tools → Security → User-Level Security Wizard**, as shown in Figure 6D.1. This will launch the window shown in Figure 6D.2.

Two choices are available to the database administrator. If the database you are trying to secure has no prior security levels installed, the selection defaults to "Create a new workgroup information file." If, on the other hand, the administrator has previously set security levels for some users and

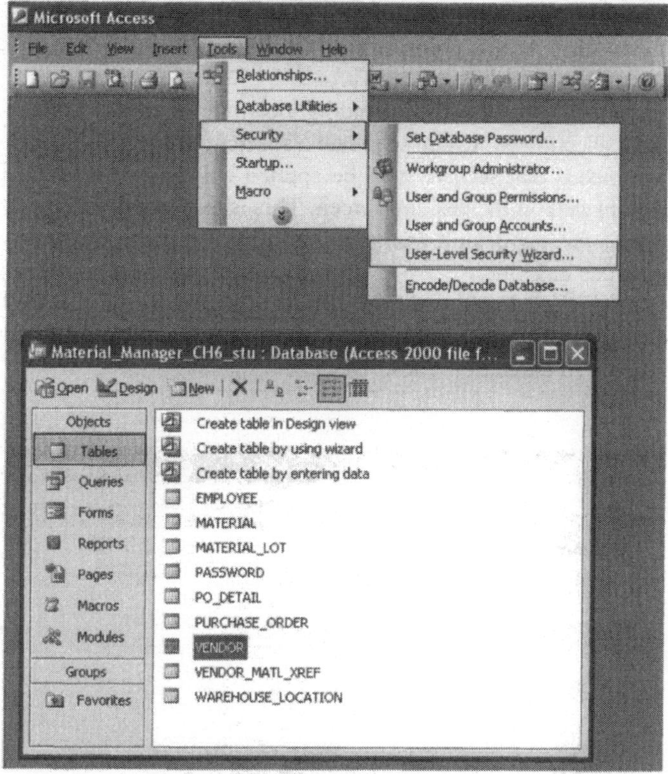

Figure 6D.1 Launching the security wizard.

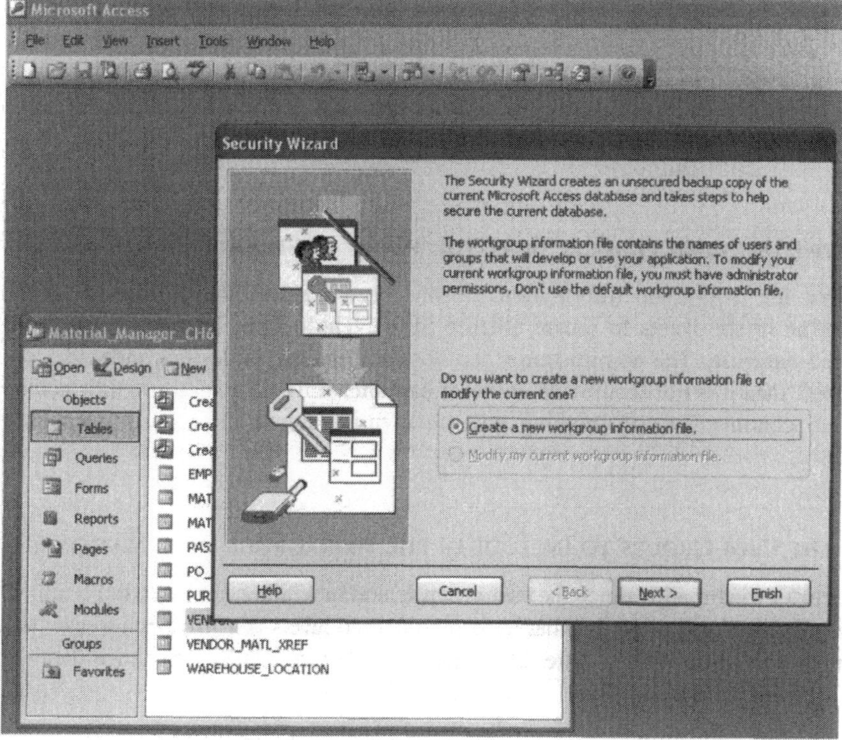

Figure 6D.2 Creating a new workgroup information file.

wants to add new users or new user groups, he or she will be allowed to "Modify my current workgroup information file." At this point, select "Create a new workgroup information file" and click "Next."

The security wizard will now launch the window shown in Figure 6D.3, with the two selections indicated. The default selection is "I want to create a shortcut to open my security-enhanced database." This selection means that security will be applied only to the currently selected database and a shortcut will be placed on the desktop screen. This is the selection you want to make. If you choose "I want to make this my default workgroup information file," the security will apply to *all* Access databases on your machine.

Note that the security wizard automatically generated the workgroup ID (WID). You can add your name and company as optional information. After selecting "I want to create a shortcut to open my security-enhanced database," click "Next."

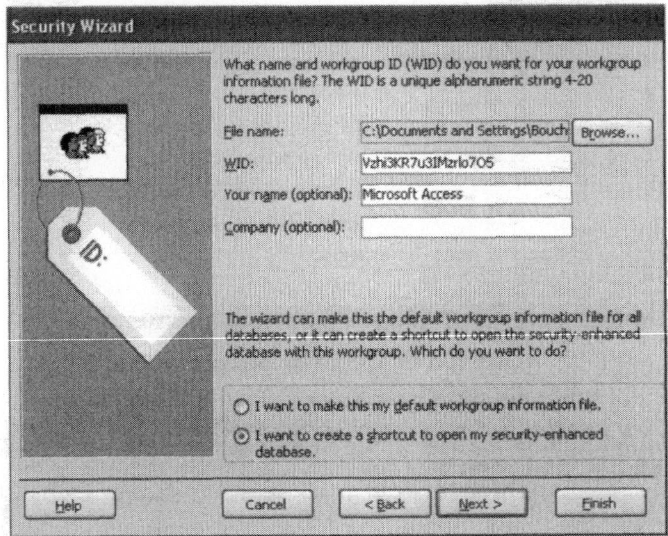

Figure 6D.3 Creating a shortcut to the secure database.

Specify the Database Objects That You Want to Secure

Now that the workgroup information file has been created, the administrator can specify the objects he or she wants to secure (Figure 6D.4). The default selection is all the objects of the selected database. The administrator can deselect objects as desired. However, if an object is deselected, then it is not secure and any user can alter it. If all objects are deselected, it makes no sense to continue because nothing will be secured. Leave all objects selected, and click "Next."

Specify the User Groups to Include in the Workgroup Information File

The security wizard will launch the user group selection window as shown in Figure 6D.5. There are seven prespecified user groups that have prespecified levels of permission to use the database. For example, "Read-only Users" are allowed to view but not create or change data. This is the

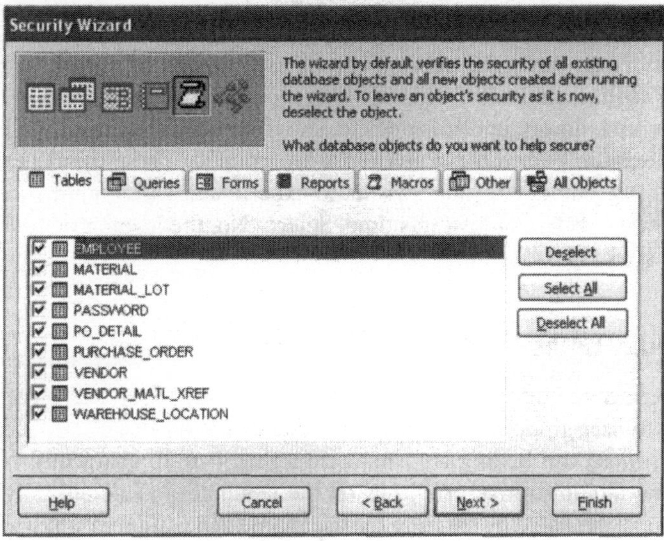

Figure 6D.4 Selecting database objects to secure.

Figure 6D.5 Selecting user groups.

default setting for this group. As we will see later, the administrator may grant additional privileges to the group, but initially the permission level is read only. On the other hand, the user group "Full Permissions" has permission to access all database objects for insert, delete, update, and read only. However, this group does not have permission to grant privileges to other users. That privilege resides with the database administrator. Select "Read-only Users," and click "Next."

Grant Permissions to User Groups

The administrator can now add permissions to user groups as shown in Figure 6D.6. For example, although we have defined a read-only user group, we can grant additional permissions to members of that group to insert, update, or delete data if we wish. Technically speaking, this group will no longer be read-only once these privileges are granted. These privileges will apply to all members of this group. Although you may want to experiment with granting privileges later, stay with the default setting of read-only at this time. Select "No, the Users group should not have any permissions," and click "Next."

Assign Individual Users and Their Passwords to User Groups

The security wizard will open the window shown in Figure 6D.7. This is the first of two steps in assigning users to user groups. The first step is to declare the user names and to assign them a password. The password can be any alphanumeric string 4 to 20 characters long. Note that the individual(s) having administrative privileges on the computer is automatically listed in the left side panel. New users are added by entering the user name and assigning a password, then clicking on the command button "Add this user to the list." Enter a user name and password, add it to the list, and then click "Next."

The second step is to assign the users to a user group. This is shown in Figure 6D.8. First select a user from the pull-down list, then place a check mark in the appropriate group. Only groups that have been previously established will be displayed. If you have been following this tutorial on your computer, you will see that "Read-only Users" and "Admins" groups are displayed. Add the user to the read-only user group, and click "Next."

At this point, the process of assigning security levels is complete. The security wizard will make a backup copy of your database. You can select the path for saving it; otherwise it will be placed in the same folder as your secure database, as shown in Figure 6D.9. Click "Finish."

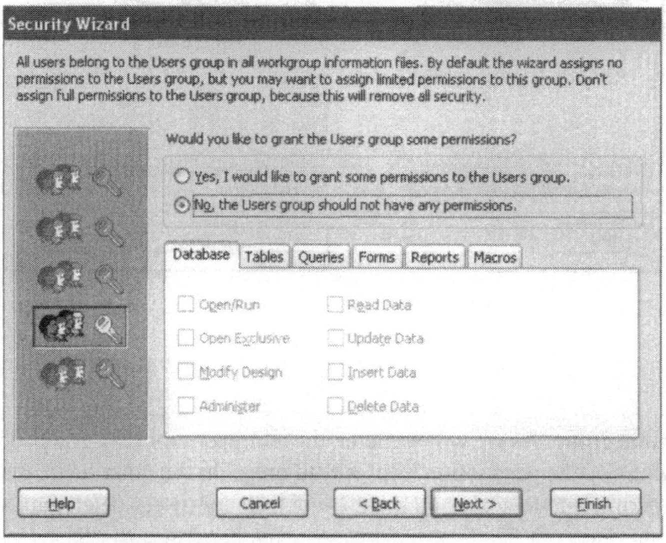

Figure 6D.6 Granting additional permissions.

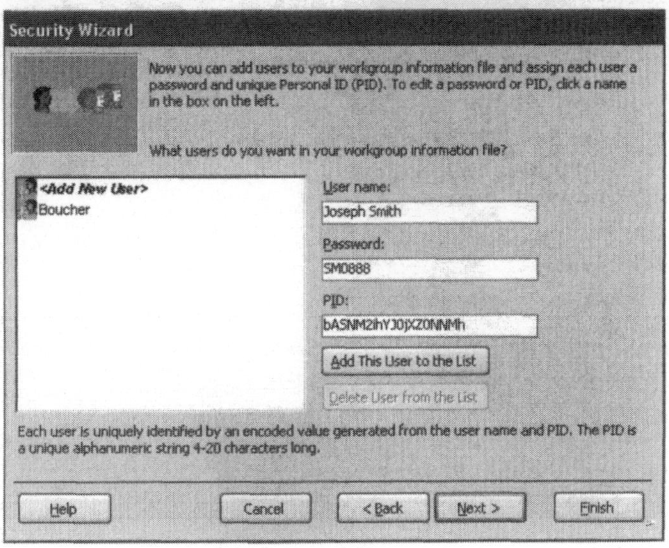

Figure 6D.7 Assigning passwords to users.

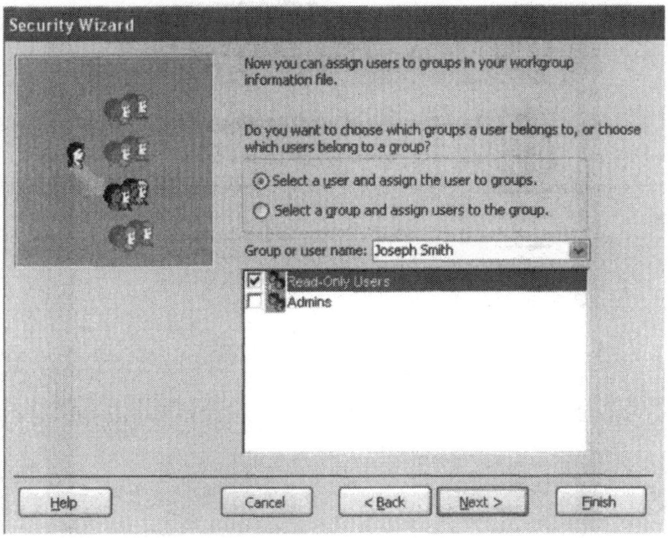

Figure 6D.8 Assigning users to groups.

The security wizard will now provide a report that includes the group IDs, user names, and user passwords for the database administrator as shown in Figure 6D.10. This is important information and should be printed and saved.

Close the database, and note that a shortcut icon had been placed on the desktop. Double-click on the shortcut, and a password screen will open. Using the user name and password that you have just created, launch the database. Note that all tables, forms, and so on are read only.

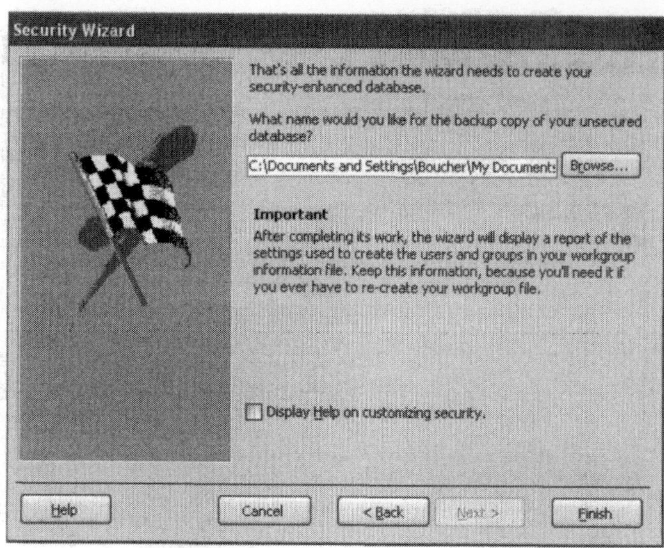

Figure 6D.9 Saving a backup database.

Security Wizard

One-step Security Wizard Report

Page: 1

Figure 6D.10 Security report.

Repeat this exercise, and "Modify the current workgroup information file" by adding a workgroup for "Full Permissions." Add a user to this workgroup. When you launch the database and sign on as the user with full permission, you will be able to insert, delete, and update data.

You can locate the workgroup information file and secure the database in the folder used for the database. If you want to remove the secure database from your machine, open the folder in Windows Explorer, delete the workgroup information file, and then the database. Go to your desktop and remove the shortcut icon.

The implementation of workgroups and users belonging to workgroups is a common method used by database administrators to control access to a database. Here we have demonstrated a way to do this using Access DBMS.

Chapter 7

Executing an Information System Design Project: A Case Study

7.1 INTRODUCTION

In the previous chapters, we presented some of the modeling techniques that are useful during the process of designing an industrial information system. In the context of an information system design project, the system analyst must decide how and when to apply these tools. The purpose of this chapter is to bring together the modeling tools discussed in the previous chapters to demonstrate their application in a case study design project.

A design project involves much more than converting paper documents into a data model, as we did in Chapter 5. Most design projects require extensive study before any attempt is made to develop a data model. Some basic questions need to be answered:

- What are the goals of the project?
- What are the boundaries of the system that is the object of the design?
- What are the needs of the prospective users?
- Who are the sources for gathering relevant information?

By going through a typical small design problem in some detail, the reader will have an opportunity to see how these and other issues influence the activities of the systems analyst. Although the activities to be described are not always entirely sequential, the approach will be formalized by suggesting a sequence of steps. These steps are summarized in phases, as shown in Figure 7.1. We provide explanations as the case study description proceeds.

7.2 PRELIMINARY STUDY AND PROBLEM DEFINITION PHASE

During the initial stages of a project, the analyst must decide what the client wants from the information system. Much of the analyst's time is consumed with understanding how the target system actually functions and how it interfaces with any existing information system. The analyst must depend on personnel familiar with the operations of the organization at various levels. This is an information-gathering stage that should result in documenting the

1. **Preliminary Study and Problem Definition Phase (7.2)**
   a) **Description of operations (7.2.1)**
   b) **Identifying system redesign objectives (7.2.2)**
   c) **Defining "as is" IDEF0 activity diagram and establishing system boundaries (7.2.3)**

2. **Design Phase (7.3)**
   a) **Identifying user information needs (7.3.1)**
   b) **Defining entities and relationships (7.3.2)**
   c) **Defining attributes and domains (7.3.3)**
   d) **Establishing global data model (7.3.4)**
   e) **Defining superclass/subclass relationships (7.3.5)**
   f) **Evaluating the need for transaction entities (7.3.6)**
   g) **Normalizing data model (7.3.7)**
   h) **Finalizing and validating "to be" IDEF0 activity diagram (7.3.8)**
   i) **Finalizing and validating IDEF1 global information model (7.3.9)**

3. **Implementation Phase**
   a) **Creating database tables and relationships**
   b) **Defining data entry methods**
   c) **Creating queries, forms, and reports**
   d) **Establishing data security and access methods**

Figure 7.1 Phases of a design project.

operations within the scope of the project in a formal way. In our methodology, the IDEF0 functional diagram is the preferred formalism. We break down the activities in this phase as (1) establishing a description of operations, (2) identifying system redesign objectives, and (3) defining "As Is" IDEF0 activity diagram, including system boundaries.

7.2.1 Description of Operations

The University Food Company manufactures shelf-stable and refrigerated packaged food products for university dining halls and supermarket brand labels under contracts with supermarkets. The company has been automating some of its business systems. For example, the company previously developed a material tracking system for controlling incoming raw materials (see Chapters 4 and 5). This tracking system establishes lot numbers for ingredients and other raw materials and maintains a record of the location of each material lot in the warehouse. Management wishes to extend the information system to include the gathering of information about the manufacture of the products as well as the control of finished goods after manufacture. You have been selected to lead this project.

The first step of the project is to study the operations involved in the manufacture of the products and to understand the general objectives of management and production personnel for the information system. Several individuals have been identified as key sources of information on the client side. These include the plant manager, the production line supervisor, the quality control manager, and the production personnel working on the line. Each of these individuals has a different perspective on the requirements and use of an information system, and before this study is complete, it will be necessary to understand the concerns of each. The following paragraphs describe how this first step might evolve.

The plant manager has instructed the production line supervisor and production personnel to brief you on the operations in the target area. To perform his briefing, the production supervisor has prepared three exhibits, shown as Figures 7.2, 7.3, and 7.4.

Figure 7.2 is the bill of materials (BOM) structure for a "typical" product manufactured by University Food Company. The product is macaroni and cheese, which is a packaged and

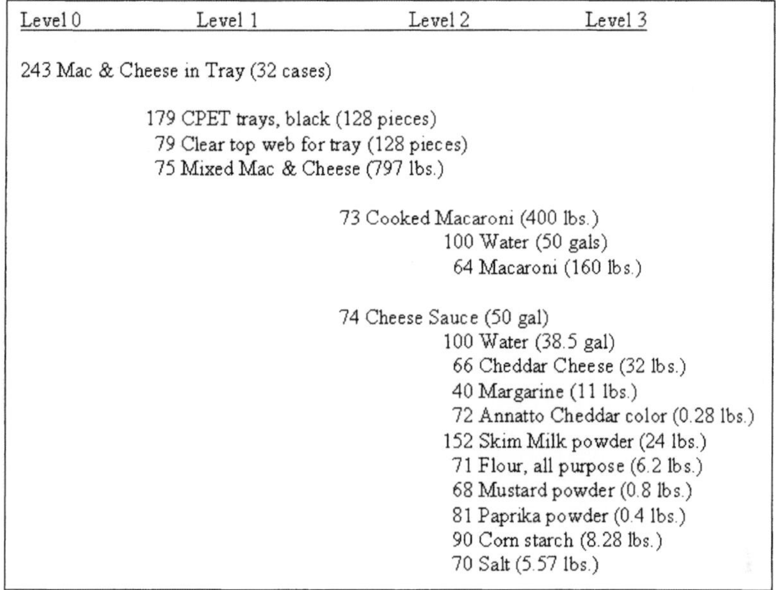

Level 0	Level 1	Level 2	Level 3
243 Mac & Cheese in Tray (32 cases)			
	179 CPET trays, black (128 pieces)		
	79 Clear top web for tray (128 pieces)		
	75 Mixed Mac & Cheese (797 lbs.)		
		73 Cooked Macaroni (400 lbs.)	
			100 Water (50 gals)
			64 Macaroni (160 lbs.)
		74 Cheese Sauce (50 gal)	
			100 Water (38.5 gal)
			66 Cheddar Cheese (32 lbs.)
			40 Margarine (11 lbs.)
			72 Annatto Cheddar color (0.28 lbs.)
			152 Skim Milk powder (24 lbs.)
			71 Flour, all purpose (6.2 lbs.)
			68 Mustard powder (0.8 lbs.)
			81 Paprika powder (0.4 lbs.)
			90 Corn starch (8.28 lbs.)
			70 Salt (5.57 lbs.)

Figure 7.2 Bill of material structure for macaroni and cheese.

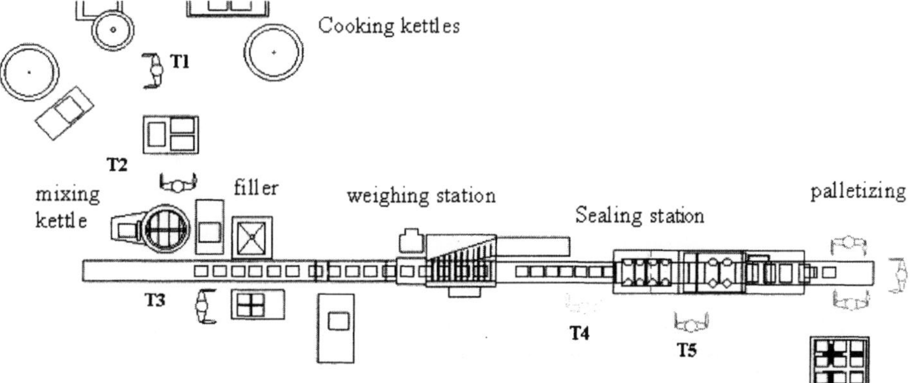

Figure 7.3 Production line layout. (Courtesy of Rutgers Center for Advanced Food Technology.)

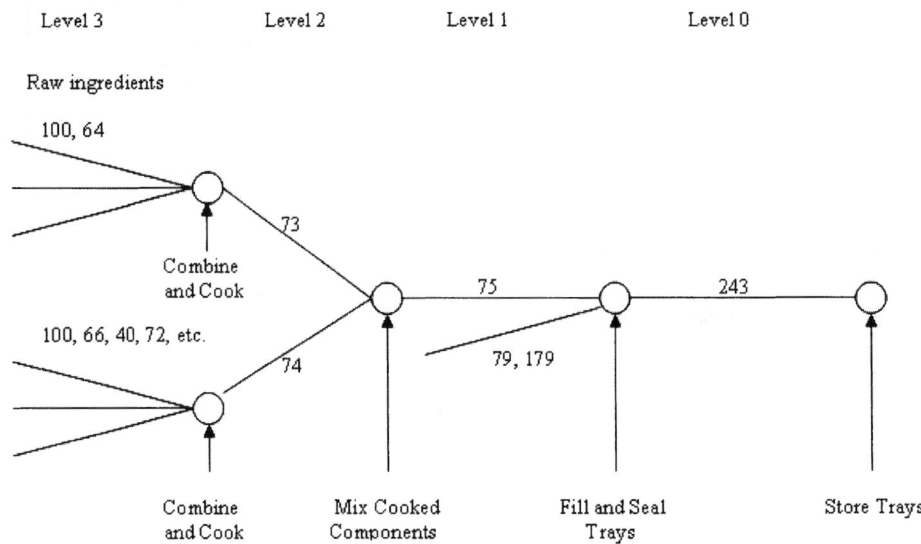

Figure 7.4 Process diagram—flow of materials in relation to production operations.

refrigerated product distributed to university dining halls. It is 1 of 20 products manufactured by University Food. As shown in Figure 7.2, there are four levels to the product BOM structure. Level 3 contains the raw ingredients, such as cheese, starch, and water that enter production at the first operations. The amounts in parentheses are the quantities combined together at level 3. These level-3 ingredients are combined into intermediate subassemblies, such as cheese sauce and cooked macaroni. Cooked macaroni and cheese sauce are the two level-2 materials. Thus, the combination of water and 160 lbs of macaroni yields 400 lbs of cooked macaroni at level 2 when the water is absorbed. At each level, a material ID uniquely identifies the material. For example, cooked macaroni is material ID 73, and cheese sauce is material ID 74. At level 1, the level-2 materials are combined into a mixture of macaroni and cheese (material ID 75). This mixture is then combined with packaging material to yield the final packaged product, shown at level 0 as Filled Mac & Cheese in a tray (material ID 243). The final product is inventoried in units of cases. Four 6.2 lbs trays of food make up one case.

The production line layout that performs the manufacturing operations is shown in Figure 7.3. Figure 7.4 is a process diagram that is also helpful in understanding the material flow in relation to the production operations. The arcs shown in Figure 7.4 indicate material in some stage of the process. The nodes in Figure 7.4 indicate the operations that are combining the material flow at each stage of production. The following is a description of the activities involved as provided by production personnel.

7.2.1.1 Combine and Cook Ingredients

Raw ingredients are brought from storage to the production line and staged in the cooking area, which is shown in the upper left of Figure 7.3. This area contains cooking stations with kettles of various capacities. The technician in this area, T1, is responsible for combining ingredients in

COOK SHEET

Product: Cheese Sauce **Revision Date: 9/09/99**
Material ID Number: 74 **Supersedes: 10/01/95**

Description:
The cheese sauce is the sauce prepared as part of the product 243, Macaroni & Cheese, from ingredients at the same proportions as listed in the cheese sauce formula.

Sauce preparation procedure: **(50 Gallons)**

1. Place water in the Groen Kettle (No. 1).
2. Turn on agitator at highest speed.
3. Add dry skim milk and reconstitute.
4. Add ingredients (corn starch, flour, salt, mustard powder, and paprika powder). Mix at room temperature until all ingredients are dissolved.
5. Add margarine and heat with agitation to 45-50 degree Celsius (115 – 125 degrees Fahrenheit), until all margarine is melted.
6. Add cheddar cheese, mix thoroughly, and add annatto cheese color.
7. Heat with agitation to a temperature of 85 – 90 degrees Celsius (180 – 195 degrees Fahrenheit) and hold this temperature for a minimum of 10 minutes.
8. Take sample to quality control for viscosity measurement.

FORMULA (50 Gallon Batch)

INGREDIENTS	In Lbs	Actual (lbs)
100 Water	308 (38.5 gal.)	
66 Cheddar Cheese	32.0	
152 Skim Milk Powder	24.0	
40 Margarine	11.0	
90 Corn Starch	8.28	
71 Flour, all purpose	6.2	
70 Salt	5.57	
68 Mustard Powder	0.8	
81 Paprika Powder	0.4	
72 Annatto Cheese Color	0.28	

Actual holding temperature ____________
Actual cook time at holding temperature ____________ Batch No. __________
Prepared by ____________________________ Date ____________

Figure 7.5 Cheese sauce cook sheet.

accordance with the formula for the product currently being produced and cooking the ingredients in accordance with the recipe. This recipe information is contained on a "cook sheet" for the product, which the technician uses as a control to guide him in preparing the batch. An example of the cook sheet is shown in Figure 7.5 for the cheese sauce (material ID 74).

The technician is also responsible for collecting certain quality control data, which he will enter on a record for each batch produced. The required quality control data include the holding time and temperature for the batch, which are recorded on the cook sheet. Although the recipe specifies the desired holding time and temperature, the actual holding time and temperature may vary from the specification in the recipe. Therefore, it is necessary to collect "actual" cooking parameters as a record of what actually happened during the production run. Also, a sample of the sauce is sent to the quality control lab for a viscosity reading. These measurements become a permanent record for the batch. A date and a sequential batch number identify each batch.

7.2.1.2 Mix Cooked Ingredients

After both the macaroni and the cheese sauce have been cooked separately for the appropriate amount of time, a technician, T2, combines the cooked ingredients into a mixing kettle, which is adjacent to the production line (left side of Figure 7.3). This is done by first draining the macaroni and then dumping the drained macaroni into the mixing kettle. Pumping the sauce into the mixing kettle is the next step. The combination of macaroni and cheese sauce is then stirred in the mixing kettle until a homogeneous consistency is obtained.

7.2.1.3 Fill Packages

A filling machine is a piston pump that draws a measured volume from the mixing kettle into a cylinder and then deposits that measured volume into the tray (see Figure 7.6). Containers move along the conveyor (shown at left in Figure 7.3). As they come under the filler, a sensor indicates their presence to the electronic control system of the filling machine, and the filler is indexed through its cycle by a programmable controller, dispensing the known volume of mixed macaroni and cheese.

7.2.1.4 Weigh Packages

As filled packages move along the conveyer, they pass over an in-line scale where they are automatically weighed (shown in center of Figure 7.3). By FDA regulation, packaged food must weigh at least as much as the net weight declared on the label. Underweight packages are detected at this point and automatically transferred off the line to a side conveyor, where they can be manually removed.

7.2.1.5 Seal Packages

After weighing, the package passes under a sealing machine that places a polymeric cover on the filled tray and seals the polymeric film to the tray using heat (see Figure 7.7). Finally, a laser jet printer prints the product ID (material ID 243, Mac & Cheese in tray), the lot number of the finished product, and the production date on the tray. A technician, T5, monitors this operation (shown at right in Figure 7.3).

Figure 7.6 Filling machine dispensing product into tray.

Figure 7.7 Sealing tray with clear plastic cover.

7.2.1.6 Store Packages

At the end of the production line (right side of Figure 7.3), technicians stack the sealed packages on pallets. The pallets of trays are taken to the warehouse for refrigerator storage. From completed production reports, the quality control department is informed of the finished product lot numbers. Quality control personnel will retrieve a sample of trays from the refrigerator to be examined for defects. When quality control completes the finished product testing, the trays will be approved for shipping, or, if the sample fails the tests, the product will be rejected and will have to be destroyed or reworked.

7.2.2 SYSTEM REDESIGN OBJECTIVES

The system redesign objectives have to do with the redesign of the information collection system and how that information is used. It does not have to do with the redesign of the physical processes of manufacture, which are fixed. The objectives of the information system design are usually expressed by the clients of the system analyst and have to be interpreted by the analyst. In this scenario, we will assume that discussions have taken place with the plant manager, quality control manager, and the production line supervisor. The following sections summarize the concerns expressed by each.

7.2.2.1 Plant Manager

"The Food and Drug Administration has just issued a new ruling that impacts the way we manage our business, particularly with regard to information requirements (see the press release in Box 7.1). It's part of this bioterrorism threat. If a terrorist introduces contaminated ingredients anywhere in the food chain, the government wants to be ready to respond by tracking down

Box 7.1
FDA Issues Rules on Food Traceability

WASHINGTON, Dec, 2004—The U.S. Food and Drug Administration (FDA) issued its final ruling requiring food companies to maintain records under the Bioterrorism Act. The ruling requires that companies that manufacture or transport food products must maintain records that identify the sources of component ingredients of a product by lot number and the customers who are the immediate recipients of the food they distribute.

The records are deemed crucial for the FDA to deal effectively with food-related emergencies, such as contamination by terrorists. The ability to trace a product back to the ingredients that have gone into it will enable investigators to get to the source of the contamination faster. The records also enable FDA to trace forward to recall tainted food that poses a danger to consumers.

The records can be kept in either paper or electronic form. Records must be available within 24 hours of an official request, or the FDA can bring civil action against the company. There are also criminal liabilities for company managements that fail to keep necessary documentation.

the source of contamination and warning the recipients of the food products. This is obviously a necessary precaution from the FDA point of view, but it's going to place some hardships on us.

"Right now we have a very good information system for managing our raw materials (Chapters 4 and 5). We've also been working on the design of a database for managing our shipping department and tracking the shipment of each lot of product to the customer (University Food Company Case [B] and Case [C]). However, we do not have a system for tracing raw materials through work in process so that we know which lots of ingredients have gone into which lots of final product. This is what the FDA wants the food industry to do. We're a small company, and this is going to add quite a bit of additional effort on our part, but we have to do it.

"In a way, we have been postponing the inevitable. There have been several newspaper articles over the past few years concerning food product recalls. Some of these recalls have occurred because of bad ingredients that were not detected in quality control or that had been improperly stored prior to production and had gone bad or been contaminated. This is not terrorism, but just accidents that can happen. If we were asked to perform a recall on a bad batch of product, whether contaminated by accident or by terrorism, it could be a nightmare. For example, if the problem were related to a lot of raw ingredients, we have no way of knowing which specific batch of product used that lot of ingredient. We only know that the lot was relieved from inventory to be used in production on a particular day. So we would have to recall and destroy an entire day's production. Also, if we suspect that a particular lot of ingredient is the cause of a problem, we would have to recall and destroy any other day's production on which that ingredient lot was relieved from inventory, even if it was used in only one production batch. We simply need better information about the flow of raw ingredients into final product.

"I am also concerned that our quality control records on production batches are hard to access. These are all paper records. In the event of a problem with a product, if we had to go back and retrieve this data for a customer or a government auditor, we wouldn't look very well organized. These new FDA regulations require that records are available within 24 hours of a government request. I think it's time for us to implement an information system that can capture and recall records of our actual production operation on a production batch basis."

7.2.2.2 Quality Control Manager

"We are way behind the rest of the industry in automating our quality control records. Our quality control records come from two sources. My technicians get samples from the line and analyze them here in the lab. We retain a paper record on each analysis. The guys on the line are responsible for collecting and recording some data. For example, if you look at a cook sheet (Figure 7.5), you can see that the technician on the line has to record the actual amounts of ingredients used in a batch as well as the cooking time and temperature. This record is his and our check that the recipe is being followed. However, I have no way of knowing that his record is accurate or whether, by error, the wrong amount of an ingredient was added. I have often thought that we should be able to compare the amounts of each ingredient that is relieved from inventory on a given day with the amount that should have been used according to the formula and the number of production batches produced on that day. This would be a double check against the accuracy of the paper record. Unfortunately, this is not an easy thing to do manually. I think the computer can help us here.

"I know that the FDA is concerned with bioterrorism events, and that will drive the specifications for this project. However, I am also concerned about how we would manage a crisis situation like a product recall or a product complaint from a customer due to accidental causes. In my opinion, most problems of this kind are not ingredient related. They are usually caused by production practices that violate quality control standards. If we have a customer complaint, it is most probably because of a quality problem related to a specific batch of our own product, not the entire day's production. If we have good quality control records for the production conditions by product batch, there should be no need to recall an entire day's production. So as part of this project, I think you should include the system of collecting quality records against production batches with the most refined degree of granularity possible. If you can easily correlate quality records and batch records using the computer, that will be a big contribution to solving quality control's problems in this area."

7.2.2.3 Production Line Supervisor

"They're asking us to do too much already. My technicians have to weigh and combine ingredients and run the line while at the same time recording quality records, such as temperature and cooking time. Now, I know that FDA wants us to record which lots of ingredients go into which batches of product, and we will have to find a way to comply with this regulation. However, in the interest of productivity, the last thing we need is more record keeping on the line.

"What would really be helpful to me is if you could come up with a way to automatically collect this information and relieve my guys from the record keeping overhead."

7.2.2.4 Interview Summary

The individuals interviewed expressed somewhat different points of view about what they would like to see done. The plant manager is primarily interested in complying with new FDA requirements—that is, being able to trace lots of raw ingredients to the batches of final product into which they went. On the other hand, the quality control manager thinks it is at least as important to automate the documentation of the quality tests during the preparation of the batches so that a routine audit of the quality tests on past batches is possible. The production line supervisor would like to accommodate these new requirements, but he does not want them to overburden his technicians. He suggests the need to automate the data collection requirements. With these varied suggested system redesign objectives, we go on to structurally document the system as it is currently operated.

7.2.3 Defining "As Is" IDEF0 Model and Establishing System Boundaries

It is during the definition of the "as is" functional model that the analyst can clearly identify the boundaries of the project in relationship to the physical processes. Each activity that is defined in the model implies a requirement for certain information or a responsibility to document certain information. Based on the description of the production operations as given in Section 7.2.1 and further observations of the activities of technicians during actual production, the analyst has constructed the hierarchical IDEF0 model shown in Figures 7.8 to 7.13. We now briefly discuss these figures.

7.2.3.1 Control Production Processes

The overall activity is described in Figure 7.8 as *control production processes*. Indeed, this activity was previously identified in Figure 4.11 as the major activity that follows the activity of

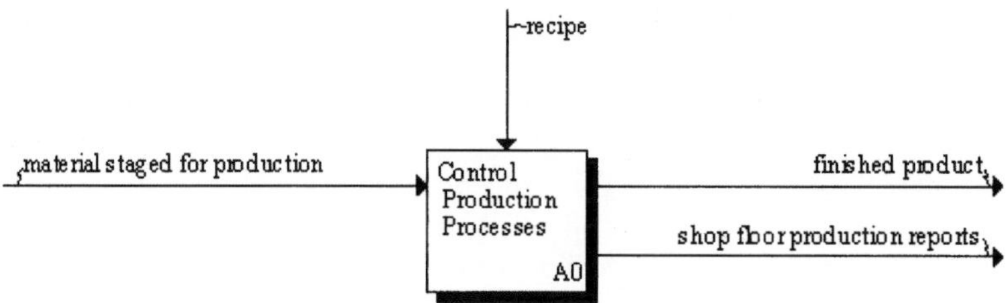

Figure 7.8 Node A0, control production processes.

control stored material. Since we are beginning a new set of diagrams in this chapter, this activity will be labeled A0. The *material staged for production* is input to this activity, which is transformed into *finished product* on the output side. In addition, the activity provides an information output, *shop floor production reports*. There are controls on the activity. The *recipe* is the complete description of how the product is made. It includes the *formula* that defines the ingredient quantities to be added in each step. The *cook sheet* in Figure 7.5 is that part of the recipe for macaroni and cheese that governs the making of cheese sauce. The complete recipe for the product also includes a cook sheet for cooking the macaroni, instructions for setting the filling machine (e.g., volume setting) and packaging machine (e.g., seal pressure for sealing the lid to the tray), information to be placed on the package label, and quality control measurements to be taken during production. Recipes can be extremely detailed. We will not dwell on these details here, but we will use the cheese sauce cook sheet (Figure 7.5) as a typical example.

There is a *production schedule* control at University Food that is generated by the production planning department. Using the backlog of orders, a schedule is prepared to service orders based on earliest due date. From the perspective of the production line supervisor, he is only concerned with two requirements of the production schedule: (1) which product to produce on the line each day and (2) which lots of raw ingredients to use in production on that day. The first requirement is provided to the supervisor by issuing him the day production schedule and cook sheets for each day of production. The second requirement is provided by production planning in the form of a *material move schedule*, which is part of the production schedule. The material move schedule tells the forklift truck driver which lots of material to withdraw from inventory on a particular day. This activity was previously documented as part of the process *Control Stored Material* in Section 4.6. For this case study, we will assume that once the production line is set up for a particular product on a given day, the same product is produced on that production line for the entire day. We will not explicitly include the production schedule control in the diagrams since it is not relevant to the objectives of the case.

7.2.3.2 Decomposition of Node A0

The decomposition of node A0 is shown in Figure 7.9. The description follows closely the description of production operations in Section 7.2.1. Activity A1, *Combine and Cook Ingredients*, is the first product preparation step on the production line. The *cook sheet* (part of the recipe) controls this activity, which yields material output (*cooked ingredients*) and information output (*shop floor production reports*).

In a similar manner, the subsequent activities of *Mix Cooked Ingredients*, *Fill and Seal Trays*, and *Store Trays in Refrigerator* represent further processing of the product. Each has a material

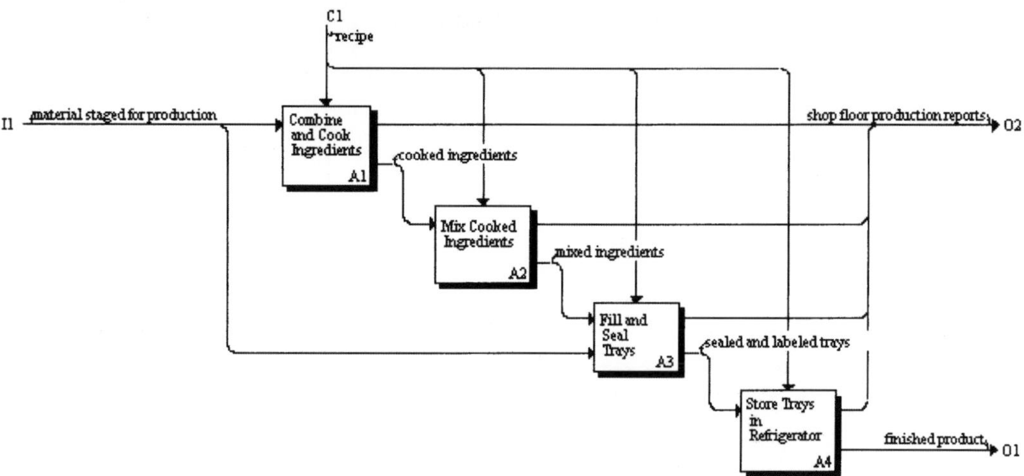

Figure 7.9 Decomposition of node A0.

input and output as well as an information output requirement. Also, each activity is guided by the instructions given in the recipe for the product.

7.2.3.3 Decomposition of Node A1

The activity *Combine and Cook Ingredients* is decomposed in Figure 7.10. In activity A11, *Combine Ingredients*, the *materials staged for production* are combined in the cooking kettle in accordance with the instructions on the cook sheet (see Figure 7.5). The technicians must complete the cook sheet record, which is an information output of the activity. In activity A12, *Cook Ingredients*, the ingredients in the cooking kettle are brought to temperature for the prescribed time in accordance with the cook sheet. Further information is added to the cook sheet record. Finally, cooked ingredients are sampled in A13, *Sample Cooked Ingredients*, and the sample is sent to the quality control (QC) lab for viscosity measurement.

7.2.3.4 Decomposition of Node A2

Figure 7.11 shows the decomposition of node A2, *Mix Cooked Ingredients*. The process begins with A21, *Transfer Cooked Ingredients*, which moves the cooked batches into the mixing kettle. In A22, *Mix Combined Ingredients*, the technician uses the recipe as a guide in setting the time and speed of the mixer for a consistent mix of the macaroni and the cheese. A temperature reading is taken at the end of the mixing cycle for quality control purposes and added to the *shop floor production reports*. Finally, a transfer hose attached to the filler pump is inserted into the mixing kettle in A23, *Insert Filler Pump*. The material will be pumped directly from the mixing kettle into the trays in the next major activity.

7.2.3.5 Decomposition of Node A3

Figure 7.12 shows the decomposed activity steps of A3, *Fill and Seal Trays*. In A31, *Fill Package*, a tray moving along the conveyor (*material staged for production*) is filled with mixed

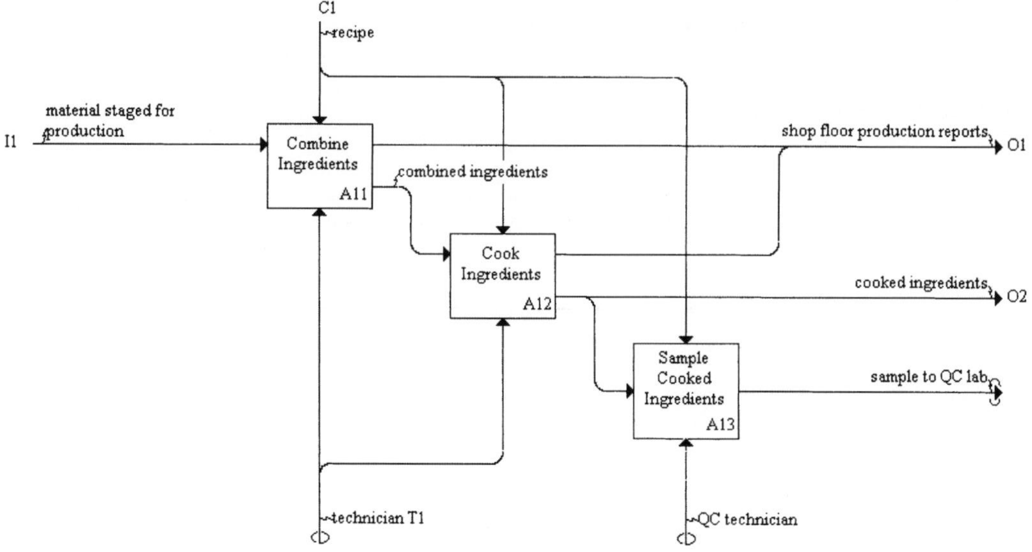

Figure 7.10 Decomposition of node A1.

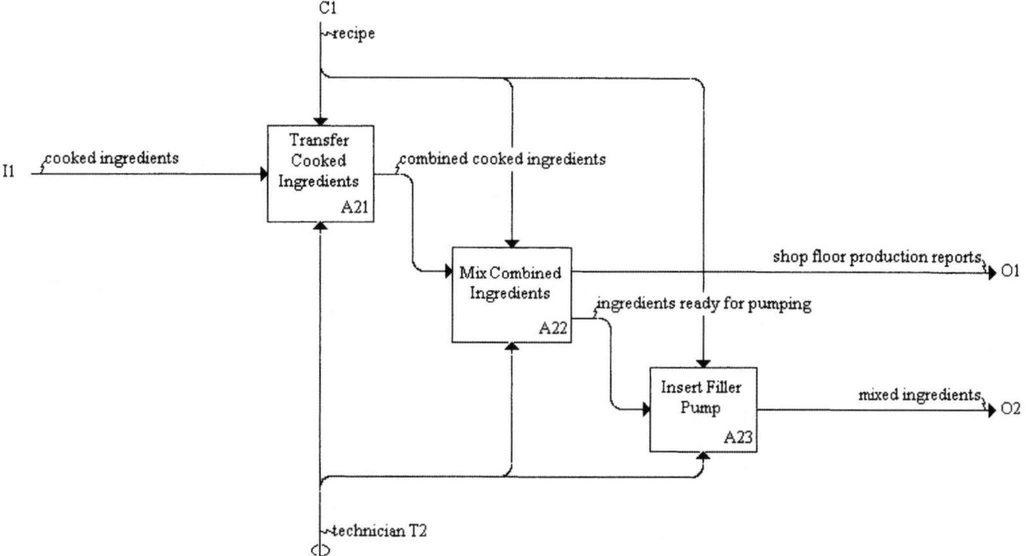

Figure 7.11 Decomposition of node A2.

ingredients. As the filled tray continues to move along the conveyor toward the sealing operation, it passes over an inline checkweigher (A32, *Weigh Tray*). Filled trays of acceptable weights move to the sealing operation (A33, *Seal Package*) where a lid is heat sealed to the filled tray. The sealed product then moves past a laser jet printer where product, lot number, and establishment information are printed on the container (A34, *Print Label Information*). The shop floor production report

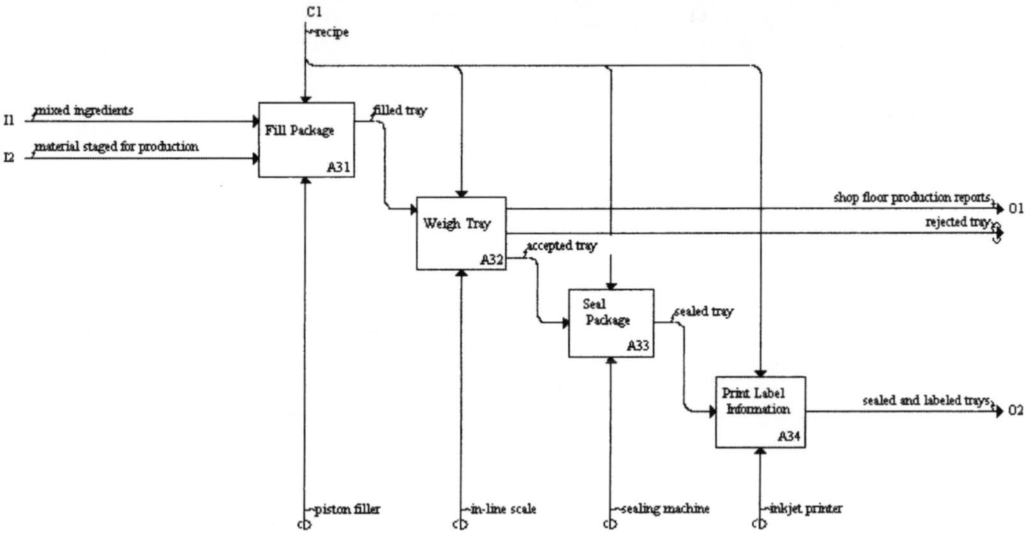

Figure 7.12 Decomposition of node A3.

is updated with information on the average weights of containers and the number of underweights (rejects).

7.2.3.6 Decomposition of Node A4

Figure 7.13 shows the decomposed activity steps of node A4, *Store Filled Trays in Refrigerator.* Trays are first stacked on pallets (A41, *Stack Trays on Pallet*) and are then transported to the warehouse (A42). When the trays are stored in the refrigerator (A43), the work-in-process status of the material ends. A record indicating the final count of trays in the batch is sent to the quality control department so that quality control tests can be performed on a sample of the product. This count is part of the *shop floor production reports* documented at node A41. Once quality control accepts the final product, sales will be notified that the finished goods lots are available for shipment against open orders. These offline quality control tests are not incorporated into the activities of *Control Production Processes.* They are part of a separate set of processes, labeled *Control Finished Goods* in Chapter 4, Figure 4.5.

7.2.3.7 Summary of "As Is" Model

The previous sections showed the IDEF0 model of the basic elements of the production process as it is currently executed. The analyst should review these diagrams with the appropriate personnel to verify that the activities have been properly documented and that everyone involved agrees that the boundaries of the system defined by the IDEF0 model are acceptable. It is possible to continue the decomposition of activities and provide a more detailed description of the tasks involved. However, we will choose to stop here, since the model has given us an acceptable level of understanding of the process.

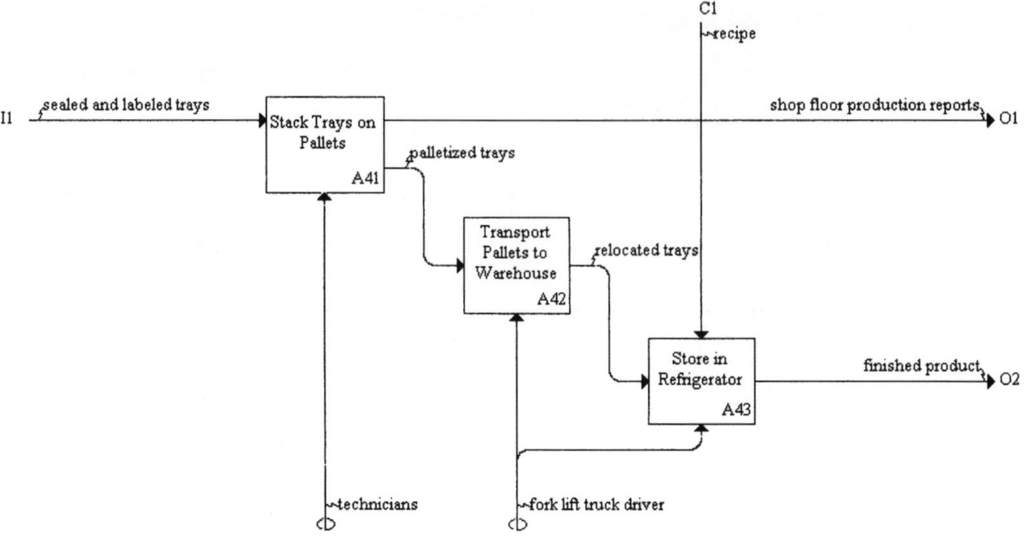

Figure 7.13 Decomposition of node A4.

This ends the preliminary study and problem definition phase. At this point, the analyst goes on to the conceptual design phase, which is the subject of the next section.

7.3 DESIGN PHASE

In the design phase, the analyst defines the entities, relationships, and attributes that are sufficient to service the needs of the users (user views). A ***user view*** is the data required by a particular user or user group, for example, quality control, sales, or shipping. This phase is independent of all physical considerations, such as the target database system, computer platform, and so forth. Since there may be many possible ways of expressing a conceptual design, the design process is somewhat creative. To establish some structure within the process, we will go through the series of steps from Figure 7.1:

- Identifying user information needs (user views)
- Defining entities and relationships
- Defining attributes
- Establishing the global data model
- Defining superclass/subclass relationships
- Evaluating the need for transaction entities
- Normalizing the data model
- Finalizing and validating a "to be" IDEF0 model
- Finalizing and validating an IDEF1X global information model

7.3.1 Identifying User Information Needs

The database design must follow user requirements. The user requirements that were identified in Section 7.2 may be summarized as follows:

1. *Plant manager.* Lot tractability from raw materials to finished goods. Here we wish to link the existing MATERIAL_LOT entity (Figure 5.12) to the finished product in which it is used.
2. *Quality control manager.* Relate quality control data to batches with which they are associated. Quality control data are collected throughout the production process at various steps in the process, which we shall call "operations." Operations include cooking, mixing, filling, and so on.
3. *Production line supervisor.* Data input should not be cumbersome. This is not a requirement that would be considered at the conceptual phase because it has to do with physical implementation of the data model and its associated user input requirements.

At this stage of the design process, it would be desirable if the clients could describe the kind of reports they would like to see as output of the information system. This would help the analyst to identify entities and attributes. At this stage, every attempt should be made to study paper documents that exist because they will contain much of the information in a format familiar to workers and management. We identified such documents as the "material move schedule" and the "cook sheet" in the previous sections. Next we will use those items and our general knowledge of the problem to define rules that govern the relationship between entities in our data model.

7.3.2 Defining Entities and Relationships

7.3.2.1 Plant Manager View

Let us begin with the first requirement and try to define relationships based on tracking raw ingredient lots into finished good lots. This might be considered the plant manager's user view. In so doing, business rules will be summarized from what we know about the case thus far. *Business rules* are statements that clarify how functions and procedures are performed in a particular enterprise. They are specific to the enterprise.

Rule 1: A material lot may go into zero, one, or more batch lots. Each batch lot will contain more than one material lot. A batch lot is simply the combining of portions of existing material lots into a new material lot that has a new lot number, for example, combining cheddar cheese, cornstarch, and so on from lots of raw ingredients into a lot of the cheese sauce.

Rule 1 summarizes the previous description of how material lots are combined in production. At each operation, material lots are combined to form new material lots until, finally, a finished product is created. This finished product is stored in inventory with its own unique material lot number.

A situation somewhat like this was presented as part of the discussion of recursive entities in Section 3.2.4. There, a bill of materials was used for illustration, where components go into subassemblies and components and subassemblies go into finished products. The current situation is similar because the material lots are being consumed into intermediate batch lots at each operation, finally yielding a finished product lot. Therefore, one way to conceive of this situation is to consider MATERIAL_LOT as a recursive entity.

As discussed in Chapter 3, an M:N recursive relationship must be decomposed into a series of 1:M relationships in order to be implemented in a database. As suggested in Section 3.2.4 (Figure 3.15), one solution to the problem is to list the components of a finished product or assembly in a related table. Were such an approach to be applied in this case, it would yield the data model shown in Figure 7.14. If the reader compares the MATERIAL_IDs in the example MATERIAL_LOT table with Figures 7.2 and 7.4, it should be clear that the BATCH table in Figure 7.14 illustrates the composition of lots for one batch of macaroni and cheese.

In Figure 7.14, each lot number is related to the lower-level component lot numbers that have gone into it in the table BATCH. Therefore, a search for the lot numbers that constitute a finished product or subassembly can be obtained with the following query:

```
SELECT COMPONENT_LOT_NO
FROM BATCH
WHERE LOT_NUMBER=(the lot number of the finished product or subassembly);
```

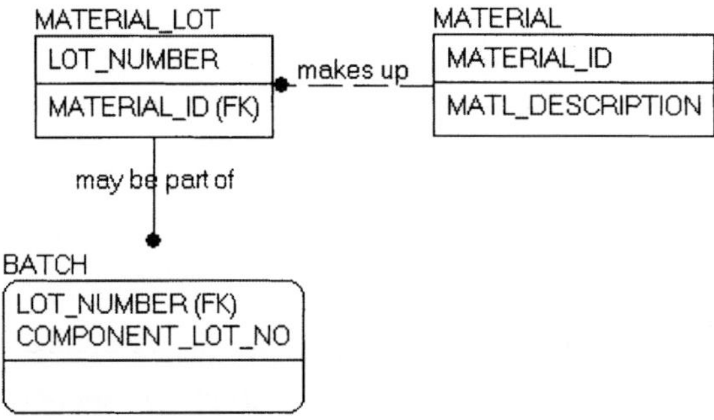

MATERIAL_LOT

LOT_NUMBER	MATERIAL_ID
10102	74
10103	73
10104	75
10105	243
1180	152
1190	64
1201	66
1206	68
1210	71
1250	40
1260	72
1281	90
1301	81
1302	70
1385	79
1400	179

BATCH

LOT_NUMBER	COMPONENT_LOT_NO
10102	1180
10102	1201
10102	1206
10102	1210
10102	1250
10102	1260
10102	1281
10102	1301
10102	1302
10103	1190
10104	10102
10104	10103
10105	10104
10105	1385
10105	1400

Figure 7.14 Candidate data model for recursive relationship.

This is a "drill down" search. It returns the components of a particular finished product lot or subassembly lot. It is consistent with the FDA requirement to "trace back . . . to get to the source of contamination." (See Box 7.1)

Alternatively, management also wants to know where each ingredient was used. For example, the information system must "enable FDA to trace forward to recall tainted food that poses a danger to consumers." (See Box 7.1) That is to say, it must be possible for the FDA to "drill up" to determine which final product lot numbers contain an offending ingredient. This can be accomplished as follows:

```
SELECT LOT_NUMBER
FROM BATCH
WHERE COMPONENT_LOT_NO=(the component lot number of the ingredient);
```

Note that these two SQL statements must be used recursively at each level in which lots are combined into other lots. For example, the drill-up search from an ingredient goes to the next level up, which is a subassembly. To get to a final product, the query must be run again with COMPONENT_LOT_NO equal to the subassembly lot number that was returned in the prior query.

Although the preceding data model may appear to solve the problem, it leaves out some important considerations that differentiate this case from the static bill of materials example discussed in Section 3.2.4. In particular, the LOT_NUMBER for the batch lots, such as cooked macaroni and cheese sauce, is assigned as the BATCH is created. This is also true for the lot number of the final product. According to Figure 7.14, in order to enter an instance of a LOT_NUMBER in the table BATCH (the M side), it must already exist in the parent entity, MATERIAL_LOT (the 1 side). Therefore, if a technician on the line wishes to create a new lot number and enter its component lot numbers, he or she must first create an instance of that number in the table MATERIAL_LOT. Such an implementation is possible, but it must be designed into the forms to be used by the technician so that an entry is first committed to the parent table and then to the child table.

Another concern is the possibility that the technician may enter an ingredient lot number (COMPONENT_LOT_NO) that does not actually exist in the inventory (MATERIAL_LOT table). Mistakes in data entry such as this occur all the time. In fact, it is desirable to ensure that the numbers entered by the technician do actually exist in the MATERIAL_LOT inventory table. The data model in Figure 7.14 does not enforce this requirement. For example, a technician creating a batch of cheese sauce (a new LOT_NUMBER) will enter a COMPONENT_LOT_NUMBER for the ingredient "Annatto Cheddar color" (see Figure 7.2). This lot number should already exist as a LOT_NUMBER in the table MATERIAL_LOT. If the COMPONENT_LOT_NUMBER is incorrectly entered in the table BATCH, the DBMS will not catch this error.

These considerations can be accommodated by a slight change in Figure 7.14. This is shown in Figure 7.15. Here we have replaced LOT_NUMBER in the BATCH table with a new attribute name, OUTPUT_LOT_NO. The OUTPUT_LOT_NO is the lot number of the new material lot that is being created. The lot number in the BATCH table of Figure 7.15 is now the LOT_NUMBER of the existing material that goes into the newly created lot. Hence, relational integrity now requires that MATERIAL_LOT.LOT_NUMBER exists before it can be used in BATCH.LOT_NUMBER. This prevents the incorrect entry of material lot numbers that do not exist. Also, BATCH.OUTPUT_LOT_NO can be created without requiring that it already exist in MATERIAL_LOT. A form designed for use in creating BATCH.OUTPUT_LOT_NO should also add the newly created lot number to the MATERIAL_LOT table. A drill-down search for the ingredients in cheese sauce lot number 10102 is accomplished as follows:

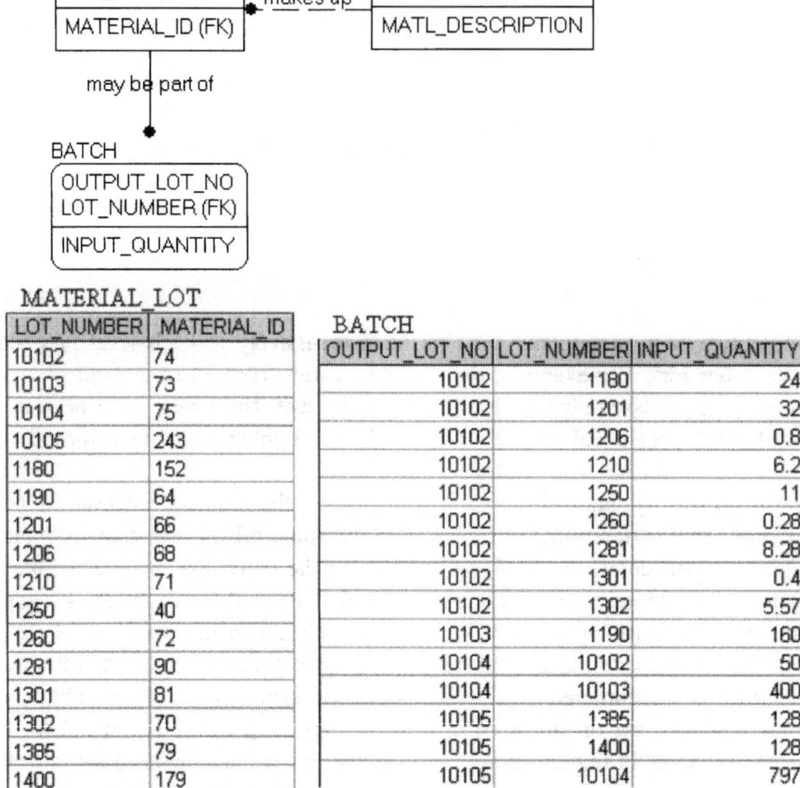

MATERIAL_LOT

LOT_NUMBER	MATERIAL_ID
10102	74
10103	73
10104	75
10105	243
1180	152
1190	64
1201	66
1206	68
1210	71
1250	40
1260	72
1281	90
1301	81
1302	70
1385	79
1400	179

BATCH

OUTPUT_LOT_NO	LOT_NUMBER	INPUT_QUANTITY
10102	1180	24
10102	1201	32
10102	1206	0.8
10102	1210	6.2
10102	1250	11
10102	1260	0.28
10102	1281	8.28
10102	1301	0.4
10102	1302	5.57
10103	1190	160
10104	10102	50
10104	10103	400
10105	1385	128
10105	1400	128
10105	10104	797

Figure 7.15 Data model to enforce existence of input lot numbers.

```
SELECT LOT_NUMBER
FROM BATCH
WHERE OUTPUT_LOT_NO=10102;
```

An upward search based on an ingredient, for example lot number 1180, can be accomplished as follows:

```
SELECT OUTPUT_LOT_NO
FROM BATCH
WHERE LOT_NUMBER=1180;
```

It could be argued that the important relationship for tracing lot usage for recall purposes is to be able to go from the lot number of the raw material, or ingredient, to the lot number of the final product. So why bother assigning lot numbers to intermediate subassemblies, like cheese sauce and cooked macaroni? There are at least two reasons. First, each subassembly may be subjected to quality control tests, such as sauce viscosity. Each of these test results should be associated with the lot number on which it was performed. Also, on any given production day there may be an excess amount of a subassembly produced and not consumed into a final product on that day. This excess will probably be inventoried for use on a subsequent day. For example, part of a batch of cheese sauce produced on one day will be combined with a new batch of cheese sauce on the next day. To trace the ingredients that went into the product on that next day, there must be a complete record of what ingredients went into the sauce on the prior day. For these reasons, generating a complete record of intermediate subassemblies is the correct way to approach this situation.

There is still one irritating feature of the data model shown in Figure 7.15. Since it only establishes connections between immediate inputs of an OUTPUT_LOT_NO, the SQL calls on the table will only allow the user to drill down or drill up one level at a time. It would be better to be able to return all the subassembly and final product lot numbers in one form or report. One way to do this, as discussed in Section 3.2.4, is to write code in a high-level language that will search the tables and recursively return each level of the relationship and display it in a table. However, writing algorithms is not what we want to emphasize in this text, so we offer another solution.

An alternative way to accomplish end-to-end tractability without writing code is to add one more attribute to the table BATCH. That attribute will be called FINISH_GOOD_LOT_NO and will be a lot number associated with the creation of an instance of a final product, such as MATERIAL_ID 243 (Mac & Cheese in a Tray, Figure 7.2). In other words, it is the final OUTPUT_LOT_NO in the series of links from ingredient to product. This modification is shown in the data model in Figure 7.16. With this attribute added to the BATCH table, the following queries will provide end-to-end tractability:

```
Drill Down:

SELECT LOT_NUMBER
FROM BATCH
WHERE FINISH_GOOD_LOT_NO=10105;

Drill Up:

SELECT FINISH_GOOD_LOT_NUMBER
FROM BATCH
WHERE LOT_NO=1180;
```

One thing to note about this solution is that the FINISH_GOOD_LOT_NO should be determined a priori—that is, it should be known when an intermediate entry for that lot is made into the BATCH table. If lot numbers are assigned as lots are created, this number will not be known until intermediate assemblies are combined at the final operation. Therefore, if all the data are collected on paper during production and entered into the database offline at the end of the day, no problem will exist. However, if the technicians are expected to input information to the database in real time, the technician must know the FINISH_GOOD_LOT_NO when he or she is creating the batch of cheese sauce. This problem shows how situation specific a database implementation can become. Recall that in IDEF0 activity A34 (Figure 7.12), the laser printer is programmed to print the finished good lot number on the package label. For this production operation, the finished product lot number is assigned at the start of the production run. Therefore, FINISHED_GOOD_LOT_NO is known a priori and no problem exists.

Although the modified BATCH table in Figure 7.16 will allow end-to-end tractability, it will not display the data to the plant manager in an elegant manner. The SQL call will simply return

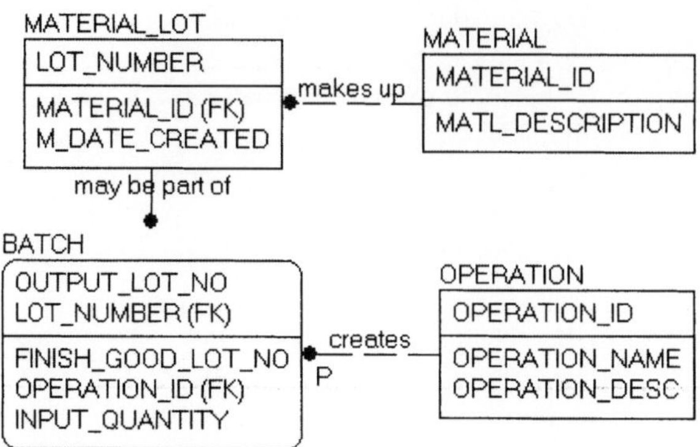

BATCH

OUTPUT_LOT_NO	LOT_NUMBER	FINISH_GOOD_LOT_NO	INPUT_QUANTITY
10102	1180	10105	24
10102	1201	10105	32
10102	1206	10105	0.8
10102	1210	10105	6.2
10102	1250	10105	11
10102	1260	10105	0.28
10102	1281	10105	8.28
10102	1301	10105	0.4
10102	1302	10105	5.57
10103	1190	10105	160
10104	10102	10105	50
10104	10103	10105	400
10105	1385	10105	128
10105	1400	10105	128
10105	10104	10105	797

Figure 7.16 Final data model.

an unordered list of ingredient and subassembly lot numbers (drill down) or a list of final lot numbers (drill up). A better formal presentation would display the results by each stage in the combining of lots. For this production system in which operations are sequential, a simple business rule does exist to make this possible:

Rule 2: A batch is created at one and only one production operation. Each production operation creates one or more batches.

This rule reflects the fact that a specific production operation (i.e., combine and cook cheese sauce) can create many different batch lots. On the other hand, the batch lot was made at one specific production operation. Figure 7.16 updates the data model based on business rule 2.

The fact that the operations are sequential in our food manufacturing system gives them an order that can be useful to us. A well-ordered display of the relationships at each level of combining the lots together can now be obtained from the following queries:

Drill Down:

```
SELECT FINISH_GOOD_LOT_NO, LOT_NUMBER, OPERATION_ID
FROM BATCH
WHERE FINISH_GOOD_LOT_NO=(the lot number of the finished product)
ORDER BY FINISH_GOOD_LOT_NO, OPERATION_ID;
```

Drill Up:

```
SELECT BATCH.LOT_NUMBER, BATCH.FINISH_GOOD_LOT_NO,
MATERIAL_LOT.M_DATE_CREATED
FROM BATCH, MATERIAL_LOT
WHERE BATCH.LOT_NO=(the lot number of the ingredient)
AND BATCH.FINISH_GOOD_LOT_NO=MATERIAL_LOT.LOT_NUMBER;
```

At this point, we have addressed the main issue raised by the plant manager. The database Lot_Trace_CH7_stu, in which the previously discussed queries are implemented, can be downloaded from the Web site. The next section reviews the needs of the quality control manager.

7.3.2.2 Quality Control Manager View

The quality control manager has emphasized the quality control data collection that takes place throughout the manufacturing process. This data collection is performed during major steps in the process, such as cooking or mixing. Table 7.1 lists some of the tests performed on macaroni and cheese product at various production operations. Quality control personnel who take a sample from the line during production perform some of these tests (e.g., gravy viscosity). There are also quality control records that are taken by technicians or recorded by equipment on the line (e.g., net weight). All of these records are associated with the product at the batch level.

There are also quality control tests that are performed on some ingredients when they arrive at the plant and before they are used in production. For example, from Table 7.1, flour undergoes a microbial count to ensure it is not contaminated. Ingredients, or raw materials, are tested on a MATERIAL_LOT basis. Tests are also performed on finished goods after production. For example, from Table 7.1, a test of seal integrity is performed on a sample of packages to ensure that the sealing machine has properly sealed the package. Although we may think of the testing of subassemblies and finished product as batch tests, both subassemblies and finished products are also material lots. Based on this observation, one can conclude that the most compact representation of quality control test results would be to associate them with the material lot entity.

Table 7.1

Some Example Quality Control Tests

Mat'l ID	Mat'l description	Oper'n ID	Operation name	Quality control text
71	Flour, all purpose	10	Incoming storage	Microbial count
74	Cheese Sauce	20	Combine & Cook	Gravy Viscosity
75	Mixed Mac & Cheese	30	Mixing	Mixing time
243	Mac & Cheese in tray	40	Fill & Seal	Net Weight
243	Mac & Cheese in tray	50	Final Storage	Seal Integrity

Rule 3: A material lot may have zero, one, or more quality control records associated with it. A quality control record is associated with one material lot.

We shall use the terminology "QC Record" to refer to a quality control record, whether it is taken by a technician on the production line or by quality control personnel in the QC lab; whether it is associated with an ingredient or a production batch material lot. An optional relationship exists because some materials may not require a quality control test, and, therefore, there will be no QC record for a material lot of that material. The implementation of this rule is shown in Figure 7.17.

Note that the QC_RECORD is not a description of a quality control test. Rather, it is the actualization of a particular test on a particular material lot. The QC_RECORD_ID is simply a unique number that specifies the incidence of a test. Therefore, a related entity is the description of the quality control test itself.

Rule 4: A quality control record requires one or more samples to be tested. Each sample yields one test result. Each sample is associated with one quality control record.

Most often an evaluation of the acceptability of a material lot is accomplished by first taking samples from the material lot. Upon testing, each sample will yield a specific result. So if a sample of size five is drawn from a material lot of flour and each sample is tested for microbial count, there will be five individual readings to record. The inspection procedure for flour may require that all five samples are within the acceptable range of the test in order to accept the material lot of flour. Therefore, the material lot has an overall test result (accept/reject) that is based on the individual test values of the samples. The implementation of this rule is shown in Figure 7.18.

Rule 5: A quality control record is associated with a quality control test procedure, which is a description of how to perform a test. A quality control test procedure may have many quality control records associated with it.

Rule 6: A quality control test procedure is associated with one or more materials. A material may be associated with one or more quality control tests.

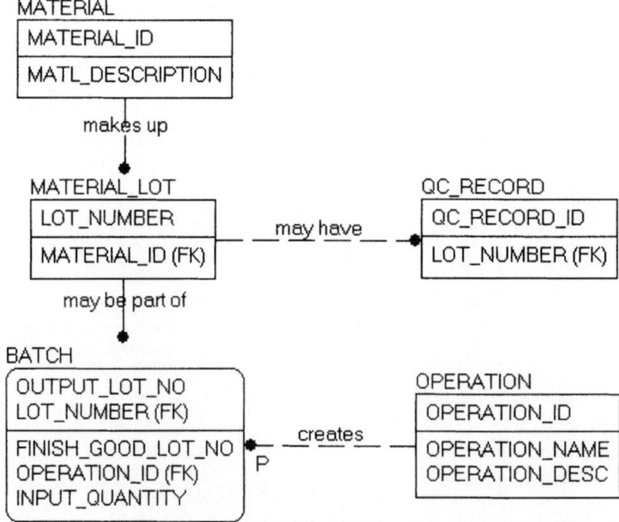

Figure 7.17 Relationship between material lot and QC test record.

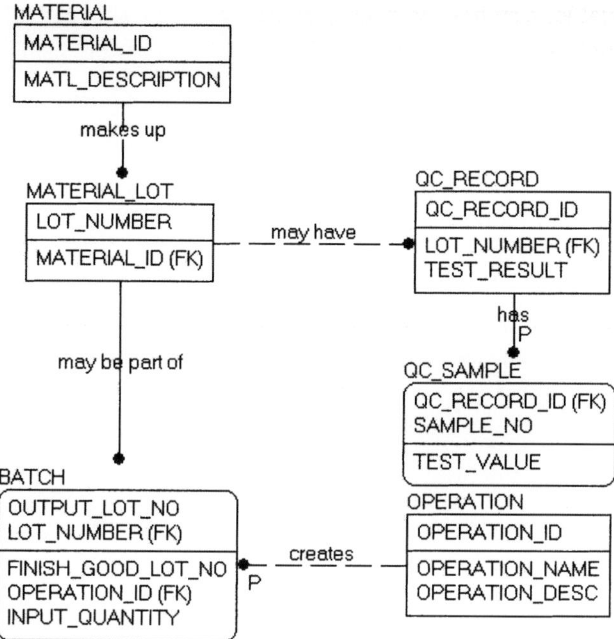

Figure 7.18 Relationship between QC test record and samples.

The QC_RECORD is a record of an actual test result. It documents an instance of a quality control test procedure being applied to a material lot. The test procedure is generic. It is just a description of how to perform a test, such as measuring viscosity. The test procedure is described in the quality control manual. A single quality control test procedure may be associated with more than one material. For example, a viscosity test procedure is performed on cheese sauce, but it may also be performed on beef gravy. The only distinction between the test will be the acceptable readings, or quality control limits of the test, for the specific material. These limits and acceptance regions are recorded in the appropriate section of the quality control manual. These ideas are captured in Figure 7.19. Note that the entity QC_TEST directs the user to a reference in the QC manual (TEST_PROCEDURE_REF). The manual itself could also be entered into the database. However, we shall stop at this level of automation.

At this point, we have addressed the quality control manager's concerns. The data model accommodates the need to document and relate each test to the material lot on which it is performed. Also, since lot tractability is also accommodated in Figure 7.19, all quality records associated with a particular end product and all the materials that went into that end product can be recalled.

7.3.3 Defining Attributes

Following the initial data model design, the analyst begins to collect more detailed information on attributes. As the identification of attributes proceeds, it may lead the analyst to reconsider the model as it is currently proposed. This can result in changes in the data model. The creation of a

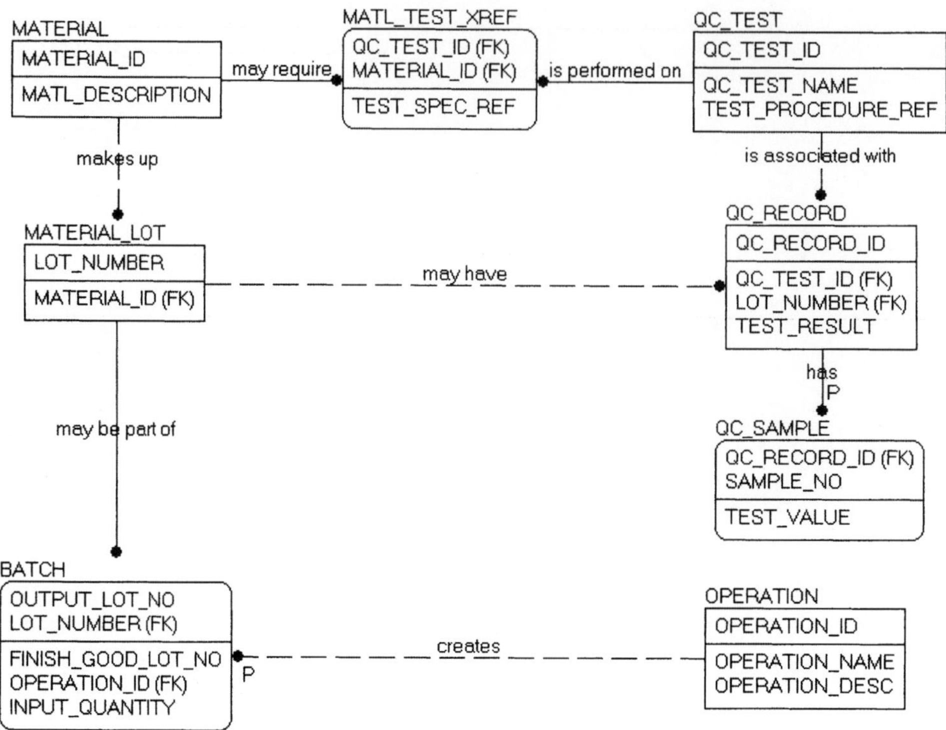

Figure 7.19 Relationship between material and QC test.

proper model is an evolutionary process. In the discussion that follows, we present attributes and their description in a tabular format.

7.3.3.1 Plant Manager View

From Figure 7.19 and corresponding information from Figure 5.13, the following attributes are proposed. Primary keys and foreign keys that are relevant to the current IDEF1X model will be identified.

Entity: MATERIAL

Attribute Name	Description	PK or FK
MATERIAL_ID	Identifier code for unique material	PK
MAT'L_DESCRIPTION	Description of material	
UNIT_OF_MEASURE	Quantity measure for material	

Entity: MATERIAL_LOT

Attribute Name	Description	PK or FK
LOT_NUMBER	Unique identifier for the material lot	PK
MATERIAL_ID	Identifier code for unique material	FK
M_DATE_CREATED	Date material lot was created in inventory	
M_QUANTITY_CREATED	Initial amount on hand at creation	
M_QTY_ON_HAND	Current amount on hand	

Entity: BATCH

Attribute Name	Description	PK or FK
OUTPUT_LOT_NO	Identifier of batch lot created	PK
LOT_NUMBER	Identifier of material (input) lot	PK, FK
INPUT_QUANTITY	Amount of LOT_NUMBER consumed	
FINISH_GOOD_LOT_NO	Identifier of end product into which batch is consumed	
OPERATION_ID	Identifier of production operation creating batch	

Entity: OPERATION

Attribute Name	Description	PK or FK
OPERATION_ID	Unique identifier of operation	PK
OPERATION_NAME	Common name for operation	
OPERATION_DESCRIPTION	Description of operation	

The attributes of the entity MATERIAL_LOT may refer to purchased materials, subassemblies, or final product. The M_DATE_CREATED is the date on which a material lot number was assigned. There are other attributes that are particular to purchased materials that were shown in the material lot entity of Figure 5.13. We will return to these shortly.

7.3.3.2 Quality Control Manager View

The QC manager has introduced the entity of the QC_RECORD, which is associated with material lots at various stages of production. Physically, the batch lot is a transient substance that is created and then disappears as it is consumed into another batch lot or into a finished good lot. It is typically not a semipermanent thing that gets stored in inventory, as in the case of raw ingredients or finished goods. The batch lot is an entity that is convenient to us in that it is the repository of work-in-process quality control data.

From Figure 7.19, we introduce the concept of the QC_TEST entity. The manner in which it is performed is given in the "Test Procedure" section of the quality control manual, which describes the various quality control tests. For our purposes here, we will assume that the QC_TEST is a generic concept for a procedure that is not associated with a specific material. The application to a specific material, such as the sample size to be taken and the acceptable limits of the test, is recorded in another section of the quality control manual under "Test Specification." Thus, in Figure 7.19, the MATL_TEST_XREF entity refers to the test specification for the particular material. Figures 7.20 and 7.21 are entries in the manual for the generic procedure for viscosity and its application to a batch of cheese sauce.

Entity: QC_RECORD

Attribute Name	Description	PK or FK
QC_RECORD_ID	Unique identifier of a record	PK
QC_TEST_ID	Unique identifier of a QC_TEST	FK
LOT_NUMBER	The lot number to which the record applies	FK
TEST_RESULT	Pass/Fail indicator	
QC_TEST_DATE	Date on which test was performed	

Entity: QC_TEST

Attribute Name	Description	PK or FK
QC_TEST_ID	Unique identifier for a type of test	PK
QC_TEST_NAME	Name of the test	
TEST_PROCEDURE_REF	Applicable QC manual reference	

Entity: MATL_TEST_XREF

Attribute Name	Description	PK or FK
MATERIAL_ID	Identifier code for unique material	PK
QC_TEST_ID	Unique identifier of QC_TEST	PK
TEST_SPEC_REF	Applicable QC manual reference	

Entity: QC_SAMPLE

Attribute Name	Description	PK or FK
QC_RECORD_ID	Unique identifier of a record	PK, FK
SAMPLE_NO	Sample identifier within the record	PK
TEST_VALUE	Result of the test on the sample	
DEFECT_LEVEL	Value indicating defect level of sample	

TEST PROCEDURE 1014: Sauce Viscosity

Equipment and Materials:

- Thermometer (EQ-19)
- Beaker (600 ml)
- Brookfield Viscometer DV-I (Eq-03)

Procedure:

1. Adjust the level of viscometer indicated by the bubble in the back of the instrument.
2. Turn power switch on.
3. Attach spindle to viscometer and turn motor switch on. Viscometer should read (0) in the air. At this point turn the motor switch off.
4. Spindle number, speed, and temperature of measurement used should be in accordance with the material test specification of the product.
5. Place 500 ml of sauce in a 600 ml beaker, stir and measure temperature.
6. At the specified temperature, insert the spindle of the viscometer in sample to the groove line, turn motor switch on and stir at specified speed for 10-15 seconds. Record reading.
7. Convert dial reading to centerpoise (cps) and record results to nearest 100 cps.

Figure 7.20 Sauce viscosity test procedure. (Courtesy Rutgers Center for Advanced Food Technology.)

MATERIAL TEST SPECIFICATION 450: Cheese Sauce Viscosity

1. Sauce is checked for viscosity according to test procedure 1014.
2. Each batch is checked for viscosity upon preparation.
3. Five samples of 500 ml each are taken from the batch.
4. Measurement is conducted at 40 degrees F using spindle #5 at 5 rpm.
5. The attribute specification for this test is as follows: Low Limit: 3000cps, High Limit: 5200 cps.
6. Samples with results falling within low and high limits will be accepted and will result in 0 defects.
7. Samples falling outside low and high limits will be rejected and will be given a score of 1 defect.
8. Batch passes test specification if number of defective samples is 1 or 0.
9. Batch does not pass test specification if number of defects is greater than 1.

Figure 7.21 Cheese sauce test specification. (Courtesy Rutgers Center for Advanced Food Technology.)

The reader will note that the entity QC_SAMPLE contains the TEST_VALUE attribute, which is the actual reading recorded in an instance of applying a QC_TEST. This can be an attribute value, such as a temperature reading in Fahrenheit or the value of a viscosity reading in centipoise. It can also be a classification value (good/bad) for a quality characteristic that is judgmental, such as in a visual examination of whether or not a seal on a finished package is acceptable.

The QC-SAMPLE entity also specifies a defect level. For example, in Figure 7.21, the value of a defect from a reading outside the viscosity limits is "1".

7.3.4 Establishing the Global Data Model

After the analyst has completed the user views, she or he must review the models with the individuals who provided the details that make up those views. It is important that there is agreement between analyst and user that the data model and attributes have captured the principal relationships of interest by the user. Once they have reached an agreement, the analyst should unite these views into a single data model for further analysis, as shown in Figure 7.22.

7.3.5 Defining Superclass/Subclass Relationships

We noted in Section 7.3.3 that each physical instance of a material, whether purchased from a vendor (ingredient) or produced (subassembly or finished good), is simply a material lot. However,

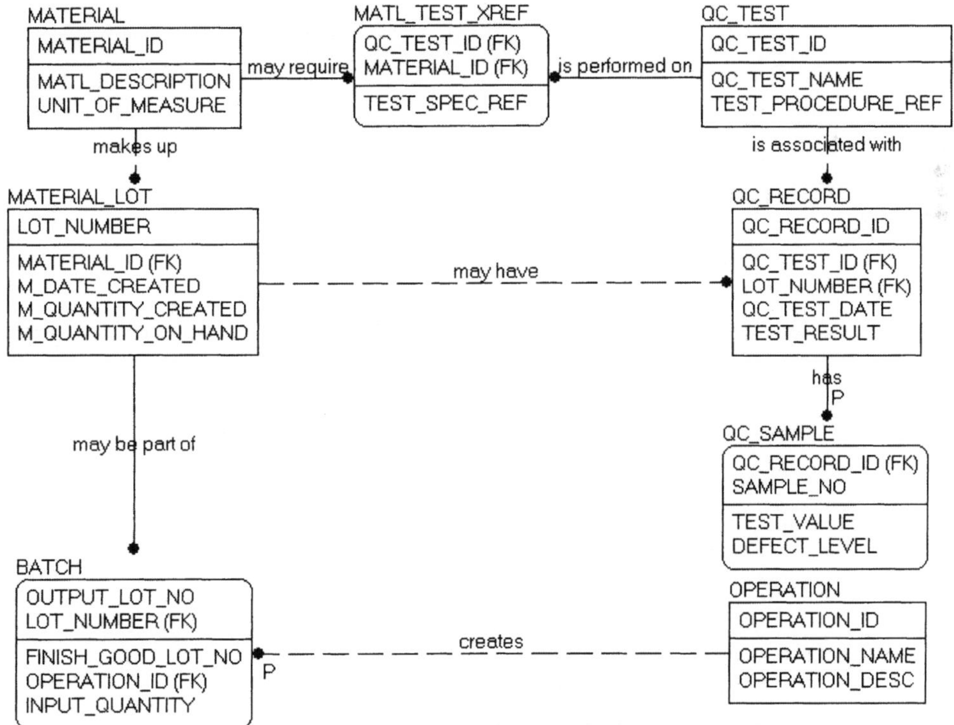

Figure 7.22 Initial global data model.

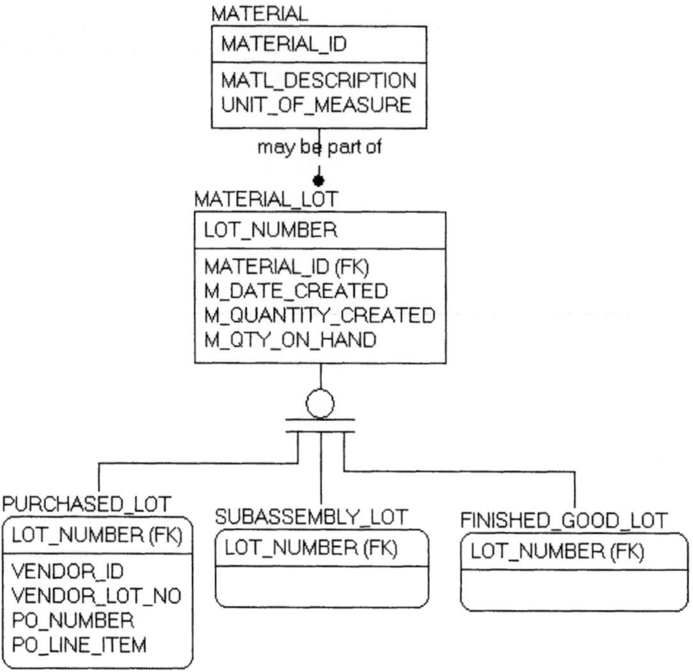

Figure 7.23 MATERIAL_LOT categorization entities.

the entity MATERIAL_LOT has attributes that distinguish between these two groupings, such as VENDOR_ID in the case of a purchased lot. Therefore, it is reasonable to establish a relationship between a superclass entity, MATERIAL_LOT, and the various subclass entities, as illustrated in Figure 7.23. New attribute tables are defined to show the division of attribute types. Note that the MATERIAL_LOT entity is a complete categorization, indicating that every instance of the superclass must be represented in the subclass. This means that the entity PURCHASED_LOT will contain all purchased materials, including items other than ingredients—for example, spare parts and supplies. A MATERIAL_LOT is a member of one and only one subclass.

Entity: MATERIAL_LOT

Attribute Name	Description	PK or FK
LOT_NUMBER	Unique identifier for the lot	PK
MATERIAL_ID	Identifier code for unique material	FK
M_DATE_CREATED	Date material lot was created	
M_QUANTITY_CREATED	Initial amount on hand	
M_QTY_ON_HAND	Current amount on hand	

Entity: PURCHASED_LOT

Attribute Name	Description	PK or FK
LOT_NUMBER	Unique identifier for the material lot	PK
VENDOR_ID	Unique identifier for vendor	
VENDOR_LOT_NO	Vendor identification for lot	
PO_NUMBER	Unique identification for a purchase order	
PO_LINE_ITEM	Line number on purchase order	

Entity: SUBASSEMBLY_LOT

LOT_NUMBER	Unique identifier for the batch lot	PK, FK

Entity: FINISHED_GOOD_LOT

LOT_NUMBER	Unique identifier for finished good lot	PK, FK

7.3.6 Evaluating the Need for Transaction Entities

In Section 5.3.3, we described the use of the transaction entity for recording the history of events leading to changes in the inventory status. Since our model includes the MATERIAL_LOT and will be united with the database shown in Figure 5.15 when implemented, which includes the WAREHOUSE_LOCATION entity, it is natural to include the LOT_TRANSACTION entity, as done in Figure 5.15.

7.3.7 Normalizing the Data Model

The steps required in normalizing a data model to third normal form were described in Section 3.5. In designing the data model, we used the IDEF1X formalism instead of the E-R modeling primitives. During model development, we have enforced 1:M relationships among entities. We have also eliminated redundancies as we proceeded. With the exception of the items discussed at the end of section 5.3.3, the model is in third normal form. Thus, this step is not necessary in this case study.

7.3.8 Finalizing and Validating a "To Be" IDEF0 Model

During the definition of the "as is" functional model, the analyst focuses on establishing the way in which current operations are executed. Prior to the implementation phase of the project, it may be desirable to redesign the processes or to add new technologies that will alter the way the processes and data collection functions are executed. Therefore, a detailed final activity diagram of the architecture may be warranted. In this case study, the activities are irreducible in the sense that they are all required in the order presented in order to manufacture the product. Therefore, we will not consider a redesign of the activity model at this time.

7.3.9 Finalizing and Validating an IDEF1X Global Information Model

The IDEF1X format is useful in this step because it gives an easy-to-read interpretation of the table structure of the database, with table names (entities), fields (attributes), and key attributes represented on the diagram. Such a diagram can now be constructed from the information given in Section 7.3. Since a database design exists for the control of purchased ingredients currently used by the enterprise (Figure 5.15), it is natural at this point to merge the two data models, as shown in Figure 7.24.

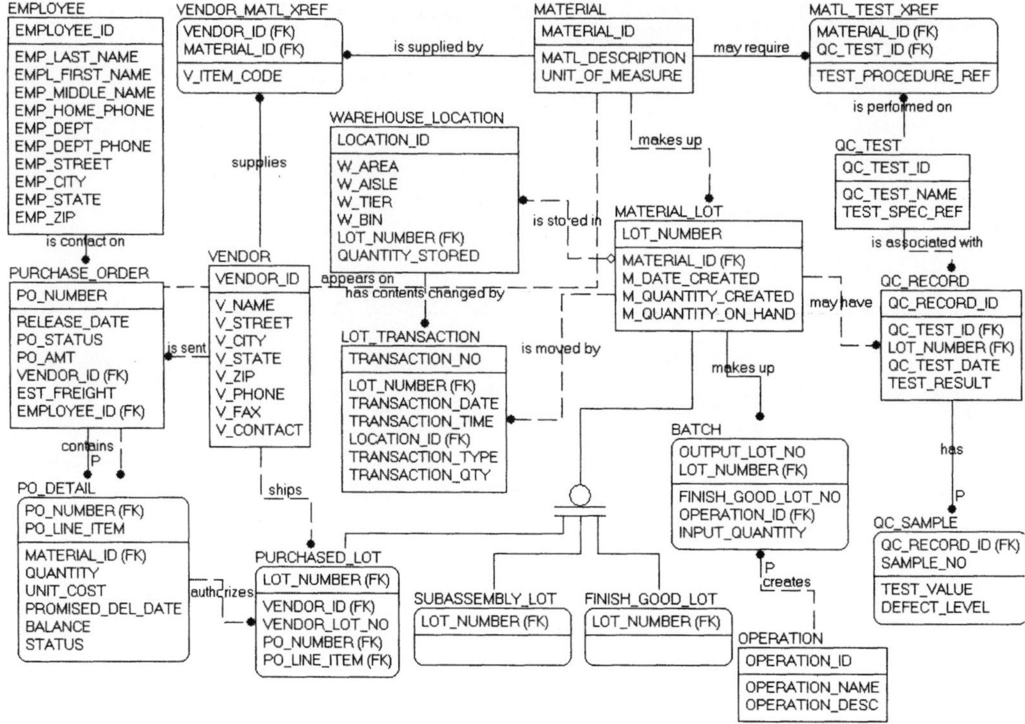

Figure 7.24 Final database design.

7.4 IMPLEMENTATION PHASE

In the implementation phase, the analyst is concerned with converting the logical database model into tables and creating data entry forms and user reports. There are four key steps to observe:

- Creating database tables and relationships
- Defining data entry methods
- Creating queries, forms, and reports
- Establishing data security and access methods

These steps were adequately covered by examples in Chapter 6. Since forms and reports can be implemented in many ways, we will not describe a typical implementation here. Rather, exercises at the end of this chapter will allow the reader to develop a set of forms and reports based on the requirements of this case study.

7.5 SUMMARY

In this chapter, we described a hypothetical case study of an information system design project. A typical design project has three distinct phases. In the preliminary study and problem definition phase, the analyst defines what the client generally wants from the system and how the target system functions. This phase concludes with a functional model that clearly defines the boundaries of the study. During the design phase, the analyst focuses on the user information needs and models the requirements for each user in a data model format. This phase ends when these individual user views are combined into a global data model. This phase produces a logical database model that can be directly converted into database tables. Finally, during the implementation phase, the tables are created and the forms and reports are designed to serve the clients' stated needs.

Information systems are not intended to be static. As organizations grow and change, the information system must be reevaluated to determine whether or not it is still serving the needs of the organization. The process of systems analysis and information system design reoccurs regularly to meet changing requirements, such as those presented in this chapter's case study. The approach to project execution illustrated in the case study provides a logical framework for executing such projects.

REVIEW EXERCISES

7.1 How does an analyst determine the boundaries of a project? How are these boundaries documented?

7.2 What is meant by a "user view," and how is a user view determined? How is it documented?

7.3 What are business rules, and why are they important in the information system design?

7.4 The FDA regulation requires total traceability of material lots throughout the supply chain. Explain how the model in Figure 7.24 will enable traceability through the supply chain from the original source to the final product.

7.5 In University Food Company Case (C), the reader was asked to develop an IDEF1X model of the shipping department data requirements. This model enabled the tracking of lots of finished product to customers, which is part of the requirements of the FDA regulation. Merge the relevant entities in the solution to University Food Company Case (C) with the relevant entities in Figure 7.24 to show complete traceability from vendor ingredients to the final customer.

7.6 The database Lot_Trace_CH7_stu is in the Chapter 7 folder of the Web site that accompanies this book. It contains tables populated with appropriate data.

(a) Prepare a management "Drill Down" report that can take a finished product lot number and determine the ingredients that went into it at each batching operation.

(b) Prepare a management "Drill Up" report that can take a material lot number and determine the lot numbers of the finished product that the material lot went into.

7.7 Using the Lot_Trace_CH7_stu database, create a form for the cooking operation that will allow the technician to enter lot numbers of ingredients according to the formula for cheese sauce. Sample data for running the application are also given in the LOT_TRACE application note on the Web site. The form should exhibit the following functionality:

- Lot numbers for ingredients added to a batch must already exist in the MATERIAL_LOT table.
- Quantities of ingredients added to a batch must be deducted from inventory when the record is submitted.
- New lot numbers created for a new cheese sauce subassembly must be unique. This can be done using an autonumber or entering a unique number.

Chapter 8

E-Business and Web-Enabled Databases

8.1 INTRODUCTION

At one time manufacturing managers scheduled and operated their plants using fairly long range planning horizons over which they forecasted demand. A typical approach was to isolate the plant operation from downstream activities, such as product distribution, by carrying buffer stocks of inventory in the plant warehouse and monitoring the level of those buffer stocks to decide when more inventory should be produced.

All that changed with advances in information technologies. The developments in computer network technology and point-of-sale data acquisition technologies, such as bar code readers, made it possible to capture real-time information at final distribution points and to incorporate this information immediately into production planning and scheduling decisions at the plant. Thus, instead of monitoring inventory levels at the plant warehouse, it was possible to see the status of inventories throughout the distribution system and to anticipate when inventory stock replenishment orders would be placed. In some cases, information that exists beyond the enterprise can be made accessible to the plant for planning purposes. For example, supermarkets have real-time inventory control through the use of bar code scanning at the checkout counters. The supplier (manufacturer) of food products may develop a relationship with the supermarket that allows the manufacturer to access the supermarket information system and extract information on current inventory levels. This allows better resupply forecasting and subsequent production scheduling decisions to be made by the manufacturer.

This vision of business relationships and information sharing has been incorporated into the concept of *supply chain management.* A ***supply chain*** is a set of related enterprises through which materials and associated information flow. A typical supply chain of a single manufacturer is shown in Figure 8.1. It has an hourglass shape, with the manufacturing enterprise at the center. The lower part of Figure 8.1 shows the physical distribution system from the manufacturing plant to the customer. Here management is concerned with customer demand and appropriate inventory levels to service that demand at minimum cost. Management problems in the distribution system also include transportation and the routing and scheduling of company-owned vehicles for delivery of the product to regional and local warehouses and customers.

The upper level of Figure 8.1 shows the input network of the supply chain. Manufacturers are customers of suppliers, who themselves may be manufacturers with their own supply chains. Problems on the input side include monitoring appropriate levels of raw materials and managing lead

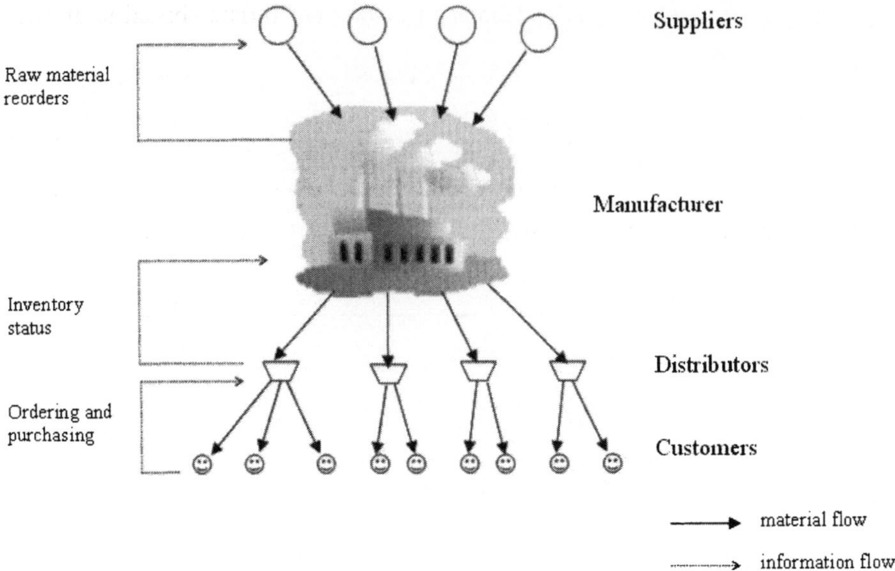

Figure 8.1 Typical supply chain network.

times between ordering and receiving shipments from suppliers so as not to disrupt production schedules due to delayed shipments.

The problems of managing the operations of the supply chain are beyond the scope of this book. Specific issues and management techniques for supply chains can be found in (Bowersox, et al., 2002; Chopra and Meindl, 2000; Simchi-Levi, et al., 2000). However, the supply chain is a useful context for the subject of e-business. Many of the problems of coordinating the operations of a supply chain arise from lack of timely information. Today, companies are moving to a tightly integrated supply chain model in which information throughout the supply chain is immediately available to all participants. The model popularized by Dell Computer is a case in point. Customers order Dell's computer products directly over the Internet, and Dell uses this ordering information to procure parts over the Internet from its extensive Web of suppliers. In this model, the connection between the plant floor and the broader supply chain is made possible by information exchange over the Internet. The term commonly used to describe this model is ***e-commerce***; here we use the associated term ***e-business***.

The purpose of information technology in e-business is to provide communication between the enterprise and its customers (**business to customer, or B to C**) and between the enterprise and its suppliers (**business to business, B to B**). The two main uses of e-business systems are intercompany transactions and information exchange. Intercompany transactions can include customer order entry, purchasing from suppliers, and electronic billing, among others. Information exchange can include inventory level information sharing, quotation of lead times, and online checking on the status of an order, among others.

Earlier electronic transactions between business partners (B to B) used a technology called ***electronic data interchange (EDI)***. EDI enabled intercompany, computer-to-computer exchange of business documents in a standard format. With the advances in computer technology and the

Internet in the late 1990s, business collaboration among enterprises shifted to Internet-based transaction processing.

Each enterprise in a supply chain partnership has its own enterprise resource planning (ERP) and manufacturing execution system (MES) and other software solutions that handle the internal business. However, in the supply chain, these systems must work together with those of other supply chain partners to achieve the business objectives of the coalition of partners in the supply chain. From the perspective taken in this book, this raises an important issue. A requirement for business partners to work together in a supply chain is that of database information exchange technologies and protocols. Database information input and retrieval can be designed as part of an Internet application. The reader will be familiar with this fact from placing orders for books and other merchandise over the Internet. Those applications are developed using a combination of hypertext markup language (HTML) for designing Web pages and some scripting language, such as VBScript or JavaScript, for interacting with the database. This popular approach is used for placing orders in B-to-C applications, which will be described in Section 8.3.

There are also cases in which an enterprise may simply want to connect to a supplier via the Internet and transfer information from its database to that of the supply chain partner. For example, Cisco Systems designs network routers that are manufactured by contractors (supply chain partners) at distant locations. Cisco wants to supply timely product specification information, including engineering changes, to the contractors. This information should immediately update the computer-aided design (CAD) database of the manufacturer and the associated database tables of the manufacturer's ERP system. One technically feasible way to do this is to require all contractors to use the same information systems and software used by Cisco Systems. In this way, changes made in one information system can be easily mapped to that of all participating partners. However, existing partner relationships do change over time and new partners may be brought into the relationship. New partners probably have already invested in information systems that would be expensive to convert, and they may have relationships with other firms that have other systems. Thus, requiring all firms in a relationship to have identical information system technology is impractical. Another way to enable the exchange of information is to adopt a single interchange format that defines the meaning of data and to move the data over the heterogeneous systems using the common format. This solution is currently being implemented using extensible markup language (XML), which is a meta-grammar that allows users to define a common language for data transactions that can be understood and parsed by diverse systems and applications. We shall discuss XML and its implementation in Section 8.4.

The starting point for studying the details of Internet applications is to review HTML, which is the formatting language for Web pages. The most common functions of this language are reviewed in the next section.

8.2 AN HTML TUTORIAL

Hypertext markup language (HTML) is a "scripting" language, as opposed to a programming language. It is used to *mark up* the pages of electronic documents with formatting commands that can be easily interpreted by a Web browser. A Web browser is a computer program that interprets HTML commands according to an (evolving) HTML standard.

Because HTML is basically a document page design tool, there are many commercial products that allow a user to directly design a Web page and automatically convert the design into HTML code so that it can be viewed on a browser. Some leading products are FrontPage (by Microsoft) and Dreamweaver (by Macromedia).

Because we are interested in implementing database queries on the Web, it behooves us to know some of the details of Web technology and HTML code. Therefore, we will review a subset of HTML and see how it fits into the overall process of making our database forms Web enabled.

8.2.1 Web Page Design Example

It is important to determine beforehand what the Web page should look like and how it should function. This is a creative process of the designer. The set of Web pages that will be used for this illustration is based on a common function in B-to-C (business-to-customer) Internet transactions, that of entering an order for a product. Two screens will be involved:

- An HTML home page with references to text and graphic files
- A form for entering customer information

The home page to be designed is shown in Figure 8.2a. The form page for entering customer information is shown in Figure 8.2b.

There are two general classifications of Web page applications that are represented by Figures 8.2a and b. These classes are "static" and "dynamic" Web pages, respectively.

A ***static Web page*** consists of a file or a set of files that are programmed in HTML. The contents of these files are not expected to change very often, and they are not capable of changing their presentation on a Web browser based on a user query. Figure 8.2a is an example of a typical static Web page.

The static Web page environment is a good example of client-server architecture. The World Wide Web (WWW) consists of a network of computers that act in two roles. The browser on a workstation acts as a client. Examples are Netscape Navigator and Windows Explorer. The server is the computer and software that handles the client's requests for a particular Web page. Typical examples of server software for Web applications are Microsoft's Internet Information Server and Sun Microsystem's Apache Web server.

Figure 8.3 depicts the basic components of the client-server architecture of a Web environment that serves HTML Web pages. The client machine establishes a connection with the Web server using a Web browser. The client requests a Web page from the server by transmitting the uniform resource locator (URL) address of the document. Then the server downloads the document (file) to the Web browser on the client machine. The Web server then closes the connection. The browser displays the document on the client machine based on the formatting instructions in the HTML file. So, basically, all the Web server does is download files on request from the client.

Dynamic Web pages are more complicated. In that environment, a client can request information that may change based on the requestor or on the timing of the request. In other words, the information is dynamic. A database search is an example of this kind of situation, as is Figure 8.2b. We will discuss the architecture for the dynamic Web page environment in Section 8.3. Here we continue with a basic description of how to build static Web pages using HTML.

8.2.2 HTML Page Tags

Again, HTML is a language for simply marking up a document page. The marking up is done using HTML *tags*. A **tag** is a coded command used to define how part of a Web page is to be displayed on a browser.

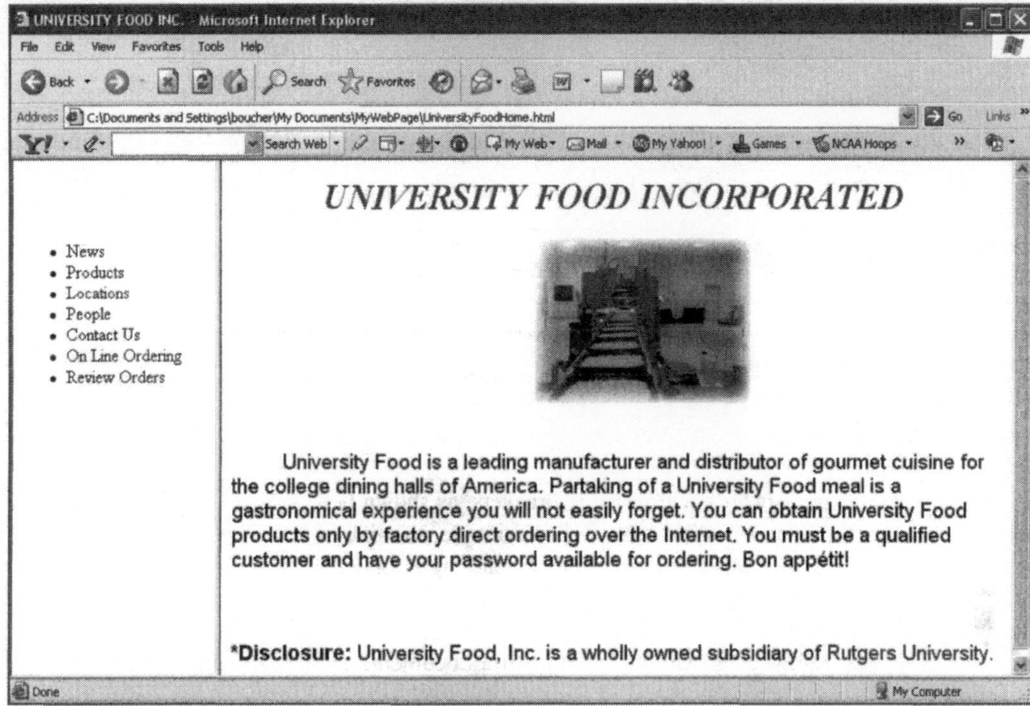

a) Homepage

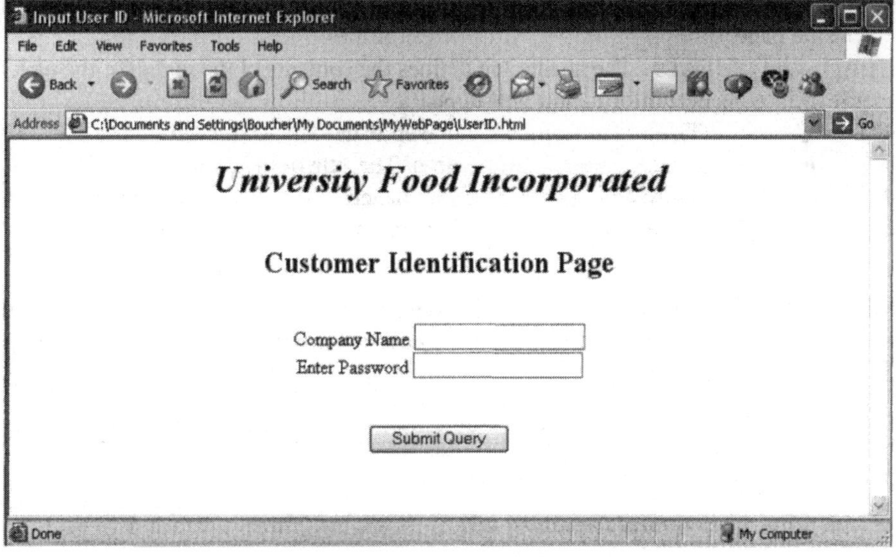

b) Customer input form

Figure 8.2 Example Web pages.

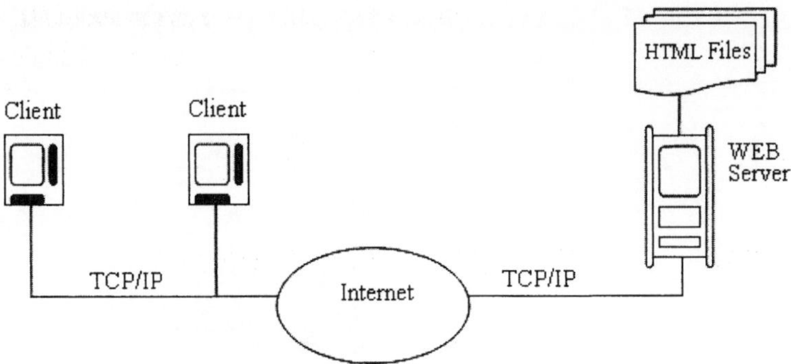

Figure 8.3 Client-server two-tier model.

Tag syntax appears as words within angle brackets, as shown here: <word>. They delineate an area of the page for a specific kind of text. For example, the title of the page is indicated by the start tag <TITLE>. A start tag usually requires an end tag to indicate the end of the assigned area. The end tag begins with a foreward slash /. Thus, the end of the area in which the title appears is indicated by </TITLE>.

Some tags are used to give overall structure to a document. These structural format tags are as follows:

- <HTML> and </HTML>. This tag set declares that the document within the tags is an HTML document. Normally, you will see these tags at the start and the end of the formatting instructions of the page.
- <HEAD> and </HEAD>. The header tag defines the contents of the region of the page where general theme (header) information will appear.
- <TITLE> and </TITLE>. The name of the Web page is placed within these tags. The title of the page is an example of header information. The title of the Web page will appear on the title bar of the Web browser when the page is opened.
- <BODY> and </BODY>. This usually follows the header and is the area where text within the browser window will appear.

HTML documents can be created in any text editor. For the examples given in this chapter, we will use Notepad, the text editor that comes with the Windows operating system.

HTML Exercise 8.1:

We wish to use the ***structural formatting tags*** of HTML to illustrate a default page based on Figure 8.2a.

1. Create a folder that you will use to save your web pages (for example, C:/MyWebPage).
2. Open the Notepad text editor ***(Accessories => Notepad)***.
3. Type the text shown in Figure 8.4.
4. Save the file as type ***"All files"*** with the File name ***UniversityFood.html*** in your folder.

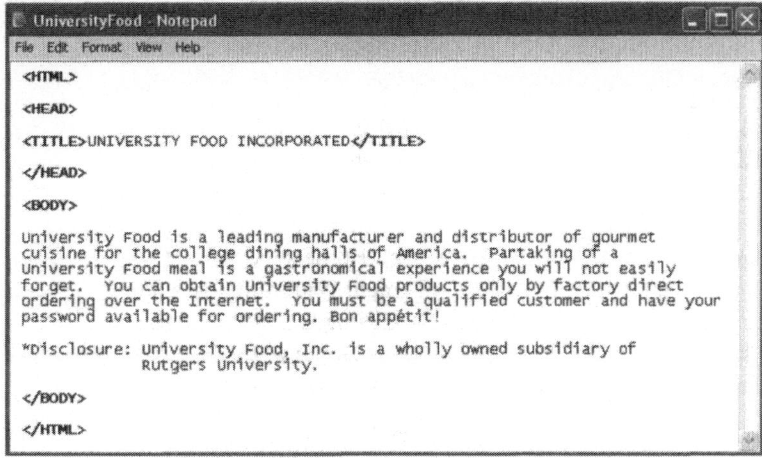

Figure 8.4 Using structural formatting tags.

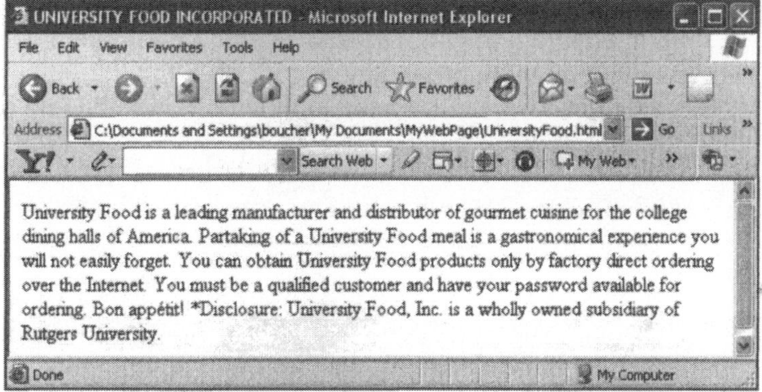

Figure 8.5 Web page created by the code shown in Figure 8.4.

Note that it is important to save the file as type "all files." If you save it as a text file, the browser will not recognize it as an HTML file.

To test the functionality of your page, open a Web browser (Netscape Navigator or Internet Explorer). Browse to the folder and open the file in your browser. It should look like Figure 8.5. The page does not look much like Figure 8.2a. For that to happen, we need to do more detailed formatting. This is discussed in the following sections.

Note that the title of the page appears in the title bar of the browser. It does not show up on the Web page. All the text that the viewer will see in the browser window must be placed between the <BODY> </BODY> tags. Also, the text is not formatted; it runs on continuously, regardless of how it was typed. To control the detailed formatting of the text, it is necessary to include formatting tags.

8.2.3 HTML Text Body Formatting Tags

Here are some of the most basic formatting tags for the text appearance on the page:

- <Hx> and </Hx>. Headings that appear in the Web browser window (placed between <BODY> and </BODY> tags) are typically larger than the other text in the body area. The easiest way to create a heading larger than the default body text size is to use the heading tags: <H1> </H1>, or <H2> </H2>, and so forth. The index indicates the relative size of the heading text, where H1 is the largest and H2 through H6 become progressively smaller.
- <P> and </P>. These tags enclose a paragraph. Within the paragraph, the text is continuous when opened in the browser window. The closing tag, </P> is optional.
-
. This is a line break. It is placed at the end of a section where a line break is desired. It is a single tag and does not require a closing tag.

To illustrate the use of these tags and to render a better appearance to the Web page, open UniversityFood.html and edit the text in accordance with the instructions of HTML Exercise 8.2. The new HTML tags are shown in bold in Figure 8.6.

> ***HTML Exercise 8.2:***
>
> We wish to add more structure to Figure 8.5 using ***basic formatting tags***.
>
> 1. Open the file ***UniversityFood.html*** in Notepad.
> 2. Add heading tags, a paragraph tag, and line breaks as shown in Figure 8.6. Save and close the file.
> 3. Using Netscape Navigator or Internet Explorer, browse to the file and open it in the browser window. It should appear like Figure 8.7.

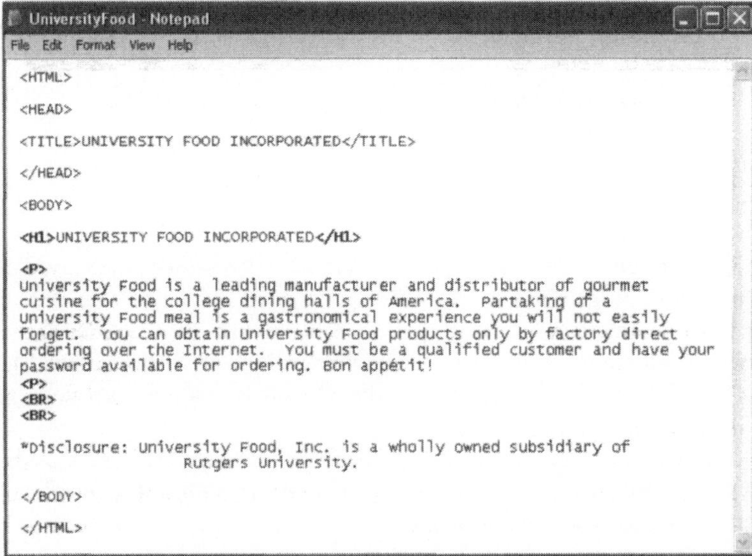

```
<HTML>
<HEAD>
<TITLE>UNIVERSITY FOOD INCORPORATED</TITLE>
</HEAD>
<BODY>
<H1>UNIVERSITY FOOD INCORPORATED</H1>
<P>
University Food is a leading manufacturer and distributor of gourmet
cuisine for the college dining halls of America.  Partaking of a
University Food meal is a gastronomical experience you will not easily
forget.  You can obtain University Food products only by factory direct
ordering over the Internet.  You must be a qualified customer and have your
password available for ordering. Bon appétit!
<P>
<BR>
<BR>

*Disclosure: University Food, Inc. is a wholly owned subsidiary of
                    Rutgers University.

</BODY>
</HTML>
```

Figure 8.6 Adding text formatting tags.

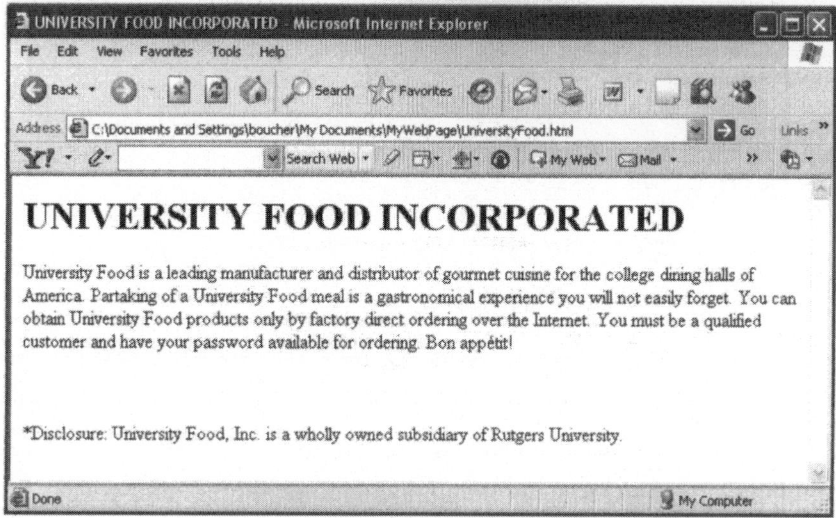

Figure 8.7 Web page created by the code shown in Figure 8.6.

Note the improved control over the layout of the page. Also, the page caption now appears as a heading of the desired size in the browser window. Heading tags within the body of the page result in a caption on the page within the browser window.

8.2.4 ALIGNMENT, POSITIONING, AND FONT CONTROL

Despite the improvement in the format of the Web page, there are still some obvious deficiencies in the appearance, alignment, and positioning of the text. The following are some additions to text formatting tags. These additions are called *attributes*. An ***attribute*** is a command, placed within the format tags, to further define a feature of the text within these tags.

- <B> and </B>. These tags indicate that the text appearing between them should be bold.
- <I> and </I>. These tags indicate that the text appearing between them should be in italic.
- <P Align="Right">, <P Align="Center">, <P Align="Left">. The command *Align* is an attribute of the P tag or the Hx tag. It specifies the alignment of the paragraph or heading. If alignment is not specified, the text defaults to <P Align="Left"> as observed in Figure 8.7.

To illustrate these tags, open the UniversityFood.html file and add the attributes as shown in Figure 8.8.

> ***HTML Exercise 8.3:***
>
> We wish to add more definition to Figure 8.7 by using ***tag attributes***.
>
> 1. Open the file ***UniversityFood.html*** in Notepad.
> 2. Add the tags for bold, italics, and align as shown in Figure 8.8. Save and close the file.
> 3. Using Netscape Navigator or Internet Explorer, browse to the file and open it in the browser window. It should appear like Figure 8.9.

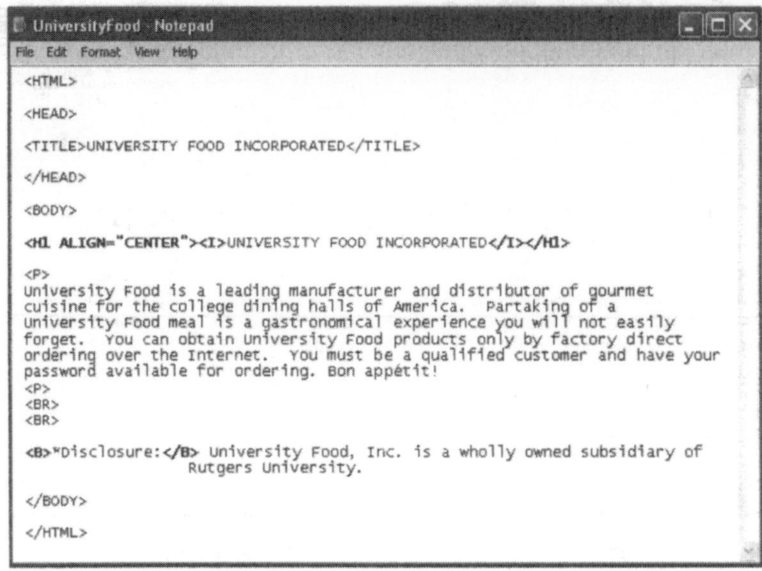

```
<HTML>

<HEAD>

<TITLE>UNIVERSITY FOOD INCORPORATED</TITLE>

</HEAD>

<BODY>

<H1 ALIGN="CENTER"><I>UNIVERSITY FOOD INCORPORATED</I></H1>

<P>
University Food is a leading manufacturer and distributor of gourmet
cuisine for the college dining halls of America.  Partaking of a
University Food meal is a gastronomical experience you will not easily
forget.  You can obtain University Food products only by factory direct
ordering over the Internet.  You must be a qualified customer and have your
password available for ordering. Bon appétit!
<P>
<BR>
<BR>

<B>*Disclosure:</B> University Food, Inc. is a wholly owned subsidiary of
                    Rutgers University.

</BODY>

</HTML>
```

Figure 8.8 Adding alignment, positioning, and font attributes.

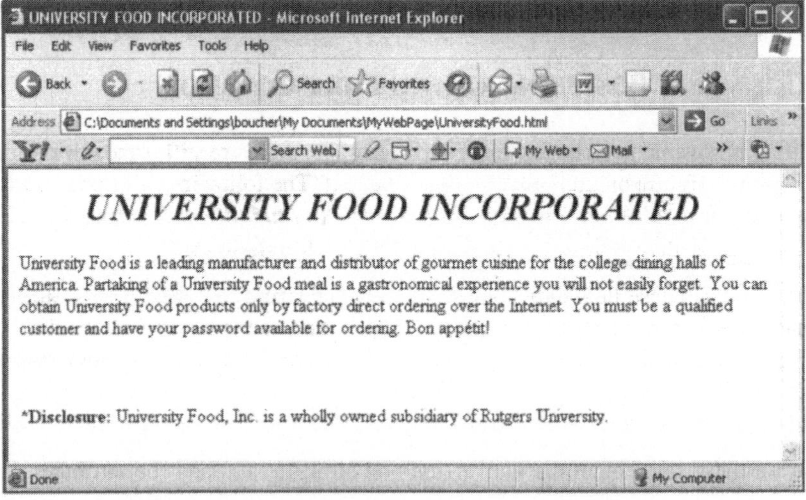

Figure 8.9 Web page design from the code shown in Figure 8.8.

8.2.5 Linking to Other Web Sites and Web Pages

The power of ***hypertext*** is the ability to link pages within a document and to link to other Web pages on other machines on the World Wide Web. A hyperlink is a command that, when executed, opens another file, browser window, or Web page. Linking to other pages and other documents on the Internet is easy using the ***anchor*** <A> tag.

The anchor tag uses a format that begins with **<A HREF=** and is followed by the page file location or the Web address of the page to which the link is directed. For example, if one is linking

to a file named "home.html" that resides within the same folder as the current page, the anchor would be as follows:

```
<A HREF="home.html">
```

If the linked page (file) is in a separate folder, the complete path to the file must be given within quotations. If the link is to another Web site on the WWW, it can be referred to by its WWW address. For example, in the case of Rutgers University, <A HREF="http://www.rutgers.edu">.

When you create a link, you do not have to display the target URL address in the browser window, although you can if you want to. For example, Figure 8.2a has a link to the Rutgers Web site that just reads "Rutgers University." This can be done by adding text after the opening anchor and adding a closing anchor as follows:

```
... subsidiary of <A HREF="http://www.rutgers.edu"> Rutgers University. </A>
```

In this format, the words "Rutgers University" are displayed as a hyperlink. Clicking on the hyperlink in the browser window with a mouse will display http://www.rutgers.edu in the Web browser URL panel, and the new Web page will be loaded.

8.2.6 Using Images

Images are popular on Web pages to gain viewer interest. Images must be created separately and stored as either .jpg or .gif files. HTML simply formats the page and indicates the placement of the image on the page. When a user opens the Web page, the image file is called and it opens in the formatted location. The following tag is used with images.

- <IMG>. This tag alerts the browser that an image should be located in the position on the screen where the tag occurs. Reference is made to the image file (.jpg or .gif) by pointing to the source file as follows:

```
<IMG SRC="ImageName.gif">
```

If the image file is not located in the same folder as the file of the Web page calling it, the IMG tag must include the complete path to the file within the quotations. The image file appearing in Figure 8.2a is named "foodpic.jpg." This file is included on the Web site accompanying this book. Copy that file into your folder MyWebPages. If the image is stored in a separate folder, the path to the file must be placed within the quotations.

Image size can be controlled by using the dimension tags to scale the image. For example, if the original image is 300 × 200 pixels, its size can be scaled to half using the HEIGHT and WIDTH specification tags as follows:

```
<IMG SRC="ImageName.gif" HEIGHT=150 WIDTH=100>
```

HTML Exercise 8.4:

We wish to add the ***image link*** and ***anchor tag*** to Figure 8.9.

1. Open the file ***UniversityFood.html*** in Notepad.
2. Add the image link and anchor tag as shown in Figure 8.10. Save and close the file.
3. Using Netscape Navigator or Internet Explorer, browse to the file and open it in the browser window. It should appear like Figure 8.11.

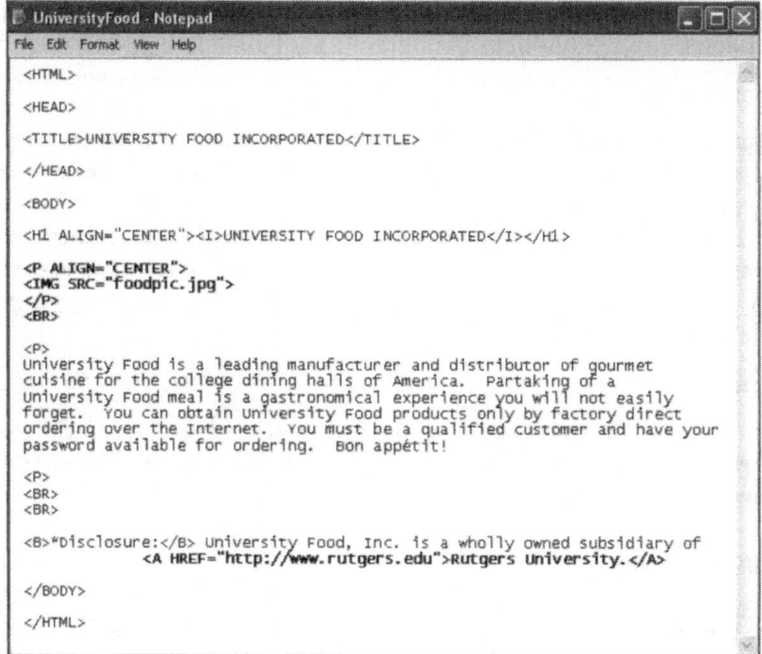

```
UniversityFood - Notepad
File Edit Format View Help

<HTML>

<HEAD>

<TITLE>UNIVERSITY FOOD INCORPORATED</TITLE>

</HEAD>

<BODY>

<H1 ALIGN="CENTER"><I>UNIVERSITY FOOD INCORPORATED</I></H1>

<P ALIGN="CENTER">
<IMG SRC="foodpic.jpg">
</P>
<BR>

<P>
University Food is a leading manufacturer and distributor of gourmet
cuisine for the college dining halls of America.  Partaking of a
University Food meal is a gastronomical experience you will not easily
forget.  You can obtain University Food products only by factory direct
ordering over the Internet.  You must be a qualified customer and have your
password available for ordering.  Bon appétit!

<P>
<BR>
<BR>

<B>*Disclosure:</B> University Food, Inc. is a wholly owned subsidiary of
            <A HREF="http://www.rutgers.edu">Rutgers University.</A>

</BODY>

</HTML>
```

Figure 8.10 Adding an anchor and an image.

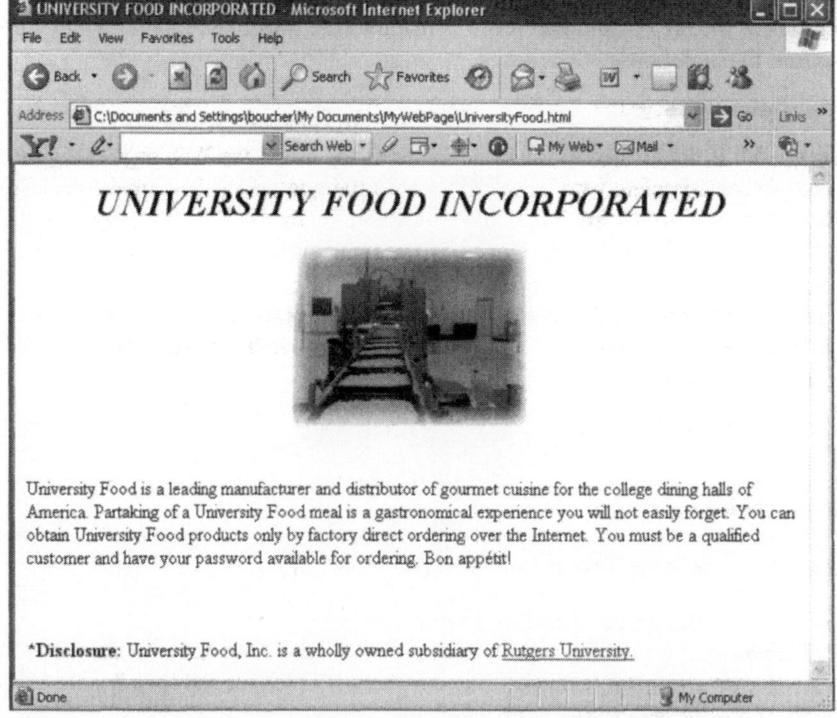

Figure 8.11 Web page created by the code shown in Figure 8.10.

8.2.7 USING CASCADING STYLE SHEETS

More control can be exercised over the appearance of a document by using **style sheets**. Style sheets have been used for many years in desktop publishing to control elements of appearance, such as font size, color, and indentation. A style sheet uses methods that are specified as arguments of a tag. The Web browser reads these arguments just as it reads any other HTML tag instruction and displays the text according to the desired style.

There are three ways of using style sheets:

- Inline
- Embedded
- External

The ***inline*** approach simply adds the argument to the tag that appears in the body of the HTML file. Consider Figure 8.11. If it is desirable to have the text beneath the image to appear indented and the text font to be 14 point Arial, the opening paragraph tag would be augmented as follows:

```
<P STYLE="text-indent: 0.75in;font: 14pt arial">
```

The keyword STYLE indicates that the instructions that follow are style methods. In this case, the text-indent method and the font method are used. When the browser encounters the STYLE keyword, it executes the methods that follow on the relevant text. Revise the HTML code in Figure 8.10 to include this style command and note the changed appearance when the file is opened in a Web browser.

Inline STYLE can also be useful for aligning text on a page. A special case is its use in conjunction with the <ADDRESS> tag that provides a single spaced address block. For example,

```
<ADDRESS STYLE="text-indent: 2.00in">NAME</STYLE></ADDRESS>
<ADDRESS STYLE="text-indent: 2.00in">STREET</STYLE></ADDRESS>, etc.
```

will provide a left-aligned, single-spaced, indented address block. The inline STYLE statement can be used with other HTML tags. For example, it can be used with block formatting tags (<BODY>, <CENTER>, etc.), list tags (<UL>, <OL>), image (<IMG>) and anchor (<A>) tags, among others.

Using the ***embedded*** style sheet format, the user places all style information in a style declarative section before the body of the text. The style declarative section is bounded by a <STYLE> </STYLE> tag set and is placed just after the header information. For example, in Figure 8.10, an embedded style sheet section could be added as follows:

```
<HTML>

<HEAD>
<TITLE>UNIVERSITY FOODS INCORPORATED</TITLE>
</HEAD>
<STYLE>
BODY {margin-left: 0.5in;margin-right: 0.5in}
H1 {color: blue}
P {text-indent: 0.5in;font: 14pt arial}
A {text-decoration: none}
</STYLE>
```

The preceding style sheet sets left and right margins for the entire body of the document. Each time the <H1> </H1> tag set is encountered, the header text will be colored blue. Each time a <P> </P> tag set is encountered, a 0.5-inch indentation will appear on the first line of the paragraph and the font will be 14 point Arial. Finally, text enclosed in anchor tags, <A> </A>, will appear without any special decoration. That is to say, no italics, bold, or underline will appear. By default, anchor tags appear with underlines and are colored blue. To alter the color, a color method can be added to the style sheet anchor tag argument.

Finally, it is possible to define the style sheet as a separate document, called an ***external*** style sheet. Under this format, each Web page will consist of two files. The first file is the .html file that defines the contents of the Web page. A second file is created that contains the style sheet. This second file is saved as a .css (cascading style sheet) file. The .css file is referenced from the .html file. When the Web browser encounters the reference while displaying the Web page, it opens the .css file and incorporates the style sheet methods contained therein into the appearance of the Web page.

Figure 8.12 is an example of an external style sheet named UFOOD.css. To link to the style sheet in Figure 8.12, the Web page (.html file) must declare the linking information in the header section of the file. This is shown in Figure 8.13 for the University Food Incorporated home page. The <LINK> tag opens the link. The arguments declare it is a link to a style sheet, with the reference filename "UFOOD.css," and is a text type file with a .css extension. Note that the HREF

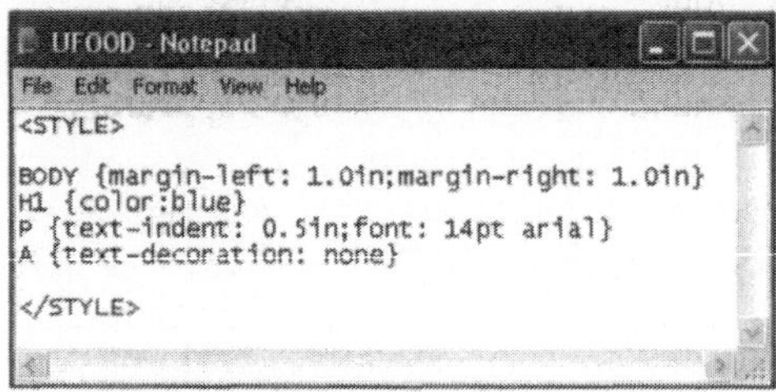

```
UFOOD - Notepad
File Edit Format View Help
<STYLE>

BODY {margin-left: 1.0in;margin-right: 1.0in}
H1 {color:blue}
P {text-indent: 0.5in;font: 14pt arial}
A {text-decoration: none}

</STYLE>
```

Figure 8.12 External style sheet.

```
UniversityFood - Notepad
File Edit Format View Help
<HTML>

<HEAD>

<TITLE>UNIVERSITY FOOD INCORPORATED</TITLE>
<LINK REL=STYLESHEET HREF="UFOOD.css" type="text/CSS">
</HEAD>

<BODY>
<H1 ALIGN="CENTER"><I>UNIVERSITY FOOD INCORPORATED</I></H1>

<P ALIGN="CENTER">
<IMG SRC="foodpic.jpg">
</P>
<BR>
<P>
University Food is a leading manufacturer and distributor of gourmet
cuisine for the college dining halls of America.  Partaking of a
University Food meal is a gastronomical experience you will not easily
forget.  You can obtain University Food products only by factory direct
ordering over the Internet.  You must be a qualified customer and have your
password available for ordering. Bon appétit!
<P>
<BR>
<BR>
<B>*Disclosure:</B> University Food, Inc. is a wholly owned subsidiary of
            <A HREF="http://www.rutgers.edu">Rutgers University.</A>

</BODY>

</HTML>
```

Figure 8.13 Web page with link to style sheet.

argument of the LINK tag is just the style sheet filename. This indicates that it is in the same folder as the file that calls it. If it is installed in a different location, the complete path to the style sheet file must be included within the quotations.

The external style sheet format allows the user to separate style settings from the main Web page document. As a result, the same style sheet can be used over and over again for different Web pages and changes to the style sheet can be made without making changes in each Web page file.

HTML Exercise 8.5:

We wish to create a ***cascading style sheet*** and reference it from UniversityFood.html.

1. Open Notepad and prepare the file UFOOD.css as shown in Figure 8.12. Save the file in your folder (MyWebPage) under Type "All files" as UFOOD.css.
2. Open the file ***UniversityFood.html*** in Notepad.
3. Add the link to the style sheet as shown in Figure 8.13. Save and close the file.
4. Using Netscape Navigator or Internet Explorer, browse to the file and open it in the browser window. Note the incorporation of the style sheet appearance.

8.2.8 USING FRAMES

Framed layouts are used to divide the browser window into multiple regions, known as ***frames***. For example, in Figure 8.2a, the left side panel is a frame, and the right side window, which is the same as Figure 8.11, is a separate frame. Only a few tags are required to implement frames:

- <FRAMESET>. This tag is used to divide the browser window into frames. There are two attributes that can be used with frameset; they are *rows* and *cols*. The first divides the browser screen into multiple rows, and the second divides the screen into multiple columns. A typical syntax is as follows:

```
<FRAMESET Cols="20%, 80%">
...
</FRAMESET>
```

This command tells the browser to divide the browser window into two columns. The first column, starting from the left, comprises 20% of the screen, and the second column comprises 80% of the screen.

The <FRAMESET> tag simply divides the browser window into multiple windows. The content of each window is separately determined by relating the window to the file that will give it content. To relate the frames to their contents, you need the <FRAME> tag:

- <FRAME>. The *frame* tag tells the Web browser where to locate the contents of the frame. Typical syntax is as follows:

```
<FRAMESET Cols="20%, 80%"> <!-make 2 column frames->
<!-Assign content to frame #1->
<FRAME SRC="List.html" FRAMEBORDER=1>
<!-Assign content to frame #2->
<FRAME SRC="UniversityFood.html" FRAMEBORDER=1>
</FRAMESET>
```

The FRAMEBORDER attribute is used to determine whether or not there will be a border around the frame. The value "1" indicates a border, and "0" indicates no border.

The preceding code includes some "comments." To document the purpose of an HTML page, it may be appropriate to add comments to the page. Comments on an HTML page are not displayed in the browser window. They are strictly there for documentation purposes. Comments are added as follows:

```
<!-This is a comment ->
```

To illustrate the use of the frameset and frame tags, two new files will be created: University-FoodHome.html and List.html. The first file will become the University Food home page and will be partitioned into framesets. The second file will become the left side panel. The reader can prepare the homepage by doing HTML Exercise 8.6.

> ***HTML Exercise 8.6:***
>
> We wish to create a container Web page with frames for holding the contents of ***UniversityFood.html***.
>
> 1. Open Notepad and prepare the file for ***UniversityFoodHome.html*** as shown in Figure 8.14. Save the file in your folder (MyWebPage) under Type "All files" as ***UniversityFoodHome.html***.
> 2. Create a new file in Notepad. Leave the file blank. Save the file in your folder (MyWebPage) under Type "All files" as ***List.html***.
> 3. Using Netscape Navigator or Internet Explorer, browse to the file ***UniversityFoodHome.html*** and open it in the browser window. It should look like Figure 8.15.

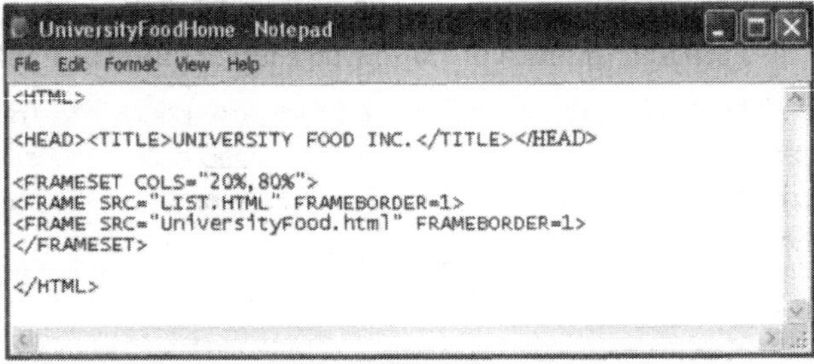

```
<HTML>

<HEAD><TITLE>UNIVERSITY FOOD INC.</TITLE></HEAD>

<FRAMESET COLS="20%,80%">
<FRAME SRC="LIST.HTML" FRAMEBORDER=1>
<FRAME SRC="UniversityFood.html" FRAMEBORDER=1>
</FRAMESET>

</HTML>
```

Figure 8.14 University food home page with column frames.

When framesets are used with anchor tags, it is necessary to direct the Web browser to the frame in which to load new content from the hyperlink. For example, List.html will eventually contain candidates for hyperlinks to other Web pages. In a typical implementation, the hyperlink will change content in the right-hand side frame, leaving the list of selection options on the left side unchanged. When a frameset is established in the container Web page, and there are hyperlinks for changing content, it is necessary to specify a name for the frame that will receive the content. If this is not done, the new content will be loaded into the frame on which the anchor tag resides.

HTML provides some keywords to control the loading of new content in a frameset. The keyword "NAME" is used to provide an identity for a frame. The name refers to the frame and not to the contents of the frame. So, for example, were we to name the right side frame of the frameset in UniversityFoodHome.html as "Rside," the appropriate code would be as follows:

```
<FRAMESET Cols="20%, 80%">
<FRAME SRC="List.html" NAME="Lside" FRAMEBORDER=1>
<FRAME SRC="UniversityFood.html" NAME="Rside" FRAMEBORDER=1>
</FRAMESET>
```

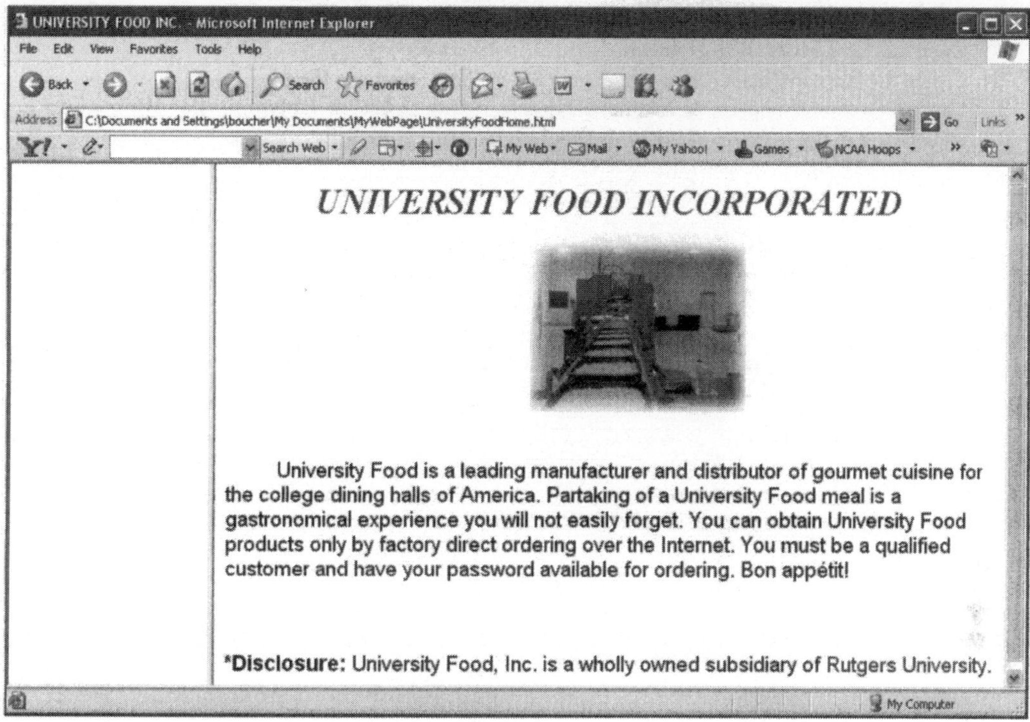

Figure 8.15 Web page created after adding the code shown in Figure 8.14.

Henceforth, the right side frame can be addressed by the name "Rside." Then, for each anchor tag, the target frame for loading new content is identified by the keyword "TARGET." For example, in Figure 8.17, the hyperlink for products could be a file that lists the various products sold by the company. The anchor tag would indicate the target frame for loading the page, for example, as follows:

```
<LI><A HREF="UniversityFoodProducts.html" TARGET="Rside">Products</A>
```

When the Web browser encounters the keyword TARGET, it knows that the argument is the name of the frame in which to load the content. The reader will have an opportunity to experiment with this idea in the end-of-chapter exercises.

8.2.9 Lists

The left panel of Figure 8.2a shows a list of user options. Each option will have its own anchor tag that is a hyperlink to another Web page. Here we will simply set up the list and leave the creation of the hyperlinks to the end-of-chapter exercises.

HTML includes tags that allow the user to set up numbered (ordered) and bulleted (unordered) lists. The tags are as follows:

- <OL> and </OL>. These tags indicate the beginning and the end of an ordered list.
- <UL> and </UL>. These tags indicate the beginning and end of an unordered list.
- <LI>. This tag is placed in front of each item to be either numbered (ordered list) or bulleted (unordered list).

The reader will create a list in HTML Exercise 8.7.

HTML Exercise 8.7:

We wish to create the left frame of Figure 8.15.

1. Open the file **List.html** in Notepad..
2. Create the file contents as shown in Figure 8.16. Save and close the file.
3. Using Netscape Navigator or Internet Explorer, browse to the file **UniversityFoodHome.html** and open it in the browser window. It should look like Figure 8.17.

```
<HTML>

<HEAD>
<TITLE>List</TITLE>
</HEAD>

<BODY>
<BR><BR><BR>

<UL>
	<LI>News
<BR>
	<LI>Products
<BR>
	<LI>Locations
<BR>
	<LI>People
<BR>
	<LI>Contact Us
<BR>
	<LI>On Line Ordering
<BR>
	<LI>Review Orders
</UL>

</BODY>
</HTML>
```

Figure 8.16 Creating lists.

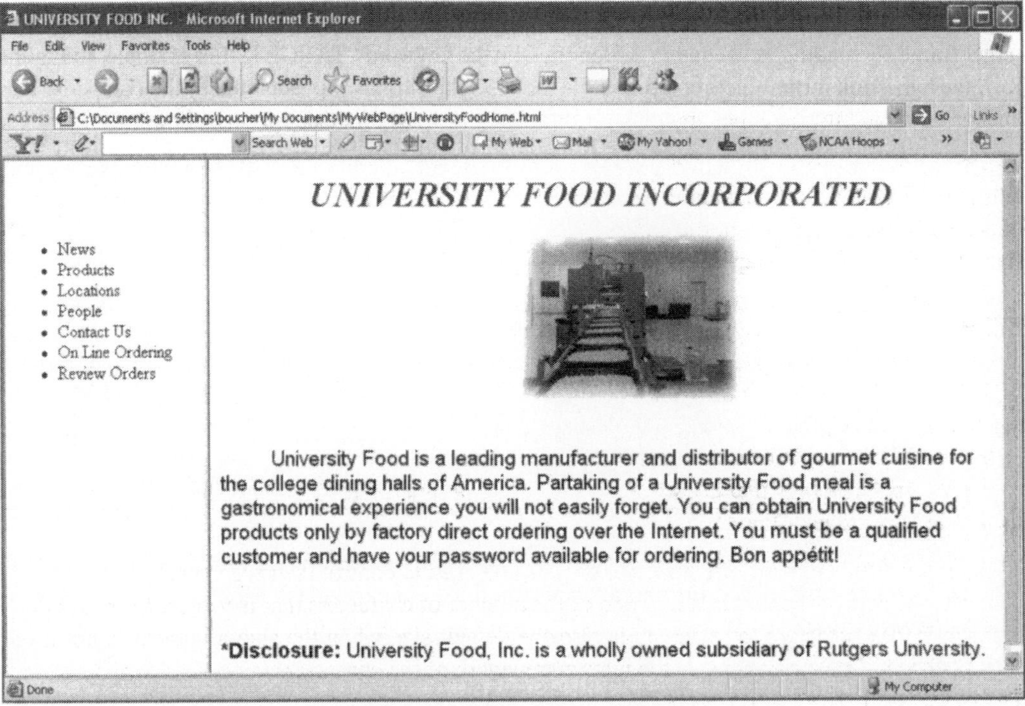

Figure 8.17 Final home Web page.

8.2.10 Forms

Forms are used to gather data from the user. Figure 8.2b is an example of a form. A form can be placed anywhere within the body of the HTML document. Forms may include text boxes, radio buttons, drop-down lists, and command buttons, among other objects. The form is usually filled out by the user and then submitted by clicking on a command button.

Since HTML documents do not include procedural programming code, it is necessary to create a separate file to handle the data being submitted on the form. This separate file may include the SQL commands to log the data into the database. Different technologies are available for implementing this. Microsoft has developed Active Server Pages (ASP) as the Windows standard technology for implementing code to handle data submitted via a Web page. The HTML page simply passes the data to the .asp file for processing. This section describes the typical components of a form object. Creating .asp files for handling the data submitted via the form will be discussed in Section 8.3.

8.2.10.1 Form Elements

The area on a form where data is entered is called a ***form element***. Typical elements include text boxes, password boxes, check boxes, radio buttons, and submit buttons. An element is created on a form object by declaring the following attributes:

- INPUT TYPE. This attribute declares the kind of element to be placed on the form. Typical TYPES include "text" (for text boxes), "password" (for password boxes), "checkbox", "radio", and "submit". All forms must include an input type "submit", which is displayed as a command button on the form. When a user clicks the command button, the input type "submit" transfers the input data to the ASP file that will process it.
- NAME. This attribute assigns a variable name to the input entered into the INPUT TYPE. The .asp file identifies the input by its variable name.

A typical form is the one shown in Figure 8.2b, which contains a text box, a password box, and a submit button. The .html file for this form is shown in Figure 8.18. Note the following structure to the form elements:

```
Element Label <INPUT TYPE="element type" NAME="input variable name">
```

The leading string, Element Label, that appears before the element type declaration is the label for the element. It will appear in the browser window as a label next to the element. The element is identified within the brackets following the label. Figure 8.18 illustrates a text box, a password box, and a submit button. Following the element type, a variable name is given. The input provided by the element will be assigned to that variable name. When the data are passed from the input element to the server for processing, the value of the input will be referred to by reference to the variable name. A Submit button has an optional VALUE argument. The VALUE argument is the label that will appear on the button. The default VALUE argument is "Submit Query."

A text box has some useful attributes that can be used to control the user input. The *SIZE* attribute sets the size of the text box in terms of the number of characters that is visible. In Figure 8.18, the size is 20 characters long, which is also the default size when the size argument is not used. The *MAXLENGTH* attribute sets the maximum length of the character string that may be entered. If the MAXLENGTH attribute is not used, it defaults to 255.

HTML Exercise 8.8:

We wish to create a form like that of Figure 8.2b.

1. Open Notepad and create the file ***UserID.html*** as shown in Figure 8.18. Save the file in your folder (MyWebPage) under Type "All files" as UserID.html. Close the file.
2. Using Netscape Navigator or Internet Explorer, browse to the file **UserID.html** and open it in the browser window. It should look like Figure 8.19.

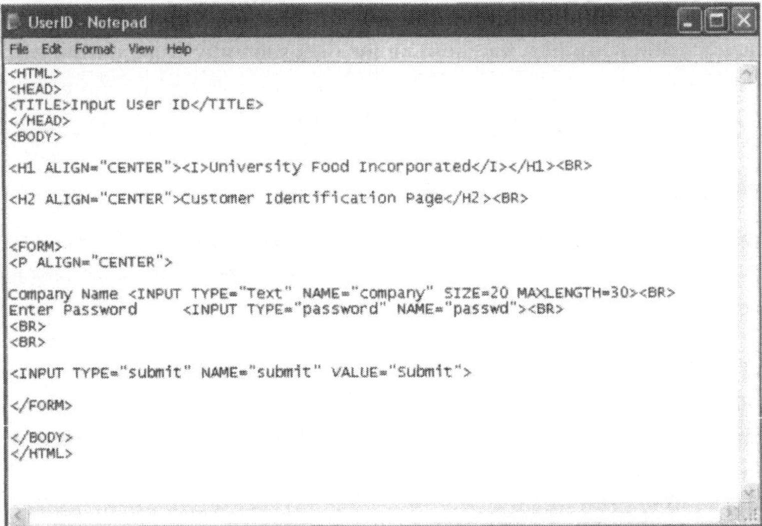


```
<HTML>
<HEAD>
<TITLE>Input User ID</TITLE>
</HEAD>
<BODY>

<H1 ALIGN="CENTER"><I>University Food Incorporated</I></H1><BR>

<H2 ALIGN="CENTER">Customer Identification Page</H2><BR>


<FORM>
<P ALIGN="CENTER">

Company Name <INPUT TYPE="Text" NAME="company" SIZE=20 MAXLENGTH=30><BR>
Enter Password     <INPUT TYPE="password" NAME="passwd"><BR>
<BR>
<BR>

<INPUT TYPE="submit" NAME="submit" VALUE="Submit">

</FORM>

</BODY>
</HTML>
```

Figure 8.18 Creating a simple example form.

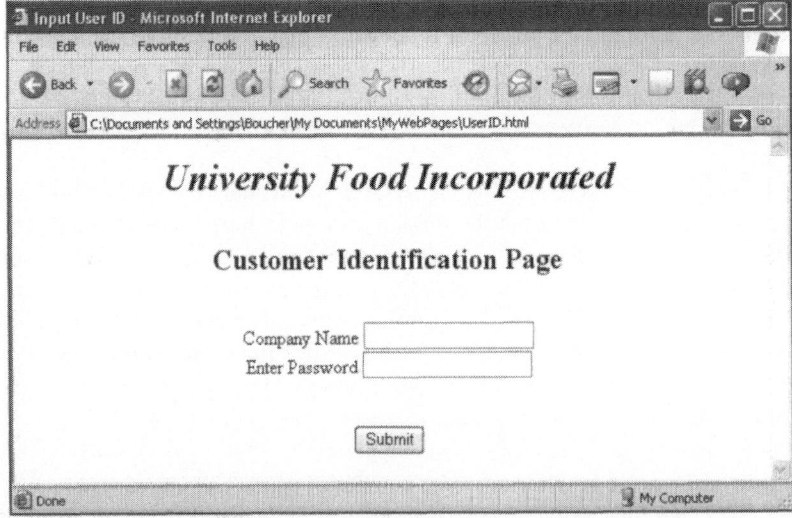

Figure 8.19 Web page created by the code shown in Figure 8.18.

Other useful form objects are the text area, radio button, check box, drop-down list, and reset button. The syntax associated with each object will now be described, and the reader will have an opportunity to implement these objects in the end-of-chapter exercises.

The *text area* allows the user to enter several sentences or paragraphs of information. The syntax is as follows:

```
Label <TEXT AREA NAME="text var name" ROWS=5 COLS=20></TEXT AREA>
```

The *ROWS* and *COLS* attributes declare the size of the display area in terms of characters. When the form is submitted, the *NAME* attribute is the variable name by which the submitted text is referenced.

A radio button allows the user to select one option from among a number of alternatives. The syntax is as follows:

```
Label_1<INPUT TYPE="radio" NAME="var name" VALUE="value_1" CHECKED>
Label_2<INPUT TYPE="radio" NAME="var name" VALUE="value_2">
```

For any group of radio buttons, the *NAME* attribute is the same. However, the value that is assigned to the name variable is different for each option. If a default option is specified at the time the Web page is loaded, that option contains the attribute *CHECKED*. A Web page may have more than one set of options displayed as radio buttons that are distinguished by different *NAME* attributes.

A check box allows the user to select more than one option from among a set of predefined options. The syntax is as follows:

```
Label_1<INPUT TYPE="checkbox" NAME="var_1 name" VALUE="value_1">
Label_2<INPUT TYPE="checkbox" NAME="var_2 name" VALUE="value_2">
```

When the page is submitted, all the values that are checked will be assigned to their respective variable names and passed on to the server.

A drop-down list allows a user to select one and only one option from a list of values. The syntax is as follows:

```
<SELECT NAME="variable name" SIZE= no. of entries shown>
<OPTION>Option_value_1 <OPTION>Option_value_2 <OPTION>Option_value_3
</SELECT>
```

The <SELECT> </SELECT> tags define a drop-down list with the options between the tags. The <OPTION> tag defines each option. Each option has a different *VALUE* that will be assigned to the *NAME* attribute when the choice is submitted. The attribute *SIZE* defines the number of values that will be shown in the drop-down window at one time. Scroll bars are automatically added for moving through the options. The default width is the width of the largest entry on the list.

A reset button allows the user to clear the data on the form and reset form elements to their default values. The syntax is as follows:

```
<INPUT TYPE="reset" VALUE="Clear Form">
```

For the reset button, the *VALUE* argument is the label that will appear on the button.

8.2.10.2 Linking a Form to an ASP File

As previously mentioned, a form will not have any functionality unless it is linked to an ASP file. The reader can test this fact by opening the form UserID in a browser window (Figure 8.19), entering data, and clicking on the "Submit" button. Nothing will happen.

Linking the form to an ASP file is done in the opening <FORM> declaration tag by adding two parameters as follows:

- ACTION. This attribute is a pointer to the .asp file that will operate on the input data.
- METHOD. This keyword calls a method that will be used to pass the form's input data to the .asp file. Here we will illustrate the GET method. Using this method, when the ASP file is called, the data are appended to the URL of the .asp file identified in the action statement. There is also a POST method, which is suitable for forms that send a large quantity of data to the .asp file. We will not discuss the POST method here.

Figure 8.20 shows the addition of these two parameters within the body tag of the form. The user should open the file UserID.html and complete the code as shown in Figure 8.20. The .asp filename "response.asp" has not yet been written, so nothing will execute at this time. We have now arrived at the point where we can begin to discuss the ASP technology and the interaction between Web pages and databases. This is the subject of the next section.

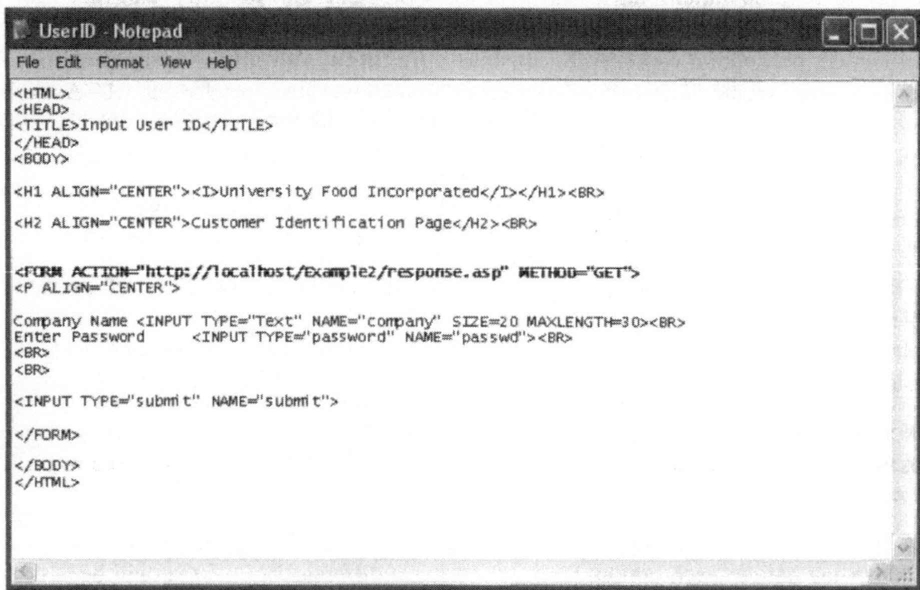


```
<HTML>
<HEAD>
<TITLE>Input User ID</TITLE>
</HEAD>
<BODY>

<H1 ALIGN="CENTER"><I>University Food Incorporated</I></H1><BR>

<H2 ALIGN="CENTER">Customer Identification Page</H2><BR>

<FORM ACTION="http://localhost/Example2/response.asp" METHOD="GET">
<P ALIGN="CENTER">

Company Name <INPUT TYPE="Text" NAME="company" SIZE=20 MAXLENGTH=30><BR>
Enter Password      <INPUT TYPE="password" NAME="passwd"><BR>
<BR>
<BR>

<INPUT TYPE="submit" NAME="submit">

</FORM>

</BODY>
</HTML>
```

Figure 8.20 Adding an action and method.

8.3 ACTIVE SERVER PAGES

HTML is used simply to format the contents of Web pages. To add interactive and dynamic properties to Web pages, Microsoft developed ***Active Server Pages (ASP)***. ASP uses a dynamic link library (DLL) component installed on the Web server to process ASP code. When a particular .asp file is required to add dynamic properties to a Web page, the .asp file is called by the Web page server. The .asp file often contains procedures in a scripting language, such as VBScript or JScript, and may contain ASP objects that have their own methods and properties.

ASP will run on either a Web server platform or a personal computer on which ***Internet Information Server (IIS)*** (Windows 2000 or higher) is installed. IIS is an add-on that comes with the

Windows Professional operating system. IIS is a middleware component. ***Middleware*** is software that acts as an intermediary between the Web server software and the target database. Appendix 8A at the end of this chapter describes how to install IIS on your laptop or desktop PC in order to process Active Server Pages. The exercises in this section assume that ISS is installed on the machine you are using to perform the exercises.

ASP can be used for database interaction over a Web browser. Using .asp files, a user is allowed to retrieve, add, delete, or edit records in a database. ASP uses server-side processing to do this. When a browser submits data that must be processed by an .asp file, such as input from a form, the server passes the data to the appropriate .asp file or method to do the processing. The .asp method returns the result to the Web server, which sends the result to the browser as HTML code.

The architecture of ASP/IIS server processing is depicted in Figure 8.21. The client submits an HTML document that has a form and some form elements for processing, such as the text boxes shown in Figure 8.19. The ACTION statement directs the processing of the data to a file with an .asp extension to indicate that the document requires further processing by the server, which was shown in Figure 8.20. When the Web server encounters the ASP file reference in the ACTION statement, it uses the application server, IIS, to process the code. In a database application, the code is typically some SQL commands. The database server returns the data, which is sent to the browser as text in an HTML document.

The architecture of Figure 8.21 is often referred to as a three-tier architecture. The Web server acts as both a client and a server. It is a server to the Web browser. It is a client to the application program, or middleware, IIS/ASP, which processes requests of the Web server. The client and server can exist on the same machine. In fact, the exercises that follow will assume that the IIS server is installed on your machine as described in Appendix 8A.

Unlike an HTML document file, an ASP file must reside in a specific directory under IIS. This directory is known as the ***root directory***, and it is found by following the path **c:\Inetpub\wwwroot**. These directories are installed when the IIS software component is installed. The user may create subfolders for .asp files within the wwwroot directory. When a browser is used to open an .asp file on the local machine, the path to this directory is **http://localhost/FolderName/filename.asp**. This is the required form of the address placed in the browser URL. The word **localhost** will direct the server to the wwwroot directory.

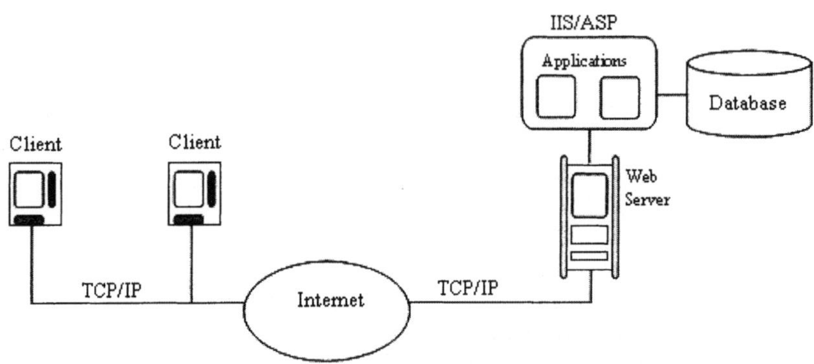

Figure 8.21 Architecture for processing dynamic HTML pages.

Using IIS/ASP as a server on a personal computer or laptop is a common way to design and debug Web pages and ASP files before uploading them to a server on the Internet. We do not address the details of setting up a Web server on a network in this book. All exercises are designed to be performed on a personal computer or laptop running IIS. There are numerous Internet service providers that can host the Web pages created in this chapter, including ASP files and databases. Many of these providers offer free space for hosting small applications. The file uploading process is specific to the service provider. The interested reader may want to look at some typical sites on the Internet, such as www.1ASPHOST.com or www.brinkster.com.

8.3.1 Adding ASP Code to a Web Page

ASP code can be added to an HTML document anywhere between the <BODY>, </BODY> tags. Two tag delimiters are used to inform the Web server that it should process the code as ASP code. They are as follows:

```
<%
   Place the ASP code here!
%>
```

The <% and %> tags mark the beginning of and the end of the ASP code. When an ASP page is saved, it must be saved with the extension .asp, even though it also contains HTML formatting code.

8.3.2 ASP Objects

ASP objects are used to control information passing between the Web server and the ASP file. There are five main objects, and each object has a set of methods. The five objects are as follows:

- *Response.* Sends information from the ASP file to the Web server for downloading to a browser
- *Request.* Retrieves information from the Web server for use by an ASP file
- *Server.* Creates objects and methods on the Web server
- *Session.* Sends session information for individual users
- *Application.* Shares information for users during an application

This section will be selective in the use of objects and methods, introducing them as needed to illustrate a point. For a more thorough discussion, the reader is referred to Ladd et al., 1999; Whitehead, 2000. To illustrate a simple ASP document, we will use the ***write method*** of the ***response object***. The syntax is Response.Write("Contents"), where Contents is a string to be passed from ASP to the server for downloading to the browser window.

An example of an ASP file used to send Web page content to the Web server for downloading to a browser is shown in Figure 8.22. Note the code within the ASP delimiters <% and %>. First, two text strings are defined: company and passwd. These are the two text string variable names we previously defined in Figure 8.18. This illustration defines these strings as constants. The **Response.Write** method is used to send the strings and variables in parentheses to the Web server. Note the use of the "&" symbol to append the variable to the string. The variable is enclosed in quotation marks. The reader can execute this example by implementing ASP Exercise 8.1.

ASP Exercise 8.1:

We wish to create an ASP file that demonstrates the use of ***Response.write*** for downloading text to a web page.

1. Establish a folder in the ASP root directory named "Example1". (C:\Inetpub\wwwroot\Example1).
2. Open Notepad and create the file shown in Figure 8.22. Save the file in the Example1 folder as a Type "All files" with the filename ***response.asp***.
3. Open a browser. Type the local host address of the root directory in the URL panel as follows: **http://localhost/Example1/response.asp** (see Figure 8.23). Hit the Enter key.
4. When the web page opens it should look like Figure 8.23.

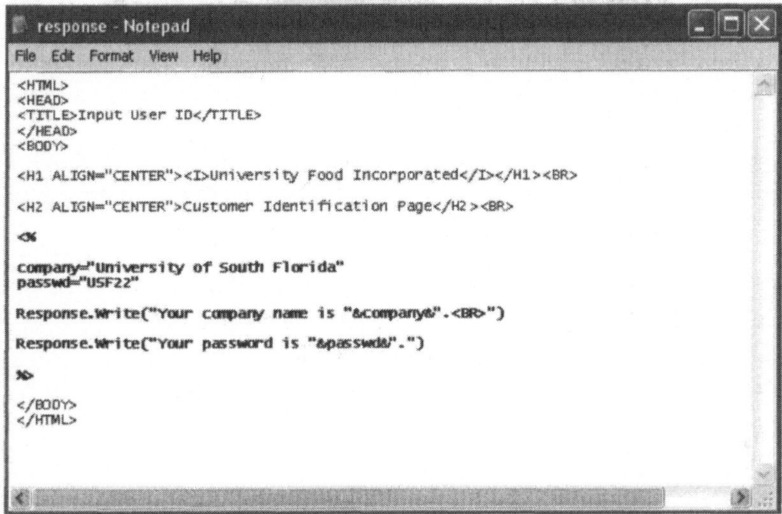

```
<HTML>
<HEAD>
<TITLE>Input User ID</TITLE>
</HEAD>
<BODY>

<H1 ALIGN="CENTER"><I>University Food Incorporated</I></H1><BR>

<H2 ALIGN="CENTER">Customer Identification Page</H2><BR>

<%

company="University of South Florida"
passwd="USF22"

Response.Write("Your company name is "&company&".<BR>")

Response.Write("Your password is "&passwd&".")

%>

</BODY>
</HTML>
```

Figure 8.22 Example ASP page.

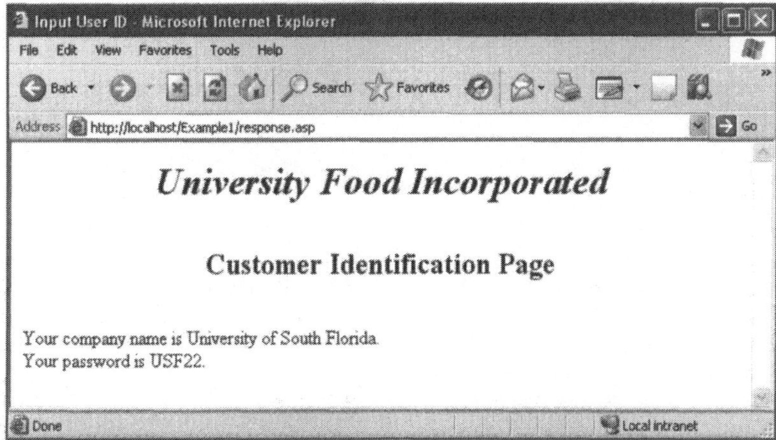

Figure 8.23 Code shown in Figure 8.22, open in a browser window.

8.3.3 Passing Data to the ASP File from a Form

Clearly, the next step is to link the HTML form (Figure 8.19) to the ASP file for processing. The method used to pass data from the browser to the server is the GET method. When the user presses the ***Submit*** button, the ***GET method*** places the URL of the .asp file (as defined by the ACTION attribute) into the browser address panel. It also appends to the address a string that contains all the input variables that will be passed to the .asp file from the text boxes and other form elements. Assuming that response.asp is the ASP filename, the syntax of the string placed in the address panel is as follows:

```
http://localhost/folder name/response.asp?the string is placed here
```

When the .asp page is called, the string is available to the .asp file for processing. The ASP program gains access to the query string using the ASP **Request** object. One of the methods of the Request object is **QueryString**. This method takes the string appended to the URL address and makes it available to the ASP program. The use of the Request. QueryString statement is illustrated in Figure 8.24, which is a new version of the file **response.asp**.

Note the syntax of the statement. The Request.QueryString statement is preceded by an equal sign, which indicates that it should simply be displayed. The QueryString method returns the string to the server, which downloads it to the browser.

Before the program UserID.html can be executed and pass the query string to the program response.asp, we must do the following. The UserID.html file should point to the location of the .asp file. Therefore, the localhost address and any path to subfolders should appear in the ACTION statement. This is shown in Figure 8.25, where the ACTION statement is modified to include the localhost address. The reader should implement ASP Exercise 8.2.

ASP Exercise 8.2:

We wish to demonstrate how an ASP file is called from a HTML file using the ***GET method***.

1. Open the file ***response.asp*** in Notepad (C:\Inetpub\wwwroot\Example1\ response.asp). Change the file in accordance with Figure 8.24. Save and close the file.
2. Open the file ***UserID.html*** in Notepad (C:\MyWebPage|UserID.html). Add the FORM ACTION statement as shown in Figure 8.25. Save and close the file.
3. Open ***UserID.html*** in a browser. Enter data in the text and password boxes and submit the query by clicking on the Submit button.
4. The ASP file should return a screen that looks like Figure 8.26.

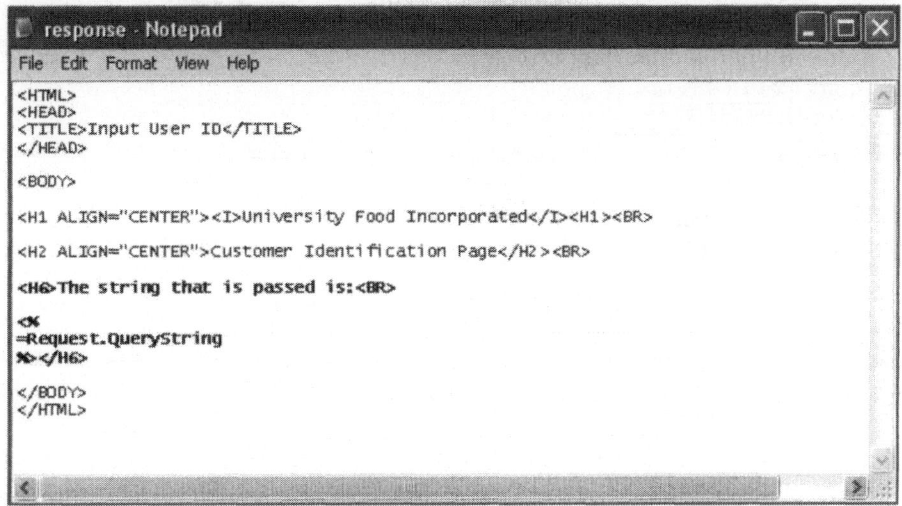

```
<HTML>
<HEAD>
<TITLE>Input User ID</TITLE>
</HEAD>

<BODY>

<H1 ALIGN="CENTER"><I>University Food Incorporated</I><H1><BR>

<H2 ALIGN="CENTER">Customer Identification Page</H2><BR>

<H6>The string that is passed is:<BR>

<%
=Request.QueryString
%></H6>

</BODY>
</HTML>
```

Figure 8.24 Using request object with QueryString method.

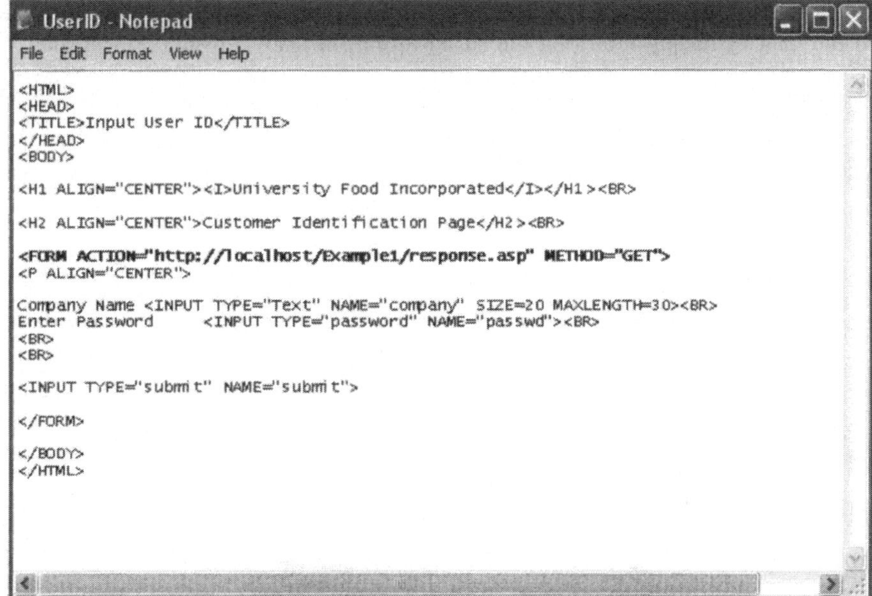

```
<HTML>
<HEAD>
<TITLE>Input User ID</TITLE>
</HEAD>
<BODY>

<H1 ALIGN="CENTER"><I>University Food Incorporated</I></H1><BR>

<H2 ALIGN="CENTER">Customer Identification Page</H2><BR>

<FORM ACTION="http://localhost/Example1/response.asp" METHOD="GET">
<P ALIGN="CENTER">

Company Name <INPUT TYPE="Text" NAME="company" SIZE=20 MAXLENGTH=30><BR>
Enter Password     <INPUT TYPE="password" NAME="passwd"><BR>
<BR>
<BR>

<INPUT TYPE="submit" NAME="submit">

</FORM>

</BODY>
</HTML>
```

Figure 8.25 ACTION statement pointing to ASP file location.

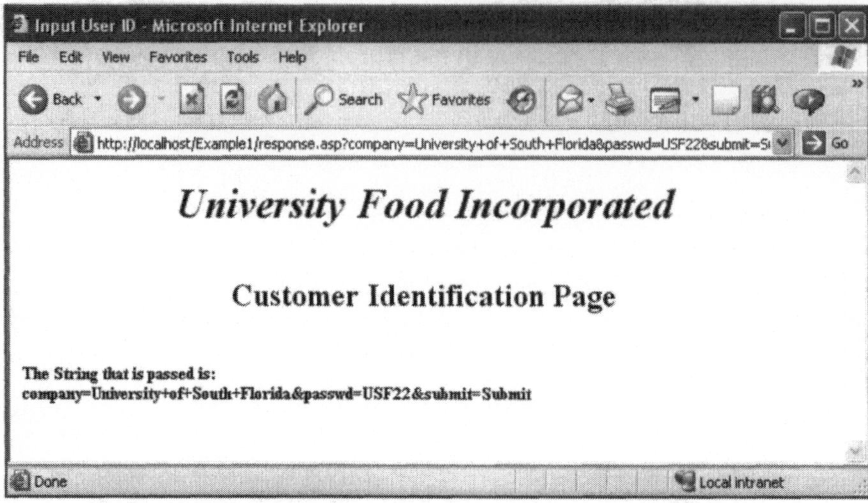

Figure 8.26 Data returned to browser from response.asp.

Note the string in the URL address panel in Figure 8.26 and compare it to the string that is returned to the browser window. The GET method has appended the values from the text box and password box to the URL of the .asp file in the address window. The .asp file (Figure 8.24) has processed the data by simply returning the string to the IIS server, which downloaded it to the browser (Figure 8.26).

The Request.QueryString statement can be used to identify each value stored as a variable. Using the variable name of each form element as an argument in the Request.QueryString statement does this. An example is shown in Figure 8.27. Here the variable names "company" and "passwd" are used in a Response.Write statement to return their values to the Web browser. The reader should implement ASP Exercise 8.3.

ASP Exercise 8.3:

We wish to demonstrate how to ***parse a query string*** for form element values.

1. Open the file ***response.asp*** in Notepad (C:\Inetpub\wwwroot\Example1\response.asp). Change the file in accordance with Figure 8.27. Save and close the file.
2. Open ***UserID.html*** in a browser. Enter data in the text and password boxes and submit the query by clicking on the Submit button.
3. The ASP file should return a screen that looks like Figure 8.28.

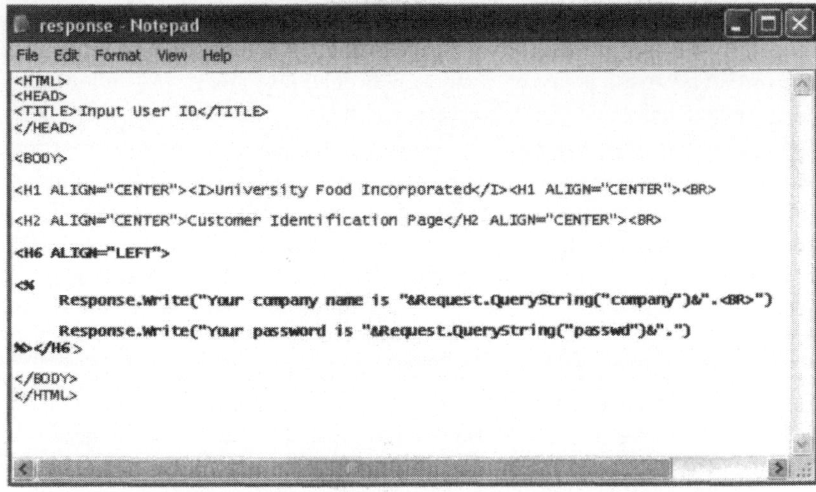

Figure 8.27 Using Request.QueryString to identify the value of a variable.

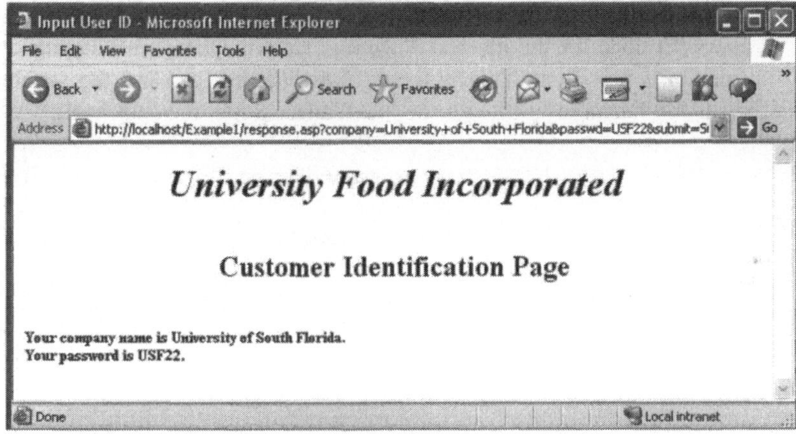

Figure 8.28 Browser window resulting from the code shown in Figure 8.27.

8.3.4 USING THE ASP SESSION OBJECT

The two previous sections have introduced key methods of the Request object and the Response object. Use of the methods Request.QueryString and Response.Write are fundamental to using ASP with HTML. There is one other object that should be described in some detail since it facilitates user interaction with multiple ASP pages. This is the ***Session object***.

Sometimes it is necessary for the client machine to interact with multiple Web pages. For example, the user may log in on one Web page and then be given access to functions that are carried out on other Web pages. For example, the user may have permission to create an order, check on the status of an order, and cancel an order. These functions are each carried out on different Web pages. As the user moves from one page to another, the server must remember the user's identification so that he or she is given access only to functions and records allowed by his or her password. The Session object allows the server to do this.

ASP creates a session when the ASP file encounters the keyword "Session." At that point, the server creates a client identification and places a cookie with that identification number on the client machine. A ***cookie*** is a hidden element stored on the client machine. The identification of the client browser is then persistent during the session and will not be lost until the session ends. The session ends if the client does not request another ASP page from the server within a session timeout interval, which can be set in the ASP file. The default timeout is 20 minutes. The session can also be ended on the server side using the Abandon method of the Session object, ***Session.Abandon***.

Figure 8.29 is the UserID.html file with which we have previously worked, except that the ACTION statement is changed to point to a new ASP file called CreateSession.asp, saved in a new subdirectory called Example4. Figure 8.30 shows the ASP file that will be called when the form data are submitted. The file CreateSession.asp uses the Session keyword to define two variables as session variables. The variable "UserName" is defined as a session variable and is assigned the value of the variable "company" from the query string. A similar assignment is made for the variable "Passwd." Any session variables become persistent and are associated with the session ID stored as a cookie on the client machine. In the second ASP component, two welcome statements are presented to the client who has just logged on. Following that, an anchor tag is provided that will send the client to another ASP page (called the NextPage).

Figure 8.31 shows the code for the file NextPage.asp. Here is where we illustrate the fact that the session information is persistent. That is to say, NextPage.asp can use the session variables created when the session began even though the NextPage.asp file does not explicitly import the information. The information is persistent as long as the session is in progress. Note that the session ID created by the server can also be retrieved using the SessionID method of the Session object. Again, the session will end if the client does not request another ASP page within the timeout period or if the server encounters a Session.Abandon statement. The reader should implement ASP Exercise 8.4.

ASP Exercise 8.4:

We wish to demonstrate the use of the ***Session object*** for retaining information during a session.

1. Open the file ***UserID.html*** in Notepad (C:\MyWebPage|UserID.html). Change the FORM ACTION statement as shown in Figure 8.29. Save and close the file.
2. Establish a new folder in the wwwroot directory names "Example4" (C:\Inetpub\wwwroot\Example4).
3. Open Notepad and create the file shown in Figure 8.30. Save the file as Type "All files" with the filename ***CreateSession.asp*** in the Example4 folder.
4. Open Notepad and create the file shown in Figure 8.31. Save the file as Type "All files" with the filename ***NextPage.asp*** in the Example4 folder.
5. Open **UserID.html** in a browser window. Enter data and submit the query.
6. The ASP file should return a screen like Figure 8.32a. Click on the anchor tag. The ASP file should return a screen that looks like Figure 8.32b.

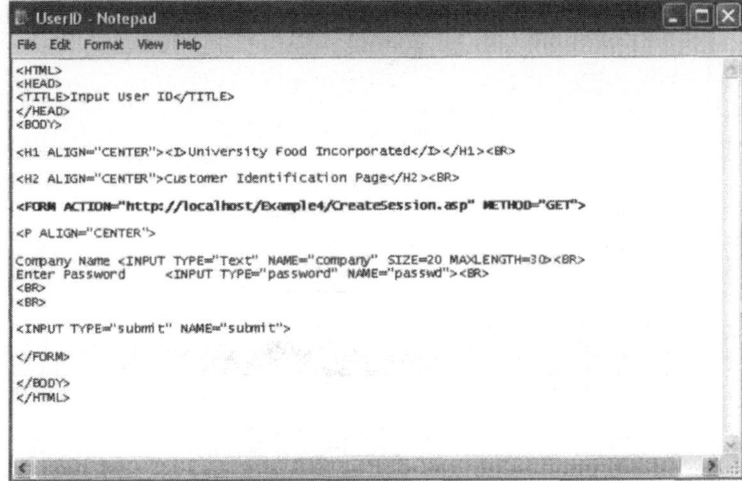

```
<HTML>
<HEAD>
<TITLE>Input User ID</TITLE>
</HEAD>
<BODY>

<H1 ALIGN="CENTER"><I>University Food Incorporated</I></H1><BR>

<H2 ALIGN="CENTER">Customer Identification Page</H2><BR>

<FORM ACTION="http://localhost/Example4/CreateSession.asp" METHOD="GET">

<P ALIGN="CENTER">

Company Name <INPUT TYPE="Text" NAME="company" SIZE=20 MAXLENGTH=30><BR>
Enter Password     <INPUT TYPE="password" NAME="passwd"><BR>
<BR>
<BR>

<INPUT TYPE="submit" NAME="submit">

</FORM>

</BODY>
</HTML>
```

Figure 8.29 Revised code of UserID.

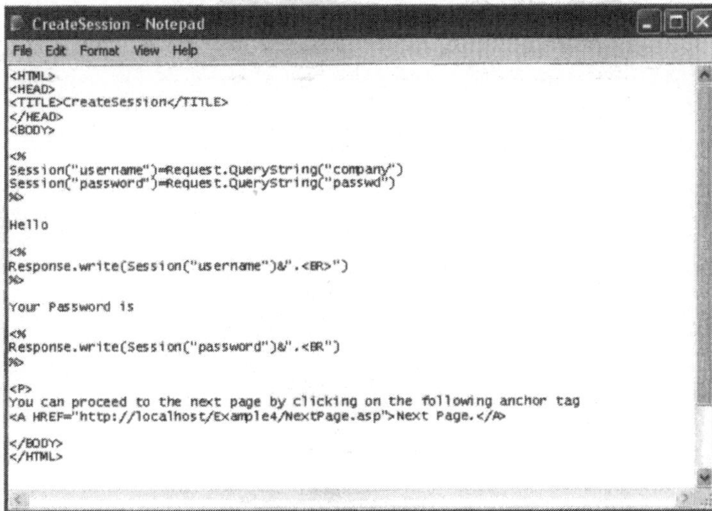

```
<HTML>
<HEAD>
<TITLE>CreateSession</TITLE>
</HEAD>
<BODY>

<%
Session("username")=Request.QueryString("company")
Session("password")=Request.QueryString("passwd")
%>

Hello

<%
Response.write(Session("username")&".<BR>")
%>

Your Password is

<%
Response.write(Session("password")&".<BR")
%>

<P>
You can proceed to the next page by clicking on the following anchor tag
<A HREF="http://localhost/Example4/NextPage.asp">Next Page.</A>

</BODY>
</HTML>
```

Figure 8.30 ASP file for CreateSession.

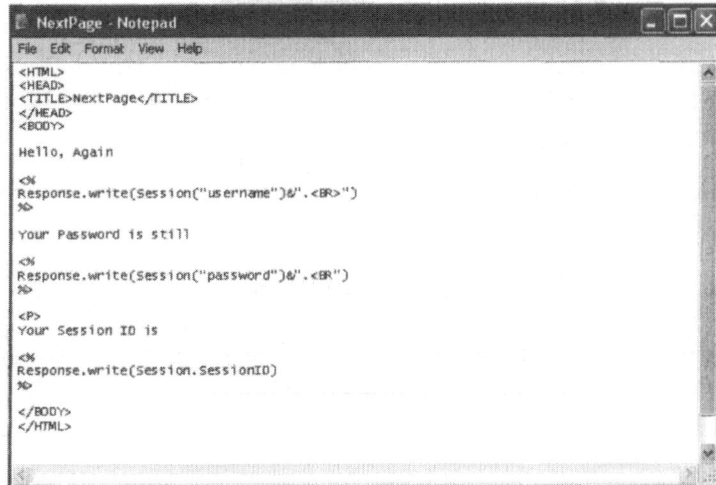

```
<HTML>
<HEAD>
<TITLE>NextPage</TITLE>
</HEAD>
<BODY>

Hello, Again

<%
Response.write(Session("username")&".<BR>")
%>

Your Password is still

<%
Response.write(Session("password")&".<BR")
%>

<P>
Your Session ID is

<%
Response.write(Session.SessionID)
%>

</BODY>
</HTML>
```

Figure 8.31 ASP file for NextPage.

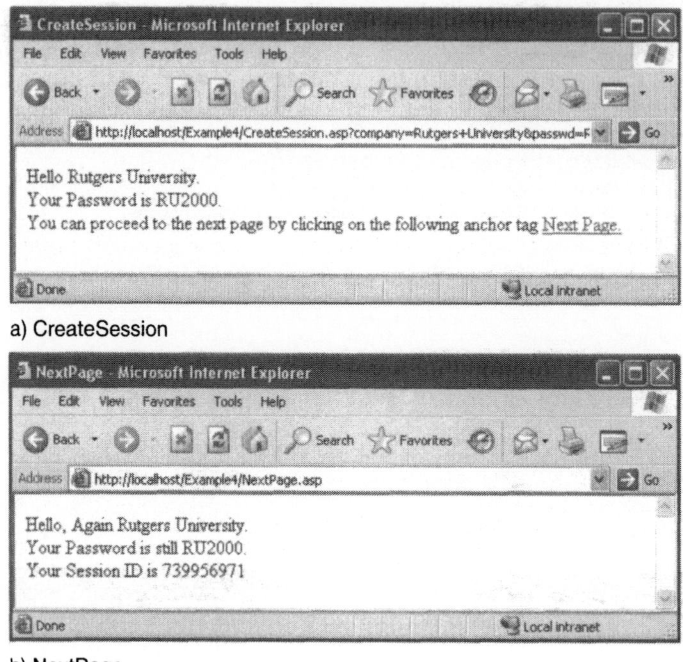

Figure 8.32 Responses to ASP Exercise 8.4.

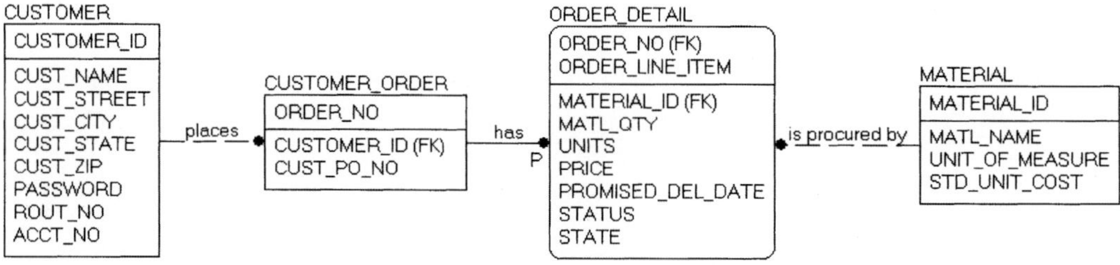

Figure 8.33 Data model for ORDERS database.

Note the Session ID returned in Figure 8.32b. This is the identifier that the server stored on the client machine when the session was created. The session remains open for 20 minutes or until the client closes the browser connection. The session object has several methods that are used for session management. Interested readers are referred to (Ladd, 1999; Whitehead, 2000).

8.3.5 Using ASP for Database Retrievals

This section describes database connectivity over the Internet. A database and populated tables that will be used for the exercises of this section are shown in Appendix 8C. Figure 8.33 shows the data model for this application. The name of the database is ORDERS.mdb, and it is on the Web site that supports this book. To execute the exercises that follow, the reader should create a new folder C:\MyDB and load ORDERS.mdb in that directory.

ASP supports connecting to and retrieving data from a database. This is done by establishing an SQL query within the ASP delimiters of an .asp file. To execute the SQL query, a connection must be established between the .asp file and the database. The Microsoft solution for executing SQL in Internet server applications is the use of ***ActiveX Data Objects (ADOs)***. ADO is a library of object models with associated methods and properties that are designed to facilitate interaction with database objects, such as tables. An application program written for a dynamic Web page may include the use of ADO to retrieve data and the use of either VBScript or JavaScript as a programming language to manipulate the data. In this chapter we will assume some familiarity with the Visual BASIC programming language, of which VBScript is a subset. The next section covers the fundamentals of using the ADO object library, after which examples of applications will be described in subsequent sections.

8.3.5.1 An ADO Tutorial

The ADO object model is Microsoft's solution for connecting an application program to a data source. ADO works with the Object Linking and Embedding Database (OLEDB). ADO can be used to link with Access using the Access Jet driver and to link with other databases such as Microsoft SQL Server and Oracle using an appropriate driver. By using ***open database connectivity (ODBC)***, a variety of other relational database management systems (RDBMSs) can also be accessed.

Access provides a library, known as the ADODB library, which contains the components of the ADO object model. The library is usually automatically installed during the Access installation. However, the user should check for the presence of the library reference on the computer before programming with ADO. This is easily done by going to the Access main menu and opening the Visual Basic for Applications (VBA) editor by choosing menu items ***Tools ⇒ Macro ⇒ Visual Basic Editor***, as shown in Figure 8.34. Once the VBA editor is open, select menu items ***Tools ⇒***

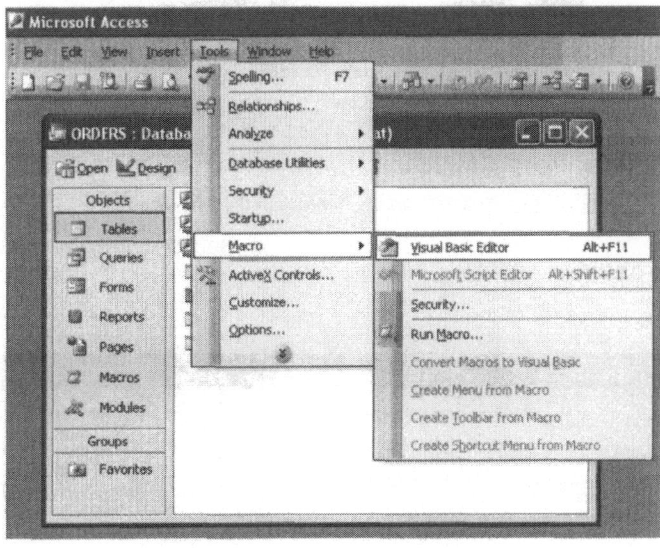

Figure 8.34 Accessing the VBA editor.

References. This will open the window shown in Figure 8.35. Make sure the ***Microsoft ActiveX Data Objects Library*** is selected by a check mark in the box on the left. If it is not, select it before closing the window.

The ADO library has several objects, but two of them are especially important in working with databases: the *Connection* object and the *RecordSet* object. These objects will be described in this section.

The ***Connection object*** is used to establish a connection to an OLE DB provider. The **provider** is a software driver that can connect to a database. If the database is an Access database, the usual driver is known as the Microsoft Jet. Two of the methods of the *Connection* object are the *Open* method and the *Close* method. The ***Open method*** is used to open the connection to the database using the provider, and the ***Close method*** closes the connection to the database. Thus, a sample program fragment used to establish a connection to a database and to open the database would be as shown in Figure 8.36.

In the first line of code, the ***Server object*** of ASP is used to create an instance of an ADODB connection with the instance name "ConnectionToDatabase." The new instance of the *Connection* object inherits all the methods and properties of the object class *Connection.* The *Server* object

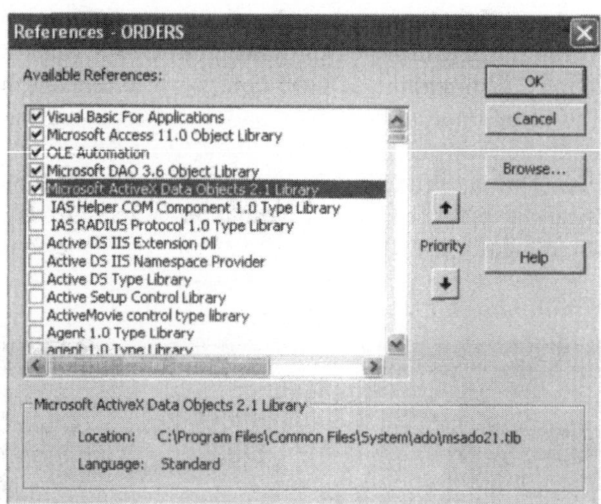

Figure 8.35 Adding libraries to the references.

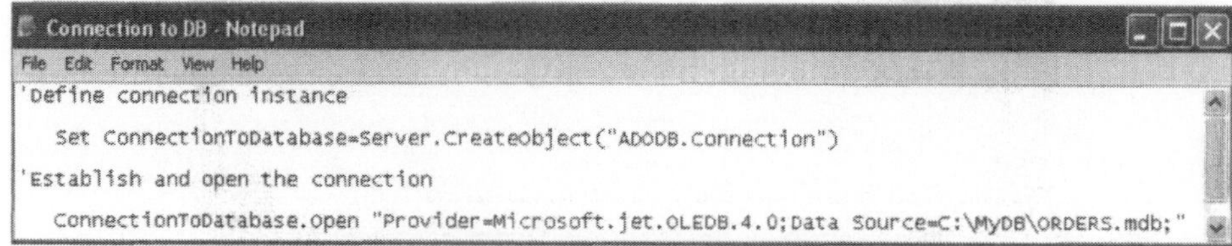

Figure 8.36 Connecting to a database.

is a component of ASP, and it provides several methods that can be used to manage the Web server. The ***CreateObject method*** of the *Server* object enables the creation of instances of object classes; for example, instances of *Connection* and *RecordSet* objects. In the second line of code, the ADO library is used to invoke the *Open* method of the connection object, followed by the provider name and the data source, including the complete path to the database. Here we call the Microsoft Jet driver and direct the connection to the ORDERS.mdb database in the MyDB folder.

An alternative way to open the connection to a database involves the use of a data source name. If the user has previously defined a ***data source name (DSN)***, as described in Appendix 8B, the second line of code could have been written as follows:

```
ConnectionToDatabase.Open "DSN=ORDERS"
```

Setting up a DSN involves defining the provider and the location of the database in advance so that the user does not have to remember those details each time the database is accessed. Hence, the DSN designation replaces those arguments in the *Open* method and is a more convenient way to program. It is necessary to have administrative privileges on the machine on which the database resides in order to set up a DSN. Appendix 8B outlines the steps involved in establishing a data source name. The exercises in this chapter specify the provider and the data source within the .asp file instead of using a DSN.

The ***RecordSet object*** is used to retrieve records from the database after the *Connection* object has been used to open the connection to the database. As in the case of the *Connection* object, an instance of the *RecordSet* object must first be created using the Server object. The *RecordSet* object has several methods associated with it, including the *Open* and *Close* methods for opening and closing the recordset. We previously encountered the RecordSet object in Section 6.4.5. After a recordset is opened, an SQL command can be issued to retrieve records from the database. A typical sequence for establishing a recordset is shown in the third and fourth lines of code in Figure 8.37.

After defining a connection to the database and opening it, a recordset is created with the instance name "MyRecordSet" in Figure 8.37. When MyRecordSet is opened, it is followed by an SQL command to retrieve records from a table. The SQL command must be followed by a reference to the connection instance to the database where the table resides. Hence, the selection is from the MATERIAL table in the ORDERS database, where the Orders database is indirectly referenced by its connection instance, ConnectionToDatabase. There may be several connections open to different databases at any point in time; thus, it is necessary to be specific about which database connection should be used to retrieve the data. Any name can be used to set up a connection instance.

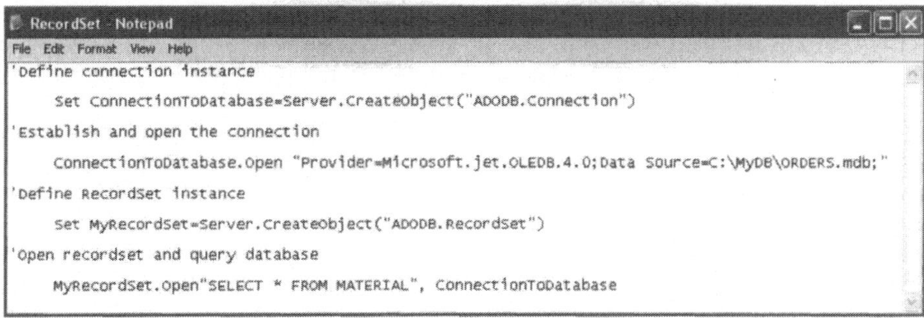

Figure 8.37 Using the RecordSet object.

Once an instance of a *RecordSet* object is opened and populated with data from an SQL query, the records in the recordset are available to be read and evaluated according to the programmed criteria. Each record, or row of the recordset, is accessed individually using the *Move* methods of the *RecordSet* object. These methods move a row pointer (called a cursor) around the recordset. The following methods apply:

- *MoveFirst.* Moves the cursor to the first record of the recordset
- *MoveNext.* Moves the cursor to the next record of the recordset
- *MovePrevious.* Moves the cursor to the previous record of the recordset
- *Move[N].* Moves the cursor N records forward (+N) or backward (–N)
- *MoveLast.* Moves the cursor to the last record in the recordset

When using the *Move* methods, one must always be aware of the limits of the number of records in the recordset. Judicious use of the beginning-of-file (BOF) and end-of-file (EOF) markers will restrict the search within the records. For example, the program code shown in Figure 8.38 sets up a loop that will move the cursor through the records until the end of the recordset file is reached. If the *Move* methods are used without constraining the search, it is always possible that the cursor will be moved outside the boundaries of the recordset. In that case, a program error will occur and further processing will cease.

In ADO, the cursor is an object that you move around within the recordset and use as a pointer to a specific record. The cursor object has several properties, but two properties are especially important in database applications. They are the *CursorType* property and the *LockType* property.

The ***CursorType property*** is specified at the time a recordset is opened. It determines whether the recordset will be used for read only or whether the recordset will allow additions. The *CursorType* is assigned a numerical value, which is interpreted as follows:

0 *Read only.* This is the default cursor type. Also, the user can only move through records in the forward direction.

1 *Read and write access.* The user can add, delete, and edit records. If the data object (table) is simultaneously being accessed by other users, changes made by others will not be visible to the user while the recordset is open. Forward and backward movement through the recordset is allowed.

2 *Read and write access.* The user can view changes made by others while the recordset is open. Forward and backward movement through the records is allowed.

3 *Read only.* Forward and backward movement through the recordset is allowed.

If the recordset is stored on a server, it may be accessed by more than one user at a time. The ***LockType property*** is used to control the order in which users are allowed to update records. As

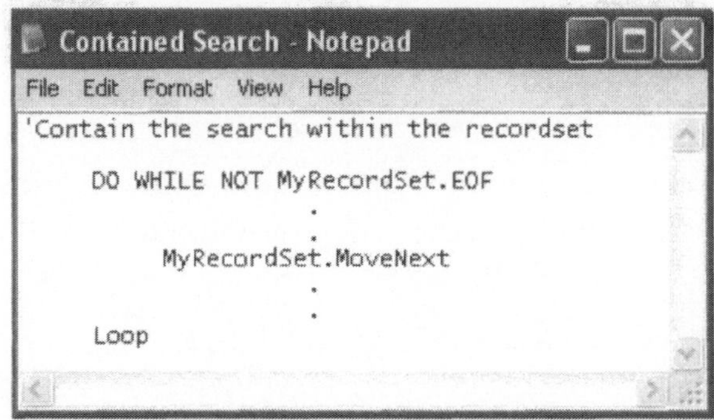

Figure 8.38 Contained recordset search.

in the case of the *CursorType* property, the *LockType* property is declared when the recordset is opened. The numerical assignments to the *LockType* property are as follows:

1 Allows read only access to other users while the recordset is open to the current user.
2 Locks a record the current user is editing. Other users will not be allowed to edit the record until the current user has finished editing.
3 Locks the record the current user is updating.
4 Locks multiple records when the current user is performing a batch update on a group of records.

The use of the *CursorType* and *LockType* properties is important if the user program is doing more than just a read-only operation. For example, the user may wish to insert a new record into the database. If the *RecordSet* object is used to insert a new record, there are two *RecordSet* methods that must be used: the ***AddNew method*** and the ***Update method***. They perform the following functions:

- *AddNew.* Adds a new record to the recordset.
- *Update.* Writes the current contents of the recordset to the database table from which it was drawn. Any changes that were made to the recordset are then recorded in the database.

To illustrate the complete cycle of opening a recordset for both reading and writing, adding a new record, and updating the database table, consider Figure 8.39, in which a new record is added to the MATERIAL table of the ORDERS database.

Note the format of the code for adding the new record. First the *AddNew* method is invoked, followed by the records that will be added to the recordset. Each attribute of the record is indicated by its attribute name, followed by the value. Finally, the Update method is invoked to write the recordset back to the MATERIAL table.

The *CursorType* and *LockType* properties are specified when the recordset is opened, where the numerical values 1 (for *CursorType*) and 2 (for *LockType*) are assigned after invoking the name of the connection object as in the fourth line of code in Figure 8.39. Hence, read and write access to the recordset is allowed, and any records currently being edited by the user are locked to other users until the editing is complete.

This section has introduced some of the basics of the ADO object model. At this point, the reader has sufficient knowledge to use ADO in designing some interesting dynamic Web pages. Applications will be discussed in the following sections.

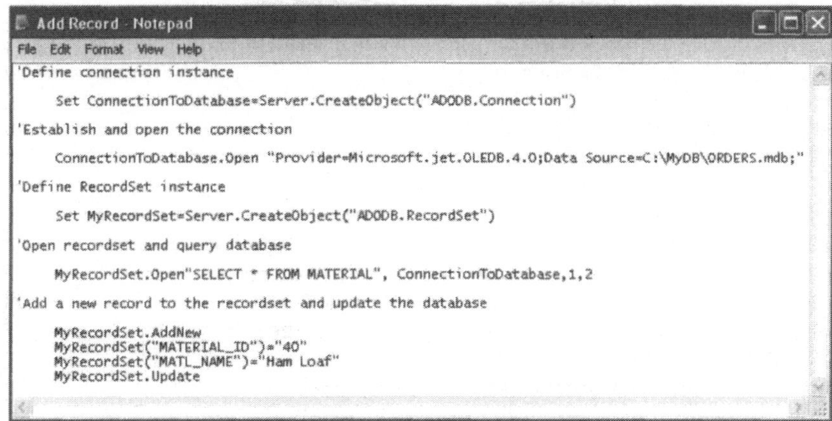

Add Record - Notepad
File Edit Format View Help
'Define connection instance
Set ConnectionToDatabase=Server.CreateObject("ADODB.Connection")
'Establish and open the connection
ConnectionToDatabase.Open "Provider=Microsoft.jet.OLEDB.4.0;Data Source=C:\MyDB\ORDERS.mdb;"
'Define RecordSet instance
Set MyRecordSet=Server.CreateObject("ADODB.RecordSet")
'Open recordset and query database
MyRecordSet.Open"SELECT * FROM MATERIAL", ConnectionToDatabase,1,2
'Add a new record to the recordset and update the database
MyRecordSet.AddNew
MyRecordSet("MATERIAL_ID")="40"
MyRecordSet("MATL_NAME")="Ham Loaf"
MyRecordSet.Update

Figure 8.39 Using the AddNew and Update methods of the RecordSet object.

8.3.5.2 A Simple Database Retrieval Example

When writing the code for an .asp file to handle a database operation, remember the following steps for implementing the code:

- Create an instance of the connection to the database.
- Define the data source and open the connection.
- Create an instance of the recordset.
- Open the recordset and execute an SQL query.
- Perform the required operations on the recordset.
- Close the recordset.
- Close the connection to the database.

An example is the simple retrieval of data to be sent to a Web page. This is shown in Figure 8.40, which is the file named "Retrieval1.asp" and is on the Web site accompanying this book. Creating and opening the connection and recordset follow the procedures described in the previous section. Assume the required operation is to return the contents of the table to the Web server for downloading to a Web page. Note the code for doing this. The ASP *Response* object and *Write* method are used. The procedure moves through each record of the recordset using a Do While loop and the *MoveNext* method such that the search is contained within the recordset. After the required operations are performed, the recordset is closed and the connection is closed. The keyword "Nothing" removes any values assigned to the object during the session.

The *CursorType* and *Locked* properties were not specified when the recordset was opened. The default values allow read-only and forward movement through the records, which are the only operations used in this exercise. The reader should implement ASP Exercise 8.5.

ASP Exercise 8.5:

We wish to demonstrate the use of ***ADO*** for connecting to a database for read only privileges and downloading contents to a Web browser.

1. Load the database ***ORDERS.mdb*** in the folder C:\MyDB. Using Windows Explorer, right click on ORDERS.mdb and select "Properties." Remove the check mark from the check box "Read Only."
2. Establish a new folder in the wwwroot directory named "DBRetrievalExamples" (C:\Inetpub\wwwroot\DBRetrievalExamples).
3. Open Notepad and create the file shown in Figure 8.40. Save the file as Type "All files" with the filename ***Retrieval1.asp*** in the "DBRetrievalExamples" folder.
4. Open a browser window and type the local host address in the URL panel as follows: **http://localhost/DBRetrievalExamples/Retrieval1.asp**.
5. The ASP file should return a screen that looks like Figure 8.41.

```
<HTML>
<HEAD>
<TITLE>A Simple Retrieval</TITLE>
</HEAD>
<BODY>
<H1>University Food Product Listing</H1>
<P>

<%
'Define connection instance

     Set ConnectionToDatabase=Server.CreateObject("ADODB.Connection")

'Establish and open the connection

     ConnectionToDatabase.ConnectionTimeout=30
     strCnn="Provider=Microsoft.jet.OLEDB.4.0;Data Source=C:\MyDB\ORDERS.mdb;"
     ConnectionToDatabase.Open strCnn

'Define RecordSet instance

     Set MyRecordSet=Server.CreateObject("ADODB.RecordSet")

'Open recordset and query database

     MyRecordSet.Open"SELECT * FROM MATERIAL", ConnectionToDatabase

'Write recordset to the Web Page

     MyRecordSet.MoveFirst
     Do While NOT MyRecordSet.EOF
          Response.Write(MyRecordSet("MATERIAL_ID"))
          Response.Write("   ")
          Response.Write(MyRecordSet("MATL_NAME"))
          Response.Write("<BR>")
          MyRecordSet.MoveNext
     Loop

'Close the recordset
     MyRecordSet.Close
     Set MyRecordSet=Nothing
'Close the connection to the database
     ConnectionToDatabase.Close
     Set ConnectionToDatabase=Nothing
%>
</BODY>
</HTML>
```

Figure 8.40 Simple database retrieval using ASP.

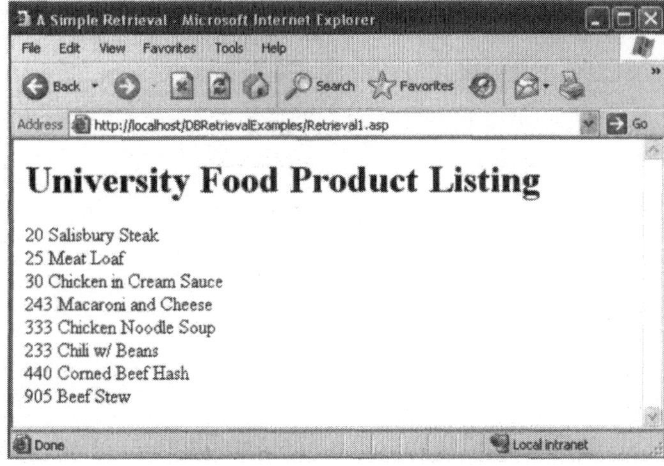

Figure 8.41 Browser window after executing the code shown in Figure 8.40.

8.3.6 Database Interaction with Forms

This section brings together the HTML forms for data input and the ASP data retrieval from the database. We will work with the login form shown previously as Figure 8.19. The file for this form, UserID.html, is reproduced in Figure 8.42. Note that it is now saved as UserID2.html, since its ACTION argument has been changed to submit the data to the ASP file named Retrieval2.asp.

The file Retrieval2.asp is shown in Figure 8.43. Using the Move method of the RecordSet object, the Do While subroutine loops through the recordset of the CUSTOMER table to see if a match is found for both input fields: company and passwd. If a match is found, a "welcome" message is displayed and a flag is set. If, after all records have been checked, the flag has not been set, an error message is displayed on the browser. The reader should implement ASP Exercise 8.6.

ASP Exercise 8.6:

We wish to demonstrate the interaction between a form for entering data and an ASP file for using that data in conjunction with a database query.

1. Open Notepad and prepare the HTML file shown in Figure 8.42. Save the file in the folder C:\MyWebPage as Type "All files" with the filename ***UserID2.html***. Close the file.
2. Open Notepad and create the ASP file shown in Figure 8.43. Save the file as Type "All files" with the filename ***Retrieval2.asp*** in the "DBRetrievalExamples" folder. Close the file.
3. Open a browser window and browse to the file ***UserID2.html***. Open it in the browser window and fill in the text boxes with an existing company name and password from the CUSTOMER table of the database ORDERS.mdb, for example, University of South Florida, and USF22. Click the Submit button
4. The ASP file should return a screen that looks like Figure 8.44.

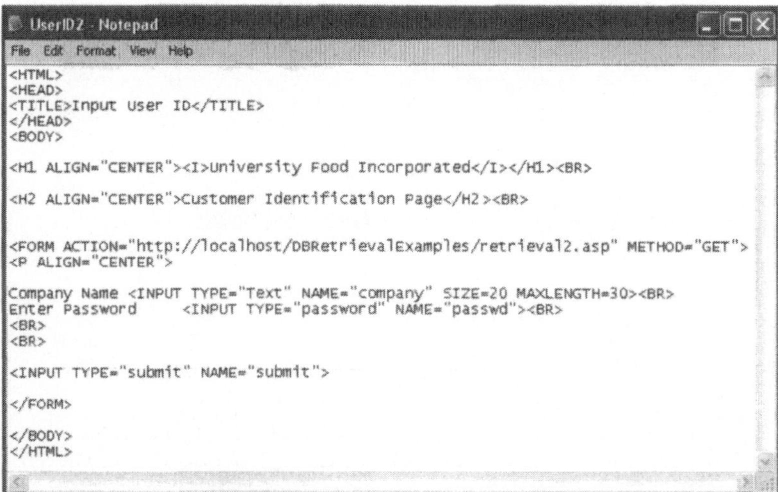

```
<HTML>
<HEAD>
<TITLE>Input User ID</TITLE>
</HEAD>
<BODY>

<H1 ALIGN="CENTER"><I>University Food Incorporated</I></H1><BR>

<H2 ALIGN="CENTER">Customer Identification Page</H2><BR>

<FORM ACTION="http://localhost/DBRetrievalExamples/retrieval2.asp" METHOD="GET">
<P ALIGN="CENTER">

Company Name <INPUT TYPE="Text" NAME="company" SIZE=20 MAXLENGTH=30><BR>
Enter Password      <INPUT TYPE="password" NAME="passwd"><BR>
<BR>
<BR>

<INPUT TYPE="submit" NAME="submit">

</FORM>

</BODY>
</HTML>
```

Figure 8.42 Code for UserID2.html.

Retrieval2.asp - Notepad

File Edit Format View Help

```
<HTML>
<HEAD>
<TITLE>A Simple Retrieval using Form Input</TITLE>
</HEAD>
<BODY>
<H1>University Food Incorporated</H1>
<P>
<%
'Define connection instance
     Set ConnectionToDatabase=Server.CreateObject("ADODB.Connection")
'Establish and open the connection
     ConnectionToDatabase.ConnectionTimeout=30
     strCnn="Provider=Microsoft.Jet.OLEDB.4.0;Data Source=C:\MyDB\ORDERS.mdb;"
     ConnectionToDatabase.Open strCnn
'Define RecordSet instance
     Set MyRecordSet=Server.CreateObject("ADODB.RecordSet")
'Open recordset and query database
     MyRecordSet.Open"SELECT * FROM CUSTOMER", ConnectionToDatabase
'Respond to Form query
  DIM flag
  flag=0

  MyRecordSet.MoveFirst
  Do While NOT MyRecordSet.EOF
  If MyRecordSet("CUST_NAME")=Request.QueryString("company") AND MyRecordSet("PASSWORD")=Request.QueryString("passwd") Then
       Response.Write("Welcome "&Request.QueryString("company")&"!")
       flag=1
  End If

  MyRecordSet.MoveNext
  Loop
  If flag=0 Then
        Response.Write("Your Login is Incorrect!  Try Again!")
  End If

'Close the recordset
     MyRecordSet.Close
     Set MyRecordSet=Nothing
'Close the connection to the database
     ConnectionToDatabase.Close
     Set ConnectionToDatabase=Nothing
%>
</BODY>
</HTML>
```

Figure 8.43 Code for retrieval2.asp.

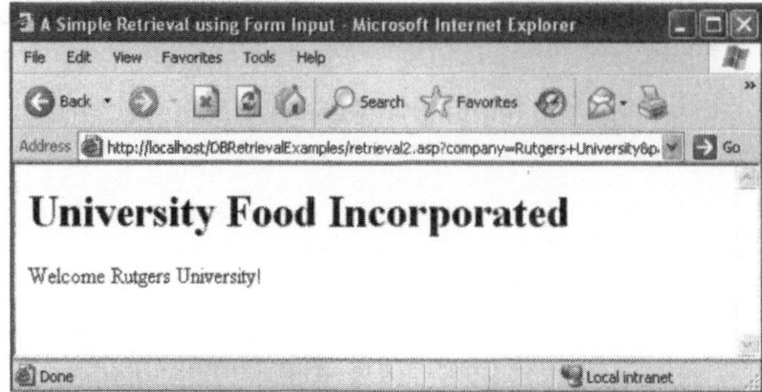

Figure 8.44 Web page after executing the code shown in Figures 8.42 and 8.43.

8.3.7 Inserting New Records into Database Tables

Inserting a new record into the database is accomplished by using the RecordSet object. An example is shown as Retrieval3.asp in Figure 8.45. There are only a few points to remember. When the table is opened, it must be specified as to what operations are allowed. This is done using the CursorType and LockType properties. Retrieval3.asp uses *CursorType* "1", which allows both read and write operations on the recordset as well as forward and backward movement through the records. The *LockType* value "2" locks the record that the user is editing, thus preventing another user from trying to update a record at the same time.

The AddNew method of the RecordSet object moves the user to a new record. After the record is added, the Update method of the RecordSet object saves the record. If the AddNew method fails because the database is read only, refer to Appendix 8D. It will explain how to eliminate this problem. The reader should implement ASP Exercise 8.7.

ASP Exercise 8.7:

We wish to demonstrate the use of ***ADO for inserting new records*** into the database.

1. Open ***UserID2.html*** in Notepad. Change the FORM ACTION statement to point to the file ***Retrieval3.asp***. Save the file as ***UserID3.html***. Close the file.
2. Open Notepad and create the ASP file shown in Figure 8.45. Save the file as Type "All files" with the filename ***Retrieval3.asp*** in the "DBRetrievalExamples" folder. Close the file.
3. Open a browser window and browse to the file UserID3.html. Open it in the browser window and fill in the text boxes with a new company name and password not currently existing in the CUSTOMER table of the database ORDERS.mdb. Click the Submit button.
4. Open the CUSTOMER table of the ORDERS.mdb database. The new record should be there.

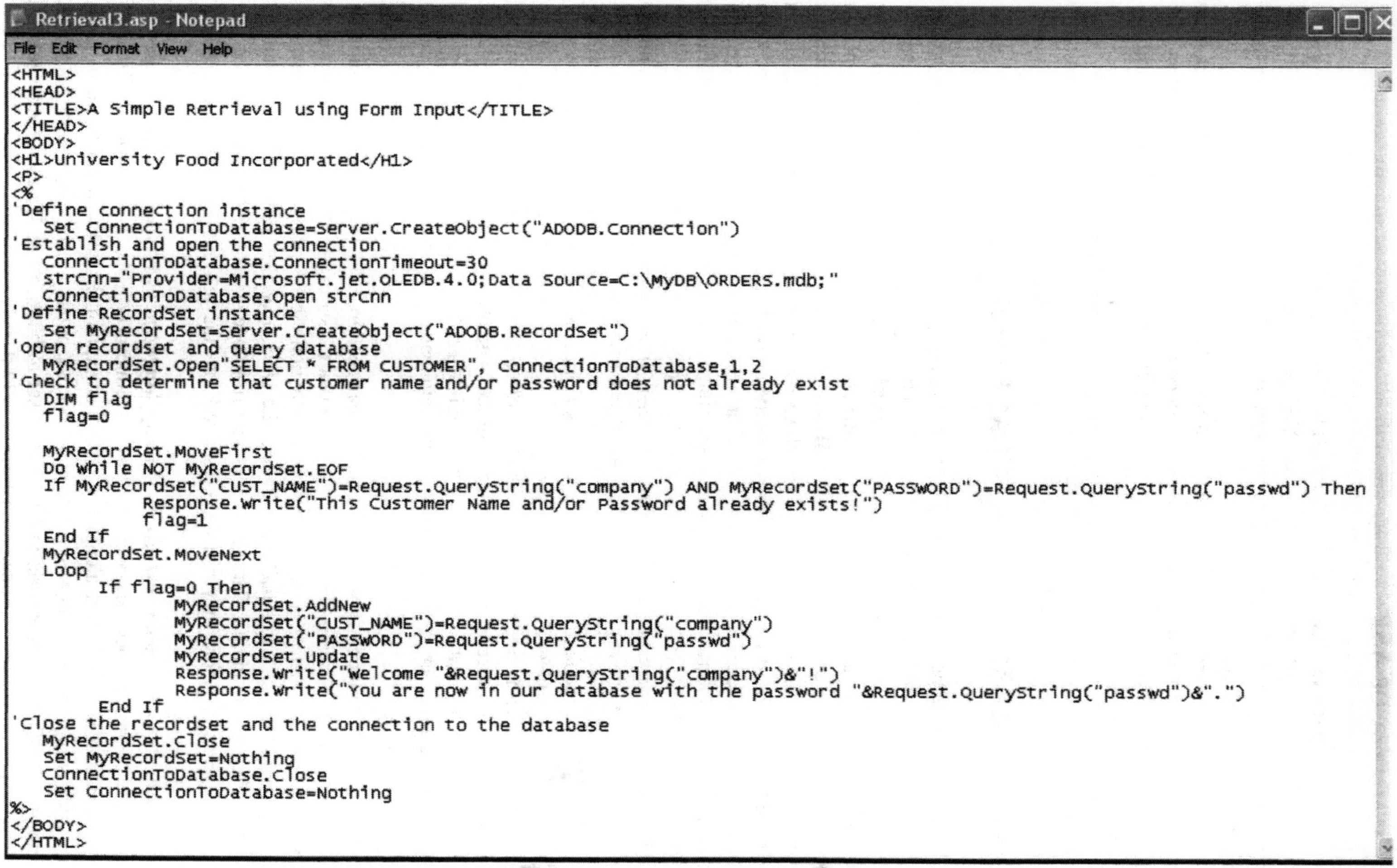

Retrieval3.asp - Notepad

File Edit Format View Help

```
<HTML>
<HEAD>
<TITLE>A Simple Retrieval using Form Input</TITLE>
</HEAD>
<BODY>
<H1>University Food Incorporated</H1>
<P>
<%
'Define connection instance
   Set ConnectionToDatabase=Server.CreateObject("ADODB.Connection")
'Establish and open the connection
   ConnectionToDatabase.ConnectionTimeout=30
   strCnn="Provider=Microsoft.Jet.OLEDB.4.0;Data Source=C:\MyDB\ORDERS.mdb;"
   ConnectionToDatabase.Open strCnn
'Define RecordSet instance
   Set MyRecordSet=Server.CreateObject("ADODB.RecordSet")
'Open recordset and query database
   MyRecordSet.Open"SELECT * FROM CUSTOMER", ConnectionToDatabase,1,2
'Check to determine that customer name and/or password does not already exist
   DIM flag
   flag=0

   MyRecordSet.MoveFirst
   Do While NOT MyRecordSet.EOF
   If MyRecordSet("CUST_NAME")=Request.QueryString("company") AND MyRecordSet("PASSWORD")=Request.QueryString("passwd") Then
            Response.Write("This Customer Name and/or Password already exists!")
            flag=1
   End If
   MyRecordSet.MoveNext
   Loop
        If flag=0 Then
               MyRecordSet.AddNew
               MyRecordSet("CUST_NAME")=Request.QueryString("company")
               MyRecordSet("PASSWORD")=Request.QueryString("passwd")
               MyRecordSet.Update
               Response.Write("Welcome "&Request.QueryString("company")&"!")
               Response.Write("You are now in our database with the password "&Request.QueryString("passwd")&".")
        End If
'Close the recordset and the connection to the database
   MyRecordSet.Close
   Set MyRecordSet=Nothing
   ConnectionToDatabase.Close
   Set ConnectionToDatabase=Nothing
%>
</BODY>
</HTML>
```

Figure 8.45 Example of adding a new record using data from an HTML form.

8.3.8 Summary of ASP

Active Server Pages was developed by Microsoft to enable database communication on a server using the Windows operating system platform. ASP provides objects and methods to control passing data between ASP files and the server. Key objects illustrated in this section include Response, Request, Server, and Session.

An ASP file can connect to and retrieve data from a database using the ADO object model library. Once data are retrieved in a recordset, they can be manipulated using a combination of the methods of the RecordSet object and VBScript. Examples were illustrated for both reading from and writing to a database.

8.4 EXTENSIBLE MARKUP LANGUAGE (XML)

In the previous sections, the use of HTML and Active Server Pages was illustrated for processing information over the Web. This approach has been successfully used in many business-to-consumer applications on the Internet, such as online ordering of products. However, it has certain limitations for exchanging information among business partners in a supply chain (business-to-business transactions). For one thing, using HTML and ASP, the server side defines the limits of the information that will be accepted from the client by the text boxes and other input element types that it provides. In that way, the server knows the meaning of each data element and knows how to handle it. However, if a client wishes to send an arbitrary message to a server, or vice versa, in some arbitrary file format, there is no way to guarantee that the receiving machine will understand the meaning of the file or be able to handle it in an application program.

Many desirable transactions among business partners in a supply chain are not easily enabled by the technologies previously discussed. Consider these examples:

- Transmitting electronic files of design drawings from a design group in one company to the manufacturing partners in another company
- Transmitting pricing and delivery quotations from a vendor to a customer and having the information automatically logged into the customer's database
- Sending inventory status of products from a point-of-sale location in a supply chain to an upstream supplier so that the supplier can anticipate the customer's inventory reorder schedule

Most of these limitations result from different partners in a supply chain having different databases, operating systems, and application software, such as middleware components. For example, one partner may use Internet Information Server running on Windows NT, while another partner uses an Apache Server and the Unix operating system. These systems do not easily "talk" to one another. Therefore, files sent from one system to the other must be reformatted or interpreted before they can be transferred into the information system of the target machine.

To overcome these problems, ***extensible markup language (XML)*** has become the de facto standard for sending business transactions over the Internet. XML uses plain ASCII text to compose a transaction document along with tags that explain the meaning of the text. Unlike HTML, XML should not be thought of as a language per se, but as a meta language (i.e., it provides a grammar, or linguistic framework, that allows a user to build his or her own language). Like HTML, the language is defined using tags. However, HTML tags just describe how a Web browser should display text—they do not describe the meaning of the text. XML, on the other hand, does not describe how text is to be displayed—it describes the meaning of the text.

When business partners employ XML, it is beneficial for them to define an agreed upon set of tags and their meaning to be used in the documents that they will transmit to each other. Using these tags and a set of rules that are defined by the XML grammar, the business partners can use the newly created XML tag language for business-to-business transactions. When the document is transmitted, the receiving computer uses the rules of the XML grammar to parse the document, extract the data, and use the data in an application program or insert the data into a database. In this way, XML has provided a general tool for use in computer-to-computer document exchange over the Internet.

In what follows, we describe the various components for defining an XML application. These include the following:

- The rules for designing an XML document. These rules are very specific and all XML documents must follow these basic rules.
- The XML document type definition (DTD) and the XML schema, which document the XML tags defined by the user for use in the XML document. The XML DTD or schema is used to validate the XML document.
- The XML processor (sometimes called parser), which is a computer program that takes an XML document and converts the contents into data structures that can be used for further processing. The processor thus handles the XML sent from another computer system and makes it readable on the local platform on which the processor resides.

Throughout the following sections, we include examples of XML documents that the reader may duplicate using Notepad or by copying the code from the Web site that accompanies this book. Several browsers include XML processors that can display XML text on a Web page. We use XML version 1.0, which can be viewed in Internet Explorer 5.0 and Netscape Navigator 4.7 and their later versions.

8.4.1 A Well-Formed XML Document

The structure of an XML document is shown in Figure 8.46. The first line is a declarative statement that gives the processor the name of the version of XML that is being used, in this case version 1.0. This is followed by the body of the document, which consists of a set of element tags and associated data. An ***XML element*** is just a tag name, which is used to identify the meaning of the data that exists between the element start tag and end tag.

One element is called the *root element*, which encloses the entire XML document. A number of other elements may be placed within the root element. This is shown in Figure 8.46 as <ELEMENT> and <SUBELEMENT> tags are placed within the <ROOT_ELEMENT> </ROOT_ELEMENT> tag set. Figure 8.46 is an example of a "well-formed" XML document. A well-formed XML document conforms to the following rules:

- There is exactly one root element.
- Elements have both start tags and end tags, which use exactly the same name. Element names are case sensitive and must be continuous (i.e., single text strings).
- Elements must be properly nested. If the XML parser encounters a start tag <x>, it must next encounter an end tag </x> or it must encounter one or more other start and end tag sets, eventually followed by the end tag </x>.

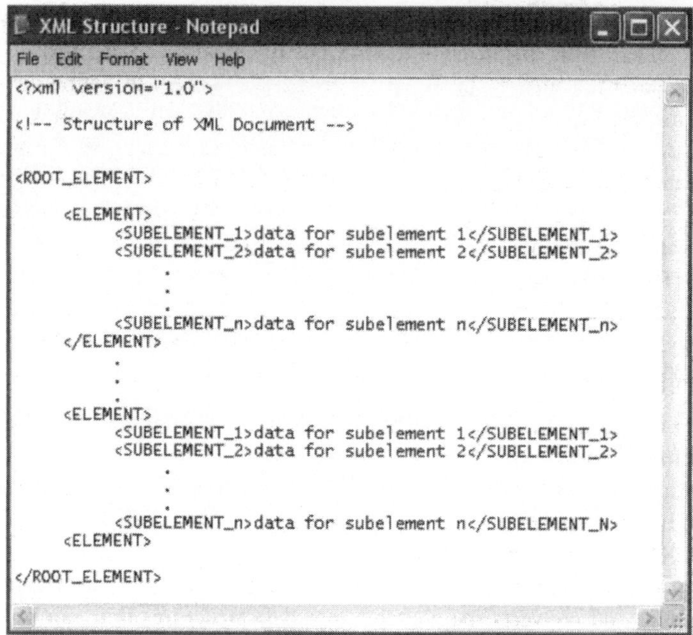

```
<?xml version="1.0">

<!-- Structure of XML Document -->

<ROOT_ELEMENT>

     <ELEMENT>
          <SUBELEMENT_1>data for subelement 1</SUBELEMENT_1>
          <SUBELEMENT_2>data for subelement 2</SUBELEMENT_2>
               .
               .
               .
          <SUBELEMENT_n>data for subelement n</SUBELEMENT_n>
     </ELEMENT>
     .
     .
     .
     <ELEMENT>
          <SUBELEMENT_1>data for subelement 1</SUBELEMENT_1>
          <SUBELEMENT_2>data for subelement 2</SUBELEMENT_2>
               .
               .
               .
          <SUBELEMENT_n>data for subelement n</SUBELEMENT_N>
     <ELEMENT>

</ROOT_ELEMENT>
```

Figure 8.46 Structure of an XML document.

The use of the declarative <?xml version = "1.0"?> is arbitrary. At the time of this writing, it is the only version of XML that has been released. No doubt it will be mandatory as new versions are released. However, if the declaration is used, it must be the first line in the file (i.e., no white spaces or text can appear before it). A document is recognized as an XML document because the filename of the document has an .xml extension.

As an example of an XML document, let us suppose that a manufacturer wishes to order materials from a vendor. Let us assume that, by arrangement among firms in the industry with their vendors, an electronic document has been designed with a set of tags that are agreed upon by all participating companies. These tags will define the meaning of information used in transferring an order from a manufacturer to its vendor. Thus, when the XML document is processed at the vendor, there is no ambiguity about the meaning of the message. Furthermore, the document can be parsed on the vendor's computer and the sales order can be automatically logged into the vendor's database. Figure 8.47 shows such a document.

The document type is known by its root element name: ORDER. Note the following characteristics of the document. First, the meaning of each tag name has been agreed upon among the parties that will use the document type. Therefore, each participant knows the meaning of the data that exist between the start and end tag pairs. Second, if the partners who establish the tags are sensible, the tag names will suggest what the meaning is; that way, any intelligent human being can read the document and deduce its meaning. For example, the document is an order. It is completely nested in a root element tag set <ORDER> </ORDER>. Finally, the document follows the established rules for nesting XML elements. Therefore, it can be easily parsed by software that extracts the text from tag elements. This, then, is the basic idea of XML as a universal grammar for designing languages to enable communication between different computer systems.

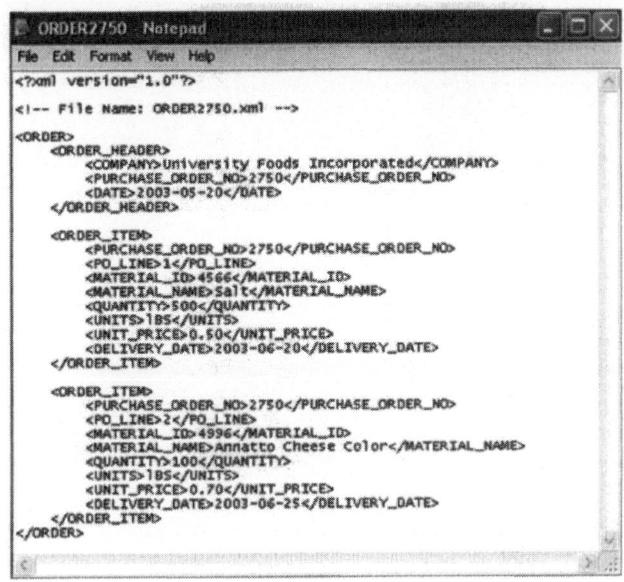

```
<?xml version="1.0"?>

<!-- File Name: ORDER2750.xml -->

<ORDER>
    <ORDER_HEADER>
        <COMPANY>University Foods Incorporated</COMPANY>
        <PURCHASE_ORDER_NO>2750</PURCHASE_ORDER_NO>
        <DATE>2003-05-20</DATE>
    </ORDER_HEADER>

    <ORDER_ITEM>
        <PURCHASE_ORDER_NO>2750</PURCHASE_ORDER_NO>
        <PO_LINE>1</PO_LINE>
        <MATERIAL_ID>4566</MATERIAL_ID>
        <MATERIAL_NAME>Salt</MATERIAL_NAME>
        <QUANTITY>500</QUANTITY>
        <UNITS>lBS</UNITS>
        <UNIT_PRICE>0.50</UNIT_PRICE>
        <DELIVERY_DATE>2003-06-20</DELIVERY_DATE>
    </ORDER_ITEM>

    <ORDER_ITEM>
        <PURCHASE_ORDER_NO>2750</PURCHASE_ORDER_NO>
        <PO_LINE>2</PO_LINE>
        <MATERIAL_ID>4996</MATERIAL_ID>
        <MATERIAL_NAME>Annatto Cheese Color</MATERIAL_NAME>
        <QUANTITY>100</QUANTITY>
        <UNITS>lBS</UNITS>
        <UNIT_PRICE>0.70</UNIT_PRICE>
        <DELIVERY_DATE>2003-06-25</DELIVERY_DATE>
    </ORDER_ITEM>
</ORDER>
```

Figure 8.47 ORDER2750.xml file in Notepad.

8.4.2 Viewing XML in a Browser Window

An XML document can be transferred between computer systems over the Internet. For example, it can be transferred as a Web page or as an e-mail attachment. It is just another ASCII text file with an .xml extension. When it arrives on the target system, it is saved as an ASCII file with the .xml extension, indicating it is an XML document. As previously mentioned, XML can be viewed in a browser window. This is demonstrated in XML Exercise 8.1.

Note that all the tags are displayed in Figure 8.48 and the format is exactly like that of the original file. The reader can confirm this by opening the file in Internet Explorer or Netscape Navigator. After opening the file, save it directly from the browser in another folder on your computer just to confirm that the file is transferable as an ASCII text file over a Web browser.

XML, version 1.0, provides a set of processing instructions that are incorporated into the Web browser for use in processing an XML file. If an XML file is sent to a browser without formatting information, the processing instructions of the browser software simply displays the file as plain text. This is what is shown in Figure 8.48.

XML Exercise 8.1:

We wish to implement an **XML document**.

1. Establish a new folder for XML documents, C:\XML.
2. Open Notepad and create the file shown in Figure 8.47 or load it from the Web site. Save the file in the XML folder as a Type "All files" with the filename ***ORDER2750.xml***. Close the file.
3. Open a browser and browse to the file ORDER2750.xml. Open the file in the browser window.
4. When the file is open it should appear like Figure 8.48.

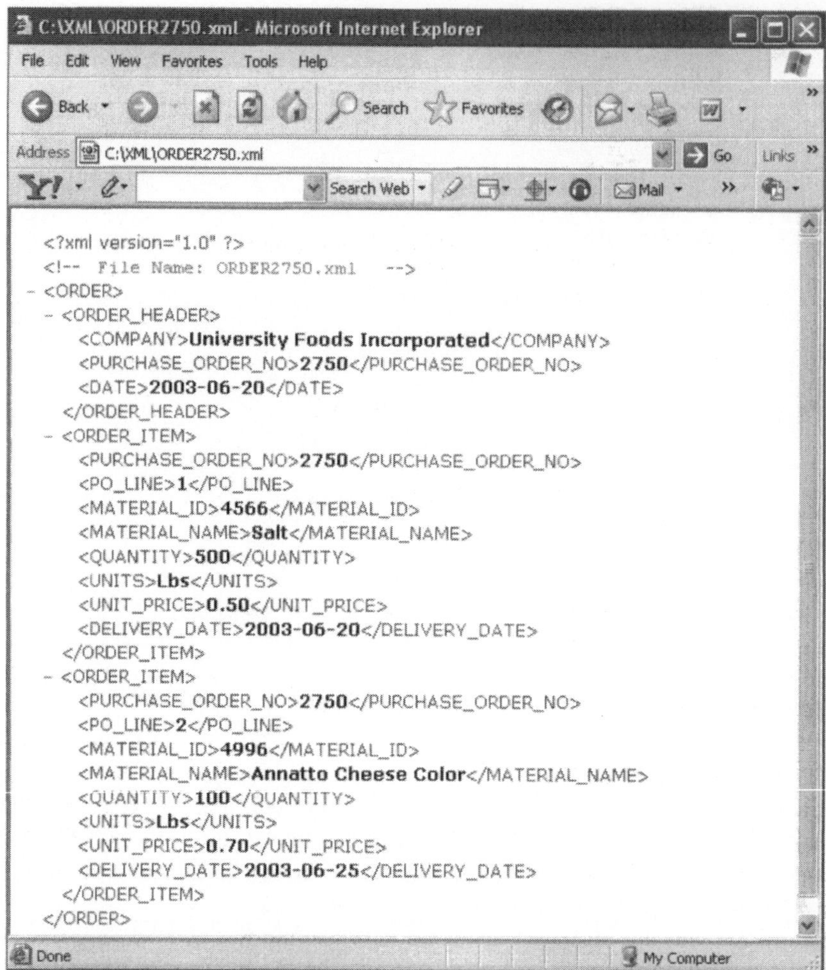

Figure 8.48 Code shown in Figure 8.47 when opened in Internet Explorer 6.0.

If it is desirable to control the format of data in the browser window and eliminate the display of tags, this can be accomplished using style sheets, a topic that was introduced in Section 8.2.7. Recall that a style sheet contains information on how the text should be displayed in a browser window and the style sheet is saved using a .css extension. The reference to the style sheet is given in the XML file. When the XML file is opened, the processor refers to the style sheet for display information. Figure 8.49 is a .css file that is created to display order information.

The style sheet of Figure 8.49 declares how each tag element will be displayed. Most of the style commands are similar to those used in Section 8.2.7. A new command type, ***display***, controls the formatting of the text on the page. When the ***block*** format argument is used, the processor

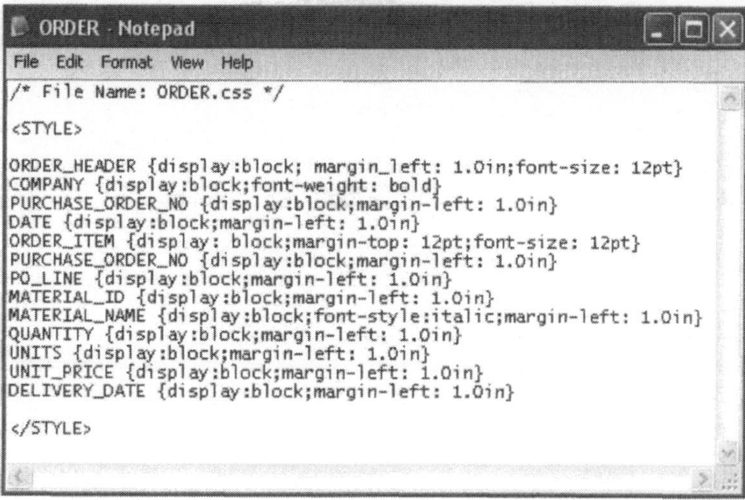

Figure 8.49 Cascading style sheet for displaying the XML order document.

inserts a line break before and after the line of text. Without a declared display type, the text of each element would simply run on continuously.

As explained in Section 8.2.7, a style sheet must be called from the program that uses it. This is done by inserting a declaration in the calling program file. For the program listing shown in Figure 8.47, the declaration can be placed anywhere between the XML version declaration and the start tag of the root element. For declaring a style sheet from an XML file, the following syntax is used:

```
<?xml-stylesheet href="path/filename.css" type="text/css"?>
```

For our example of Figure 8.49, we use the filename "ORDER.css." If this file is placed in the same folder as the file that calls it, the path argument is not necessary. Otherwise, the complete path to the file should be specified.

XML Exercise 8.2:

We wish to implement a **style sheet** for viewing XML files of the type ORDER.

1. Open Notepad and create the file shown in Figure 8.49 or load it from the Web site. Save the file in the XML folder as a Type "All files" with the filename **ORDER.css**. Close the file.
2. Open the file ORDER2750.xml in Notepad. Add the style sheet reference before the root element (see Figure 8.51). Close the file.
3. Open a browser and browse to the file ORDER2750.xml. Open the file in the browser window.
4. When the file is open it should appear like Figure 8.50.

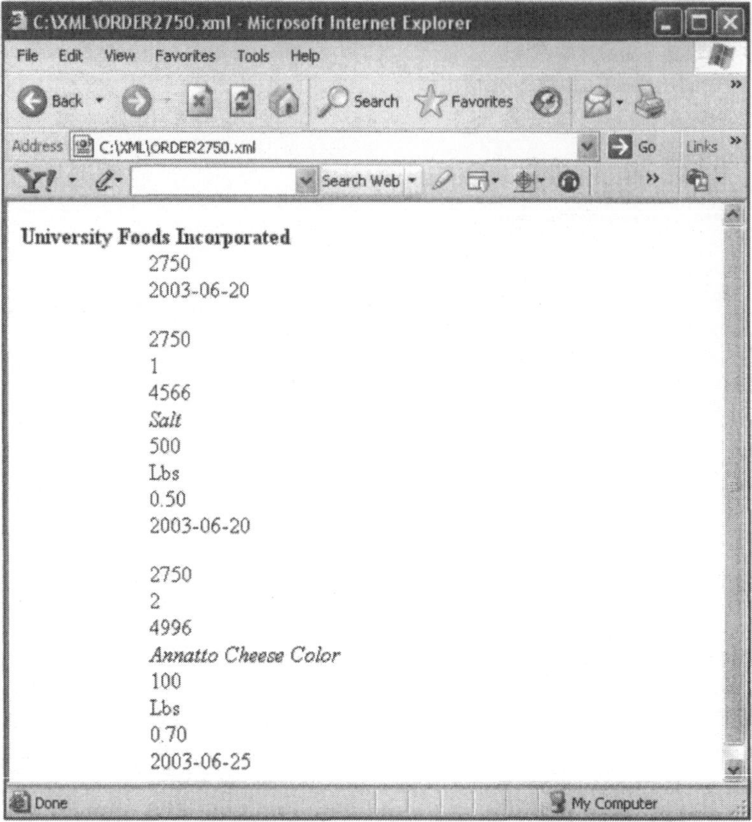

Figure 8.50 ORDER2750.xml file using style sheet ORDER.css.

8.4.3 Document Type Definitions

The XML document created in Section 8.4.2 is a "well-formed" XML document. This means that it conforms to the structural rules for creating an XML document. A distinction is made between a well-formed document and a "valid" document. A ***valid XML document*** is one that is well formed and conforms to the specifications of a document type definition (DTD) or an XML schema. When an XML document is to be validated, it can be done using either a DTD or schema. Only one method is used. This section will discuss validation using DTD. The following section will discuss the XML schema.

To understand the role of a DTD, let us suppose that a manufacturer sends an order to a vendor as shown in Figure 8.51. The document is well formed and complies with the rules of XML. However, note the use of the tag set <ID> </ID> to enclose the material ID. If the participants in the design of the ORDER type XML document have agreed on the use of the tag <MATERIAL_ID>, as in Figure 8.47, then the text enclosed within the <ID> tag set will have an ambiguous meaning. Furthermore, any user application program that is written to extract the contents of the element <MATERIAL_ID> will ignore the contents of <ID> and will not pass it on for further processing or for logging into a database.

The DTD is a specification for an XML document type that declares all the elements that will be allowed to appear in the document. If <MATERIAL_ID> is specified in the DTD, but <ID>

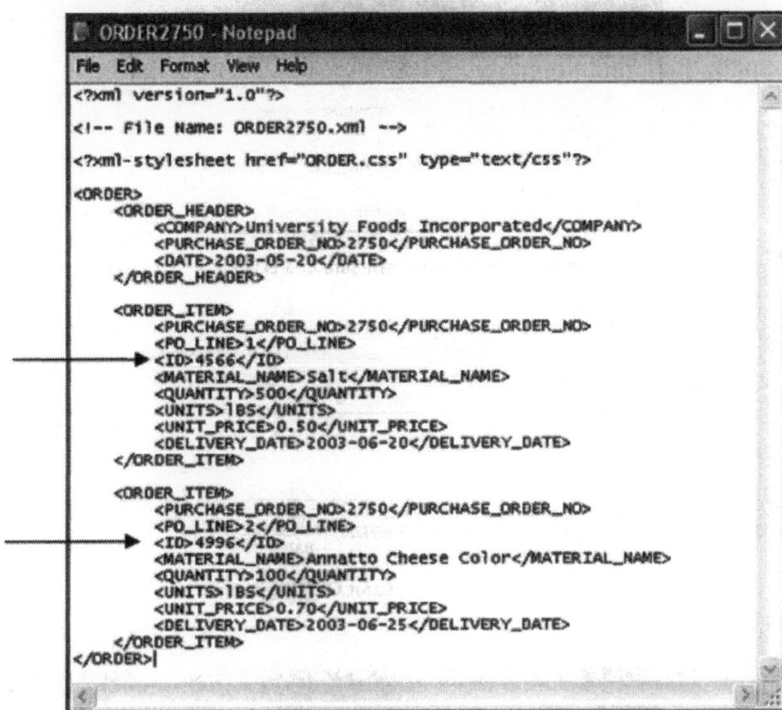
```
<?xml version="1.0"?>
<!-- File Name: ORDER2750.xml -->
<?xml-stylesheet href="ORDER.css" type="text/css"?>
<ORDER>
    <ORDER_HEADER>
        <COMPANY>University Foods Incorporated</COMPANY>
        <PURCHASE_ORDER_NO>2750</PURCHASE_ORDER_NO>
        <DATE>2003-05-20</DATE>
    </ORDER_HEADER>

    <ORDER_ITEM>
        <PURCHASE_ORDER_NO>2750</PURCHASE_ORDER_NO>
        <PO_LINE>1</PO_LINE>
        <ID>4566</ID>
        <MATERIAL_NAME>Salt</MATERIAL_NAME>
        <QUANTITY>500</QUANTITY>
        <UNITS>lBS</UNITS>
        <UNIT_PRICE>0.50</UNIT_PRICE>
        <DELIVERY_DATE>2003-06-20</DELIVERY_DATE>
    </ORDER_ITEM>

    <ORDER_ITEM>
        <PURCHASE_ORDER_NO>2750</PURCHASE_ORDER_NO>
        <PO_LINE>2</PO_LINE>
        <ID>4996</ID>
        <MATERIAL_NAME>Annatto Cheese Color</MATERIAL_NAME>
        <QUANTITY>100</QUANTITY>
        <UNITS>lBS</UNITS>
        <UNIT_PRICE>0.70</UNIT_PRICE>
        <DELIVERY_DATE>2003-06-25</DELIVERY_DATE>
    </ORDER_ITEM>
</ORDER>
```

Figure 8.51 ORDER document type with illegal tag.

is not, the file of Figure 8.51 will create an error when the processor checks it against the DTD. When a user programs an XML document and declares the DTD that validates the document, it can be validated using a processor that supports DTD before it is sent on to the target computer. Hence, a DTD is a higher level protection against sending incorrect XML documents.

A DTD can be added to an XML file by placing it right after the version declaration. A DTD can also be written as a separate file, with a .dtd extension, and referenced from the XML file.

The general format used in a DTD is shown in Figure 8.52. The document begins with the declaration of the document type. The document type refers to the root element of the document. So, for example, in Figure 8.51, the file ORDER2750.xml is of the document type ORDER. One can imagine that it is the 2750th order sent by University Food Incorporated, or that it is PO#2750. University Food will want to keep a file of all the ORDER type documents it sends, and using the PO number as the filename is a natural way to do so. The document type declaration is followed by a set of brackets that enclose specifications of each element. The specification includes the keyword "!ELEMENT," followed by an element name and a content specification declaration. The content specification refers to the type of content that can appear between the start tag and end tag of the element.

The structure of an XML document, with its nested tags, is easily pictured as an inverted tree diagram. For example, Figure 8.53 shows the tree structure of the ORDER type XML document. The highest node shows the root element of the document, the ORDER tag. There are two indented second-level elements: ORDER_HEADER and ORDER_ITEM. The ORDER_HEADER and ORDER_ITEM tags are nested within the ORDER tag. Finally, the COMPANY, PURCHASE_ ORDER_NO, and DATE tags can be nested within the ORDER_HEADER tag. Similarly, the ORDER_ITEM tag can have nested elements as indicated in Figure 8.53. It is typically the terminal

```
<!DOCTYPE root_element_name
     [
       <!ELEMENT element_1_name content_specification>
              .
              .
              .
       <!ELEMENT element_n_name content_specification>
     ]
>
```

Figure 8.52 General structure of a DTD file.

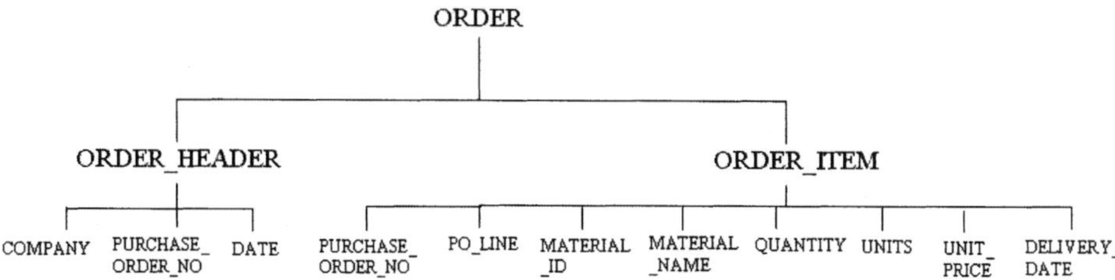

Figure 8.53 Tree diagram for ORDER document.

```
<!DOCTYPE ORDER
    [
      <!ELEMENT ORDER (ORDER_HEADER, ORDER_ITEM+)>
      <!ELEMENT ORDER_HEADER (COMPANY, PURCHASE_ORDER_NO, DATE)>
      <!ELEMENT ORDER_ITEM (PURCHASE_ORDER_NO, PO_LINE, MATERIAL_ID, MATERIAL_NAME, QUANTITY, UNITS, UNIT_PRICE, DELIVERY_DATE)>

      <!ELEMENT COMPANY (#PCDATA)>
      <!ELEMENT PURCHASE_ORDER_NO (#PCDATA)>
      <!ELEMENT DATE (#PCDATA)>
      <!ELEMENT PO_LINE (#PCDATA)>
      <!ELEMENT MATERIAL_ID (#PCDATA)>
      <!ELEMENT MATERIAL_NAME (#PCDATA)>
      <!ELEMENT QUANTITY (#PCDATA)>
      <!ELEMENT UNITS (#PCDATA)>
      <!ELEMENT UNIT_PRICE (#PCDATA)>
      <!ELEMENT DELIVERY_DATE (#PCDATA)>
    ]
>
```

Figure 8.54 DTD for ORDER document type.

nodes of the tree that show the elements that will have data content. The higher nodes of the tree typically show elements that will serve as containers for other elements.

The tree structure serves as a convenient way to organize the DTD. Figure 8.54 is a typical DTD that we can apply to the ORDER document type. The content specification of each of the elements is declared within parentheses. For example, the element ORDER must contain the elements ORDER_HEADER and ORDER_ITEM, in that specific order. The + sign following the element ORDER_ITEM is a modifier, which indicates that the child element may appear one or more times. The element ORDER_HEADER does not have a modifier, indicating it must appear only once. This specification is consistent with the example, ORDER2750.xml, shown in Figure 8.47. The first-level child elements also have their content described in terms of the elements that will be nested within them.

The lowest level elements, those that contain the data, have their contents specified as #PCDATA. The keyword PCDATA stands for "parsed character data." This means that the content specification of the element is only character data; no child elements are allowed. Since the processor will "parse" the tree to retrieve this data, it may not contain any reserved characters, such as angle brackets (<,>), square brackets ([,]), or ampersands (&).

Figure 8.54 is an example of a DTD for validating a document of the type ORDER. It is not the only way that a DTD could be written, but it is the most rigid specification for the case being considered. DTDs can be constructed with optional content specifications in which the content may take on different element or data types. A DTD can specify *attributes* that may be attached to elements for providing additional information. In this book, we do not explore the whole range of XML constructs and the possible content specifications for DTDs. Rather, we are interested in conveying the basic ideas and principles involved. The study of XML can be the subject of an entire course and the interested reader is referred to the bibliography at the end of the book for further reading (Ladd and O'Donnell, 1999; Young, 2002).

8.4.4 The XML Schema

Like a DTD, an ***XML schema*** is a document that also defines the content and the structure of a class of XML documents, such as the ORDER class. A schema is an alternative way to validate an XML document. In the evolution of XML, the use of the schema for validation is replacing the DTD. Going forward, most browsers and other programs that parse XML documents will use the schema. This is the case because schemas offer certain advantages over DTDs. The most important advantage is that they provide a much more precise control over the structure and content of the document. All the power of schemas will not be explored in this book. Rather, we will examine the basic structure of a typical schema, leaving the reader to refer to the references at the end of this book for more information.

A schema is similar to an external DTD since it is stored in a separate file and referred to when the XML document is opened. The schema is based on a set of rules and a particular set of tag definitions called the ***XML Schema Definition Language***. Each class of XML documents may have a corresponding schema with the same name as the XML document class and an extension, .xsd. To illustrate a schema for the hypothetical XML document class ORDER, refer to Figure 8.55.

The schema is itself an XML document that is written according to the rules defined by the World Wide Web Consortium (W3C); thus, the reference in Figure 8.55 to the XML Schema Definition Language (xsd) at www.w3.org. A specific XML document written according to the schema definition language is an instance of the schema.

Schema elements are declared using the ***xsd:element*** keyword followed by the element name. The ORDER schema shows the nested structure of the elements in the ORDER document type. The root level element and its subelements are shown first. Note that the allowed frequency of occurrence of subelements is specified by the minOccurs and maxOccurs constraints.

The XML schema definition specifies element types as either simple or complex. A ***simple element type*** permits the element to contain only character data. A ***complex element type*** may contain child elements or more than one data element. The ORDER element contains two child elements and is a complex element type. The <xsd:sequence> tag begins a description of the order of subelements within the element tag.

The ORDER_HEADER element and ORDER_ITEM element follow the same format. Both show the sequence of subelements within each tag set. Each subelement is a data element and is given a defined data type. Hence, when the ORDER document is parsed and the data are inserted into tables, PURCHASE_ORDER_NO will be a character string and QUANTITY will be a decimal number. The xsd:date data type requires a specific format in the XML document. The format is YYYY-MM-DD, as shown previously in the ORDER.xml file of Figure 8.47. Any other date format is not allowed.

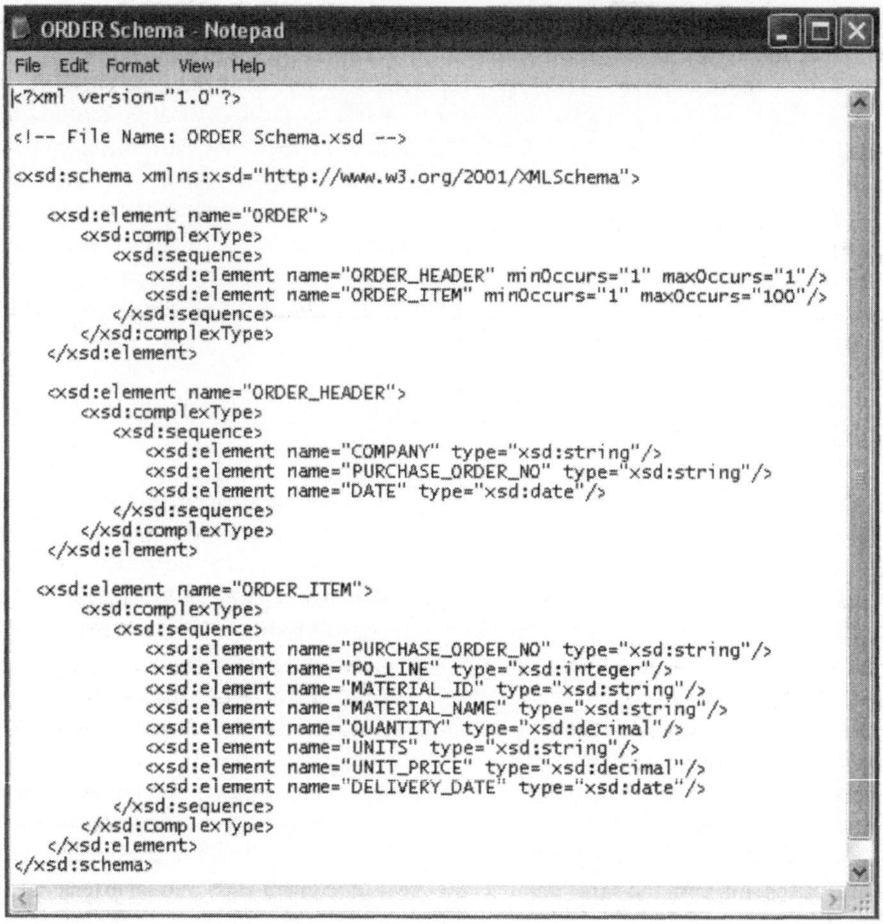

ORDER Schema - Notepad

File Edit Format View Help

```
<?xml version="1.0"?>

<!-- File Name: ORDER Schema.xsd -->

<xsd:schema xmlns:xsd="http://www.w3.org/2001/XMLSchema">

   <xsd:element name="ORDER">
      <xsd:complexType>
         <xsd:sequence>
            <xsd:element name="ORDER_HEADER" minOccurs="1" maxOccurs="1"/>
            <xsd:element name="ORDER_ITEM" minOccurs="1" maxOccurs="100"/>
         </xsd:sequence>
      </xsd:complexType>
   </xsd:element>

   <xsd:element name="ORDER_HEADER">
      <xsd:complexType>
         <xsd:sequence>
            <xsd:element name="COMPANY" type="xsd:string"/>
            <xsd:element name="PURCHASE_ORDER_NO" type="xsd:string"/>
            <xsd:element name="DATE" type="xsd:date"/>
         </xsd:sequence>
      </xsd:complexType>
   </xsd:element>

  <xsd:element name="ORDER_ITEM">
      <xsd:complexType>
         <xsd:sequence>
            <xsd:element name="PURCHASE_ORDER_NO" type="xsd:string"/>
            <xsd:element name="PO_LINE" type="xsd:integer"/>
            <xsd:element name="MATERIAL_ID" type="xsd:string"/>
            <xsd:element name="MATERIAL_NAME" type="xsd:string"/>
            <xsd:element name="QUANTITY" type="xsd:decimal"/>
            <xsd:element name="UNITS" type="xsd:string"/>
            <xsd:element name="UNIT_PRICE" type="xsd:decimal"/>
            <xsd:element name="DELIVERY_DATE" type="xsd:date"/>
         </xsd:sequence>
      </xsd:complexType>
   </xsd:element>
</xsd:schema>
```

Figure 8.55 Example XML schema.

Each data element will appear one time within their parent element, ORDER_HEADER or ORDER_ITEM. If minOccurs and maxOccurs arguments are not specified, a default value of "1" is assumed. Therefore, in this case it is not necessary to define the number of occurrences at the data element level. In cases where it was necessary to define the number of occurrences at the data element level, the data type definition would be followed by the minOccurs and maxOccurs arguments. Since each data element will occur exactly once in each repetition of a parent tag of the ORDER document, NULL values will not be allowed. If, for example, the UNIT_PRICE were not always known at the time of ordering and a NULL value were desirable, the UNIT_PRICE element tag would appear as follows in the schema:

```
<xsd:element name="UNIT_PRICE" type="xsd:decimal" minOccurs="0"
maxOccurs="1"/>
```

This means that there could be some instances of the ORDER_ITEM element that will not include a UNIT_PRICE element.

The XML schema language provides the user with additional constructs that allow her or him to be specific about the content of the XML document. For example, when an xsd:decimal data type is assigned to an element, it is possible to assign the range of acceptable values. It is also possible to assign enumeration values for xsd:string data types in cases where the data content will be from a fixed set of possible strings. These and other features of the schema language are beyond the range of topics in this general introduction; the reader is referred to the references at the end of the book.

XML Exercise 8.3:

We wish to create an ***XML Schema*** for the type ORDER.class.

1. Open Notepad and create the file shown in Figure 8.55 or load it from the Web site. Save the file in the XML folder as a Type "All files" with the filename **ORDER Schema.xsd**. Close the file.
2. The Schema will be used in conjunction with the ORDER class XML file in the next exercise.

8.4.5 Processing XML Files

Once an XML document has arrived on the target system, it can be handled in a variety of ways. As previously shown, it can be opened in a Web browser for viewing and checking for correctness and then manually entered into the enterprise database, if that is appropriate. Alternately, an application program can be written to parse the document and extract the data for entry into a database.

DBMSs have been automating the process of importing XML documents into databases. Oracle, SAP, and other enterprise databases now have a variety of facilities for managing XML documents. With Access 2003, importing and exporting XML documents have been reduced to a few simple steps. This section describes the process of importing an XML document using Access 2003.

As previously shown in Figure 8.55, an XML schema reads like a database format. Each main element tag can be interpreted as an entity (table), and each data element can be interpreted as an attribute (field) with the assigned data type indicated by the type argument in the schema. Hence, a program can be written that parses an XML file according to the XML schema and creates tables and fields having the labels given in the schema. When an XML document is imported into the database, the instances of data associated with the element names can be entered into the appropriate tables and fields. This is precisely how XML documents are imported into databases.

Fortunately, the facilities for importing and exporting XML are now integral to DBMSs. Consider the scenario in which a company sends an order to a supplier as an XML file over the Internet. It may be sent as an attachment to an e-mail message. Such an order would most probably go from the purchasing agent of the customer to the sales department of the vendor. The vendor would take the attachment and put it in a folder. If the instance of the order adheres to the schema and the enterprise database tables use the same element names as the schema, the vendor can open the enterprise database and import the order file directly into the appropriate table. Alternatively, if the sales department wishes to check an order before appending it as a new record in the enterprise database, the XML file could be imported into a temporary table and moved to the enterprise database later.

We will assume that the second protocol is used. The process of importing the XML file into a database will have two steps:

- The XML schema will first be imported into the database, thus establishing the table structure for importing the data.
- The XML document will be imported into the database, thus populating the tables that were established by the schema.

The reader should implement XML Exercise 8.4 to confirm thses steps.

XML Exercise 8.4:

We wish to demonstrate the *importation of an XML document* to an Access database.

1. Ensure that the **ORDERS.mdb** database is in the folder C:\MyDB. If not, load it at this time.
2. Ensure that the files **ORDER2750.xml** and **ORDER Schema.xsd** are in the folder C:\XML. Here we assume that the order has been received and placed in this folder.
3. Open Microsoft Access and the ORDERS.mdb database.
4. To import the schema, from the file menu select **File=>Get External Data=>Import** as shown in Figure 8.56. This will open the **Import** window.
5. Choose the XML file type and browse to the folder in which the XML documents reside (C:\XML), as shown in Figure 8.57.
6. Select the ORDER Schema and click the Import command button. This will open the **Import XML** window, as shown in Figure 8.58. This window shows the table structure that will be created from the schema. Note the table ORDER, which is the root level tag and does not contain any data elements. This table is created by the Access import program although it is not required and can be deleted afterward. Click on the OK button and the tables will be created from the schema. If there is a failure importing the schema, an error message will indicate why. The reader should observe the tables in the ORDERS.mdb database and note the existence of the tables ORDER_HEADER and ORDER_ITEM. Open the tables ORDER_HEADER and ORDER_ITEM and observe that they have been established and are currently empty. The table ORDER was created from the root directory tag and can be deleted.
7. Follow the same process to import the XML ORDER document type, except choose the ORDER2750.xml file. When you reach the Import XML window, click on the Options command button. Here you will be given three options for importing the data elements, as shown in Figure 8.59. The import options are as follows:

    - Structure – This option will establish a set of tables based on the XML document format and the element tags. Schema information is not used. The default data type for all data elements is xsd:string (text).
    - Structure and Data – This option will import the ORDER2750.xml file without regard to the existing tables established by the schema. It will establish and populate the tables based on the interpretation of the structure of the XML document. Since it does not use schema information, it will default all data types to string.
    - Append data to existing table(s) – This option assumes the existence of the tables that have the same table names as the element tags and the same field structure as the data tags. Such a table could be created manually or using the schema file, as we have previously done.

    Use the "Append data to existing table(s)" option, and click OK. The transfer should now be complete. Open the ORDER_HEADER table and the ORDER_ITEM table and ensure that the data has been properly transferred.

The two new tables may be considered to be intermediate tables in the ORDER database. Once the order is accepted, it can be moved into the enterprise tables, CUSTOMER-ORDER and ORDER_DETAIL. The XML document could have been constructed so that it would exactly match the format of the enterprise database, but it is usually more reasonable to examine the contents of the file before loading it into the main enterprise database. The transfer of data between tables and the creation of data for new field elements (such as ORDER_NO) can be done using an application program or using forms and some of the techniques that were described in Chapter 6.

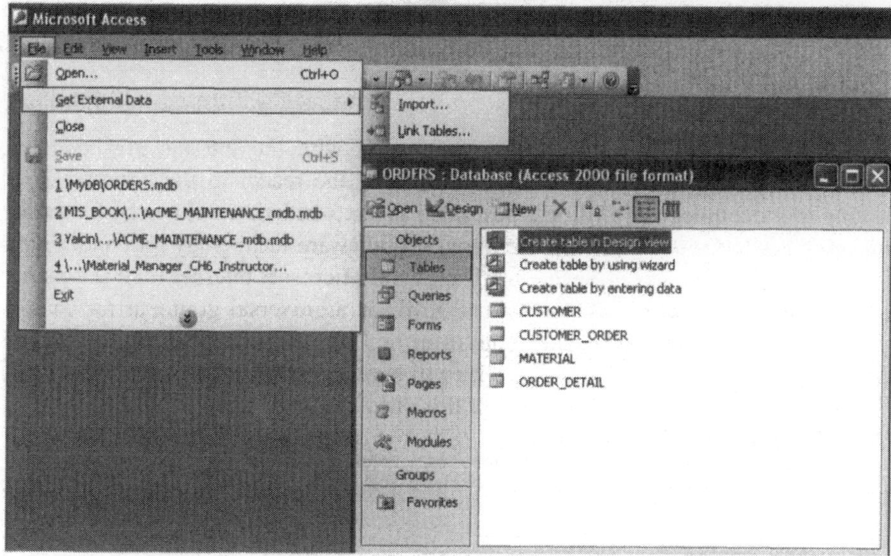

Figure 8.56 Launching the XML import.

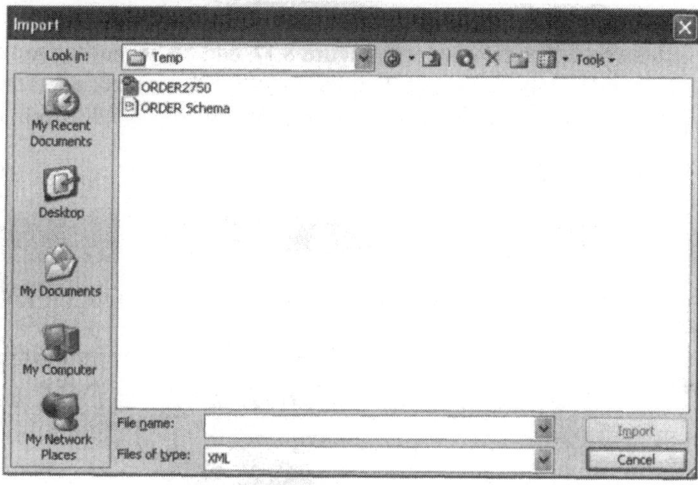

Figure 8.57 Selecting the XML schema.

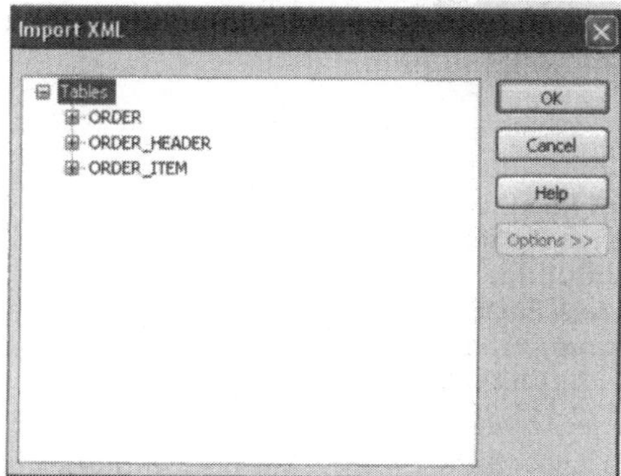

Figure 8.58 Importing the schema table structure.

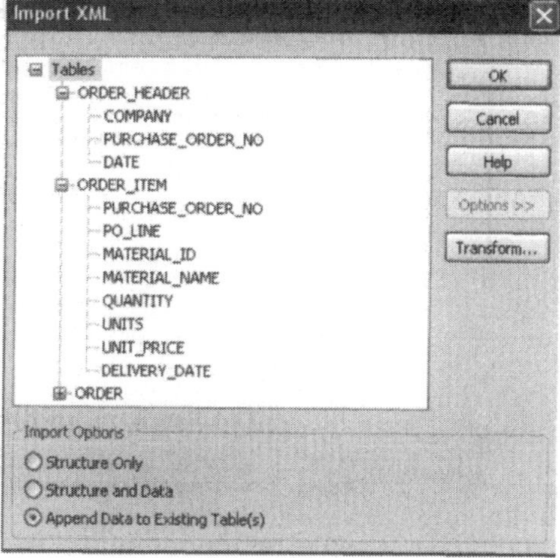

Figure 8.59 Options for importing the XML document file.

8.5 SUMMARY

The application and use of databases in conjunction with the Internet are becoming very common features of commerce. This chapter introduced the reader to the basic techniques for implementing this technology. HTML was described as the foundation technology for writing Web pages and including forms for handling user input. Middleware technology for handling database interaction was illustrated using Active Server Pages and Microsoft Internet Information Server as a typical example. Finally, the fundamentals of XML as a universal grammar for enabling dissimilar computer systems to talk to each other over the Internet are explained. The potential use of these technologies is considerable. The reader will have an opportunity to explore several applications in an extensive case study at the end of this chapter.

REVIEW EXERCISES

8.1 Using the information in Section 8.2, create your own personal Web page. Include your educational background, interests, hobbies, sports, and so on. Add an image if you have a digital picture.

8.2 From the page UniversityFoodHome.html, Figure 8.17, add an anchor tag to the "Contact us" line item on the modified unordered list (List.html). When this link is selected, it should open the following Web page. The anchor tag, Here, should return the Web browser to the homepage.

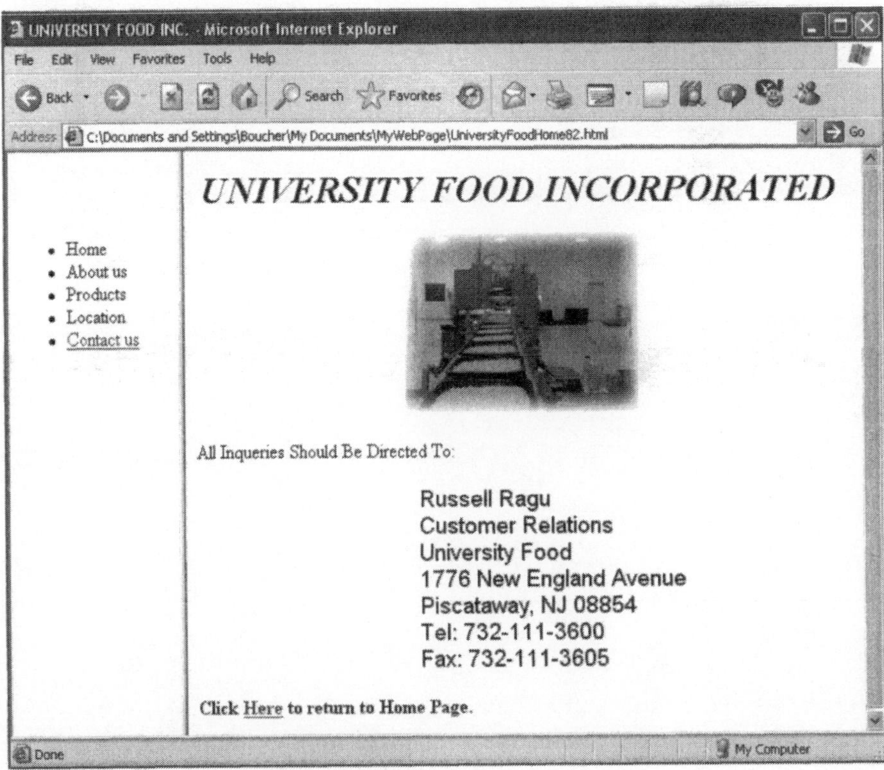

8.3 USF Rent-a-Car (Review Exercises 2.2 and 3.5) has decided to implement a Web page as a convenience for customers to make reservations. The design is shown next. Customer information is entered in text boxes. Car class, location, and date fields are implemented as drop down lists. Discount information is implemented as a radio button selection. At this time there is no ASP file to handle the submission. Prepare the form elements on a Web page as shown here.

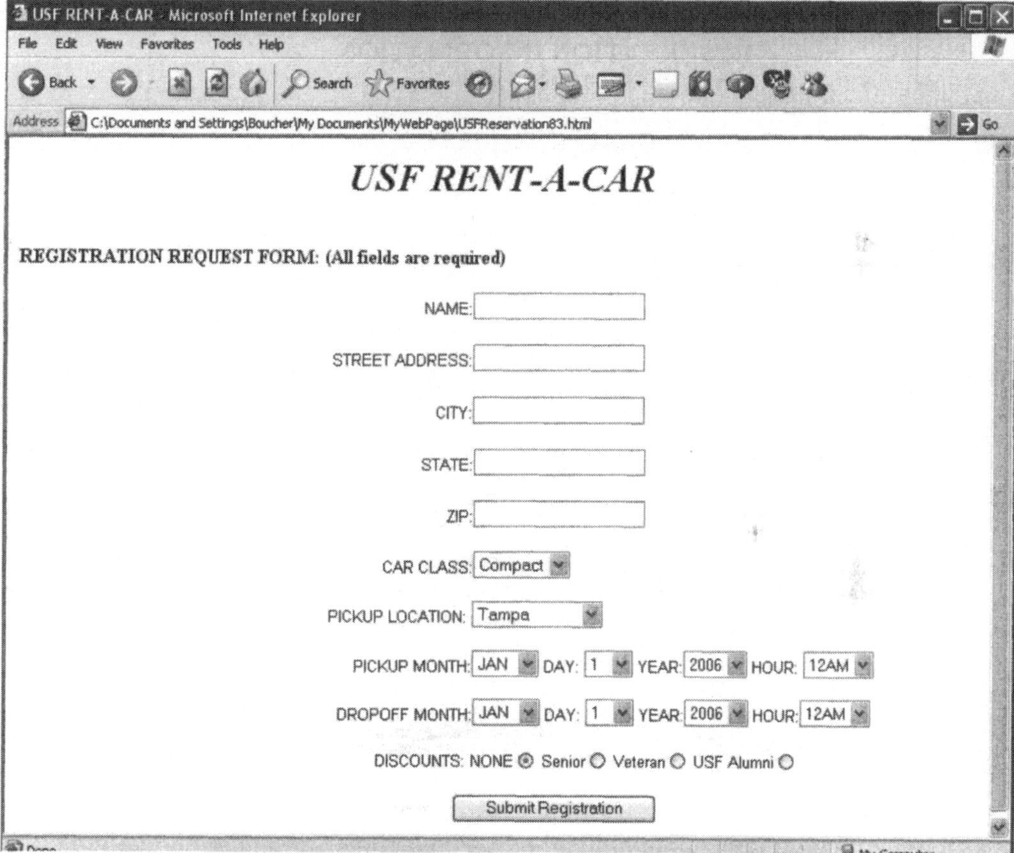

8.4 Using the USF Rent-a-Car reservation page of Review Exercise 8.3 as input, create an ASP file that will take the input and return all the form element values back to the customer. As an example, a typical Web page returned to the customer should be designed as shown here:

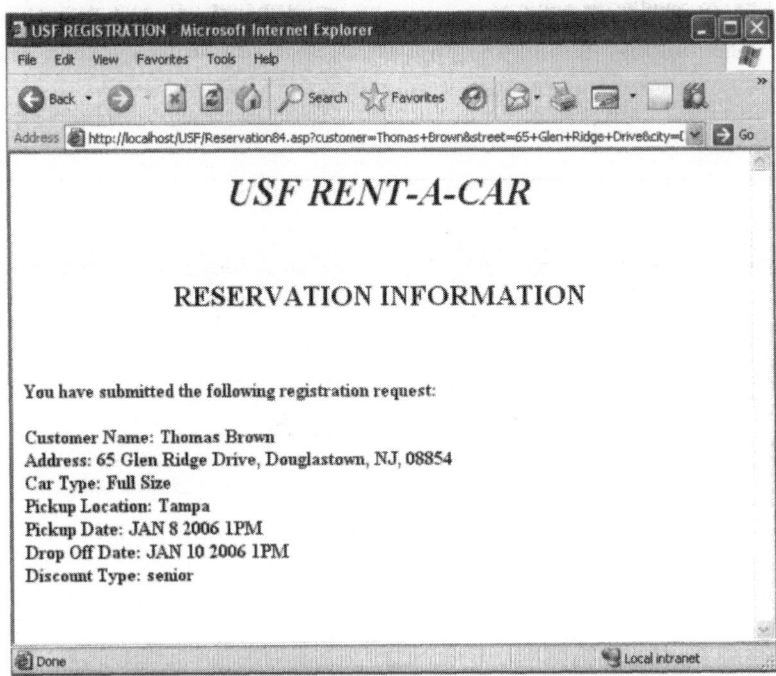

8.5 Create a database, USF.mdb, and a table RESERVATION. The table should have all the attributes shown on the input form of Review Exercise 8.3. Make the primary key Reservation_no, with a data type AutoNumber. The auto number (Reservation_no) will index by 1 each time a new record is entered. Create an ASP file that opens the database and inserts the new reservation entered from the form of Review Exercise 8.3. Finally, if all operations are successful, the server should return a page that looks like the one that follows:

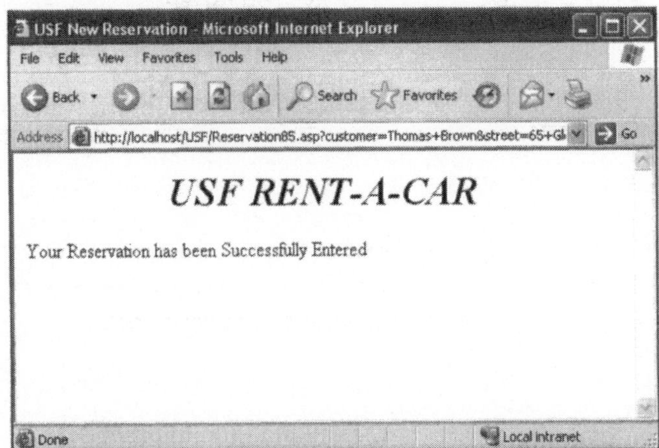

8.6 Perform the requirements of Review Exercise 8.5 except return the reservation number to the customer. This is done by retrieving the last AutoNumber created and writing it back to the customer on the Web page shown in Review Exercise 8.5.

8.7 The management of USF Rent-a-Car wants each branch location to send the files of reservations to the corporate office at the end of each day. This is to be done using XML.

(a) Create an XML document that can be used to submit elements of the RESERVATION class from a branch location to the central office. Create the RESERVATION class from the RESERVATION table used in Review Exercises 8.5 and 8.6.

(b) Create an XML schema that clearly defines the tags, their elements, and data types.

(c) Using Access, the XML document, and the XML schema, simulate the transmission of data from the branch location to the corporate office. Follow the method described in Section 8.4.5.

CASE STUDIES

8.8 University Food Case (E)

University Food Incorporated wants to develop a Web site that will allow the following functionality:

- New customers can register and have their credit checked.
- Existing customers can purchase products.
- Existing customers can check the status of the orders that they previously placed.

In the exercises that follow, each of these functionalities will be implemented. The database, ORDERS.mdb, and tables that will be used are shown in Appendix 8C and can be downloaded from the Web site.

Exercise 1

The home page, UniversityFoodHome.html, is shown in Figure CS8.1. Use the database tables shown in Appendix 8C to define the names of data elements on the forms. If a new customer selects the link ("here"), the anchor tag will send the user to a file named NewCustomer.html, shown in Figure CS8.2. Create the .html file for Figure CS8.1.

Exercise 2

The first step for the user is to register. When a customer selects the anchor tag "here," the NewCustomer.html file will be loaded. This is shown in Figure CS8.2. All registration elements are required. The elements "routing number" and "bank account number" are credit elements that are used to transfer payments upon billing for orders that are shipped. They are used to check the current status of the customer's credit. To align the text boxes on the screen, use the inline style sheet approach and the "margin-left" method. Create the file for Figure CS8.2. Link the Web page shown in Figure CS8.2 to the anchor tag on the Web page shown in Figure CS8.1. Save all files, including foodpic.jpg, in a folder within the root directory, wwwRoot.

Exercise 3

When the new customer submits a registration, an ASP file must handle the data and do the following:

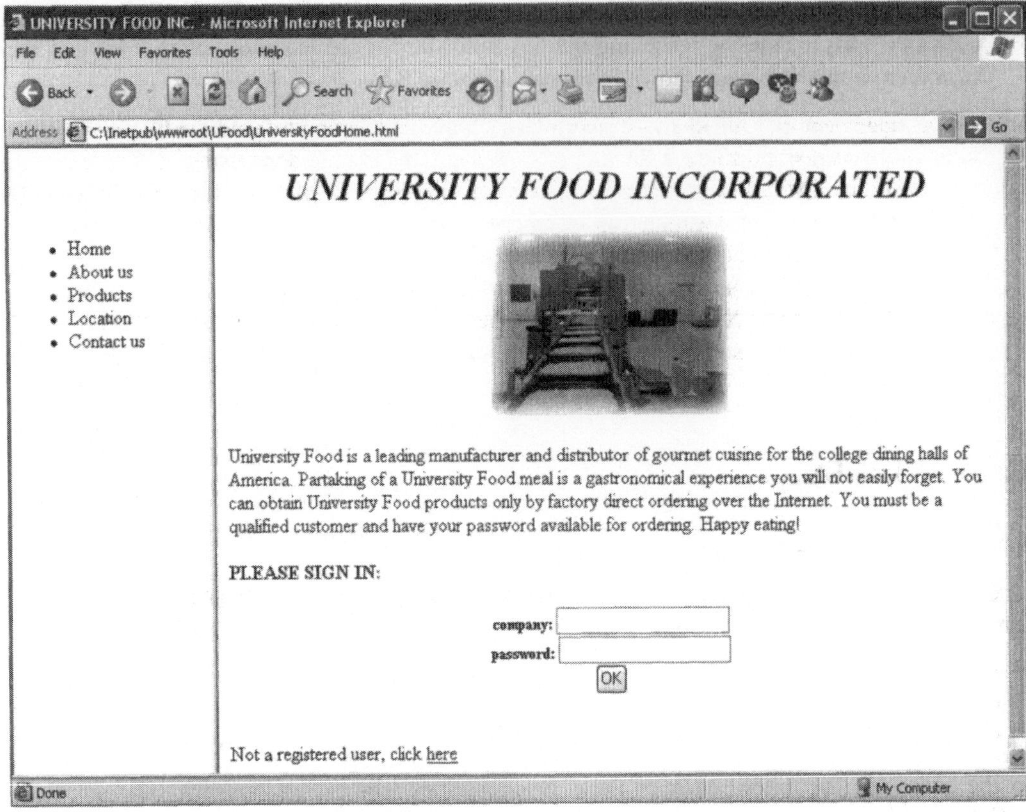

Figure CS8.1 University Food home page.

- Check to see that all data elements have been entered and that, as far as possible, they have been checked for a correct format.
- Check to see that the "create password" and "confirm password" entries are the same.
- Check to see that the requested name or password has not already been assigned to another customer.

You should program the following responses:

- If all three checks are true, enter the data into the database table CUSTOMER and return the screen shown in Figure CS8.3 to the customer. The anchor tag "here" should bring the customer back to the home page (Figure CS8.1).
- If all data elements are not complete, return the screen shown in Figure CS8.4 to the customer. This should be done before going on to checking the password. The anchor tag "here" should bring the customer back to the registration page (Figure CS8.2).
- If all the data elements are entered but the "create password" and "confirm password" entries are not identical, return the screen shown in Figure CS8.5 to the customer. The anchor tag "here" should bring the customer back to the registration page (Figure CS8.2).
- If all the data elements are entered and the "create password" and "confirm password" are identical, but the name or the password has previously been assigned, return the screen shown

UNIVERSITY FOOD INC. - Microsoft Internet Explorer

File Edit View Favorites Tools Help

Address C:\Inetpub\wwwroot\UFood\UniversityFoodHome.html

UNIVERSITY FOOD INCORPORATED

- Home
- About us
- Products
- Location
- Contact us

NEW CUSTOMER REGISTRATION: (All fields are required)

COMPANY NAME:

STREET ADDRESS:

CITY:

STATE:

ZIP:

ROUTING NUMBER:

ACCOUNT NUMBER:

CREATE PASSWORD:

CONFIRM PASSWORD:

Submit Registration

Done My Computer

Figure CS8.2 New customer registration page.

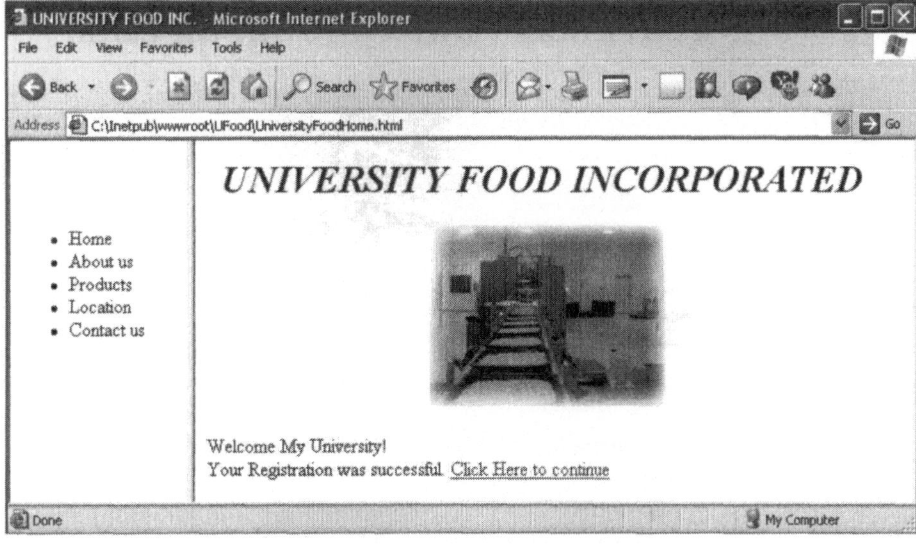

Figure CS8.3 Successful registration page.

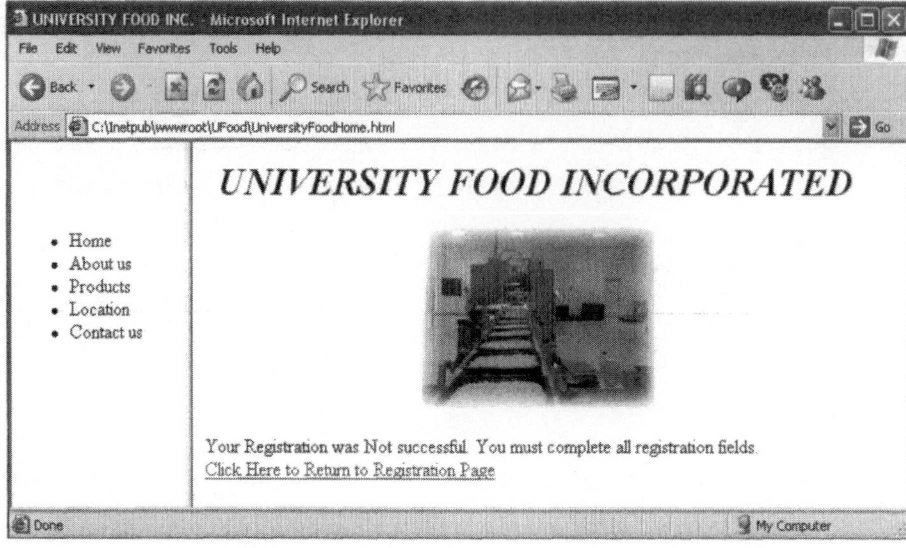

Figure CS8.4 Unsuccessful registration, data missing.

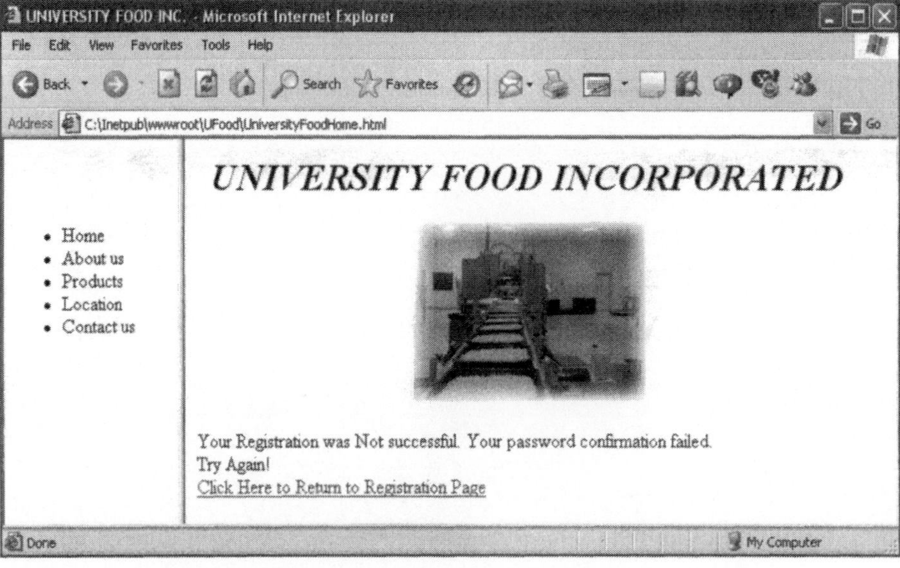

Figure CS8.5 Incompatible passwords failure.

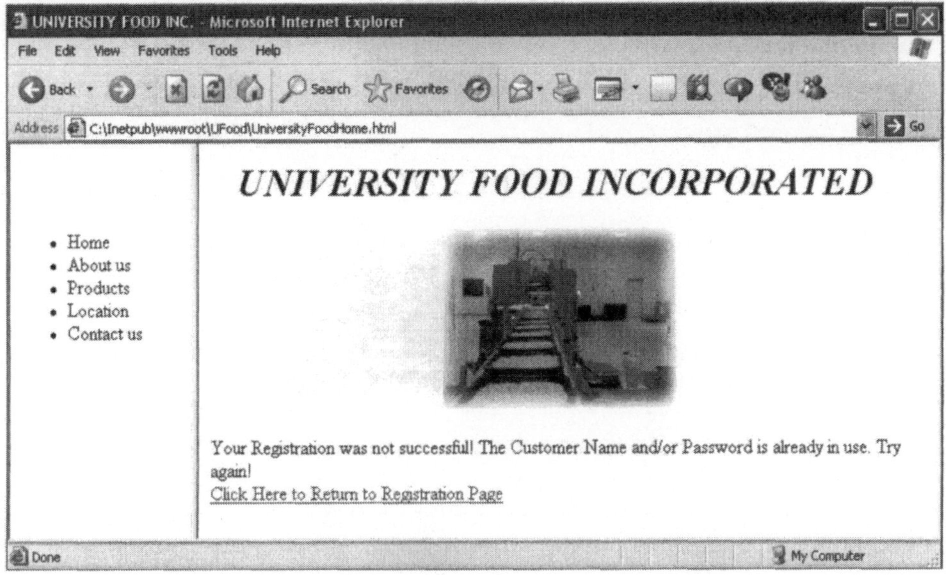

Figure CS8.6 Password already in use.

in Figure CS8.6 to the customer. The anchor tag "here" should bring the customer back to the registration page (Figure CS8.2).

These HTML and ASP files should also be put in the newly created folder under the root directory, wwwRoot.

Exercise 4

When an existing customer uses the page shown in Figure CS8.1 to log in, the login must be checked against the CUSTOMER table. The possibilities are as follows:

- The login is correct.
- The customer name is incorrect.
- The customer name is correct but the password is not correct.

You should program the following responses:

- If the login is correct, return Figure CS8.7, which allows the user to continue on with the session now that he is logged in. If the user clicks on the anchor tag of Figure CS8.7, Figure CS8.8 will be loaded into the browse window.
- If the customer name is incorrect, return a page to the customer that indicates the company name is not in the database. Try again! Provide an anchor tag that allows the customer to return to the home page (Figure CS8.1).
- If the company name is correct but the password is not correct, inform the customer that the password is not correct. Try again! Provide an anchor tag that allows the customer to return to the home page (Figure CS8.1).

There ASP files should also be put into the same folder as the other files under the root directory, wwwRoot.

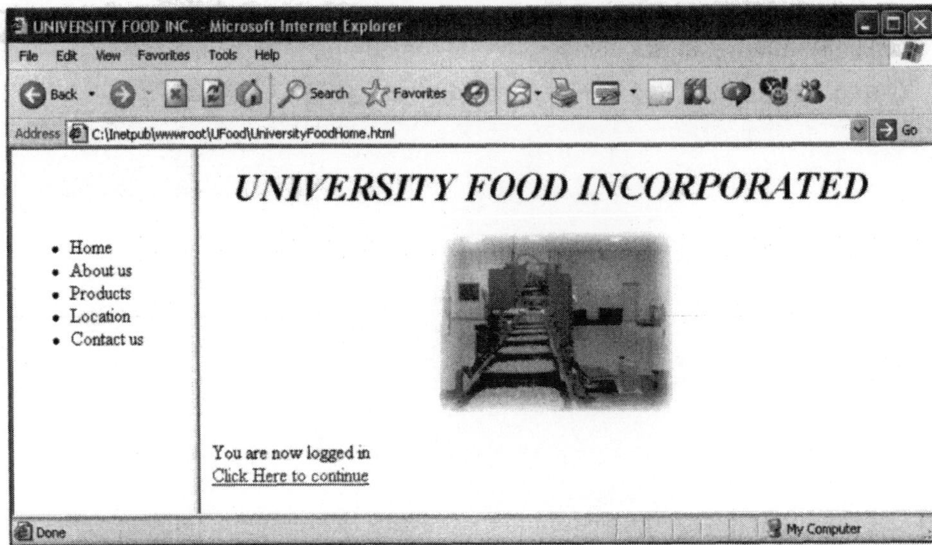

Figure CS8.7 Login correct.

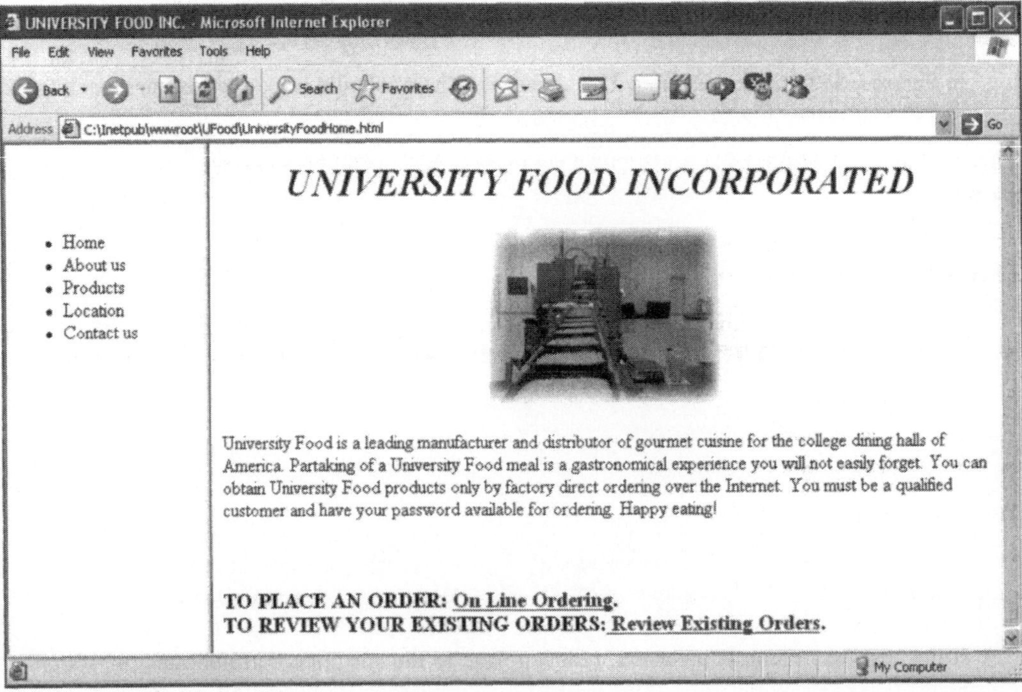

Figure CS8.8 Select option page.

Exercise 5

When a customer chooses online ordering (Figure CS8.8), Figure CS8.9 will be presented. When the user places the order, a Web page should be returned that informs the customer that the order has been accepted and should give the customer the order number for the customer's records. Provide an anchor tag that allows the customer to return to the option page (Figure CS8.8).

Exercise 6

If the customer selects "Review Existing Orders" from the option page (Figure 8.CS.8), a page should be returned that shows all the customer orders by displaying the purchase order numbers, as shown in Figure CS8.10.

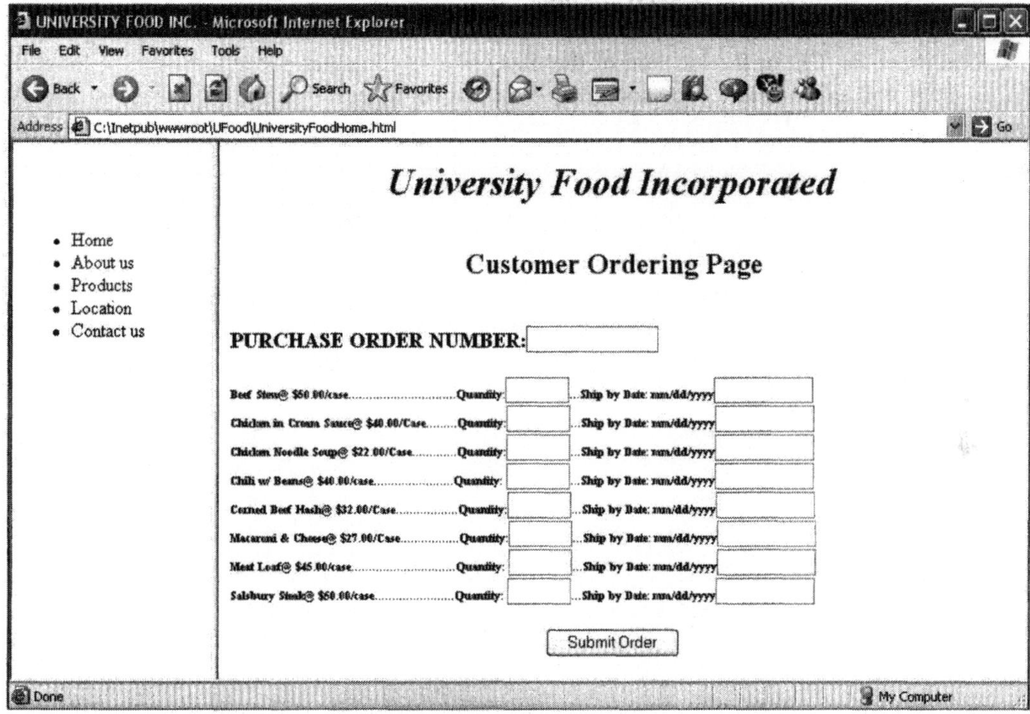

Figure CS8.9 Order entry page.

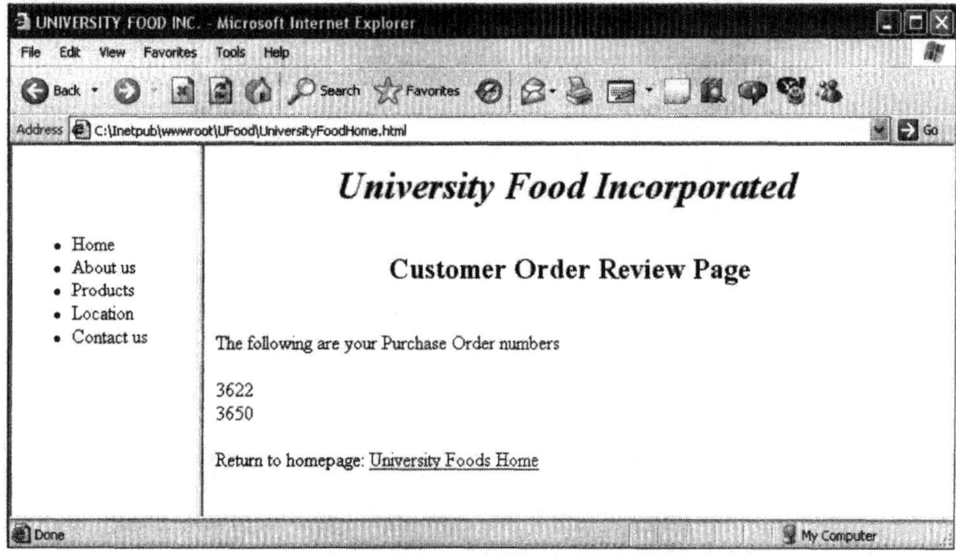

Figure CS8.10 Order review page.

APPENDIX 8A INSTALLING INTERNET INFORMATION SERVER SOFTWARE

Microsoft Internet Information Server (IIS) is a software component that allows the computer on which it is installed to host multiple web pages. It is an add-on that is included with the Windows 2000 or higher Professional CD-ROM. To install IIS, go to the Control Panel and choose Add or Remove Programs. Select ADD or Remove Windows Programs and you will be prompted for the location of the CD or the location of the files on your computer. Once the files are located, Figure 8A.1 will appear. Here you choose the Windows component to install. If it has not already been installed, check Internet Information Services (IIS). Click on Next and continue with the installation.

Once the installation is complete, reboot the computer. When the computer has restarted, check the presence of two important services that were installed by IIS. This is done by going to the Control Panel and clicking on Administrative Tools and then Services. If the installation was done correctly, there are two services that should have been automatically started when the computer was rebooted. They are "IIS Admin" and "World Wide Web Publishing." If they are not running, either there was an error in the installation or a previously installed service is conflicting with IIS. For example, the presence of another database server on the same operating system can cause a conflict. If both services are running, you should be able to program Active Serve Pages (.asp) files and run them.

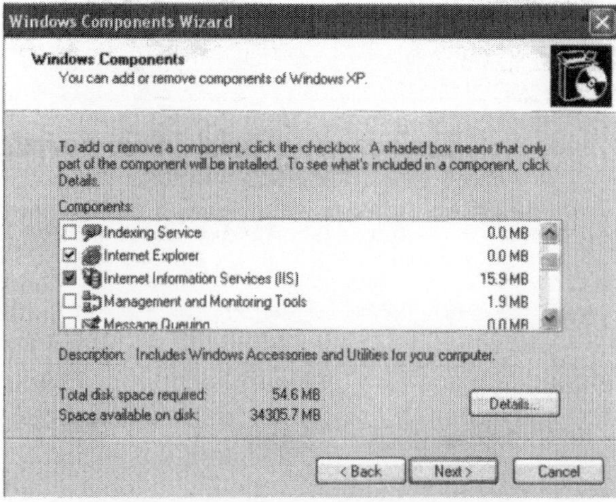

Figure 8A.1 Windows components window.

APPENDIX 8B CREATING A DATA SOURCE NAME

The purpose of a data source name (DSN) is (1) to identify the kind of DBMS that the database is running on and (2) to establish the path to the folder in which the database resides. You only have to create the DSN once, even if you make changes to the database tables, as long as you do not change the database name or its location. The DSN is created in the ODBC Data Source Administrator, an administrative tool of the Windows operating system. The DSN must be created on the machine (server) that is hosting the database and the web page. In the following description, we assume that the operating system on the server is Windows 2000 or Windows XP. To establish a DSN, the user must first log in as the administrator.

To create the DSN, go to Control Panel and open "Administrative tools." Under Administrative tools, double-click on "Data Sources (ODBC)." ODBC stands for "Open Database Connectivity." This will open the ODBC Data Source Administrator, as shown in Figure 8B.1.

In the ODBC Data Source Administrator window, click on the System DSN tab, then click on the Add command button. A new window titled "Create New Data Source" will appear. In this

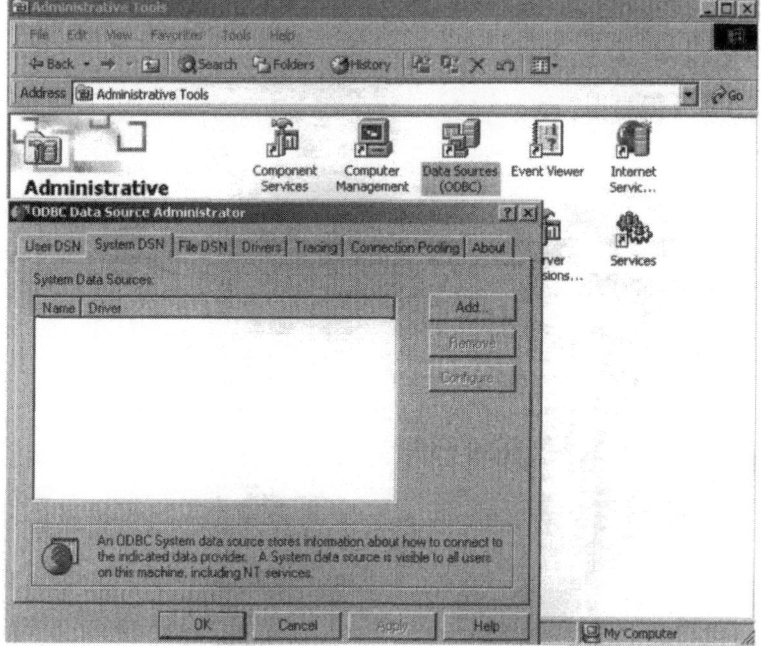

Figure 8B.1 ODBC data source administrator.

window you select the driver for the DBMS used in creating the database. Each DBMS has its own file structures and its own unique way of executing a SELECT command or an INSERT command. The driver allows any generic SQL statement to be properly interpreted for the DBMS in use. Choose "Microsoft Access Driver (*.mdb)," as shown in Figure 8B.2. Click on Finish.

After clicking on Finish, the ODBC Microsoft Access Setup window will appear. Here the user provides the data source name that will be referred to by application programs each time the user connects to the database. The DSN can be any name, including the name of the actual database with which it will be associated. We have named the database ORDERS.mdb. In this illustration, we will use the same name for the DSN (i.e., ORDERS) without the .mdb extension, as shown in Figure 8B.3. After typing the Data Source Name in the appropriate text box, click on the Select command button. This will open the Select Database window. Here the user chooses the .mdb database name to which the associated DSN name will point to at the time a connection is made. The reader should choose ORDERS.mdb, as shown in Figure 8B.4. After closing all open ODBC windows by clicking on the OK buttons, the relationship between the DSN and the database is established. Henceforth, any connection to DSN ORDERS on this machine will connect the application program to the Microsoft Access database ORDERS.mdb.

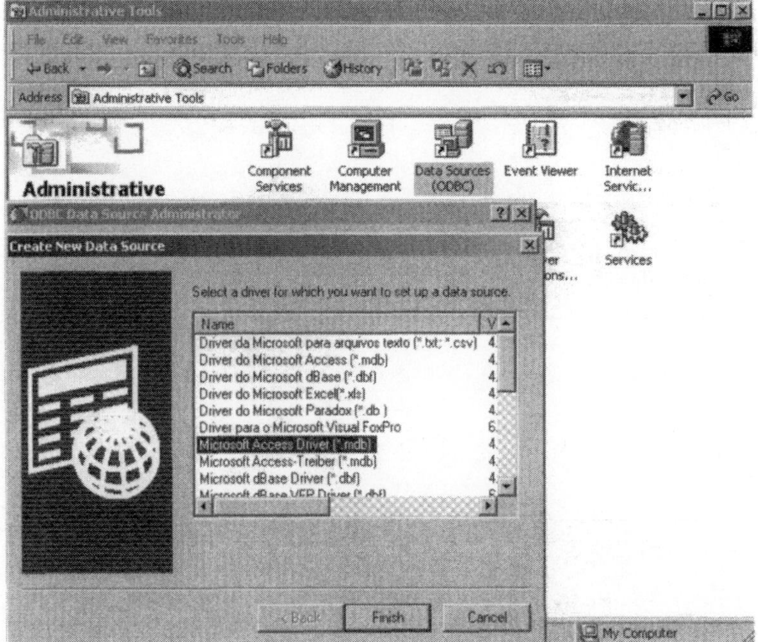

Figure 8B.2 Create new data source window.

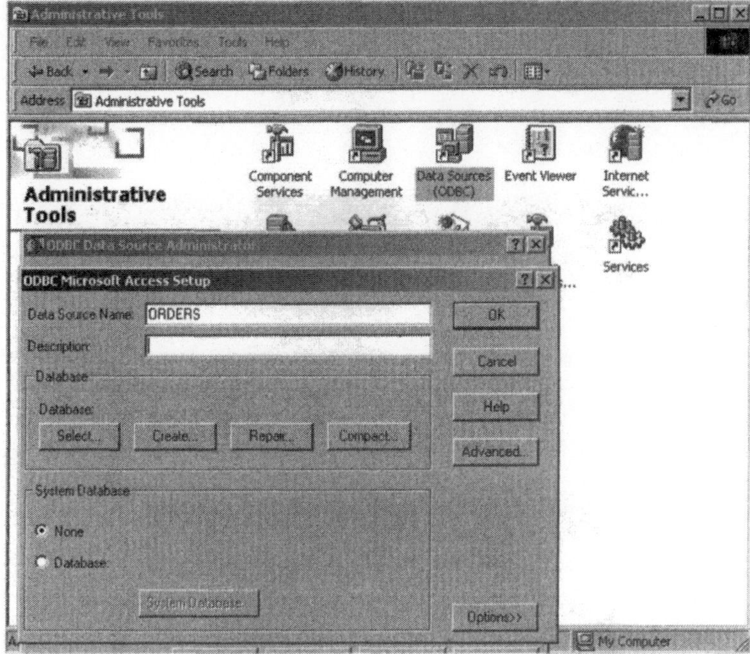

Figure 8B.3 Establishing the DSN in ODBC Microsoft Access setup window.

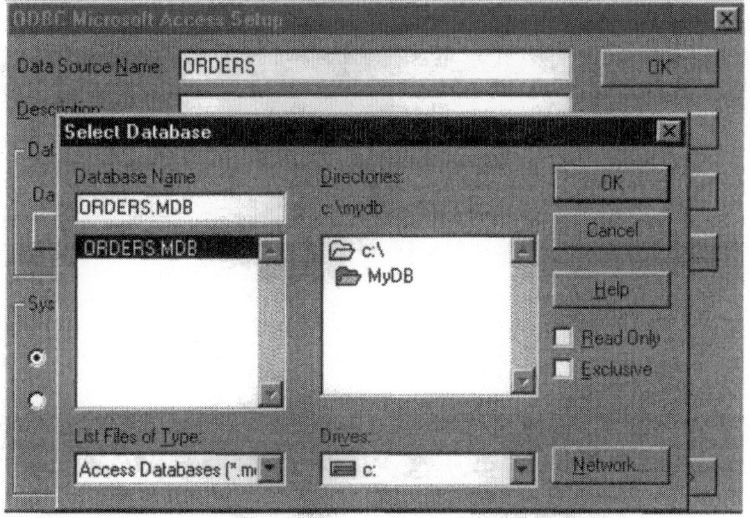

Figure 8B.4 Establishing the link between DSN and the database.

APPENDIX 8C TABLES OF THE ORDERS.MDB DATABASE

CUSTOMER : Table

CUSTOMER_ID	CUST_NAME	CUST_STREET	CUST_CITY	CUST_STATE	CUST_ZIP	PASSWORD	ROUT_NO	ACCT_NO
1	Rutgers University	94 College Ave.	New Brunswick	NJ	08854	RU2000	0809456878923	27152715809
2	University of South Flori	4202 Fowler Ave.	Tampa	Fl	33620	USF22	0999883764827	23345343232
3	New Jersey Institute of	500 Martin Luther Kin	Newark	NJ	07103	NJ1000	8548439384938	23387494505

CUSTOMER_ORDER : Table

ORDER_NO	CUSTOMER_ID	CUST_PO_NO
100	1	3622
101	1	3650
102	2	2400
103	2	2456
104	3	NJI2340

ORDER_DETAIL : Table

ORDER_NO	ORDER_LINE_ITEM	MATERIAL_ID	MATL_QTY	UNITS	PRICE	PROMISED_DEL_DATE	STATUS	STATE
100	1	20	500	Cases	$50.00	1/20/2004	CLOSED	SHIPPED
101	1	25	200	Cases	$45.00	1/25/2004	OPEN	IN PRODUCTION
101	2	243	200	Cases	$27.00	1/25/2004	OPEN	IN PRODUCTION
102	1	30	100	Cases	$40.00	1/26/2004	OPEN	ON ORDER
103	1	243	300	Cases	$27.00	1/26/2004	OPEN	ON ORDER
104	1	20	200	Cases	$50.00	1/27/2004	OPEN	ON ORDER

MATERIAL : Table

MATERIAL_ID	MATL_NAME	UNIT_OF_MEASURE	STD_UNIT_COST
20	Salisbury Steak	Cases	$50.00
233	Chili w/ Beans	Cases	$40.00
243	Macaroni and Cheese	Cases	$27.00
25	Meat Loaf	Cases	$45.00
30	Chicken in Cream Sauce	Cases	$40.00
333	Chicken Noodle Soup	Cases	$22.00
440	Corned Beef Hash	Cases	$32.00
905	Beef Stew	Cases	$50.00

Data Types:

Table Name	Attribute	Data Type
CUSTOMER	CUSTOMER_ID	AutoNumber – Long Integer
	CUST_NAME	Text
	CUST_STREET	Text
	CUST_CITY	Text
	CUST_STATE	Text
	CUST_ZIP	Text
	PASSWORD	Text
	ROUT_NO	Text
	ACCT_NO	Text
CUSTOMER_ORDER	ORDER_NO	Number - Integer
	CUSTOMER_ID	Number - Integer
	CUST_PO_NO	Text
ORDER_DETAIL	ORDER_NO	Number - Integer
	ORDER_LINE_ITEM	Number - Integer
	MATERIAL_ID	Text
	MATL_QTY	Number - Decimal
	UNITS	Text
	PRICE	Currency
	PROMISED_DEL_DATE	Date
	STATUS	Text
	STATE	Text
MATERIAL	MATERIAL_ID	Text
	MATL_NAME	Text
	UNIT_OF_MEASURE	Text
	STD_UNIT_COST	Text

APPENDIX 8D CORRECTING WRITE ERRORS WHEN USING MICROSOFT JET CONNECTION

When using the Microsoft JET connection to access the database from a Web page, the following problem is sometimes encountered. The security permissions setting on the database or folder in which the database resides is limited. When you copy a database from the Web site that accompanies this book, it may initially be "read only" until you change the access level on the database. You do this by browsing to the database file using Windows Explorer, right-clicking on the database file, and choosing "Properties." In the Properties window, remove the check mark from "read only." Click on OK, and the database file will be read/write when you open it in Access.

Even if you have established write permission to the database when it is opened in Access DBMS, you can still have problems writing to it from an ASP file. When the Web browser tries to send data for entry into the database, the Web server sees the Web browser as an Internet user. If the Internet user does not have write permission, the Web server will return an error statement that the database is "read only," even though you can write to it from Access DBMS. The **Internet user** has the server account **IUSR**. When this error occurs, this server account may need to be given write permission to the database folder and the database file. This appendix illustrates how to give IUSR write privileges.

Follow these steps to resolve this issue:

1. Sign on the computer as the administrator. Use Windows Explorer to find the folder that contains the database file(s). Right-click on the project folder and select "Properties."

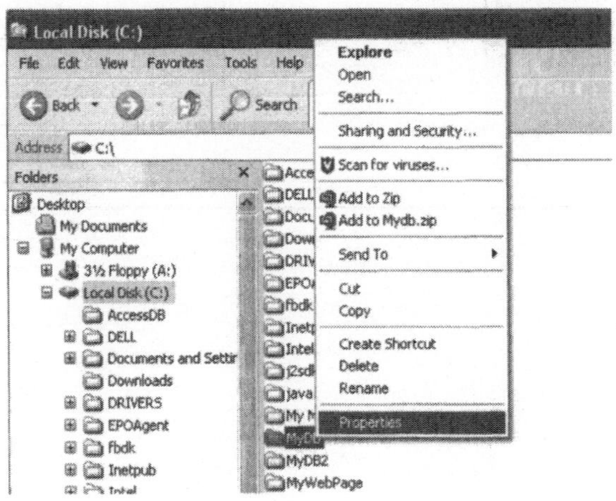

2. Uncheck the Read-only property.

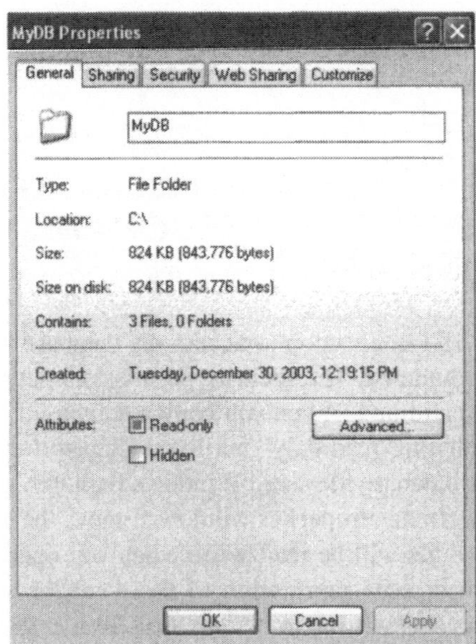

3. Click on the Security tab in the Properties window. (If you don't see a Security tab, close the Properties window, open any folder, select Tools → Folder Options → View, and uncheck the option "Use simple file sharing.")
4. In the security tab, click Add, which will open the Select Users and Groups window as shown next. This is where you will select Internet User for write permissions.

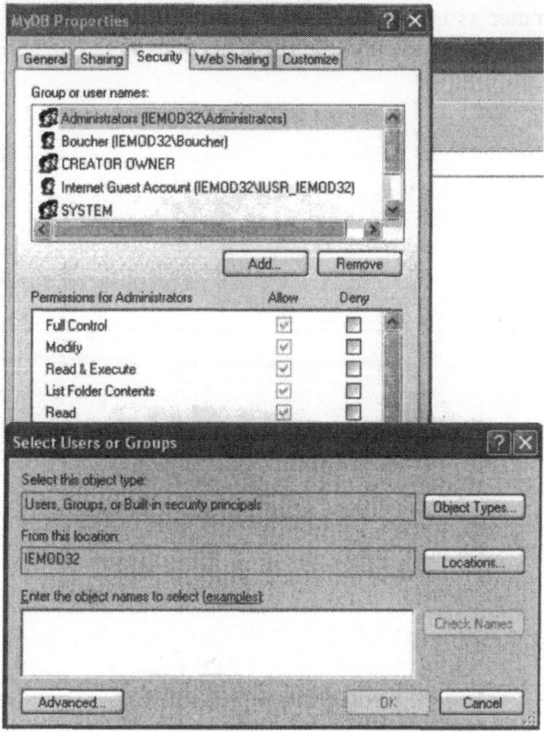

5. Click on Advanced in the Select Users or Groups window. Then click on Find Now in the Select Users or Groups window. This window will then display a list of users as shown:

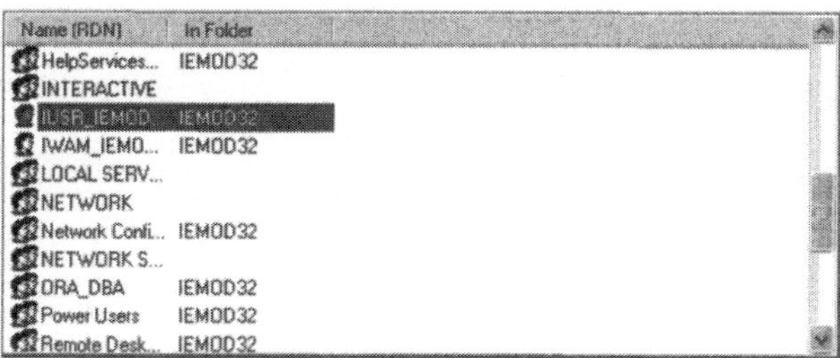

6. Find the user IUSR_<COMPUTERNAME>. Click on OK. The user will then appear in the memo box of the Select Users or Groups window under "Enter the object names to select." At this point, you have selected Internet User. Click on OK and you will return to the Database Properties window.

7. The last step is to give write privileges to that user. Select the Write checkbox in the Allow column to assign write permission to the IUSR account as shown here. Click on OK and you are finished.

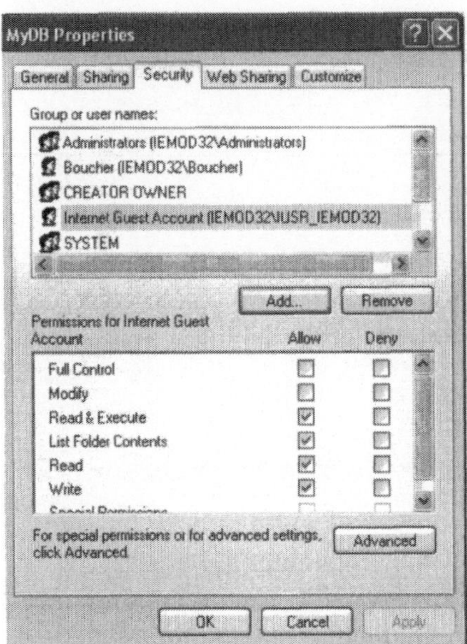

Chapter 9

Unified Modeling Language

9.1 INTRODUCTION

Chapters 4 and 5 emphasized a particular process for the conceptual-level design of an information system. This process includes an activity-centric design component (IDEF0 or DFD) and a data-centric design component (IDEF1X). In Chapter 6, we showed how they fit together in designing forms for the information system users in the enterprise. The methods followed in Chapters 4 and 5 are considered traditional in the sense that they have been around for a couple of decades and are widely used in the information system development community. However, since the early 1990s or so, new approaches have emerged and have achieved some level of success due in part to the introduction of new modeling frameworks and new enhancements that are not part of the traditional framework. Some examples are the Zackman framework (Zackman, 1987; Billo et al., 2006) and the Unified Modeling Language (Jacobson et al., 1999; Roques, 2001; Rumbaugh et al., 1999).

One of the most significant of the newer approaches owes its genesis to developments in software engineering, namely ***object-oriented (OO)*** software development. The object-oriented approach stresses the encapsulation of data and procedures within segments of software code called objects. Instead of separating programming code from data files used by the code, the two are bound together in a module, the object. Objects are instantiations of classes, which are generic templates of the objects. The object has the characteristics of the class from which it is created. In the course of earlier chapters (Chapters 6 and 8), we introduced some concepts from object-oriented programming. In fact, many of the conceptual underpinnings of Microsoft Access and its supporting programming language, Visual Basic for Applications (VBA), are based on classes and objects.

A specific set of object-oriented design tools that evolved from the trend in object thinking is known as ***Unified Modeling Language (UML)***. UML, like the IDEF framework, is a set of tools developed to assist analysts in uncovering the important features of a design project, finally arriving at a set of models that will be used to design, document, and implement the project, whether it is an information system or other software development project. In database design projects, UML is often thought of as a tool for object-oriented databases. As we shall see, the modeling components of UML have special features that complement the object-oriented paradigm. However, even if the target database management system (DBMS) is a relational database, the UML design tools can be used.

Although UML is attracting much interest among information system professionals, its use is in its infancy. It is our observation that, at the time of this writing, traditional structured analysis methods, such as IDEF and data flow diagrams, are still widely used. Hence, in this chapter we introduce UML and compare it to IDEF in order to orient the reader to this emerging technology.

9.2 OBJECT-ORIENTED DESIGN CONCEPTS

To introduce UML modeling formalisms, some background in object-oriented terminology is required. In traditional programming, code is written as functions that operate on input data and (usually) provide some output to a program. Functions are called from a main program in some sequence to accomplish the work required by the program. Among the functions required may be one to open a database or perform a structured query language (SQL) query on the contents of a database. The overall flow of control lies in the main program, which calls the required functions sequentially. This form of program design has traditionally been called "structured programming."

Consider a sensor that provides a temperature reading to a software program running on a computer, such as that shown in Figure 9.1a. The computer program may have a function called

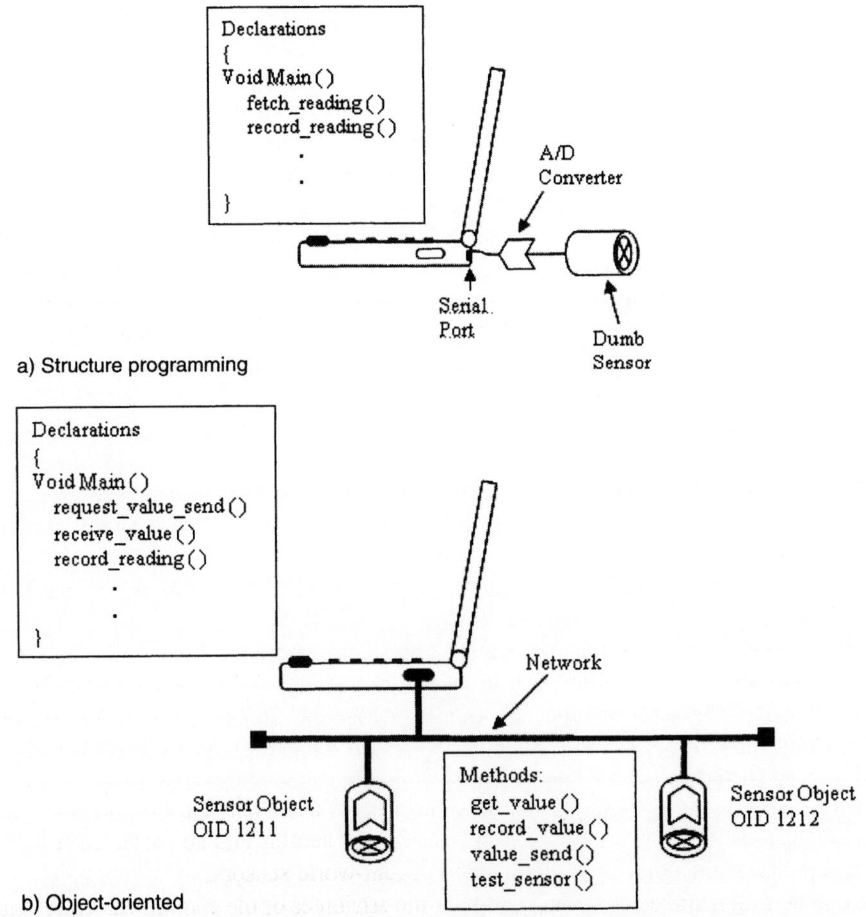

Figure 9.1 Illustrative comparison of (a) traditional versus (b) object-oriented system design.

fetch_reading(), which reads the current value of the sensor. The sensor data are always present at the port of the computer for reading whenever the function is called. The program resident on the computer is a series of functions that collect the data and record it or do some computation with it. The sensor just converts an environmental condition into a voltage that is then converted to a series of bits that a computer can read. The term "dumb sensor" is appropriate since the sensor has no oversight of its own functionality.

Object-oriented thinking uses a different paradigm. In ***object-oriented design (OOD)***, program modules are designed and built to imitate the behavior of real-world objects. For example, a sensor is a real-world object that collects sensory data in a real-world system. In OOD, the software that defines the functionality of the object is encapsulated within the object. The object "sensor" has its own program logic and its own thread of control. In a physical device, such as a sensor, this usually requires a microprocessor that is resident within the object. This arrangement is illustrated in Figure 9.1b. In this implementation, the computer (an object itself) requests a reading of the current value of the sensor. The sensor, accepting the request, runs a software component (called a method) to read the current value and then calls another software component (another method) to send the value to the computer. The thread of control is distributed to each object instead of being centralized in one main program. The sensor may also have methods that allow it to diagnose problems in its functioning, for example, diagnosing reading errors that may occur due to electronic malfunctions. In that case, the term "smart sensor" seems appropriate since the sensor object has oversight of its own functionality.

The object-oriented approach creates software in modules (objects). Each object is an independent entity of the system in which it resides. It has a set of tasks it can perform (the methods) and encapsulated data that it can access, maintain, and use. In summary, object-oriented design helps create autonomous structures (the objects) that represent real-world entities and have the ability to interact with other objects.

The motivation for OOD methods in the software development community is economic. If software can be designed as objects, the objects can be used and reused to design systems in a plug-and-play fashion. The programmer no longer has to deal with writing detailed code. Rather, the system is designed by making appropriate connections between objects that must communicate with each other. This reduces development time and enhances system reliability.

There are a number of concepts in object-oriented design that, when taken together, define whether or not an application is truly object-oriented. Here we briefly review some of these concepts.

The concept of class is an important feature of object-oriented design. A ***class*** is a collection of objects with shared structure and behavior. In OOD, the object is a specific individual member of a class. Class is the generic description of the collection of objects. So, for example, class would be the abstract concept of a tree, whereas a specific tree in your backyard, a member of that class, is an object.

With regard to Figure 9.1b, the sensors shown in the example are objects. On the other hand, Figure 9.2 shows a hierarchy of the sensor class. The most generic concept, SENSOR, is shown at the top. But there are also subclasses. Sensors can be divided into BINARY class, which measures only two states (for example, above a critical temperature and below a critical temperature), and CONTINUOUS class, which measures an analog value (for example, –10 to 200 degrees Fahrenheit). These subclasses can be further divided into other subclasses, such as continuous temperature sensor, continuous position sensor, continuous velocity sensor, and so forth. Each of these class descriptions refers to a concept of a collection of real-world sensors.

The object is an instance of the class and has the attributes of the class of which it is an instance. One of the attributes is its object ID (OID), which is a unique identification of the object. No two

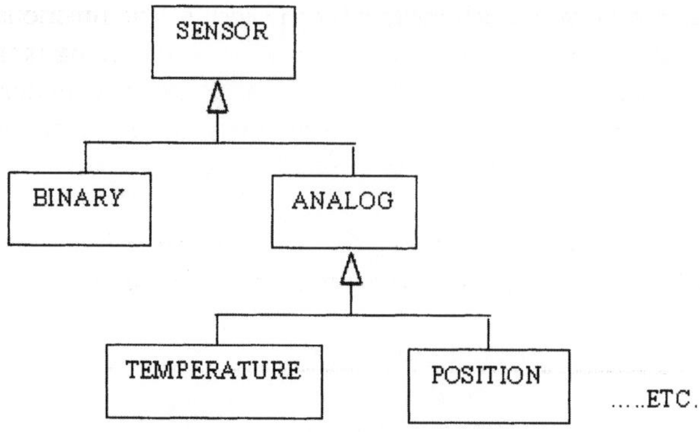

Figure 9.2 Sensor class hierarchy.

objects of the class can have the same OID. The OID has something in common with the concept of a social security number for the class of humans called U.S. residents.

The concept of ***inheritance*** has a special meaning in OOD. Inheritance is used to describe a class and its subclasses. For example, the superclass SENSOR has two subclasses, BINARY and ANALOG. BINARY and ANALOG are subclasses of SENSOR, and they inherit some methods and attributes from the superclass. However, there may be some methods and attributes of the superclass that are not inherited by the subclass. An object, on the other hand, is a realization of a class. An object of the TEMPERATURE class—for example, OID 1211 in Figure 9.1b—always has all the attributes and methods of the TEMPERATURE class.

The concepts of class and object have many real-world counterparts. For example, a cat is a subclass of another abstraction, animals (which is itself a class). However, the cat in your house, Felix, is a unique individual and a real thing. Felix is an object.

An ***attribute*** is a characteristic of an object and class. The continuous sensor class may have an attribute called measurement_range, which indicates the domain of the sensor reading. An object of that sensor class—for example, temperature sensor OID 1211—will have that same attribute as one of its characteristics and a ***value*** for the attribute, say –10 to 200 degrees Fahrenheit. Similarly, the cat class has an attribute color. The object in your house, Felix the cat, has an attribute color with a value black. Classes have attributes, but no values. Objects of a class have the attributes of the class and values for those attributes.

We previously encountered the term "attribute" in the entity-relationship (E-R) model. In that case, the attribute values are the values assigned to instances of the entity set. Although there are some distinctions that we will discuss later, the attribute value of an object is conceptually similar to the attribute value of an entity.

Objects operate under their own thread of control. An object evolves through a series of ***states***, which describe the dynamic behavior of the object. Consider, for example, the temperature sensor OID 1211 object. When there is no power to the sensor, it is in a dormant state. When the sensor is under power, it may be reading temperature (one state), idle or doing nothing (another state), or sending temperature values to the computer on request (yet another state). The states are a representation of the various phases the object passes through during its operating cycle. Similarly, Felix the cat is usually in the sleeping state, but when awake he could be in the eating state, moving state, and so on. These analogies are quite general and are a part of the description of an object's behavior.

An object has a set of software functions that it can execute. The functions are often associated with a state of the object, but this is not necessary. The important thing is that an object can use these functions whenever it is appropriate. In the OO world, these functions are called ***methods***. For example, temperature sensor OID 1211 in Figure 9.1b has a method called get_value( ). In this method, it simply assigns the current value of the temperature reading to a variable. Perhaps this is done periodically—for example, every second. We say that the sensor ***invokes*** the method every second. An object's methods may be invoked by the object itself. Felix the cat has methods that may be self-invoked. For example, the purr method or the meow method comes to mind.

An object's methods may also be invoked by other objects. Figure 9.1b shows the relationship between the computer and the sensor objects. The computer may request that the most current reading be sent to it from the sensor, request_value_send( ). This would invoke the sensor object's value_send( ) method, a software function that passes the current value from the sensor to the computer, usually over a network. Felix the cat has methods that can be invoked by other objects. For example, the scream method is invoked when you (another object) step on his tail.

All of this interest in object-oriented programming is encouraged by the ease with which a programmer can create an instance of a class. Once a class is defined and programmed in software with its attributes and methods, any number of objects can be created as instances of that class. Each object will have the attributes, methods, and state behavior of the class. By assigning values to the attributes (including the system assigned OID), the object becomes a unique implementation of the class. Therefore, software system design can be performed by creating objects and connections among objects that need to communicate with each other in the system. In the OO world, there is no need to focus on the details of programming an individual object since that was done when the object classes were created. The designer can focus on the system, how the objects are situated in the system, and how the objects communicate.

Object-oriented programming was a motivating force for the development of design architecture tools for OO design projects. This design architecture is known as Unified Modeling Language. Though based on OO concepts, UML is finding use in software and database development even when it is not based on OO technologies, like C++, Java, and object-oriented databases. Therefore, we present UML as an emerging alternative to traditional conceptual modeling techniques and illustrate its application for object-oriented design as well as more traditional information system design projects.

9.3 UML DESIGN FORMALISMS

UML supports nine application specification design models in the form of UML diagrams. They can be classified as either structural diagrams or behavioral diagrams. The ***structural diagrams*** define the static relationships among components of the architecture. The ***behavioral diagrams*** define the dynamics of components of the design. The categorization of diagrams is as follows:

- *Structural.* Class diagrams, object diagrams, component diagrams, and deployment diagrams
- *Behavioral.* Use case diagrams, sequence diagrams, collaboration diagrams, state chart diagrams, and activity diagrams

Some of these diagrams are more closely associated with software development than information system design. We shall focus on the most relevant design diagrams. However, a brief definition of each diagram type is as follows.

1. *Class diagram.* A graphic representation of a collection of model elements, such as classes, types, and relationships. It includes a conceptual model of database design elements.

2. *Object diagram.* Shows objects and their relationships at a point in time. It is a special case of a class diagram showing instances of classes and their relationships.
3. *Component diagram.* A component is a physical piece of implementation of a system, such as a module of software code. Components can be connected to other components through interfaces. A component diagram is a diagram that shows the organization and dependencies among component types.
4. *Deployment diagram.* When components are used in a system, they are deployed as component instances. A deployment diagram shows the component instances and the objects that are associated with them. Therefore, a deployment diagram shows instances of a component, while a component diagram shows the definition of component types.
5. *Use case diagram.* Use case is a function that the system can perform when it interacts with outside actors. The concept of an "actor" is an abstraction of a person or thing outside the system that interacts directly with the system—for example, a database user. The use case diagram shows the functional interaction between actors and the system when performing a user function.
6. *Sequence diagram.* A sequence diagram shows the interaction of the actor with the objects in the system and the timing of the interactions. It gives a dynamic view by time-sequencing the flow of messages among objects.
7. *Collaboration diagram.* A collaboration is an arrangement of objects and links that interact to implement a behavior, such as a use case. The collaboration diagram illustrates that interaction.
8. *State chart diagram.* A state machine is a sequence of states and transitions between states that an object goes through in response to events in its life cycle. A state chart diagram shows the state machine, including nested component states, in a simple diagrammatic form. State chart diagrams depict the dynamics of objects.
9. *Activity diagram.* An activity is an execution of a process, which could be within the system or a real-world function that interacts with the system. An activity diagram shows a graph that models the activity. Activity diagrams are very useful in modeling business operations and workflows.

It is evident from this list that UML is rich in tools for documenting a software design project or an information system design project. In fact, several vendors of computer-aided software engineering (CASE) tools for UML exist, and some of them are cited in the bibliography at the end of this book. As previously stated, in this chapter we will focus on a subset of UML diagrams that are useful in information system design, and we will compare them with the traditional methods discussed in Chapters 4 and 5. The relevant subset includes use case diagrams, sequence diagrams, activity diagrams, state chart diagrams, and class diagrams.

9.4 ARCHITECTURE DESIGN USING UML

UML supports software development and use throughout the life cycle of the software application. According to UML proponents, development begins with the elicitation of requirements, proceeds to specification and logical modeling (analysis), then to architectural modeling (design) and implementation and coding, and finally to testing and maintenance.

In traditional architecture specification, as in the case of IDEF, elicitation of requirements, design, and analysis is done from the top down based on functional decomposition, combined with separate data model design. UML permits functional analysis and system (database) specification to evolve simultaneously with the goal of identifying objects and object classes, their relationships, and their methods. Once an object class is identified, it can be worked on (encoded) independently of other classes in the system.

Although there is no hard and fast rule for using UML in identifying objects and classes, it is usual to first uncover ways in which the system is expected to interact with its environment by creating use cases. The use cases should suggest objects, their attributes, and their methods. Once objects are identified, more specificity can be added using sequence diagrams, which model the interaction among objects and the timing of that interaction. The execution of code by an object implies the existence of a method. More detailed understanding of object methods is gained by modeling the behavior of an object as an activity diagram and as a state chart diagram. These show the various activities of the system involved in execution of the method.

With respect to databases, objects are synonymous with entities. However, in addition to having attributes, objects also have methods. The entity set, called a class diagram in UML, brings together the concepts of object, attribute, and method and provides the logical database design.

The following sections focus on these diagrams for modeling an application. An application for the sale of theater tickets will be shown first, to be followed later by an example from Chapters 4 and 5.

9.4.1 Use Case Diagram

A ***use case diagram*** is a description of how users interact with the system. To develop a use case diagram, the analyst must identify three entities:

- The system boundary and interfaces
- The actors who use the system
- The use cases (i.e., the functions that the actor calls upon the system to perform)

Consider Figure 9.3, which shows a use case diagram for a hypothetical online theater ticket sales system. The diagram shows the three components. The use case is depicted as an oval with the use case name within the oval, in this example "make sale." The system is shown as a box that makes explicit the boundaries between the use case and its environment.

The ***system*** is that which is being designed. The analysts use UML models to document their design. For the purposes of this textbook, the system is usually an information system. An information system can have many components, of which the database is perhaps the most important. But, in addition, it may have sensors and data acquisition devices, such as bar code scanners. It may also have intelligent agents that serve as intermediaries between the user and the database.

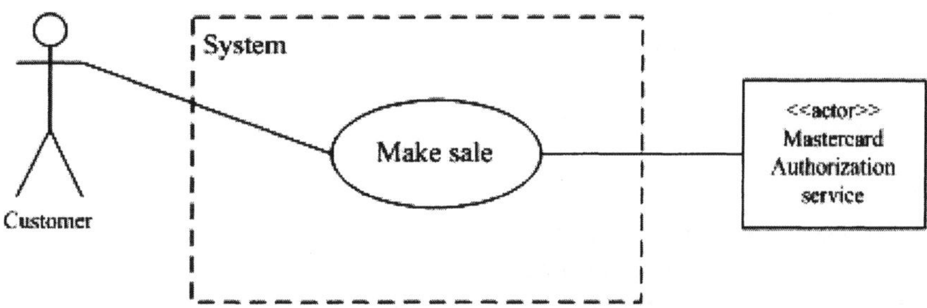

Figure 9.3 Make sale use case diagram.

An ***agent*** is a software component that is situated in some environment, can perceive its environment, and is capable of autonomous action to meet the design objectives of the application (Farahvash and Boucher, 2004; Weiss, 1999). If you have called Amtrak recently to obtain a train reservation, you probably talked with the reservation agent, Julie (1-800-USA-RAIL). Julie is an intelligent software agent with sophisticated voice recognition capabilities that allow it to hold a conversation with a customer. For Amtrak, Julie is part of the software system that is the focus of the design and would be included within the system box on a use case diagram.

In the case of online theater ticket sales, we assume the existence of a Web server with forms and software to serve as an interface. We further assume VBScript or JScript code that accesses the database and communicates with the actors. We will refer to this interface as "Ticket Agent." Ticket Agent is part of the software system for theater ticket sales and would be included within the system box on a use case diagram.

Finally, there are the actors. The ***actors*** are external to the system but interact with it. A common way of showing an actor when it represents a person type is to use a stick figure, with the role of the actor shown below the stick figure. This is illustrated as the "customer" in Figure 9.3. An actor should be described in terms of the "role" it plays. The actor may represent many different groups of people, but it is the logical role it plays when interfacing with the system that is important. Returning to the Amtrak example, the customer interacts with the system (which includes Julie) when he or she calls for a reservation. So the theater ticket use case diagram could also apply to Amtrak ticket sales. However, suppose that Julie were replaced with a real ticket agent. The customer would telephone the real ticket agent and ask for a reservation. The ticket agent would then use the computer keyboard to interface with the system in order to key in the reservation for the customer. The customer would not interact with the system. For this case, the actor in Figure 9.3 would be the ticket agent, not the customer. So the use case diagram considers the actors that are *directly interfacing* with the system.

Another way to represent an actor, particularly if it is not a person, is by using a box with the word <<actor>> in guillemets. In the case of online theater ticket sales, a credit card is usually used to buy the ticket. One of the actors that interact with the system is the authorization service of the credit card company. The credit card authorization service is another automated system (belonging to MasterCard in this illustration). However, to the theater ticket sales system, it is external and, therefore, an actor.

The use case diagram of Figure 9.3 tells a story. One of the uses of the system is to make a sale. There are two actors (external entities) involved. One is the customer and the other is the credit authorization entity. When documenting a use case, it is important to write an explanation of the use case that will help convey the message of the diagram. This document can be in the form of a text file that describes the activities that take place in executing the use case. These descriptions are called scenarios. Scenarios will be discussed in the following section.

A use case diagram can contain several use cases, depending on the various functions to be supported by the system. Another use case has been added in Figure 9.4. This is the case in which customers go online and query the system for the availability of seats for a performance. This is often done without following through to purchase a seat. For this use case, the customer is the only actor involved.

9.4.2 Sequence Diagram

A use case diagram is a broad description of the functions to be performed by the system and the external users (actors) of those functions. To obtain a more refined understanding of the various

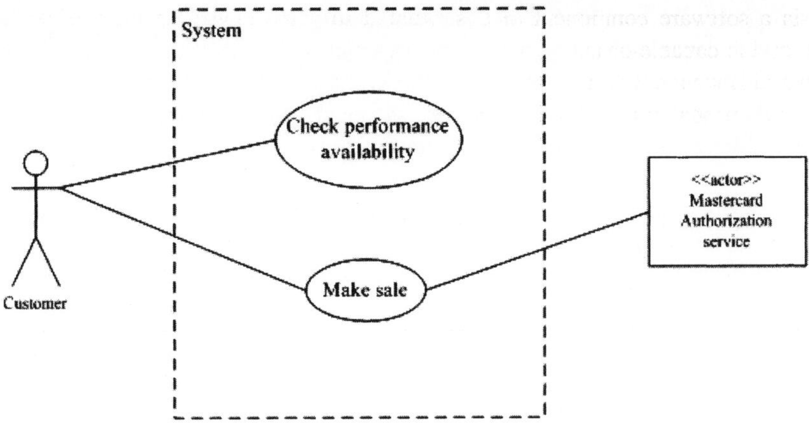

Figure 9.4 Theatre ticket use case diagram.

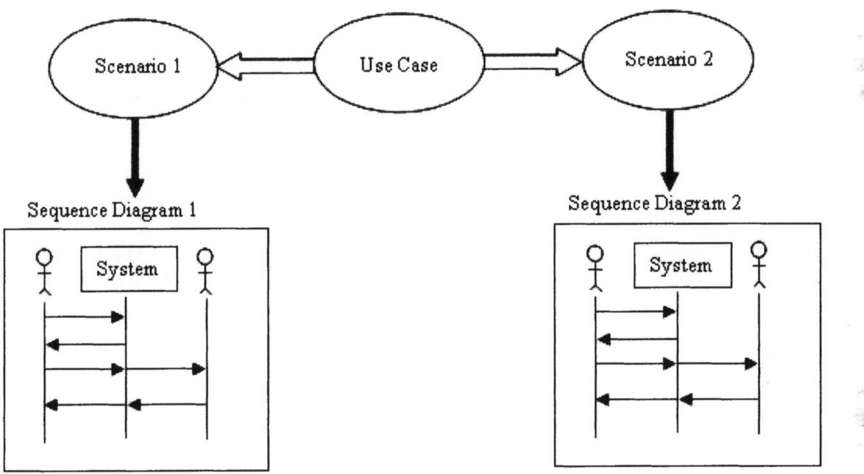

Figure 9.5 Transition from use cases to sequence diagrams.

possible interactions that may occur between the actors and the system during the execution of a use case, it is necessary to construct scenarios. A ***scenario*** is a description of a sequence of actions that illustrates the execution of a use case instance. In most use cases, several possible scenarios may occur. For example, in trying to make a ticket sale, it is possible that the sale will be made without complication or that some complication will occur, for example, the submitted credit card information could be invalid. The normal use case, in which all goes well, is one scenario; the case of the rejected credit card is another scenario. Figure 9.5 illustrates the fact that out of one use case there may be one or more possible scenarios.

One way to document a scenario is to write a text file description of what will happen under a given scenario. The text description should document all the steps that the actors will perform and the associated system responses. For example, consider a "normal" scenario in which a ticket sale is made. A possible description is as follows.

Scenario 1: Make Sale Use Case Normal Scenario

1. The customer requests a performance and seating preference from the ticket sale system.
2. The system verifies whether the performance and seating are available.
3. The performance and seating are available. The system offers the ticket to the customer.
4. The customer accepts the ticket.
5. The system temporarily places the seating for that performance "on hold" while it processes the sale.
6. The system requests credit card information from the customer.
7. The customer supplies credit card information.
8. The system requests authorization to debit the credit card from the MasterCard authorization service (we will assume that MasterCard is the only acceptable form of payment).
9. The system receives the authorization approval.
10. The system confirms the sale to the customer and requests final acceptance of the transaction by the customer.
11. The customer accepts the transaction.
12. The system bills the price of the ticket(s) to MasterCard.
13. MasterCard accepts the charge.
14. The system changes the status of the ticket from "on hold" to "sold."
15. The system sends a final confirmation to the customer, which includes the sale transaction number that the customer will use to pick up the ticket at the box office prior to the performance (we don't mail tickets).

Obviously, in this scenario, several twists and turns could have been introduced to upset the "normal" transaction sequence. The requested seating might not have been available (step 2), the credit card might have been invalid (step 9), or the customer might have changed his or her mind and canceled the transaction (step 11). These deviations would all have led to different scenarios.

Once a typical scenario is documented, it is straightforward to convert it to a sequence diagram. The relationship between scenarios and sequence diagrams is illustrated in Figure 9.5. A ***sequence diagram*** is a two-dimensional chart that displays the interaction between actors and the system across the horizontal direction and the sequencing, or timing, of that interaction in the vertical direction. The sequence diagram is basically a pictorial model of a scenario that was previously documented as text, as was done with scenario 1.

Figure 9.6 is the sequence diagram of scenario 1. The objects are arranged across the top of the diagram. In this case, they include the customer, system, and MasterCard authorization service. The vertical dashed lines beneath the objects are called *lifelines*. The lifeline shows the duration, or persistence, of the object during the scenario. The arcs with arrows indicate the direction of flow of the communication between objects. Some system arrows are self-loops because the system is communicating with itself. The precedence of the arrows from top to bottom shows the order (sequence) in which the communication transactions occur. This is why it is called a sequence diagram.

Review the text description for scenario 1 and compare the 15 steps involved in scenario 1 with the sequence diagram shown in Figure 9.6. You will find that they correspond to the 15 arcs on the sequence diagram. For a particular use case, there can be one or more sequence diagrams based on the possible scenarios that exist. You will be asked to create other scenarios and sequence diagrams for the "make sale" use case in exercises at the end of this chapter.

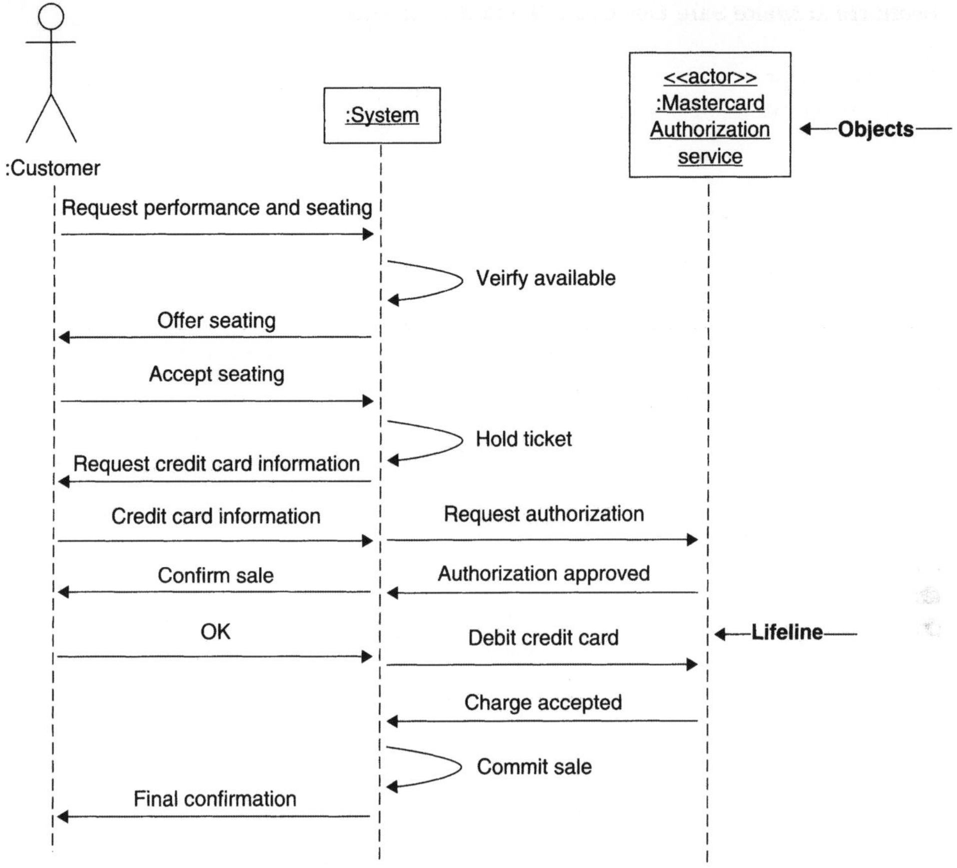

Figure 9.6 Sequence diagram of scenario 1.

Figure 9.6 is a fairly high-level view of a sequence diagram. Embedded in the SYSTEM object are lower-level objects. Sequence diagrams can be opened up to uncover more objects and show more detail. In fact, this is part of the discovery process that occurs during the UML modeling phase, which intends to uncover object classes that can be coded independently of one another.

Figure 9.7 gives a more detailed view of the system by introducing the two main object classes of which the "system" is composed. When a user goes to a Web page for online sales, the customer enters a dialog with a system software component that resides on the Web page server. This software component deals with gathering information from the customer. You created similar software when you completed the ASP exercises in Chapter 8. Let us refer to this software module as a Ticket Agent object. It is one component of what we have referred to as the System object in Figure 9.6. Information on performances, seating, ticket prices, and sales is kept in a database, which is another object class. The Ticket Agent object stores and retrieves information by sending SQL queries to the database server. Thus, in Figure 9.7, System object is broken down into two components: Ticket Agent and Database Server. The self-loops of Figure 9.6 are shown in Figure 9.7 as queries (functions) and responses between the ticket agent and the database server. Figure 9.7 is a more detailed view of scenario 1 as presented in the sequence diagram of Figure 9.6.

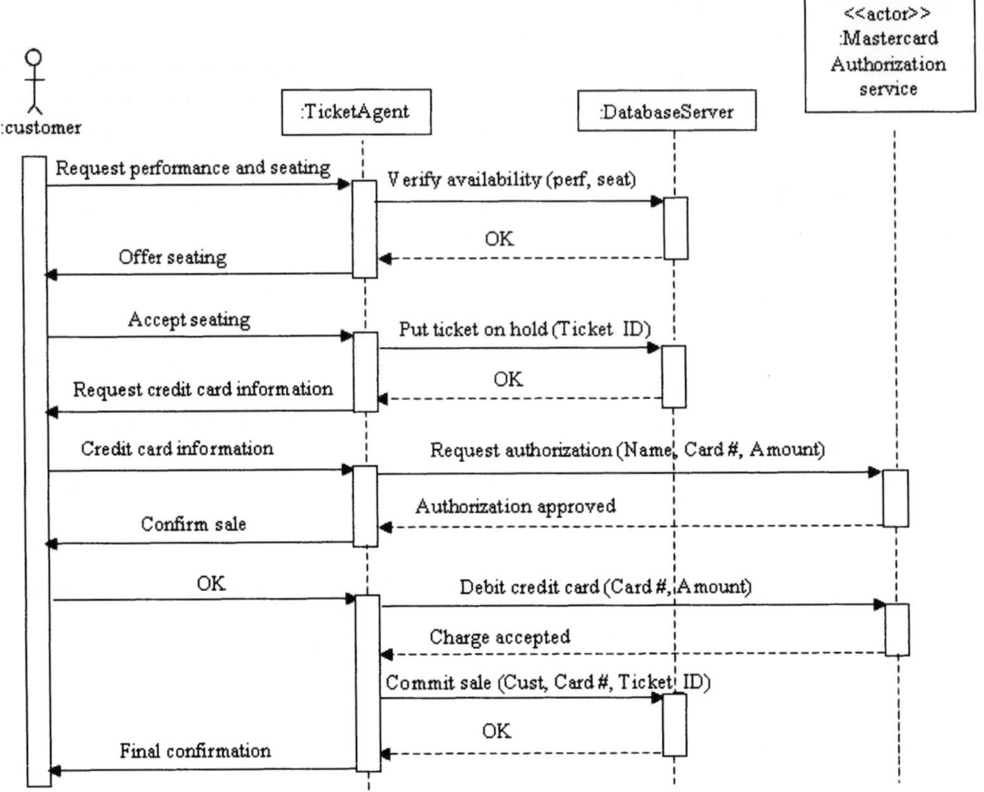

Figure 9.7 Sequence diagram with more details of scenario 1.

Additional enhancements have been added in Figure 9.7. The vertical rectangle boxes along the lifeline are activation boxes. ***Activation boxes*** indicate the beginning and end of the activation of a procedure. The procedure usually begins with a message between objects, shown as an arrow having a solid line. The procedure ends with a return arrow to the initiator. Messages may be communications or they may be method calls. If a method call returns a value, this is shown as an arrow with a dash line. Nested procedures are also possible, as shown in Figure 9.7. For example, when Customer requests performance and seating information, Ticket Agent initiates a nested method call (verify availability ( )). The nested call ends when a time value is returned (OK). Ticket Agent then offers the ticket to the customer. At that point this procedure ends.

If an analyst were to have a sequence diagram for each of the possible scenarios of the "make sale" use case, she or he would then know all the possible inputs and outputs of the ticket agent object, including all the database queries required in response to customer requests. From this information, the analyst could begin writing the software for the Ticket Agent class. Once the Ticket Agent class is designed and encoded as a module, it could be reused in any similar project for online ticket sales.

Figure 9.7 could be further elaborated by introducing entity objects (tables) that will exist within the database. However, we shall stop at this level of abstraction for the time being. The basic features of the sequence diagram are sufficiently illustrated in Figures 9.6 and 9.7.

9.4.3 Activity Diagram

One of the behavioral design formalisms of UML is the activity diagram. An ***activity diagram*** is basically a flowchart that documents the flow of control between activities of the system. An ***activity*** is a state of the system in which some transaction or data computation takes place. It is represented by a box with rounded corners that has the name of the activity within it, as illustrated in Figure 9.8. Figure 9.8 is an activity diagram that represents the flow of control of scenario 1 and some other possible scenarios of the use case "make sale." The activity diagram begins with a solid circle, the starting point. This leads to the first activity, "verify seating available," and continues to other activities. The activity diagram has one or more terminal points, which are shown as circles with a solid dot.

In modeling the flow of control, activities are usually followed by decision points, shown as diamonds. A decision point models branching behavior when the flow of control is dependent on the outcome of the activity in the state prior to the decision point. For example, the first activity, "verify seating available," can have two mutually exclusive results: the seating is available or it is not available. The labels on the output arcs of the decision point accommodate both cases in the flow of control. If the seating is available, the next activity performed by the system is "offer seating." However, if the seating is not available, the next activity is "inform the customer of failure," followed by a termination, or canceling of the transaction. Presumably, the customer can try again with another seating request.

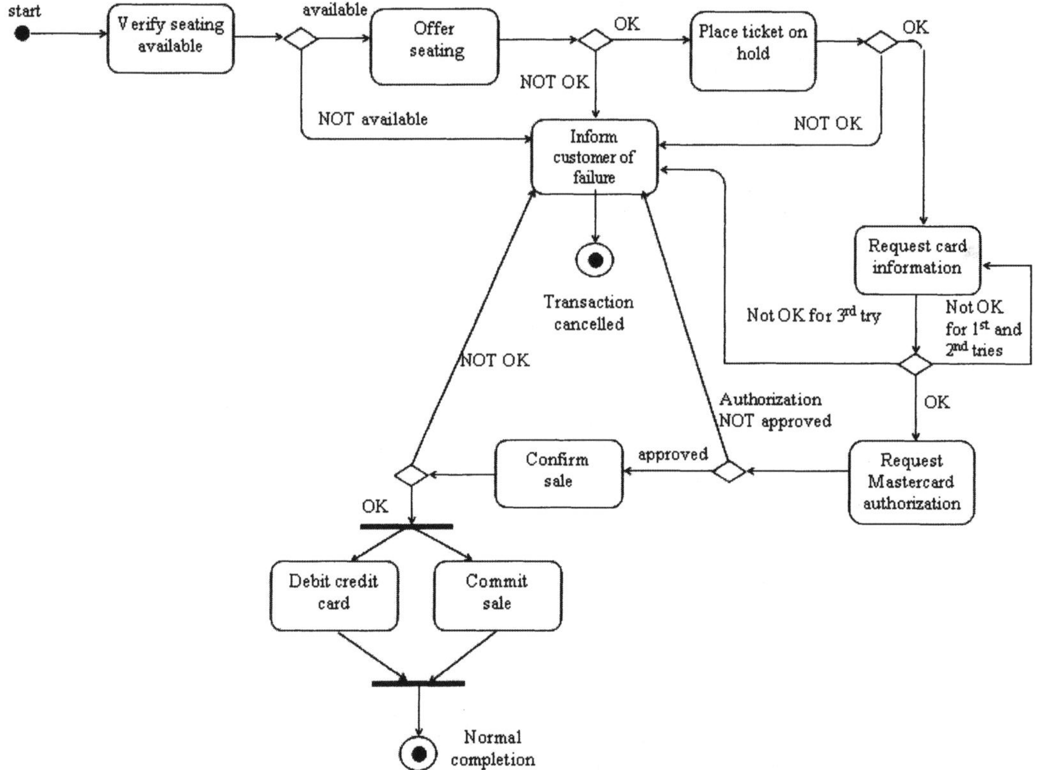

Figure 9.8 Activity diagram of some combined scenarios.

The directed arcs show the direction of the flow of control. All arcs in an activity diagram are directed arcs. From Figure 9.8, one can trace the execution of scenario 1, the normal scenario for making a reservation. The string of activities of scenario 1 is as follows: start → verify seating available → offer seating → place ticket on hold → request card information → request MasterCard authorization → confirm sale → commit sale and debit credit card → normal completion. Note the following details about this string of activities:

1. It begins with the first activity *within the system*. In the sequence diagram shown in Figure 9.7, the first event within the system is labeled "Verify availability."
2. It includes all transactions and operations of scenario 1 *carried out by the system*.
3. Each activity incorporates one or more sequence transactions, (e.g., "Request MasterCard authorization" incorporates the sequence "Request authorization" and "Authorization approved" of Figure 9.7).

Activity diagrams look somewhat like flowcharts in procedural programming. However, unlike flowcharts, activity diagrams can model concurrency (i.e., the execution of more than one activity simultaneously). Consider the last two activities of Figure 9.8. When a sale is confirmed, the activity diagram enters a fork, shown as a bar. A **fork** has one input arc and more than one output arc. This leads to the activation of more than one activity in parallel. When the simultaneous and independent activities are complete, they enter a join. A **join** has two or more input arcs and one output arc. The activity that follows a join can begin when all input activities are complete.

The activity diagram of Figure 9.8 includes scenario 1 and other possible scenarios. Consider, for example, the following string of activities: start → verify seating available → inform customer of failure → transaction canceled. This, too, is a possible scenario of the use case "make sale." This scenario can be modeled in a sequence diagram such as Figure 9.9. In fact, when an activity diagram is used to document a use case, the activity diagram should cover all of the scenarios that might occur in the execution of a use case.

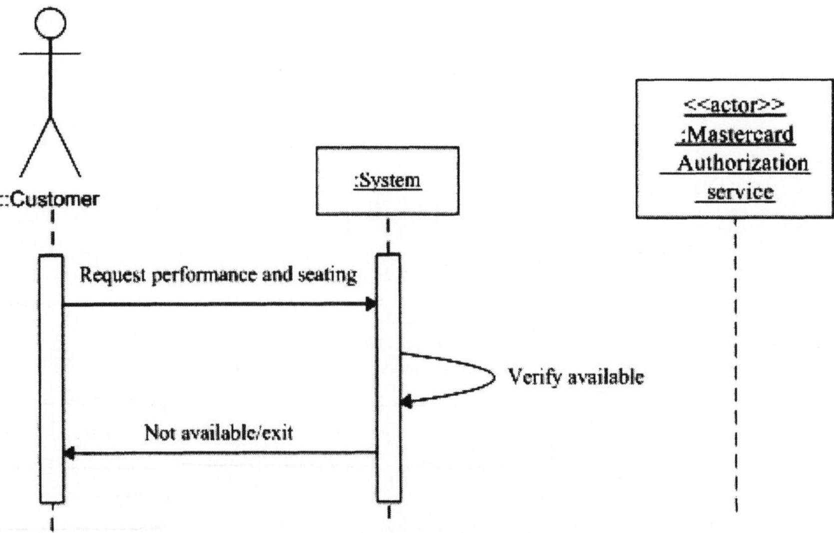

Figure 9.9 Sequence diagram for scenario 2.

9.4.4 State Chart Diagram

A ***state chart diagram*** is a dynamic model that shows the various states that an object can be in and how the object evolves through those states over time as a result of events that occur. The state chart can capture behavior at various levels of the system, including overall system behavior, class behavior, and detailed behavior of specific objects.

UML state chart diagrams are a variant of a traditional model of computer behavior known as a finite state machine. A ***finite state machine*** is composed of three elements: states, transitions, and event conditions. These elements are shown in Figure 9.10, which is a high-level state chart diagram of a ticket object during the execution of events in the Ticket Sales system. A state, represented as a box with rounded corners, is a stage of an object's life cycle that has a finite duration. In Figure 9.10, three states of the ticket sales system are modeled. In the first state, the ticket sales for a performance are being opened and tickets are being created. In the last state, the ticket sales for a performance are being closed. The interim state is the period during which the ticket sales are taking place. These three states are a fairly high level view of the behavior of an instance of a Ticket object class during its life cycle.

The transitions are the connecting arcs between states. A transition is labeled with an event condition that identifies the event that triggers the transition. When an object is in a particular state, the output transition of that state is **armed** (ready to be fired). If a transition is armed and the event condition on an output transition is true, the transition will fire and the object's state changes from the input state of the transition to its output state. The event condition is labeled Event Name/ [Action]. The **event name** is the trigger (condition) that causes the transition to fire; the optional **action statement** is an action that results from the event. Unlike a state, which may be occupied over a period of time, an event and the firing of its transition occur at an instant in time. Transitions can also be self-loops of a state.

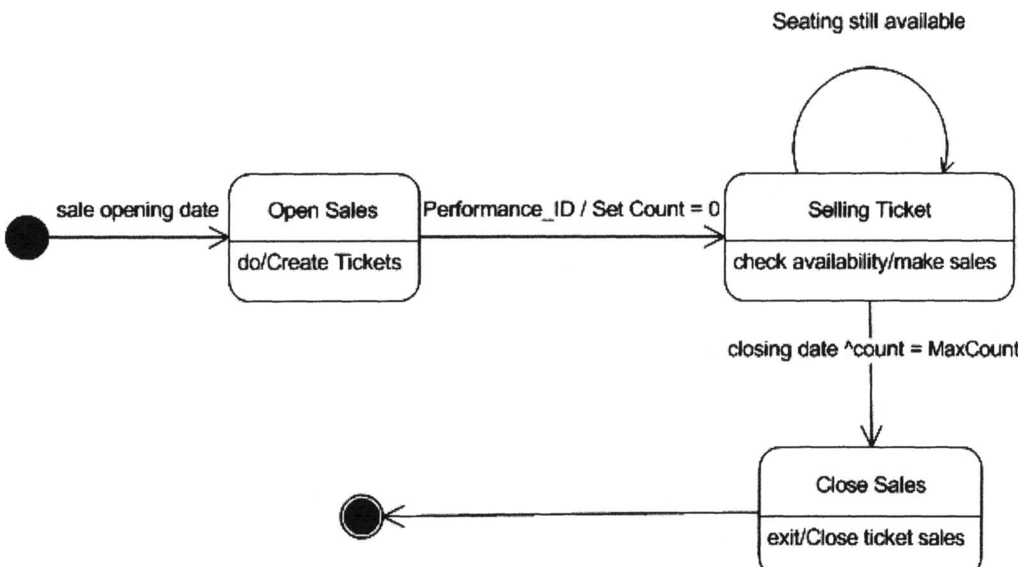

Figure 9.10 State chart diagram of instances of the ticket sales system.

Note that Figure 9.10 models the behavior of the ticket sales system. With regard to the behavior of objects of the system, there will be many such objects executing their state chart models. Let us focus just on the Ticket object class. A ticket is a claim of the owner on a seat at a specific performance. The initial event that begins the execution of the state chart is the opening date of the sale. Control first passes to that state in which ticket objects are being created and sales are opened. This activity provides an output event identifying the performance to go on sale and setting the count of total tickets sold to zero. The self-loop on the state "selling tickets" indicates continuing ticket sales. This continues until one or both of two events occur: the closing date *or* tickets are sold out. Either of these events will transition the state chart to the "close sales" state, and the state chart will terminate. Implicit in this diagram is the rule that if tickets are sold out prior to the end-of-sale date, they will remain sold out. In other words, all ticket sales are final.

A state chart diagram can model more details of object behavior. The UML standard has adopted Harel state charts (Harel and Politi, 1998), which specify modeling of layers of object behavior. For example, in Figure 9.10, significant object behavior is masked in the single high-level state "selling tickets." Figure 9.11 exposes more details of the "selling ticket" state by showing the state machine of the ticket object. Here we see that, after creation, a ticket can be in one of three states. Immediately after creation, it is in the "available" state. When a ticket is in the available state *and* the event "place on hold" occurs, the ticket transitions to the "on hold" state. Note that the event "place on hold" is generated by the Ticket Agent object during the "make sales" use case (see sequence diagram, Figure 9.7). Also, the same event appears as the third activity of the activity diagram in Figure 9.8.

Similarly, when a ticket sale is committed, the state of the object transitions from "on hold" to "sold." The state chart shows that during the period of sales, the states of the ticket objects are changing from "available" to "sold." Presumably, when the closing date for ticket sales for a performance occurs, tickets will be in one of two states: available or sold. Note that the events of the activity diagram in Figure 9.8 that result in a cancellation of a reservation after a ticket is placed

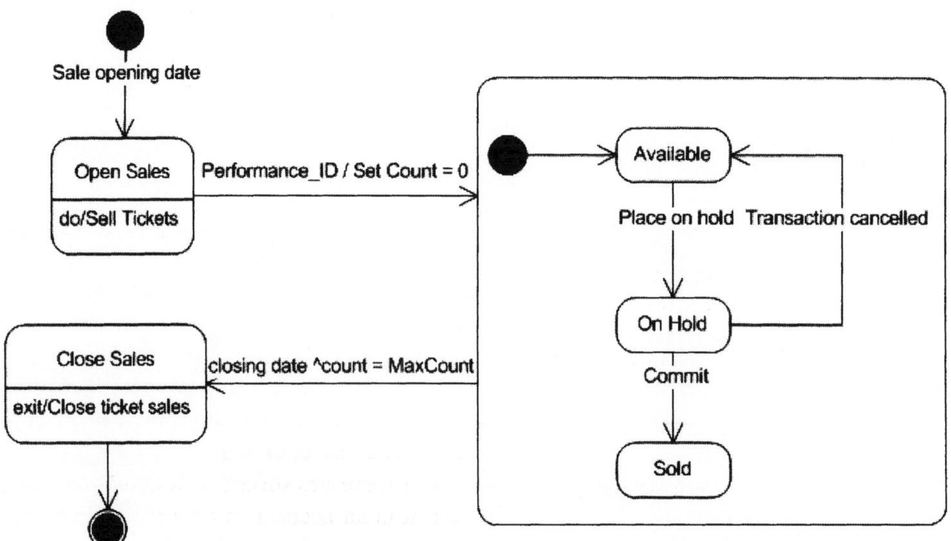

Figure 9.11 State chart diagram showing ticket object states during sales.

"on hold" but before it is "sold" are captured in Figure 9.11 as the transition labeled "transaction canceled." This transition takes the ticket back to the "available" state.

In the illustrated use case, sequence, activity, and state chart diagrams of the ticket sales system, we have not tried to capture the models associated with canceling a sale (returning a ticket) after the sale has been made. Implicitly, we have imposed a business rule that states "all ticket sales are final." The reader will have an opportunity to develop a "cancel sale" use case in the end-of-chapter exercises.

9.4.5 Class Diagram

One purpose of the step-by-step construction of UML diagrams is to identify objects and to group objects into classes. The ***class diagram*** is the most important UML model as it defines the object types that can be created by the system. These would include software objects such as the ticket agent and database objects such as ticket entities.

Earlier in this chapter, we stated that UML could be used whether or not the information system design is object oriented or based on the relational model. So let us first apply our knowledge of the relational model to the "make sale" use case. After constructing use case, sequence, activity, and state chart diagrams, the analyst will have some familiarity with the requirement specifications of the case. This should suggest data elements that are required for the data model. So we will begin the discussion of class diagrams by first suggesting a data model based on relational database design rules. This will then be contrasted with the object-oriented class diagram.

From the sequence diagram of Figure 9.7, some concepts have been identified: performance, seating, ticket, credit card, and sale. Considering a relational database design, we can declare the structural relationship between these concepts as follows:

1. A Ticket represents a claim on a Performance and a Seat.
2. Each Performance is associated with many Tickets.
3. Each seat (which we shall call a Location) is associated with many Tickets, since the same seat is used for more than one Performance.
4. A ticket Sale is made for one or more Tickets.
5. A Credit Card is used in each sale.

These structural relationships suggest the relational data model shown in Figure 9.12 in IDEF1X format. Both PERFORMANCE and LOCATION have a one-to-many relationship with TICKET. For a given performance, there will be many tickets, but each ticket is for one and only one performance. A location has many tickets since the same location is used over and over again for different performances. We have chosen to make Price an attribute of TICKET under the assumption that the same seating location for different performances may have a different price. The SALE entity set documents the sale information once a ticket is sold. Note that Transaction_ID is a foreign key of TICKET. The same Transaction_ID can occur for more than one ticket indicating the sale of more than one ticket during the same transaction. Tickets not yet sold will have a NULL value for Transaction_ID in the TICKET table. All tickets are created on the opening day of sales, as documented in the state charts shown in Figures 9.10 and 9.11. The Status attribute of TICKET is used to keep track of whether or not a ticket is available, on hold, or sold.

Figure 9.12 could be modified in several ways and there are different designs that could be considered, but for the purposes of this explanation, it is an acceptable data model of the ticket sales database as we understand it from the previous diagrams. This could indeed be the outcome of the analysis of data elements in the sequence diagram from the "make sale" use case if we were

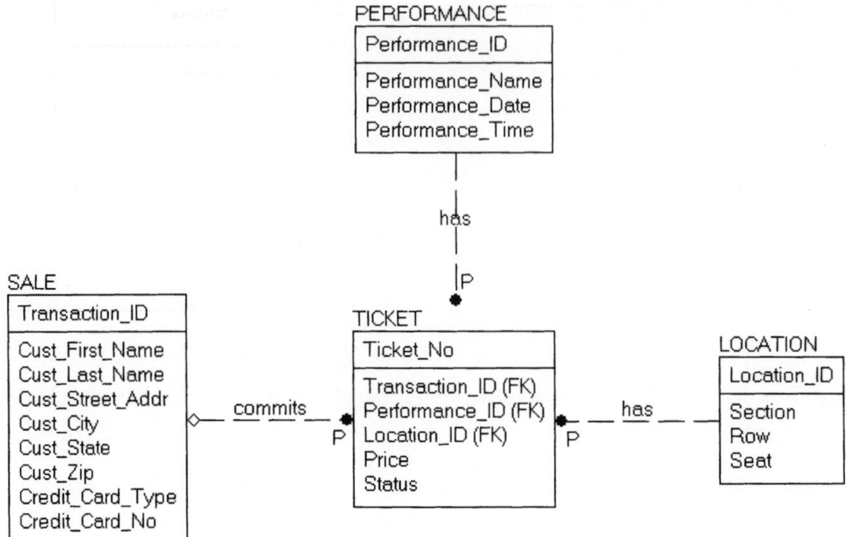

Figure 9.12 Relational model of ticket sales database.

ClassName
Attribute1:string Attribute2:date Attribute3:integer
Method1() Method2 ()

Figure 9.13 Components of a class.

to implement the data requirements in a relational database. In that sense, UML is useful in documenting the requirements of an information system and may be preferred to other modeling techniques such as IDEF or DFD, based on the taste of the analyst. Figure 9.12 will be used to contrast the relational data model with the object-oriented class diagram.

9.4.5.1 Components of a Class

The correspondence between the relational data model world and the OO data model world is rather casual as they each have a different philosophy, emphasis, and implementation requirements. However, it is useful to draw parallels between the two to explain the UML class diagram. Roughly speaking, what is referred to as an "entity set" in relational data models is analogous to a "class" in OO data models. A class diagram is composed of classes, which are the templates for sets of similar objects that have the same attributes and behavior.

In a class diagram, each class is represented by a box with three distinct sections, as shown in Figure 9.13. The name of the class is given in the top panel. This is a descriptive name and is analogous to the name of the entity set in relational data models. The center panel contains the attributes of the class and their data types. Object-oriented models support the usual data types of

Sale
CustName:string CustAddress:string CreditCardType:string CreditCardNumber:string
CreateSale ()

Ticket
Performance:string Location:string Price:currency Status:string
VerifyAvailable () PlaceOnHold () Commit () Release ()

Performance
Name:string Date:date Time:date
CreatePerformance ()

Location
Section:string Row:string Seat:string
CreateLocation ()

Figure 9.14 Ticket sales classes.

string, integer, date, and so forth, as well as user-defined data types. Attributes of the OO data model are analogous to the attribute names in relational data models. So far a class looks much like an entity set. However, the third panel contains the behavior of the class in the form of operations that the class can perform on its data—that is, on the values of its attributes. These operations, which are functions that can be called (invoked), are sometimes referred to as "operations" and sometimes as "methods." We will continue to use the term "methods." As stated in the beginning of this chapter, an object combines both data (attributes) and behavior (methods). The method can be invoked by other objects or may be invoked by the object itself. In the relational model, data are operated on by SQL commands from the user or from a user program. Data and functions that operate on data are separate. The operations and associated code are not built into the entity set (table of the database). In OO data models, the operations that are performed on the data are part of the object definition and reside in the object itself.

The four entity sets in Figure 9.12 are shown as classes in Figure 9.14. Note the declaration of the data type for each attribute. Primary keys are not used to uniquely identify an object because each object of a class will automatically be assigned a unique object ID (OID) by the system when the object is created.

Methods are shown in the lower panels of the classes. The method "Commit ( )" in the Ticket class, when invoked, results in the change of status of the Ticket object from "on hold" to "sold." This behavior was described in the state chart in Figure 9.11. It is invoked by the function call "Commit sale ( )" that appears at the bottom of the sequence diagram of Figure 9.7. The Ticket class has an attribute "Status" that is assigned the value of the state of the ticket (available, on hold, sold).

9.4.5.2 Associations

An ***association*** in a class diagram is analogous to a relationship in a traditional E-R or IDEF1X diagram. Associations are shown as lines connecting classes. The cardinality of the association is shown at each end of the line using the notation "<lower limit>..<upper limit>." For example, in Figure 9.15, the association between Performance and Ticket indicates "1..1" on the Performance side and "1..*" on the Ticket side. Each performance is associated with one or more tickets, where "*" is the generic "more than." For a cardinality of zero, one, or more, the notation would be "0..*", and for exactly 2, it would be "2..2."

The naming convention is usually a verb phrase and points in both directions. Therefore, in Figure 9.15, a sale is for one or more tickets and one or more tickets are purchased via a sale.

If the association is a parent-child relationship, the connecting arc has a diamond on the parent side. An example taken from prior chapters is the association between PurchaseOrder and PODetail. Such an association is shown in Figure 9.16.

The diamond on the PurchaseOrder side indicates that it is the parent in the relationship. The diamond can be either hollow or solid. If the association is optional to the parent, then the diamond is not filled in (hollow). When the association is mandatory to the parent, the diamond is solid. For the case of a mandatory association, deletion of the parent object cannot occur without prior deletion of the child object.

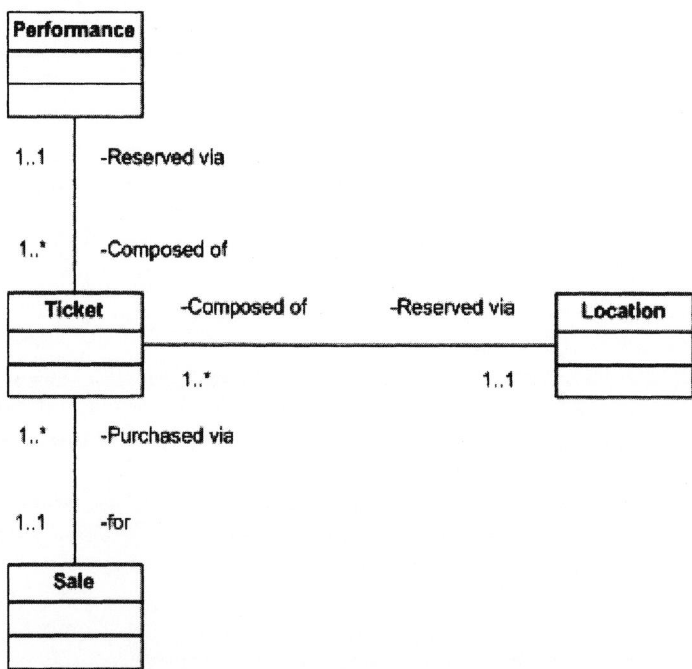

Figure 9.15 Associations of ticket sale classes.

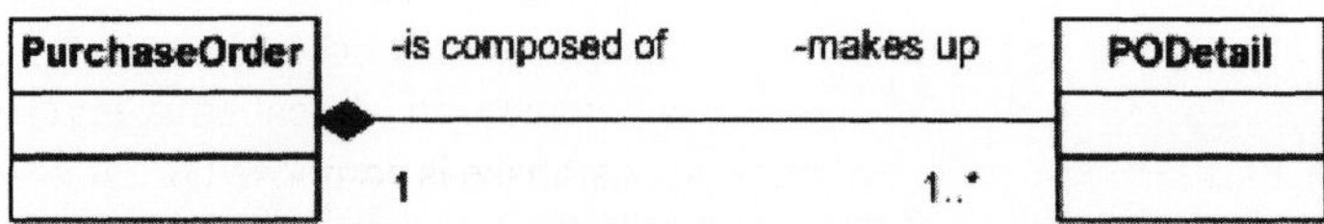

Figure 9.16 Mandatory parent-child association.

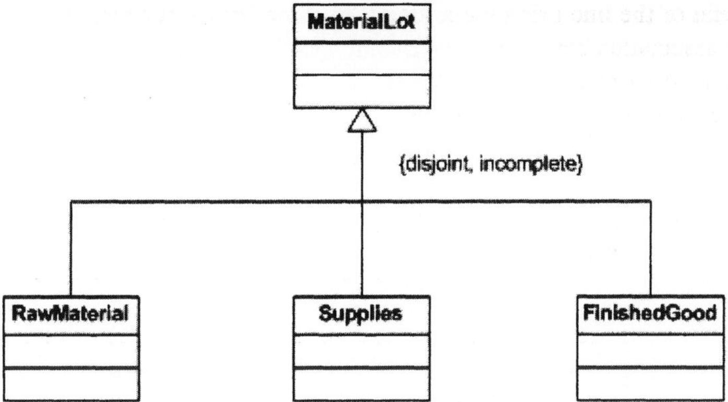

Figure 9.17 Example of a generalization in UML.

9.4.5.3 Generalization

Class diagrams also support superclass/subclass relationships, which are called supertype and subtype generalizations in UML. This is shown in Figure 9.17 using the familiar MaterialLot entity of Figure 3.21. The subtype classes are connected to the supertype classes using arcs with an arrow pointing to the supertype. Each occurrence of a subtype has an "is a" association to the supertype. The coverage of the subtypes is indicated by a label on the arc. For example, {disjoint} means that there is no overlapping between subtypes and {incomplete} means that there are material lots of the supertype that are not represented in the subtype, for example, subassemblies. So the actual situation can be described by combinations of {overlapping, disjoint} and {incomplete, complete}.

9.4.5.4 Comparison Summary

In this section, we have covered some fundamentals of class diagrams. There are more features to the class diagram that we will not cover here; the interested reader is referred to the bibliography at the end of this book. We have suggested that the class diagram could be used as though it were another representation of a relational data model. Although it could be used for that purpose, the class diagram has particular relevance to object-oriented design since it includes the methods of the class, which do not apply to the relational database tables. In the following section, we contrast an object-oriented implementation with a relational implementation for the ticket sale system.

9.4.6 Logical System Design and Implementation

The way in which one "characterizes" the use case, sequence, and class diagrams is a guide to how the system is to be implemented. For the Ticket Sales system, we characterized the operations, or methods, as being distributed among the following classes: TicketAgent, Ticket, Sale, Performance, and Location. If the context of the application were a typical relational database implementation, the methods assigned to Ticket, Sale, Performance, and Location would have no meaning. They would have no meaning because a relational database is composed of tables, not objects with methods. The only meaningful operations would be those programmed into TicketAgent, which is a software component and not a database table.

We will use the Ticket Sales system to illustrate the differences in logical design between the use of a relational DBMS versus objects in the implementation. In this section, first we propose a design based on a conventional implementation assuming a relational database. This will be followed by an object-oriented proposed design.

In Chapter 8, we illustrated the use of a Web page for interacting with a database. By structuring the elements on a Web page and the functionality available using the Submit button, we controlled all the operations that are allowable via the Web page, including the SQL queries of the database tables.

The Web page or, more specifically, the set of ASP files that runs the SQL queries is an interface that can be designed to be roughly equivalent to the TicketAgent. In a relational database design, our TicketAgent will run all the operations on the database centrally, as opposed to a true object-oriented design where operations are usually decentralized.

Figure 9.18 is a sequence diagram that gives a "generic" representation of the user interaction with the system over a Web browser. The objects identified for the system are the Web page, Web server/middleware, and the database. The object descriptions include *stereotypes.* In progressing from analytical models to design, Jacobson suggested the use of stereotypes that enable representation of how a message sent by an actor traverses the presentation, application, and business layers (Jacobson, 1992; Jacobson et al., 1999). Stereotypes add more descriptive context to the model. Stereotypes are shown as text strings surrounded by guillemets (<< >>). He identified three stereotypes for this purpose:

<<Boundary>>. A class stereotype used to model the interaction between the system and its actors

<<Controller>>. A class stereotype used to represent the coordination and control of other objects

<<Entity>>. A class stereotype used to model information

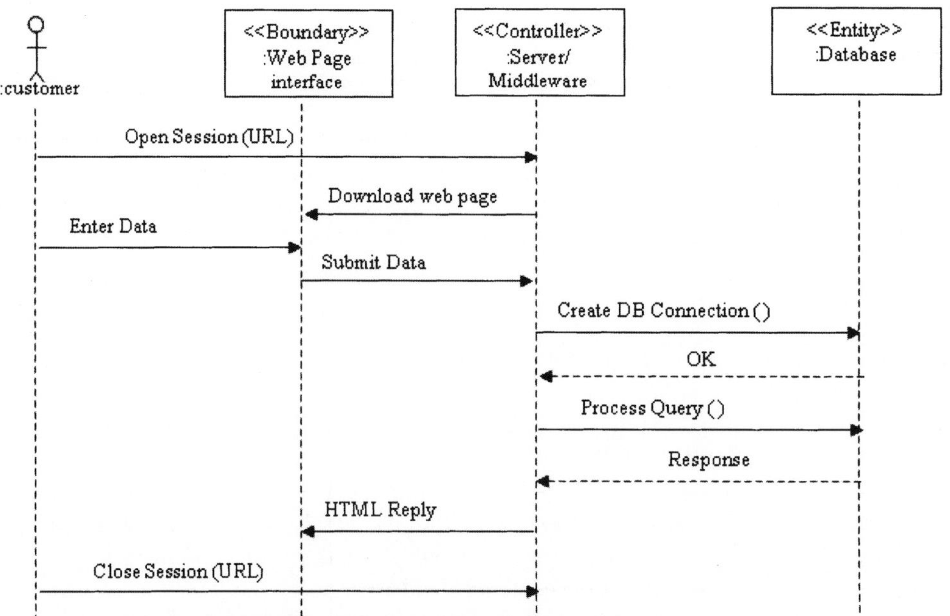

Figure 9.18 Web page application sequence diagram.

As illustrated in Figure 9.18, the Web page is of the boundary class, and the middleware component, which controls the interaction with the relational database, is of the controller class. The generic sequence diagram of Figure 9.18 is self-explanatory. The user opens a session by accessing the Web server, which downloads a Web page. Data entered via the Web page is submitted to the middleware component on the Web server, which opens a connection to the database and controls the execution of the query. Chapter 8 provided examples of how this is done. Finally, responses are downloaded to the Web page, and the user closes the session. You will be asked to augment Figure 9.18 for more specific cases in the end-of-chapter exercises.

Figure 9.19 carries this design into an extension of a class diagram called a ***view of participating classes (VOPC)***. Here we show the relationship between the controller class and the key entities of the database participating in the Ticket Sales case. When a sale is made, the Web server/middleware runs "CeateSale ( )," which is an ASP file that populates the Sale table with a new record. It then runs "Commit ( )," which changes the status of the tickets sold from "on hold" to "sold" and also enters the Transaction ID of the sale. The other functions of TicketAgent are as previously described, and all functions are centralized and executed via submit buttons and are controlled via the Web server and ASP files. This is a conceivable implementation for a relational database described by UML models.

Unique identifiers are rarely shown in object-oriented models. If the behavior of a class requires locating a specific occurrence of another class, it is useful to indicate unique identifiers. Stereotypes can be used to designate attributes that are unique identifiers. This is done by preceding the attribute

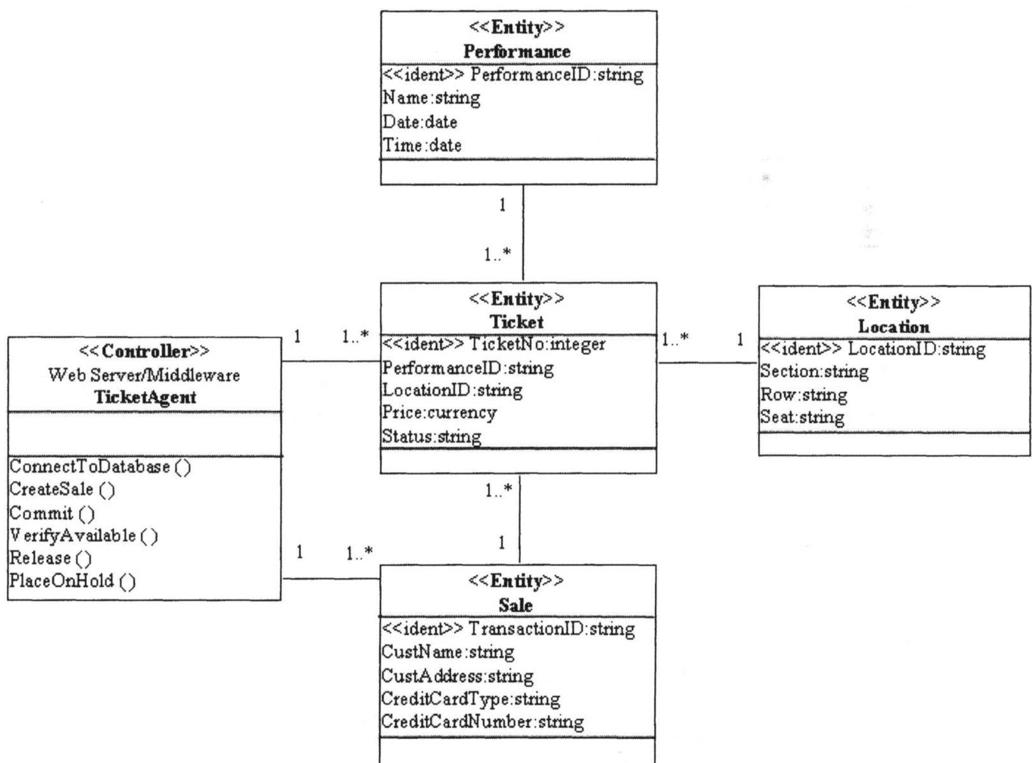

Figure 9.19 View of participating classes.

name with the designation <<ident>> . Since Figure 9.19 is based on the use of a relational database, we have included the primary keys as unique identifiers.

Let us continue with the same example and consider the decisions that have to be made in an object-oriented design implementation. The thought process is somewhat different because the design architecture is driven by a different set of principles. One principle is that the object classes should be designed to make their use as general as possible. Consider, for example, a button class. To maintain generality, the button class should be designed to be used anywhere—on a lamp, a television, a computer, and so on. It is unnecessary for the button to have any knowledge of the object to which it is being connected through an association. On the other hand, the lamp object must know about the existence of the button because it uses it to turn itself on. Therefore, the association is from the lamp to the button and not the other way around. By not assigning a lamp attribute to the button, we maintain its general usage by other object classes.

Now consider two objects from the Ticket Sale case, a Performance object and a Ticket Object. These objects will be created from the Performance class and Ticket class. Three possible ways of implementing these objects are shown in Figure 9.20. In Figure 9.20a, the Ticket object knows about the existence of the Performance object because one of its attributes is Performance. Conversely, the Performance object has no knowledge of Ticket because it has no attributes that refer to Ticket. For the situation of Figure 9.20a, we say that Ticket has a one-to-one association to performance because each ticket object references one and only one Performance object through the attribute Performance, whose value is the OID of the Performance object. On the other hand,

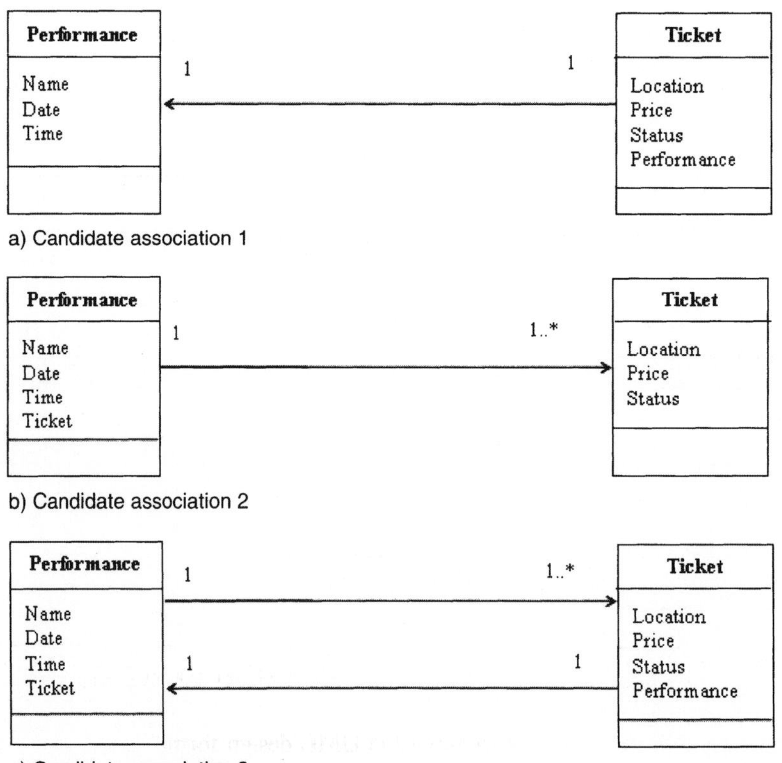

Figure 9.20 Three candidate associations.

Performance has no relationship to Ticket. We can explicitly show the direction of the association using an arrow on the association arc pointing in the direction of the association. This restricts the navigability of associations according to the direction of messages.

Now consider Figure 9.20b. In Figure 9.20b, Performance knows about the existence of Ticket. Here, one of the attributes of Performance is Ticket. In object-oriented design, an attribute can contain more that one value. This is referred to as **multiplicity** and allows a set of values to be assigned to an attribute. In other words, the OIDs of all the tickets to a performance (a set of tickets) can be an attribute of Performance. On the other hand, in Figure 9.20b, Ticket does not know about Performance. Therefore, in this configuration, there is a one-to-many association from Performance to Ticket and no association from Ticket to Performance.

In Figure 9.20c, Performance carries an attribute with information about Ticket, and Ticket has an attribute that refers to Performance. For this situation, there is a one-to-many association from Performance to Ticket and a one-to-one association from Ticket to Performance.

Unlike a relational database model built on the rules of the relational architecture discussed in earlier chapters of this book, an object-oriented model allows many possibilities for associations depending on the specific situation. In general, if an object must access another object in order to perform some function, it must have knowledge of that object. The attribute that contains the value(s) of the other object(s) are a pointer to that (those) is object(s). It is analogous to the concept of foreign key(s).

For the methods that we have defined in the Ticket Sales case, there is no reason for Performance to have knowledge of Ticket since no operation (method) was defined that requires it to communicate with Ticket. Therefore, Figure 9.20a is sufficient for this case. As the requirements of the system evolve, it may be necessary to use a different candidate association. This is part of the design process in OOD since the design of the data elements does not have the same structural rules as a relational database model.

Figure 9.21 is a View of Participating Classes for the Ticket Sale case implemented as objects. Note that Ticket has a one-to-one association with Location, and Location has no association with Ticket. Sale has a one-to-many relationship with Ticket since a Sale object has an attribute Ticket, which is the set of ticket OIDs that belongs to the Sale object.

Figure 9.21 also indicates how the system's methods function. For example, when the TicketAgent object completes a sale, it invokes the "CreateSale ( )" method of the Sale object by sending the message "Do CreateSale ( )" to the Sale object. The TicketAgent also invokes a change in the status of the ticket just sold by sending the message "Do Commit ( )" to the appropriate Ticket object(s), which runs its method "Commit ( )." The other functions have a similar interpretation.

If you compare Figure 9.21 with Figure 9.19, the difference between a relational database implementation and an objectoriented database implementation should be apparent. In the objectoriented implementation, operations on data and the data are encapsulated within the object. Operations (methods) may be invoked from the outside, but they are performed by the object owning the data.

Finally, Figure 9.22 extends the sequence diagram to show interactions between TicketAgent and the data objects for scenario 1 based on the object model of Figure 9.21.

9.5 CASE STUDY: UNIVERSITY FOOD RECEIVING DEPARTMENT

The purpose of this case study is to apply the UML design formalism to a situation that we previously modeled in IDEF formalism. In so doing, it is possible to compare and contrast the two approaches. Here we will apply it to the models encountered in Chapters 4 and 5.

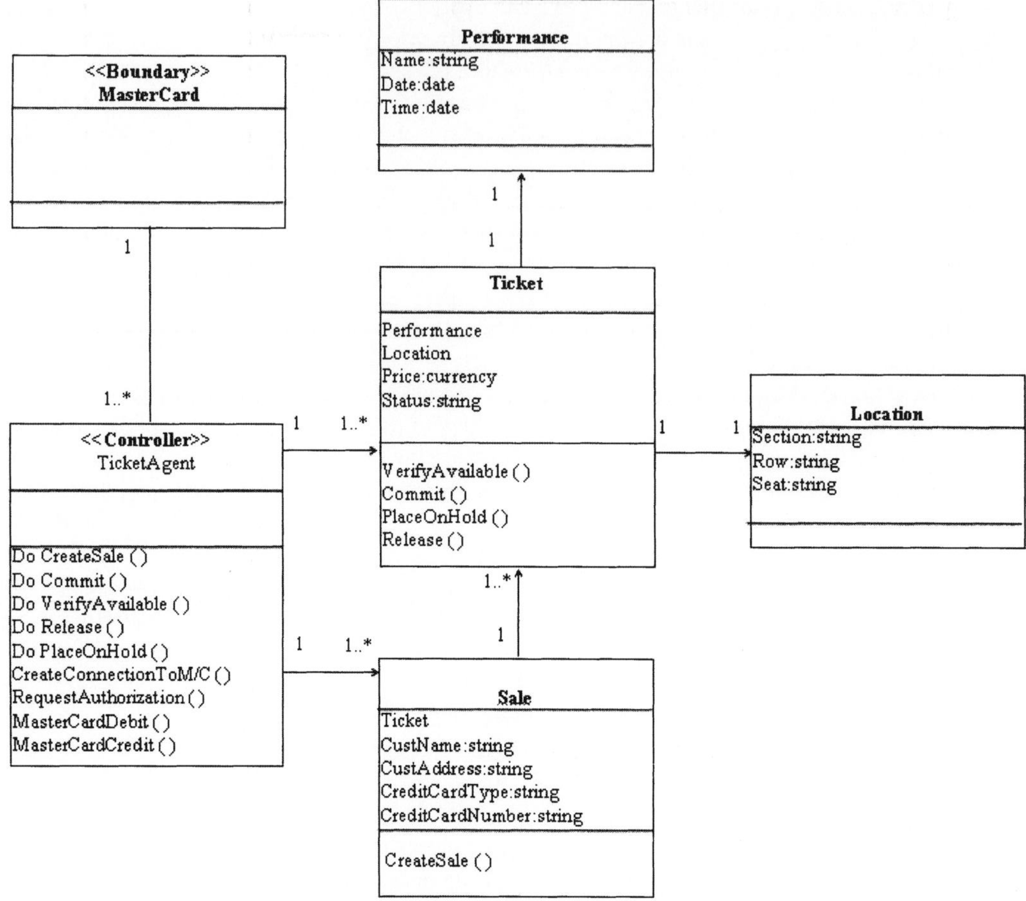

Figure 9.21 Object design view of participating classes.

9.5.1 Use Case Diagram

The scenario described in Section 4.4.4 and Figure 4.12, "Confirm Validity of Shipment," is used in this example. Recall that the receiving clerk first compares the shipping document (bill of laden) with the referenced purchase order number. If the material is accepted, the purchasing agent is informed via the receiving report, which documents the accepted materials from the shipment.

Figure 9.23 shows the use case "Confirm Validity of Shipment," as well as some other use cases having to do with the life cycle of a purchase order. Two actors interface with the system (database) when using the purchase order in various use cases. The purchasing agent creates the purchase order. This includes all of the header information in the Purchase Order table and the detail information in the PO_Detail table. After being created, the purchase order is printed and sent to the vendor.

No further use is made of the purchase order information until material is delivered. At that time, the receiving clerk compares the shipment information to the purchase order (Confirm Validity of Shipment) and prepares a receiving report or otherwise informs the purchasing agent of the

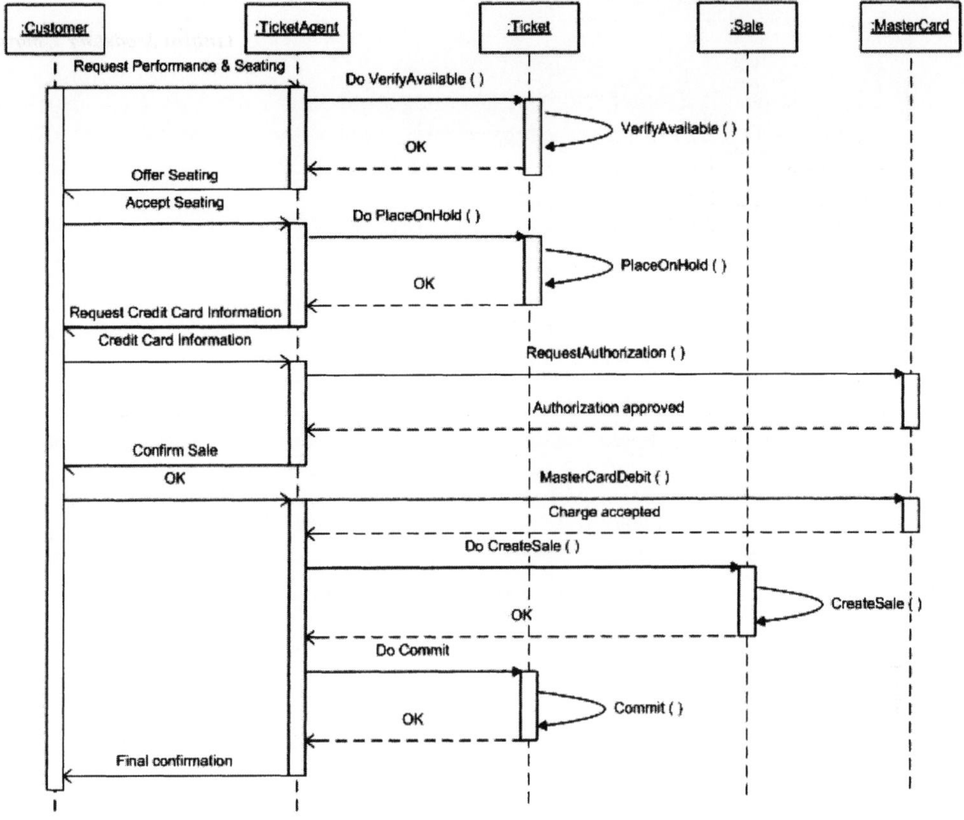

Figure 9.22 Sequence diagram with data objects.

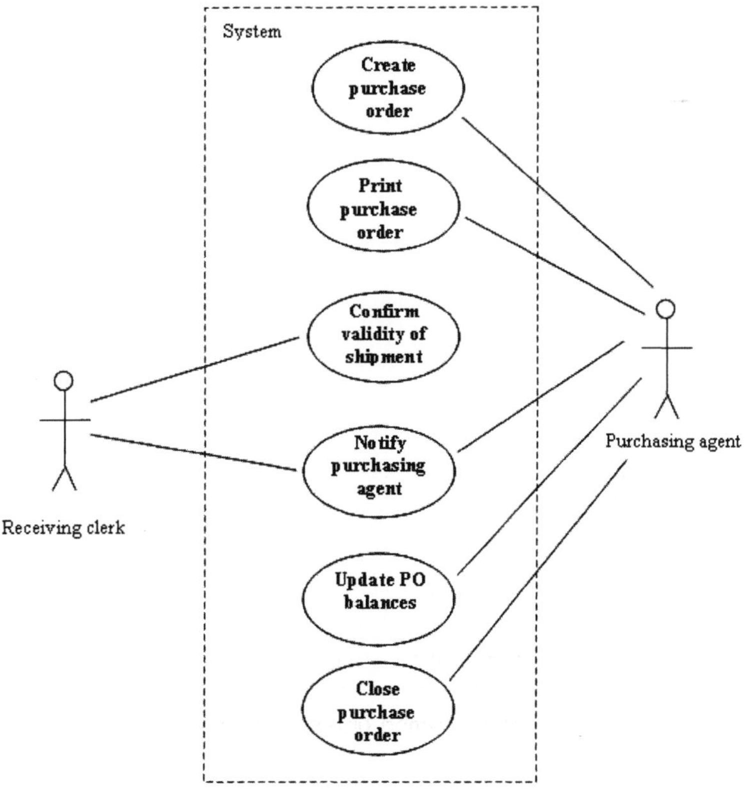

Figure 9.23 Use case diagram for University Food.

delivery against the purchase order (Notify Purchasing Agent). The purchasing agent will then change the purchase order balances by subtracting the amount received by line item from the current balance. Finally, when all line items have been delivered, the purchasing agent will close the purchase order.

There are some parallels and differences to be noted between a use case and the IDEF functional diagram. The use case diagram only models interactions between the user of the system and the system (database). IDEF0, on the other hand, includes functions that operate on material flows outside the database as well. The UML actor is analogous to the IDEF0 mechanism (i.e., it models the role of the outside entity involved in making the function happen).

9.5.2 Sequence Diagram

Figure 9.24 is the sequence diagram for the normal success scenario of the use case "Confirm Validity of Shipment." The receiving clerk must first login to the DBMS. Then the purchase order number is submitted via a user form, and a confirmation of the PO number and vendor ID are returned to be compared with the bill of laden. The line items are also retrieved for comparison with the shipment details. Depending on how the forms are implemented, two queries can be used or a single query can return a master/detail record for the entire purchase order. Regardless of implementation details, the order of reviewing content is as suggested in the sequence diagram. Finally, the receiving clerk closes the form.

There is no analog to the sequence diagram in the IDEF formalism. The closest model is the Functional/Entity Interaction Matrix illustrated in Figure 6.1. Unlike the sequence diagram, it only shows the type of action (insert, update, delete, read-only) that may occur between functions (mechanisms) and the entities of the database. It does not show details of scenarios and the order of interaction.

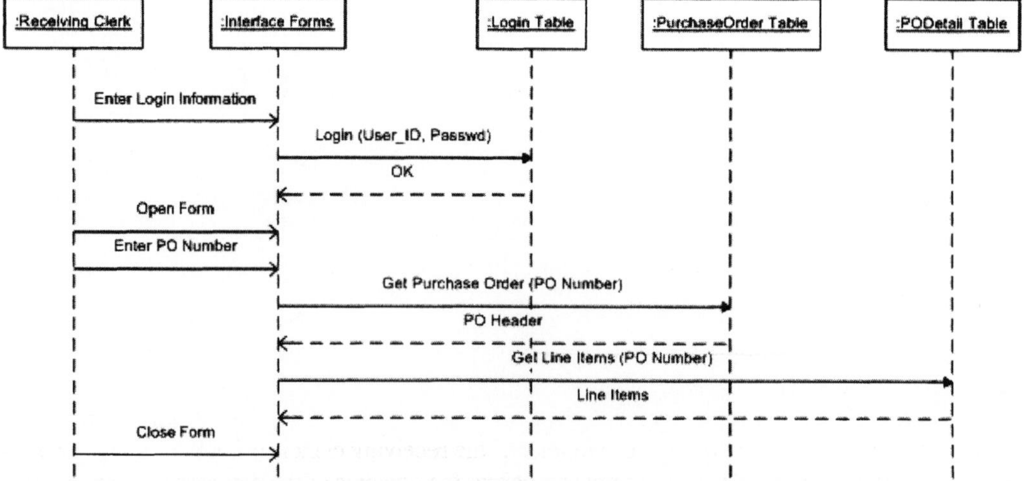

Figure 9.24 Sequence diagram for confirm validity of shipment use case.

9.5.3 Activity Diagram

The activity diagram represents the various scenarios of the use case from the system perspective. Unlike the Ticket Sale example, there are no agents or other software objects in the "Confirm Validity of Shipment" use case. Decisions on whether or not to accept materials are made by the receiving clerk, who is external to the system. This is reflected in Figure 9.25.

After logging in, the receiving clerk enters the PO number from the bill of laden. The system (database) either returns the purchase order from the database or indicates that no record with that PO number is found. A false result terminates the activity. A true result is followed by an examination of the PO detail. Since a purchase order will always have a detail, the record will always be returned. From the system point of view, this is followed by a termination of activity. Now consider the possibility that the PO detail may not correspond to items on the bill of laden. The receiving clerk has a decision to make (i.e., whether to accept or not accept the shipment). But this is a decision that is *made outside the system.* So no decision node is included for that decision in Figure 9.25. If the designers had decided to include a software object within the system that would take the information on the bill of laden and make the decision as to whether or not to accept the shipment, then such decision nodes would be appropriate to the model of Figure 9.25. So the activity diagram reflects only decisions made by the system.

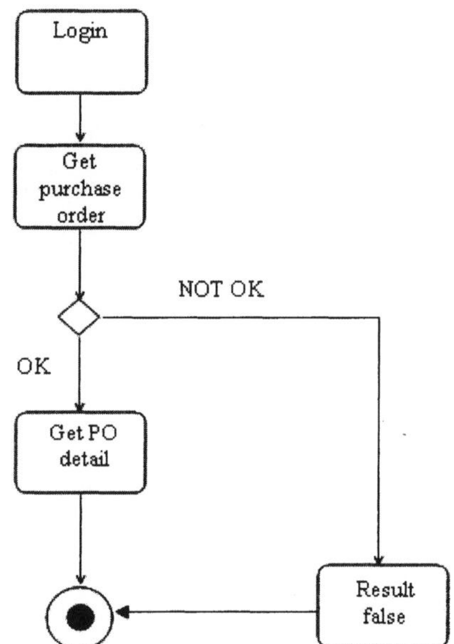

Figure 9.25 Activity diagram for confirm validity of shipment.

9.5.4 State Chart Diagram

In the use case "Confirm Validity of Shipment," the receiving clerk is using the system for read-only purposes. That is to say, he is simply comparing the contents of the database with an external document, the bill of laden. Hence, the values of the attributes of the purchase order are not changed during this use case. *Since the attribute values are not changed, there are no changes of state.* Therefore, from the point of view of the system, no state chart is specifically associated with this use case.

On the other hand, as indicated in Figure 9.23, the purchase order is involved in several use cases. For example, it is created in the first use case (Create Purchase Order), its quantity balances are updated for each line item in another use case (Update Purchase Order Balances), and, finally, it is closed (Close Purchase Order). The reader will have an opportunity to model the state behavior of the purchase order in an end-of-chapter exercise.

9.5.5 Class Diagram

From Chapter 5, Figure 5.15, we already know what the relational database model will look like. For this case study, we are assuming that the relational data model will be used. Here we will employ some of the features of a class diagram.

Figure 9.26 shows four entities that are involved in the purchase order form that was created in Chapter 6. The entities are basically as they appear in Figure 5.15, with some different notations. Here we have added a stereotype <<fk>> to indicate foreign keys. A parent-child relationship exists between PurchaseOrder, and PODetail, and the relationship is mandatory. There are no methods shown in the classes since these classes are being used to represent relational database entities.

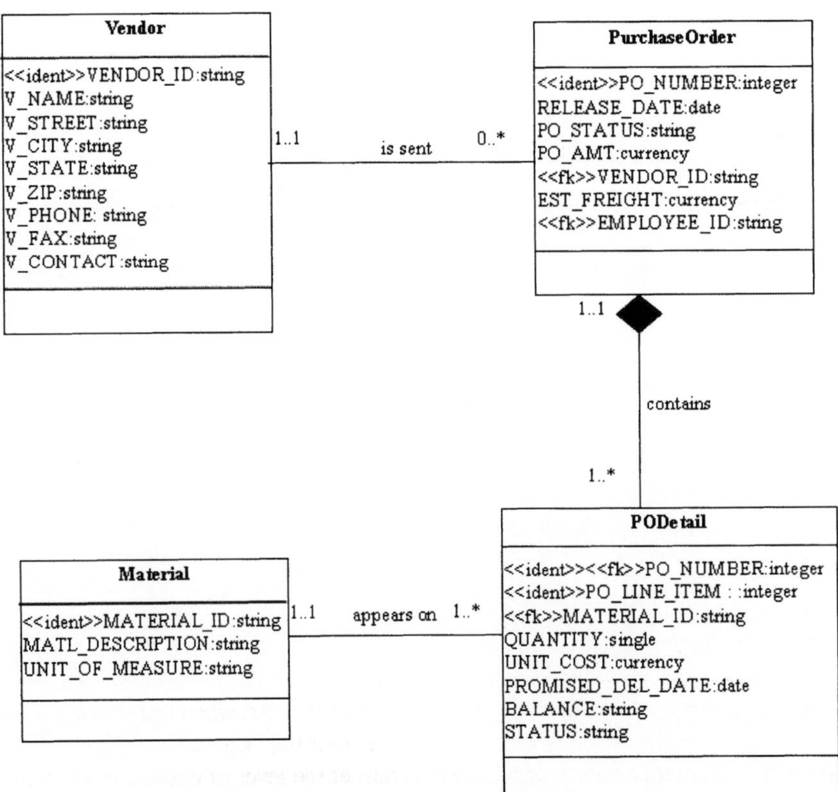

Figure 9.26 Class diagram representation of relational model.

9.6 SUMMARY

Techniques for the conceptual modeling of information systems and software projects have been evolving as the underlying technology for implementation has been changing. There are several techniques to choose from. In earlier chapters, we emphasized traditional approaches. UML is a recent development based on object-oriented thinking, and it has gained some measure of popularity because many design projects that use databases also use software components such as Web pages, intelligent agents, and other software objects. UML enables the analyst to document a more complete design by showing all the interactions among software components, including the database or data objects.

Ideally, UML should be used with objec-oriented databases. However, object-oriented databases have not replaced the relational database in business applications. Instead, it has found its use in specialized applications such as computer-aided design, which visualizes information based on real-world objects such as mechanical components and subassemblies. Despite the current minimal penetration of object-oriented databases, UML is used to specify software components that interface with relational databases and can be used to specify many details of system operation that are not captured in traditional approaches to information system design such as IDEF and DFD.

The knowledge base for the proper use of UML is more detailed than what this chapter has covered. This chapter is introductory in nature and meant to convey the basic ideas underlying UML and to contrast the UML approach with the traditional techniques used throughout most of this book. Refer to the bibliography at the end of this book for more details.

REVIEW EXERCISES

9.1 Consider the use case "Check performance availability" in Figure 9.4. Prepare a sequence diagram like Figure 9.7 for the normal execution of that use case. Prepare an activity diagram that also considers other possible scenarios, like Figure 9.8. Include a text description of the steps of each scenario.

9.2 Using the activity diagram in Figure 9.8 and the sequence diagrams in Figures 9.7 and 9.9, list the steps of other scenarios that can occur in the "make sale" use case. Diagram some of these scenarios as sequence diagrams. (*Hint*: There will be eight additional scenarios.)

9.3 The theater is considering a change in its policy that all sales are final. In particular, it is going to allow cancellations of a sale up to 24 hours after a sale is made. The customer requests a cancellation by submitting the Sale Transaction ID, name, and address on the cancellation page of the Web form. The system interrogates the database to confirm the information and that the transaction is less than 24 hours old. If true, the system asks the customer to confirm that he or she wants to cancel the sale. If the customer confirms the cancellation, the MasterCard account is credited and the Status attribute of the Sale object is changed from Active to Canceled. The system confirms to the customer that the sale has been canceled. Also, the Ticket object status changes from "committed" to "available." The cancellation of a sale requires crediting the MasterCard (reversing the debit) and writing a cancellation to the Sale and Ticket objects. All tickets involved in the sale become uncommitted when the sale is canceled.

(a) Write a text description of a "normal" scenario for the "Cancel sale" use case.
(b) From the scenario, create a sequence diagram at the level of detail shown in Figure 9.6.
(c) Create an activity diagram. Include other logical scenarios in the activity diagram.
(d) Revise the ticket object's state chart diagram (Figure 9.11) to include the new use case.

(e) Create a new state chart showing the behavior of the Sale object.
(f) Update the class diagram (Figure 9.21) to include the new attributes and methods required for this use case.

9.4 Consider the use case diagram of Figure 9.23. The purchase order object goes through two relevant states, one when it is created (OPEN) and another when it has completed its use (CLOSED). Draw a state chart diagram that shows the state behavior of the Purchase Order object.

9.5 Consider the Web page application in the sequence diagram shown in Figure 9.18 and the application described in Section 8.3.7, where a customer is entering a new customer name and password. As indicated in Figure 8.45, there exists a success scenario and a failure scenario.
(a) Prepare a sequence diagram for each scenario.
(b) Prepare an activity diagram for this use case.

9.6 Review Exercise 3.5 (USF Rent-a-Car) provides a fairly detailed explanation of a use case, which shall be called "Rent Car." Review the case, and use the information given to do the following:
(a) Construct a use case diagram that includes the "Rent Car" use case and two other use cases from the problem description: "Make Reservation" and "Return Car."
(b) Prepare a sequence diagram for two scenarios: one in which the reservation has been previously made when the customer executes the "Rent Car" use case and one in which the reservation was not made prior to picking up the car.
(c) Prepare a state diagram for the object CAR, which is one of the classes in this use case.
(d) Using the class diagram format shown in Figure 9.26, develop a class diagram based on the relational data model.

Chapter 10

Workflow Management Systems

10.1 INTRODUCTION

The work on business process reengineering in the 1970s led to the evolution of workflow concepts. Since then, workflows have been a subject of ongoing development in the traditional areas of business processes, such as office automation, health care, telecommunication, and manufacturing. A number of different workflow management systems are available as commercial products or research prototypes that support the traditional areas of business process modeling and coordination, document and image management, and emerging areas such as business-to-business and business-to-consumer interactions. In this chapter, we introduce the basic concepts associated with workflows, focusing on database transaction management in workflow management systems and workflow process definition modeling.

The ***Workflow Management Coalition (WFMC)*** defines a ***workflow*** as a computerized facilitation or automation of a business process, in whole or part, during which documents, information, or tasks are passed from one participant to another for action according to a set of procedural rules. Workflows generally represent processes and activities, which are composed of well-defined tasks. These tasks are related and dependent on one another and are executed either by humans or by processes such as application programs or database management systems. As an example of workflow, consider the patient management processes within hospital management and administration. A workflow of this kind may consist of several tasks such as entering the patient data into a database, obtaining information on earlier visits and medical history, ascertaining insurance information, entering the medical attendant's diagnostics, prescribing treatment medicine, assessing cost, and billing the patient.

Automated workflow management facilitates mapping out business processes, including describing the roles of various entities involved with the workflow, rules for their interaction through the execution of tasks and activities, and movement of documentation. It also implements process models coupled with an enterprise's software applications, shared databases, and e-mail systems to ensure a monitored and supervised information flow through the organization. The application of automated workflow concepts to business processes through organizational analysis and documentation leads to better specification and higher quality of both standard and ad hoc processes. Better reactivity, flexibility, and integration of organizations are achieved due to reduced turnaround times and easier redesign of business processes to adapt to market needs. Workflow systems also link legacy information systems with new business processes, leading to better integration within an organization.

Automated workflow management systems may not be suitable for every organization or every type of process. Workflow management systems are difficult to define and implement for unstructured processes that may be continuously changing. Automated workflow management implementations are more effective in organizations with processes composed of well-defined and structured tasks.

10.2 CLASSIFICATION OF WORKFLOWS

While there is no generally recognized classification of workflow systems, based on the dynamics and structure of the business processes involved, workflows can be broadly categorized as *transaction* and *ad hoc* workflows.

10.2.1 TRANSACTION WORKFLOWS

Transaction workflows entail complex and high-volume processes that are central to the operations within an organization. Generally, these processes are conducted in a structured environment, engaging various departments within the organization integrating legacy and new distributed information systems. Examples of transaction workflows for a manufacturing enterprise include purchasing, production scheduling, and quality control. For an insurance company, loan processing, insurance underwriting, and claims processing are examples of transaction workflows.

10.2.2 AD HOC WORKFLOWS

Many business tasks and activities do not use extensive processes and procedures and have goals and deliverables whose steps and dynamics are unstructured and difficult to predict. ***Ad hoc workflows*** usually involve creative and high-level knowledge workers who may decide the sequence of the activities without the help of workflow systems. The documents and information may be shared with others through e-mail or specialized applications such as spreadsheets or design software. Ad hoc workflows usually are not used for mission-critical business processes; however, some project management capabilities for task scheduling and deliverables may be required. Therefore, ad hoc workflow systems attempt to provide some sort of control for assuring that the tasks are accomplished on time and the deliverables are timely and acceptable. The activities related to product design and marketing are examples of ad hoc workflows. In carrying out the workflow for such activities, there are deadlines and deliverables, but the individual responsibilities and sequencing of tasks can change from one execution to another.

10.2.3 CLASSIFICATION BASED ON WORKFLOW TECHNOLOGY

Considering the transport mechanisms that route work items, workflow systems are divided into four categories: (1) production (or database-based) workflow systems, (2) message-based workflow systems, (3) suite-based workflow systems, and (4) Web-based workflow systems. The choice of the type of workflow technology to implement is largely based on the nature of the processes that are involved.

Production or database-based workflow systems support a central data repository including electronic images of documents and a database management system that supports version control, check-in, and check-out functionalities. The work is routed to different actors within the enterprise who must all have access to the central repository.

Message-based workflow systems (also referred to as administrative workflows) rely on existing e-mail systems to deliver work as file attachments. These are more appropriate for ad hoc workflows related to administrative processes that lack the rigid structure found in transaction workflows. The low cost associated with these types of systems makes them a good candidate for implementation in companies that are at the early stages of workflow automation.

Suite-based workflow systems offer the type of flexibility associated with the processes at the high levels of an organization. These systems integrate office applications such as word processing, spreadsheets, and e-mail. The users can create documents on their desktop and route them to different actors in the organization for review, editing, or comments.

Web-based workflow systems utilize the popularity of the WWW to route documents to actors that are geographically distributed. Since these are enterprise-wide solutions, a significant amount of preplanning is required.

10.3 WORKFLOW MANAGEMENT SYSTEMS

A ***workflow management system (WFMS)*** is a tool that integrates humans, computer systems, information resources, and organizational processes to provide a unified solution. Hence, the requirements of a WFMS are more challenging than those for a database management system (DBMS). The database management system might constitute a part of the WFMS, which additionally involves other users and application tasks that are nontransactional in nature. To standardize the requirements of real-world applications, the WFMC developed a **workflow reference model** as shown in Figure 10.1. The model outlines the architectural representation of a WFMS. According to the workflow reference model, an entire WFMS is centered around a *workflow engine*, which is responsible for enacting task execution, monitoring workflow state, and evaluating conditions related to task dependencies.

A WFMS consists of several other functional components such as the following:

- Process definitions tool
- Workflow enactment service
- Administration and monitoring tool
- Interface to interoperate with client application

The *process definition tool* is used to specify and analyze the workflow process definition. Process definition contains the information regarding the tasks that are to be carried out, the component operation and primitives within the tasks, their starting and completion conditions, and rules and dependencies for navigating between tasks. In general, the process definition tool developed at design time includes the following:

- Formalism for modeling and specification of workflows
- Specification of the task and information associated with it
- Specification of business rules (dependencies and constraints)
- Analysis of the workflow model for desirable properties

The *workflow enactment service* provides a run-time environment, which handles the control and execution of the workflow. A *workflow instance* represents a single execution of a workflow with its own data and independent control requirements as it progresses toward completion. Similarly, a *task instance* is the representation of a single execution of a task within a workflow instance,

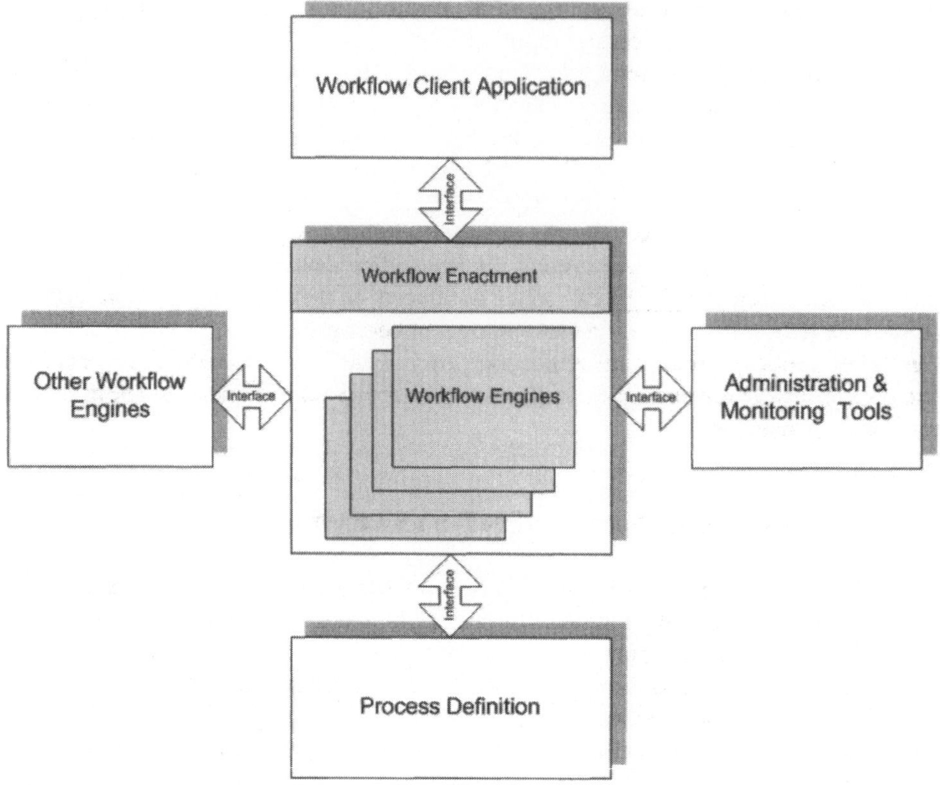

Figure 10.1 Workflow management system reference model.

which contains several task instances. For example, a customer making a cash deposit to his or her account using an ATM is a workflow instance where several instances of tasks such as *verify customer identity* and *credit account* task instances are executed. In general, the execution of a workflow includes enforcing all intertask dependencies and testing for workflow safety.

Administration tools provide functions such as managing users, roles, and security policies. *Monitoring* tools are used for tracking and reporting workflow states and data generation during workflow execution. All of these components have application interfaces that provide standard means of communication between components and the workflow engine.

10.4 WORKFLOW BASICS: TASKS, TASK STRUCTURES, AND TASK DEPENDENCIES

The most important concept in workflows is a task. A ***task*** is a logical unit of work that can be processed by a processing entity. The tasks that are of particular interest to us are database transactions. Consider the online theater ticket sale application introduced in Chapter 9. From a workflow perspective, selling a seat for a performance is a business process that requires the coordinated execution of multiple tasks. For example, the activity "Commit Sale" is a task processed by a software application that updates the status of seats as sold in the enterprise's central database.

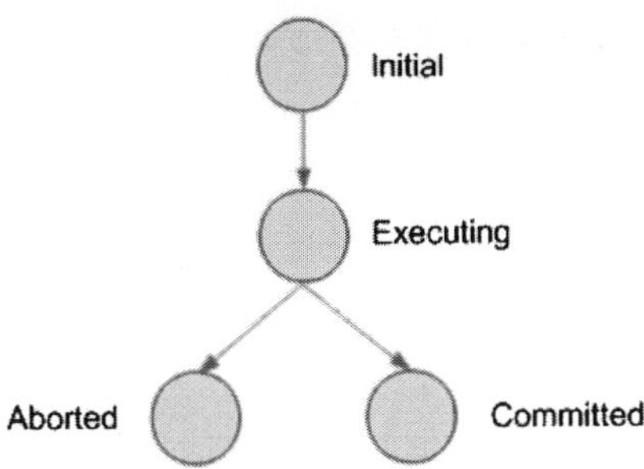

Figure 10.2 Typical task structure.

During its execution, a task can be in one of its externally observable states. Consider the task structure shown in Figure 10.2. A task before it begins execution is in its *initial* state. Once a task begins, it is in its *executing* state until it either commits and reaches its final state, *committed*, where all the operations in the task have been completed successfully and their effects are permanently stored in the system, or it aborts reaching the final state, *aborted*, which signifies the failure of the execution of a task and all effects of the task are eliminated as if it had never been executed.

The exact set of states that belongs to a task is determined by the type of the task. A task can be *transactional* or *nontransactional* in nature. Transactional tasks have four externally visible states: *initial*, *executing*, *committed*, and *aborted* as shown in Figure 10.2. Typically, tasks executed by information system software applications that do not involve humans, such as the "Commit Sale" task, are considered transactional tasks. Human activities are usually considered to be the nontransactional tasks. The externally visible states of a nontransactional task are *initial*, *executing*, *failed*, and *done*, where the failed and done states correspond to the aborted and committed states of the transactional task structure.

10.4.1 Task Structure

A task's internal structure also depends on the characteristics of the system in which the task is executed and the properties of the processing entity responsible for the execution of a task. In the workflow environment, a transactional task executes a sequence of operations and then requests a commit or abort. If a commit fails, then the task is aborted. This type of structure is a *one-phase commit structure* as depicted in Figure 10.2. However, this is not always the case as transactional tasks can have a *two-phase commit task structure* as shown in Figure 10.3. A two-phase commit first enters prepared to commit (*precommit*) state after completing its execution; if an external controller decides to commit, the task is guaranteed to commit, otherwise it is aborted. The two-phase commit task can also abort from its executing state as in the one-phase commit task structure and unlike the transition connecting the precommit state to the aborted state, this transition is *uncontrollable* and triggers if the processing entity fails to execute the operations in the task.

Consider the sequence of activities described by "Verify Seating Available," "Offer seating," and "Place Ticket on Hold" in the online ticket agency example in Figure 9.8 reproduced in this chapter as Figure 10.4. The combination of these activities can be captured in a two-phase commit

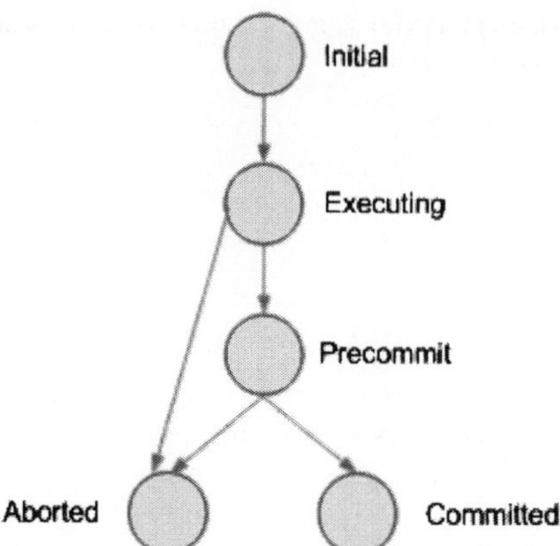

Figure 10.3 Two-phase commit task structure.

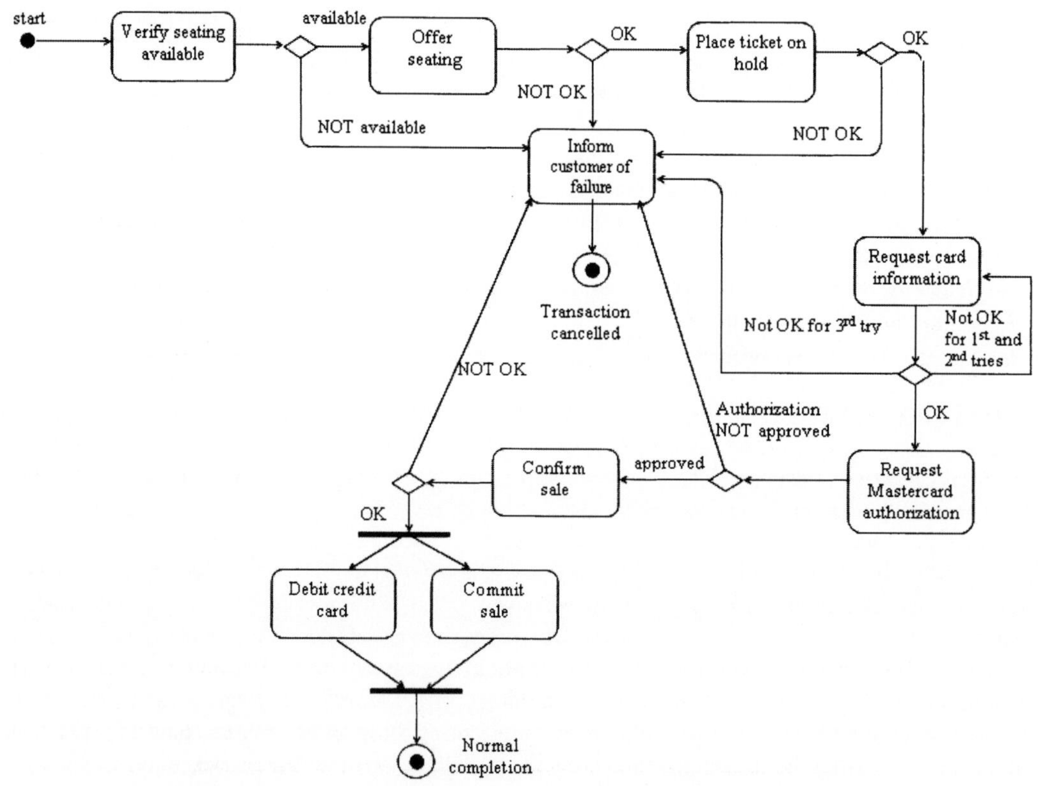

Figure 10.4 Activity diagram of online ticket sales, reproduced from Figure 9.8.

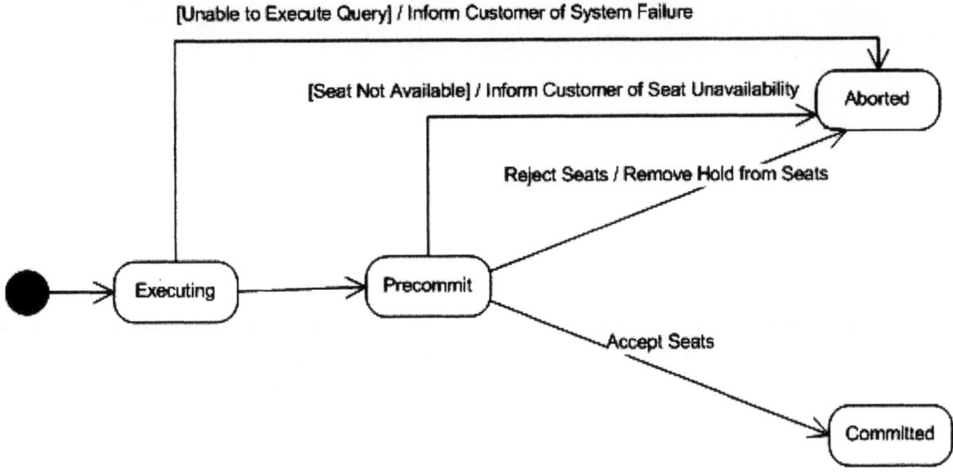

Figure 10.5 Two-phase commit task structure for the reserve seat task.

task structure as a "Reserve Seat" task. In the online ticket sales workflow, a customer may reach a Web page and enter information regarding his or her seating requirements. As soon as the customer submits the request (by clicking a button on the Web page perhaps), the task is initiated. During its executing state, a query is run to determine if any seat(s) matching the customer's preferences is available. If the query cannot be successfully completed, the task aborts from the executing state. Otherwise, it reaches the precommit state. At this state, if the query has returned no available seats, the task is aborted without any permanent changes to the database and the customer is informed of the outcome. If there are available seats, these seats are temporarily put on hold and offered to the customer who at this point may accept or reject the offered seats. If the seats are rejected, the task is aborted with no affects on the database. If the seats are accepted, then the task commits saving the changes to the status of the seats in the database to reflect the fact that they are put on hold. In this example, the response of the customer accepting or rejecting the seats acts as an external controller, which decides whether the task will be committed or aborted after it successfully completes execution. A diagram of the two-phase commit task structure for the "Reserve Seat" task is shown in Figure 10.5.

10.4.2 Task Dependencies

The collections of tasks that make up a workflow are executed in accordance with an enterprise's business rules. The activity diagram shown in Figure 10.4 depicts several activities and the interrelationships among them for the online ticket agency example in Chapter 9. The diagram indicates that the credit card information of a customer is requested only if the "Place Ticket on Hold" activity is successful. This may be the result of a business rule that requires that the credit card information from a customer is required after a ticket for that customer is put on hold. Similarly, a sale is not committed unless the customer has an authorized credit card. In workflows, the business rules that govern the execution of tasks are described by intertask dependencies. In this section, we will use the *ACTA* (which means "actions" in Latin) formalism, originally developed to define the interrelations among transactions in database systems. The intertask dependencies associated with transactional workflows can be broadly categorized into three types:

1. *Control-flow dependency.* A control-flow dependency between two tasks T_i and T_j specifies the condition under which T_j is allowed to enter state st_j based on the state st_i of T_i. Under these conditions, task T_i is the *parent* and task T_j is the *child.*
2. *Value dependency.* A value dependency specifies task dependencies based on the output value generated by certain tasks.
3. *External dependency.* These dependencies are due to some external factors such as time. These are also termed as temporal dependencies.

10.4.2.1 Types of Control-Flow Dependencies

Based on their precedence order and incompatible states, control flow dependencies are classified into three general types.

Strong Causal Type

This type of a dependency generally implies the necessary condition of a relation. If such a dependency exists between two tasks T_i and T_j, then it implies that T_j can enter state st_j only if T_i enters state st_i. The dependencies that belong in this category are as follows:

- *Begin dependency.* Task T_j cannot begin execution until task T_i begins.
- *Begin on commit dependency.* T_j cannot begin executing until T_i commits.
- *Begin on abort dependency.* T_j cannot begin executing until T_i aborts.
- *Serial dependency.* T_j cannot begin executing until T_i either commits or aborts.
- *Terminating dependency.* T_j cannot commit or abort until T_i either commits or aborts.

Weak Causal Type

This dependency specifies the sufficient condition for the relationship. The dependency can be interpreted as T_j and must enter state st_j if T_i enters state st_i. The dependencies that belong in this category are as follows:

- *Strong commit dependency.* If task T_i commits, then task T_j commits.
- *Forced commit on abort dependency.* If task T_i aborts, then task T_j commits.
- *Abort dependency.* If task T_i aborts, then task T_j aborts.
- *Exclusion dependency.* If task T_i commits and task T_j has begun executing, then task T_j aborts.

Precedence Type

This dependency enforces a condition where T_i must enter state st_i, before T_j enters state st_j, if both st_i and st_j are to occur. The dependencies that belong in this category are as follows:

- *Commit dependency.* If both task T_i and T_j commit, then the commitment of T_i precedes the commitment of T_j.
- *Weak abort dependency.* If task T_i aborts and task T_j has not yet committed, then task T_j aborts. In other words, if task T_j commits and task T_i aborts, then the commitment of T_j precedes the abortion of T_i.
- *Weak begin on commit dependency.* If T_i commits, T_j can begin executing after T_i commits.

To demonstrate the use of intertask dependencies in the workflow process definition, consider the online ticket sales business process. The set of tasks in this business process is as follows: (1) *Reserve Seat*, which combines the "Verify Seating Available," "Offer Seating," and "Place Ticket on Hold" activities as described in Section 10.4.1; (2) *Authorize Payment*, which combines "Request Card Information" and "Request MasterCard Authorization"; and (3) *Confirm Sale*, (4) *Process Payment*, and (5) *Commit Sale* tasks corresponding to the "Confirm Sale," "Debit Credit Card," and "Commit Sale" activities, respectively. We will consider the following business rules, which govern the execution of these tasks:

- *Authorize Payment cannot start until Reserve Seat commits.* This rule indicates a begin on commit dependency between the "Authorize Payment" and "Reserve Seat" tasks.
- *Confirm Sale cannot start until Authorize Payment commits with approved credit card information.* This rule indicates a begin on commit dependency as well as a value dependency, which requires an approved credit card between the "Confirm Sale" and "Authorize Payment" tasks.
- *The credit card information for a customer for any reservation can be tried at most three times for approval before it is considered not approved.* This is an external dependency indicating that the "Authorize Payment" task must commit within at most three tries.
- *Commit Sale and Process Payment cannot start until Confirm Sale commits.* This rule indicates a begin on commit dependency between "Commit Sale" and "Confirm Sale" and "Process Payment" and "Confirm Sale" tasks.
- The activity diagram of Figure 10.4 does not explicitly indicate a relationship between the "Commit Sale" and the "Process Payment" tasks. However, lack of coordination between these tasks may result in undesirable situations. For example, a ticket may be sold to a customer without properly processing a payment if the "Process Payment" task aborts and "Commit Sale" task commits. *The online ticket agency may want to ensure that a sale is not finalized until the credit card is debited.* This type of business process requirement requires careful specification of control-flow dependencies. The synchronization of these two tasks will be revisited in the next section when we introduce formal modeling of control-flow dependencies.

Figure 10.6 summarizes the dependencies discussed here. Compared with the activity diagram shown in Figure 10.4, groups of activities are represented by a single task (e.g., Reserve Seat task represents Verify Seating Available, Offer Seating, and Place Ticket on Hold activities). In addition, some activities such as "Inform Customer of Seat Unavailability" are not shown in Figure 10.6. This activity is considered as an *action*, which we will further discuss in the next section when formal workflow models are considered.

While activity diagrams are useful in capturing the overall execution of activities in business processes, specification of workflows based on task structures and intertask dependencies allows formal modeling tools such as Petri Nets, state charts, and other discrete event system modeling tools to be used in the verification of desirable properties of the workflow process definition including proper termination, nonblocking, and correctness. In the next section, we demonstrate the use of state charts, one of the Unified Modeling Language (UML) diagramming tools, to create formal models of workflows and compare their expressiveness with activity diagrams.

10.5 MODELING WORKFLOWS USING STATE CHARTS

State chart diagrams were previously introduced in Section 9.4.4 of Chapter 9 as a diagramming tool that shows the various states that an object can be in and how the object evolves through its

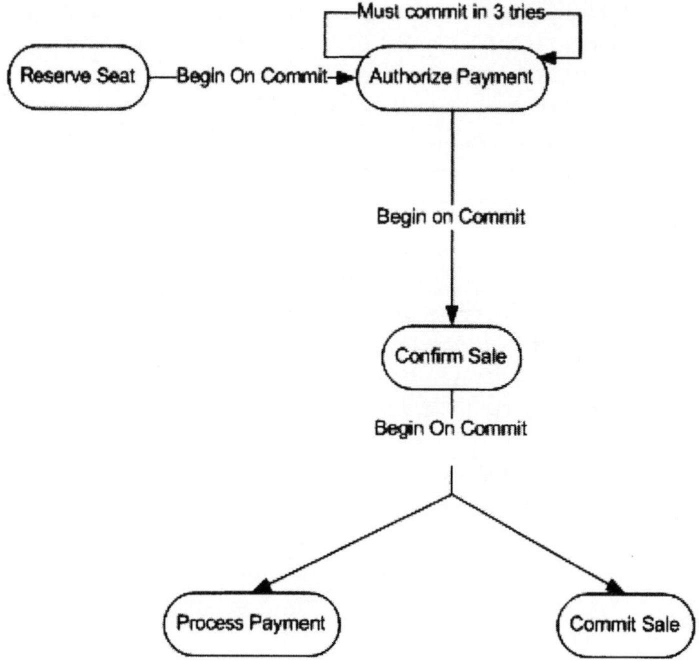

Figure 10.6 Intertask dependency diagram for the online ticket sales workflow.

states over time as a result of external and internal events that occur in a dynamic system. A workflow system is also a dynamic system where the execution of the tasks within the system changes the state of the workflow and must be coordinated based on the dependencies specified in the process definition.

State charts offer several advantages over activity diagrams in modeling of workflows. Similar to activity diagrams, graphical elements are used to describe the complex behavior of large-scale systems. In addition, through hierarchical decomposition, state charts allow the users to focus on different levels of detail as required. State charts also possess formal semantics, which allow for verification and validation of workflow models that are otherwise not possible based on workflow models created using activity diagrams. The basic constructs of state charts useful in modeling complex workflows are discussed next.

10.5.1 Hierarchy

The hierarchy associated with state charts allows the users to view the system at different levels of refinement where a set of states are contained within a single state. For example, in Figure 10.7, states X and Y are contained in state Z. When the system is in state Z, then the system can be in state X or Y but not both, representing an *OR-decomposition*.

An *AND-decomposition* indicates that if a system is in an AND-state, then the system is in all of its substates. The AND-state is shown by a dashed line separating the boundaries of the states. For the state chart in Figure 10.8, when the system is in state Z, it is both in states S and V. As soon as the system reaches state Z, both the transitions from the initial states of S and V are trig-

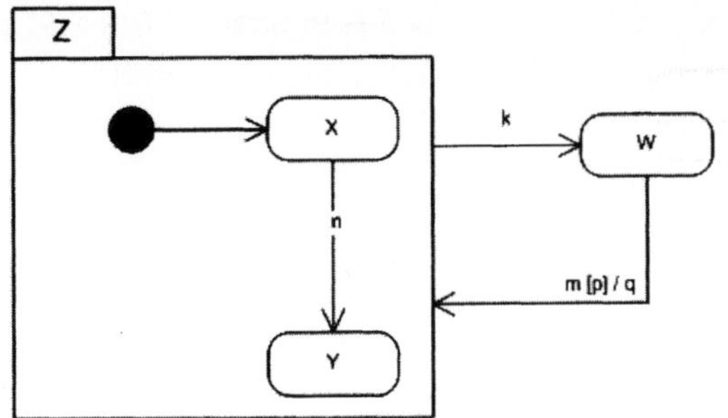

Figure 10.7 State chart OR-decomposition.

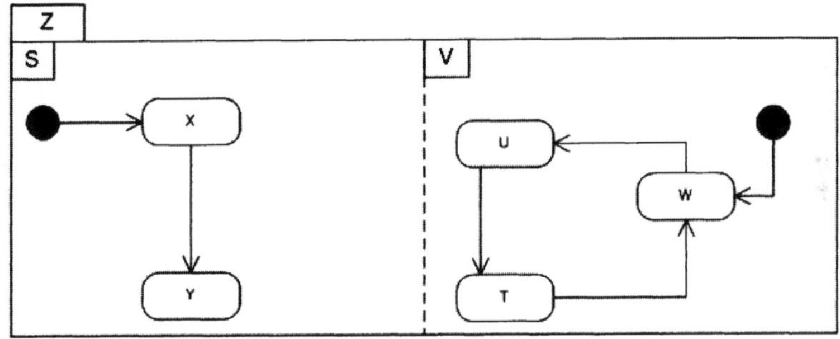

Figure 10.8 State chart AND-decomposition.

gered and move the system to state X and W. The expressiveness of the AND-decomposition greatly decreases the number of states in complex workflow models.

10.5.2 Dynamics

The transitions in state charts are labeled according to ***ECA rules*** which follow the syntax "*E[c]/A*" where "E" is an *event*, "c" denotes a *condition*, and "A" represents an *action*. A transition may be associated with all, some, or none of the three. A transition is triggered immediately when the system is in the source state of the transition and the transition is not labeled by any event, condition, or action, moving the system to the transition's target (output) state. Similarly, if a transition is labeled by "E," then it is triggered when event "E" occurs, if it is labeled by "E[c]," the transition is triggered when event "E" occurs and condition "c" is satisfied. Finally, if a transition is labeled "E[c]/A," when event "E" occurs and condition "c" is satisfied, the transition is triggered and action "A" is executed before the system reaches the transitions output state. For example, in Figure 10.7, event "k" will take the system to state W, whether the system is in state X or Y. When the system is in state W and condition "p" holds, event "m" will execute action "q" and take the system to state Z. In state Z, the system will default to state X since state X is connected to the initial state with a transition that automatically triggers from the initial state.

10.5.3 Modeling Control-Flow Dependencies using ECA Rules

The conditions associated with the transitions in state chart diagrams can be used to model the control-flow dependencies between tasks. In this section, we demonstrate the modeling of the three types of control-flow dependencies outlined in Section 10.4.2.1.

The strong causal type dependencies imply that task T_j can enter state st_j only if task T_i enters state st_i. These dependencies can be modeled by associating a condition that requires that task T_i is in state st_i with a transition that leads to state st_j. These dependencies also imply a precedence relationship, where task T_i enters state st_i before T_j enters state st_j. Consider the begin on abort dependency which requires that T_j cannot begin executing until T_i aborts. The ECA rule for this dependency is shown in Figure 10.9.

The weak causal type dependencies imply that task T_j must enter state st_j if task T_i enters state st_i. These dependencies are used to coordinate the aborting and committing of two tasks by associating a condition with the transition leading to the complementary state st_k (aborted and committed are complementary states since a task cannot terminate in both states). Figure 10.10 illustrates the task structure of task T_j, which has an abort dependency with task T_i; namely, if task T_i aborts then task T_j aborts. Notice that before task T_j is committed, the condition that task T_i has not aborted is checked. If the condition is false (task T_i has aborted), task T_j will not be allowed to commit and terminate in its complementary state aborted. Weak causal dependencies do not imply a precedence requirement—namely, task T_j can abort or commit before T_i has terminated without violating the requirements of the Abort dependency.

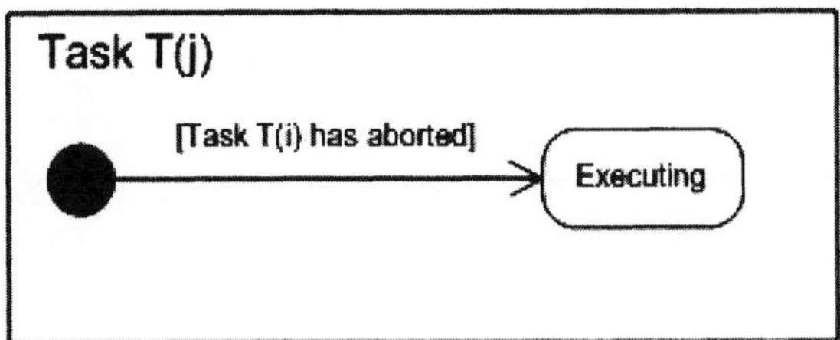

Figure 10.9 Begin an abort dependency.

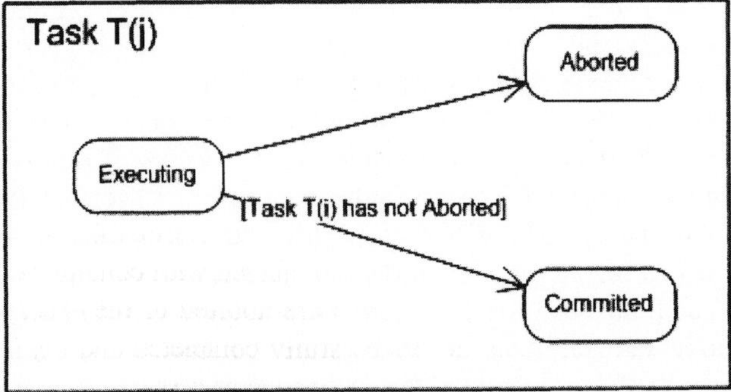

Figure 10.10 Abort dependency.

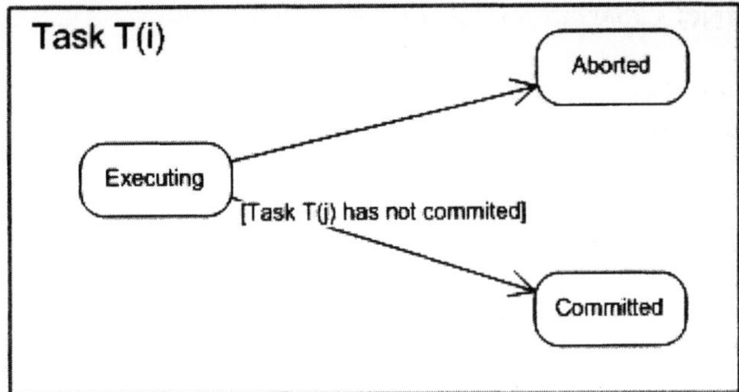

Figure 10.11 Commit dependency.

Precedence type dependencies imply a condition where T_i must enter state st_i before T_j enters state st_j if both st_i and st_j are to occur. This dependency is slightly different than the strong and weak causal relationships in the fact that a condition is associated with the parent side of the dependency (task T_i) rather than the child side (task T_j). Figure 10.11 shows the state chart diagram for the Commit dependency between task T_i and task T_j. The condition that task T_j is not committed is checked before task T_i is committed since committing task T_i after task T_j is not allowed by the precedence dependency. Other combinations of terminal states between tasks T_i and T_j, such as Task T_i aborts and Task T_j commits in any order, do not violate the requirements of the Commit dependency.

10.6 ILLUSTRATIVE EXAMPLE USING STATE CHARTS TO MODEL WORKFLOWS

The general approach to modeling workflows using state chart diagrams follows a top-down approach, in that a high-level model of the workflow is first constructed whereby the states of the state chart diagram represent the tasks in the workflow model. The high-level state chart diagram for the online ticket sales workflow is shown in Figure 10.12. The "Process Payment_AND_Commit Sale" is a composite state representing an AND-decomposition, which indicates the simultaneous execution of the "Process Payment" and "Commit Sales" tasks initiated after the "Confirm Sale" task commits.

A workflow instance of the online ticket sales workflow is initiated when a "Reservation_req" (Reservation Request) event triggers the transition from the Initial state to the "Reserve Seat" state, which represents the "Reserve Seat" task in the workflow. If the "Reserve Seat" task instance commits, then the workflow moves to the "Authorize Payment" state initiating an instance of the "Authorize Payment" task. If "Reserve Seat" task instance aborts, then the workflow instance is terminated. The other transitions in the diagram are labeled with conditions that either terminate the workflow instance if any of the task instances are aborted or the execution of the workflow instance is continued as task instances are successfully completed and committed. At this high-level view, the internal structures of tasks and dependencies associated with these internal structures are hidden.

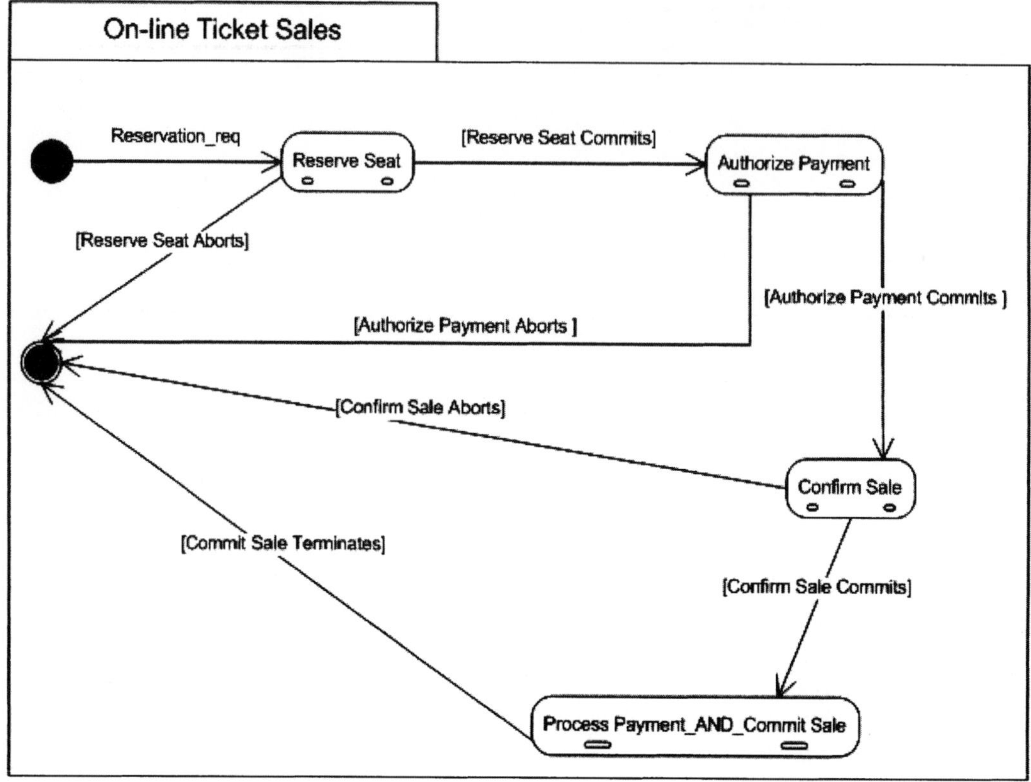

Figure 10.12 Top-level state chart diagram for the online ticket sales workflow.

The "Reserve Seat" state is a *composite state*, representing an OR-decomposition, which has internal states and transitions representing the "Reserve Seat" task's structure as shown in Figure 10.13. When the "Reserve_req" event is executed, the transition triggers and defaults to the initial state of the "Reserve Seat" task, which immediately triggers the transition between the Initial and the Executing states of the task. The internal transitions of this task were discussed in Section 10.4.1. The uncontrollable transition depicting the failure of the processing entity for a typical two-phase commit task, shown in Figure 10.4, is omitted in the diagram in Figure 10.13. Control-flow specification in the presence of uncontrollable events requires sophisticated analysis methods and will be discussed in Section 10.7. The workflow tasks in this example are assumed to be executed by processing entities, which guarantee execution of the tasks without failure.

It is important to note how some of the activities in the activity diagram are captured as actions associated with transitions. For example, in the Precommit state, if the query based on the customer's request returns no available seats, the "[seat not available]" condition holds, the system moves to the Aborted state executing the "Inform Customer of Seat Unavailability" action which presumably involves displaying a message box or a Web page describing the status of available seats. The decision of the customer to accept or reject the requested seats acts as the external controller in the two-phase commit task structure and is modeled as external events which decide whether the "Reserve Seat" task will be committed or aborted. As the top level model indicates, if the "Reserve Seat" task is aborted, the transition connecting the Aborted state and the Final state of the online ticket sales workflow triggers, terminating the workflow instance. The terminal state of the workflow instance is outside of the rectangle corresponding to the internal states of the "Reserve Seat" task. If the "Reserve Seat" task commits, then the system moves to the "Authorize Payment" state.

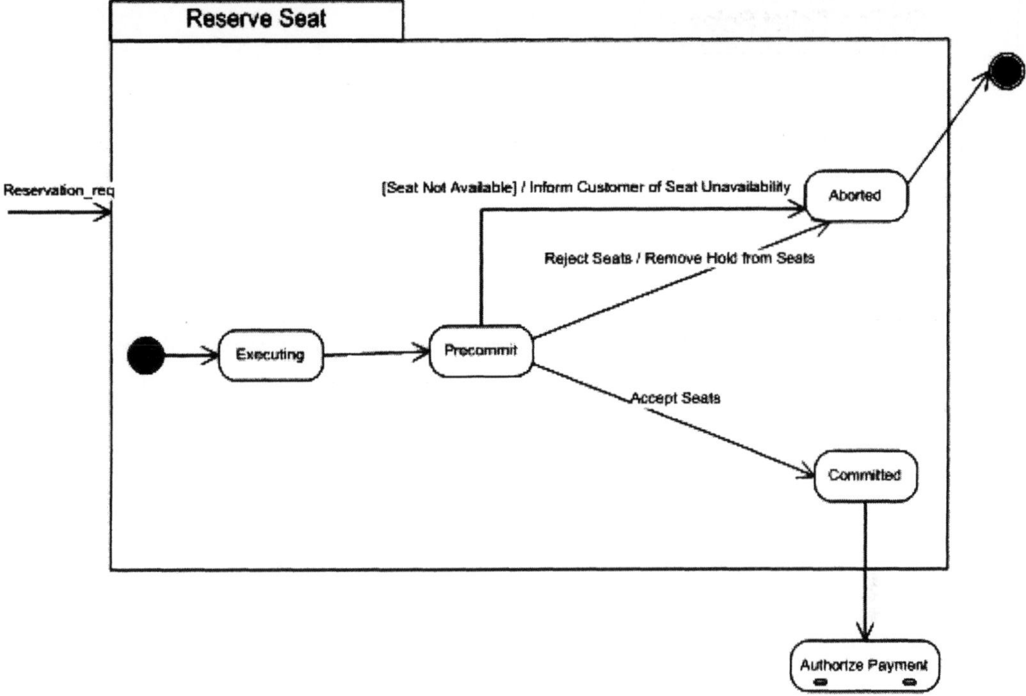

Figure 10.13 Low-level state chart diagram for the reserve seat task.

The lower-level state chart diagram of the "Authorize Payment" task, shown in Figure 10.14, illustrates how external and value dependencies are captured in state chart diagrams via the use of conditions. When the "Authorize Payment" task reaches the Precommit state, the output of the execution as "Approved" or "Not Approved" is evaluated (value dependency). If the credit card is approved, the task commits and the transition targeting the "Confirm Sale" task is triggered. If the card is not approved, the system checks the value of a counter (external dependency), which keeps track of the number of times the "Authorize Payment" task instance in this workflow instance is executed without approval. If the value of the counter is less than 3, the counter is incremented and the task is reinitiated by the transition connecting the Precommit state with the outer layer of the "Authorize Payment" task. If the counter's value is 3 and the card is not approved, the customer is informed and the task instance aborts terminating the workflow instance.

The lower-level state chart diagram for the "Confirm Sale" state is shown in Figure 10.15. Presumably during the Executing state of this task instance, a Web page is displayed where the customer reviews the seat assignments and the pricing for the requested seats as well as the payment information. If the customer agrees with the purchase, he or she may click a button and confirm the sale; otherwise the customer may click another button to reject the sale if any of the information is not accurate. Once the task instance reaches the Precommit state, the response from the customer is evaluated (value dependency). If the sale is confirmed, the task instance reaches the Committed state and triggers the transition, which takes the workflow instance to the "Process Payment AND Commit Sale" state. If the customer rejects the sale (sale not confirmed), the hold status on the seats is removed and the task instance reaches its Aborted state from which the workflow instance terminates.

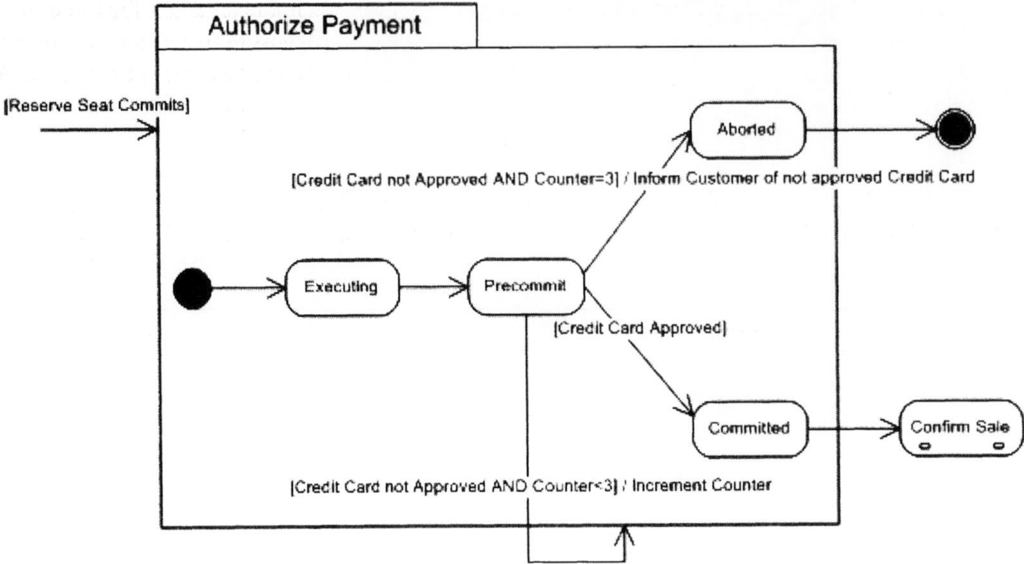

Figure 10.14 Low-level state chart diagram for the authorize payment task.

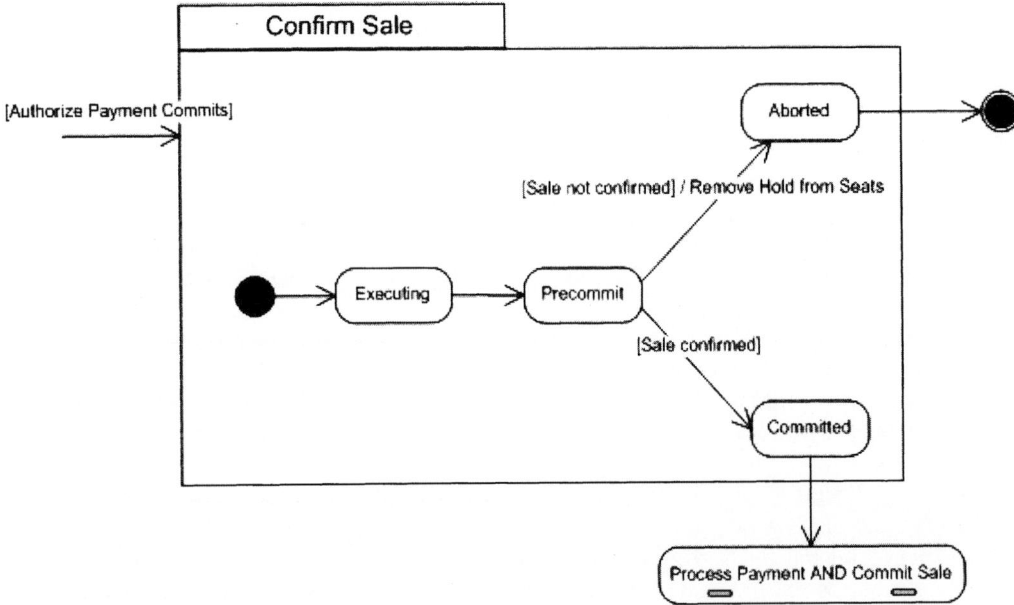

Figure 10.15 Low-level state chart diagram for the confirm sale task.

The lower-level state chart diagram for the "Process Payment_AND_Commit Sale" task is a composite state representing an AND-decomposition. The state represents two tasks, "Process Payment" and "Commit Sale," which are executed simultaneously. If no dependencies are specified between these tasks, they can terminate in any order and in any of their terminal states. For example, the "Process Payment" task may commit, and the "Commit Sale" task may abort, resulting in charging a customer without selling a seat. This is not a desirable situation for the online ticket sales agency. Let us consider how the online ticket agency can ensure that a sale is committed if a customer's credit card is debited. The first requirement is that the "Process Payment" task must terminate before the "Commit Sale" task, which can be accomplished by a Terminating dependency. In addition, a Strong Commit dependency will ensure that if the "Process Payment" task commits, then the "Commit Sale" task commits. Figure 10.16 illustrates how these dependencies are described in a state chart diagram. The transition labeled with the condition "[Confirm Sale commits]" targets the outer layer of the orthogonal state, and when triggered, it initiates both tasks. The condition "[Process Payment has terminated]" is the result of the Terminating dependency and will not allow the "Commit Sale" task to commit or abort until the "Process Payment" task commits or aborts. The Strong Commit dependency, which is a weak causal type dependency,

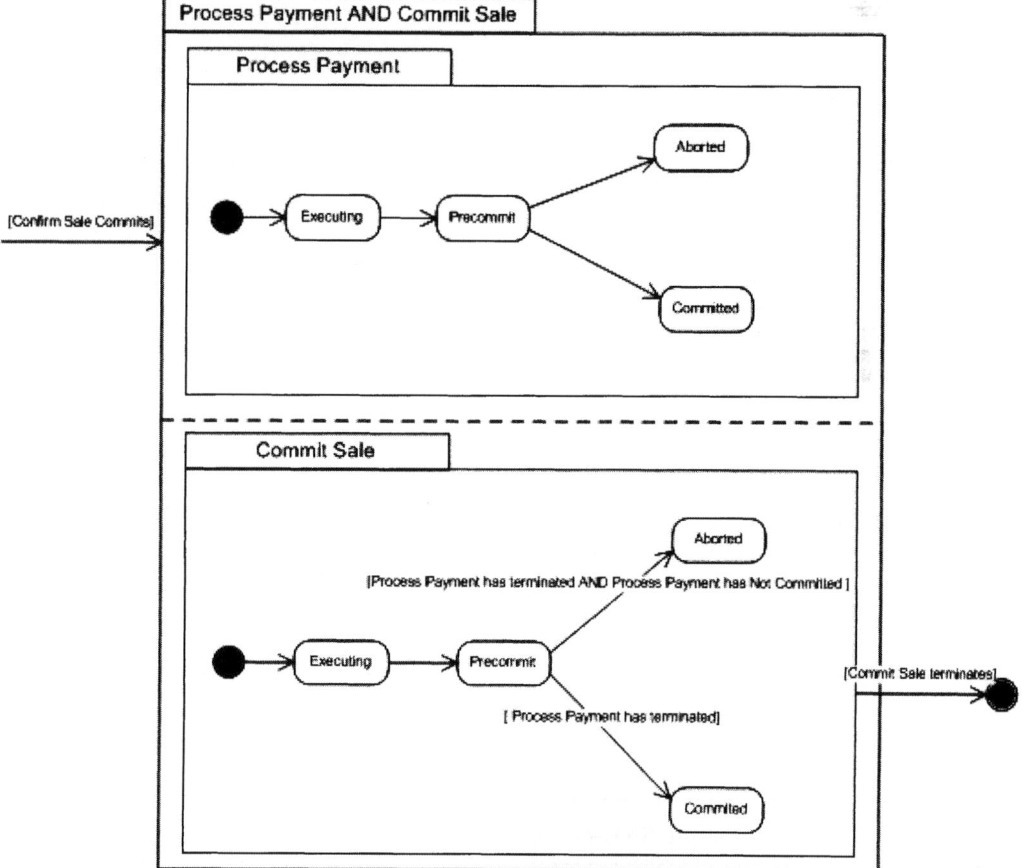

Figure 10.16 Low-level state chart diagram for the process payment AND commit sales state with terminating and strong commit dependencies.

is implemented by the condition "[Process Payment has Not Committed]," which is checked before the "Commit Sale" task is aborted. If the condition is false ["Process Payment" task has committed], then this transition cannot be fired, forcing the "Commit Sale" task to commit. Once the "Commit Sale" task reaches one of its terminal states, the workflow instance is also terminated.

Figure 10.17 shows the low-level state diagram for the complete online ticket sales workflow, which integrates the low-level state charts of all the composite states in the top-level diagram.

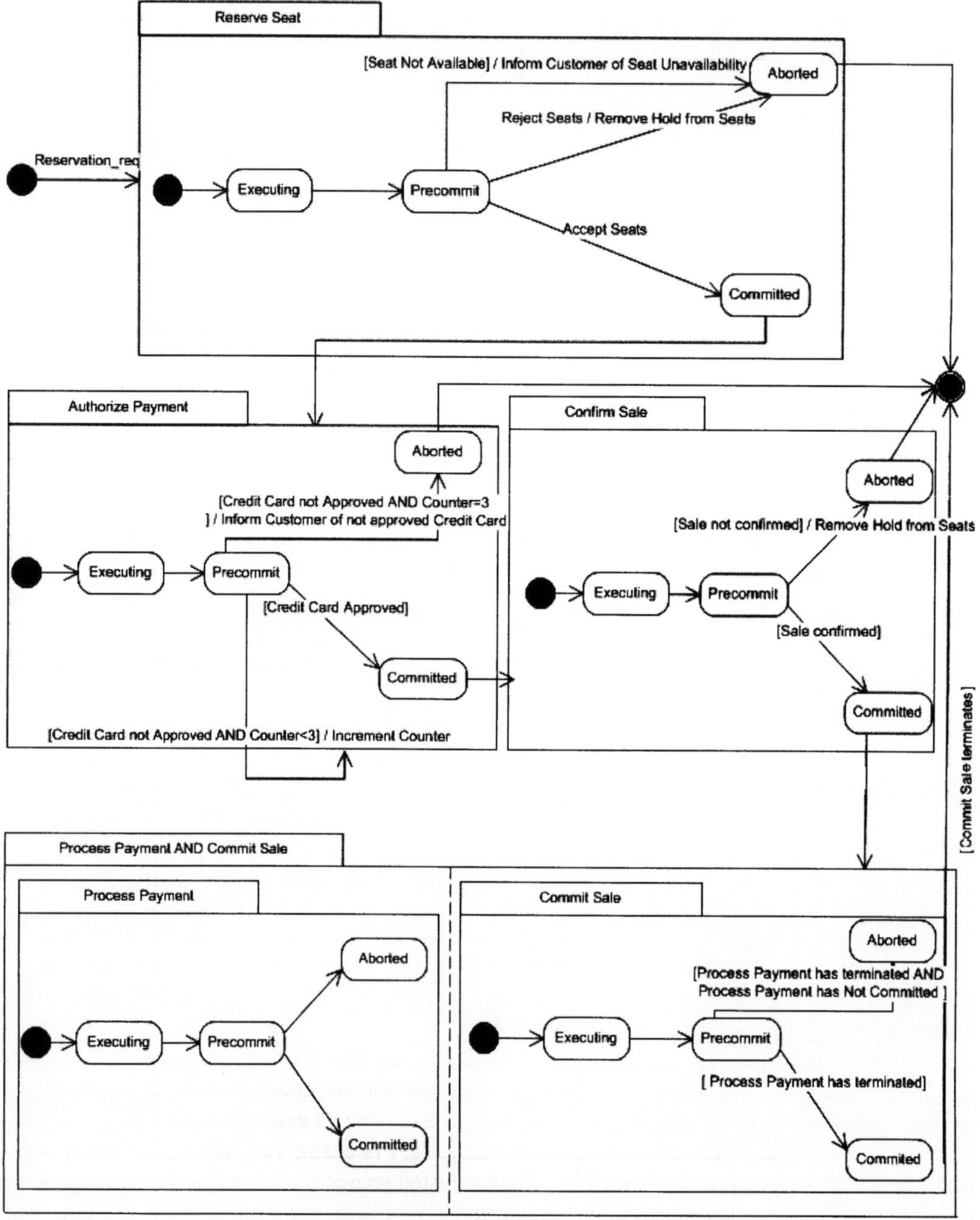

Figure 10.17 Low-level state chart diagram for the online ticket sales workflow.

10.7 ANALYSIS OF WORKFLOW PROCESS DEFINITIONS

State charts and other formal modeling tools, such as Petri Nets, have been extensively used in the analysis and verification of desirable properties of the workflow process definitions, including consistent dependency specification, safe termination, nonblocking, and correctness. Formal discussions of these approaches are beyond the scope of this text. To motivate an informal discussion of the analysis of process definitions, consider the tasks "Process Payment" and "Commit Sale" and the Terminating and the Strong Commit dependencies between these tasks. The discussion in the preceding example considered task structures where the processing entities are guaranteed to complete the execution of the tasks. The uncontrollable transition from the Executing state to the Aborted state of the two-phase commit task structure is not considered (see Figure 10.3). In such an environment, the Terminating dependency is sufficient to guarantee that the "Process Payment" task would be committed before the "Commit Sale" task terminates since the "Commit Sale" task is guaranteed to reach its Precommit state and wait until the "Process Payment" task terminates. Therefore, the Terminating and the Strong Commit dependencies between these tasks are sufficient to guarantee proper termination of these two tasks.

Consider the task structure where both tasks follow the typical two-phase commit task structure including the uncontrollable transitions from the executing states to the aborted state as depicted in Figure 10.18. This situation may lead to several undesirable terminal state combinations based on just the Terminating and the Strong Commit dependencies specified in the preceding example:

- Both tasks start executing and the "Process Payment" task aborts from its Executing state, and the "Commit Sale" task commits (since the only condition required to commit the "Commit Sale" task is termination of the "Process Payment" task) resulting in a committed seat without payment. This situation is a result of incorrect dependency specification.
- Both tasks start executing, and due to a problem with the processing entity, the "Commit Sale" task aborts from its executing state prior to the termination of the "Process Payment" task. While this violates the Terminating dependency between these tasks, this is an uncontrollable transition and triggers when there is a failure associated with the processing entity. This situation is a result of improper controller design.

The incorrect dependency specification between these two tasks can be resolved by adding an Abort dependency in addition to the Terminating and Strong Commit dependencies between these two tasks. Incorporation of this dependency is left as an end-of-chapter exercise. Correcting the improper controller design error is more involved. In this small example, one simple solution may be obvious to the reader—that is, preventing the start of the "Commit Sale" task until the "Process Payment" task terminates. In this manner, the Terminating dependency would not be violated.

The reader should also observe that the presence of uncontrollable transitions causing these tasks to be uncontrollably aborted requires considering the redesign of this workflow since some scenarios may require crediting the customer's payment or taking seats off hold if the sale is not committed.

The proper design of workflow process definitions and their associated workflow engines, which control the execution of workflow tasks, is a complex and critical process. While informal analysis of very small workflows may uncover inconsistencies as described earlier, methods for formal and systematic analysis of large-scale workflows are necessary to ensure functional and correct workflow management system implementations. The interested reader is referred to the works of Adam (1998), Darabi (2006), Van der Aalst (2002), and Yalçın (2005) for further discussion of formal methods used in the analysis of workflow process definitions.

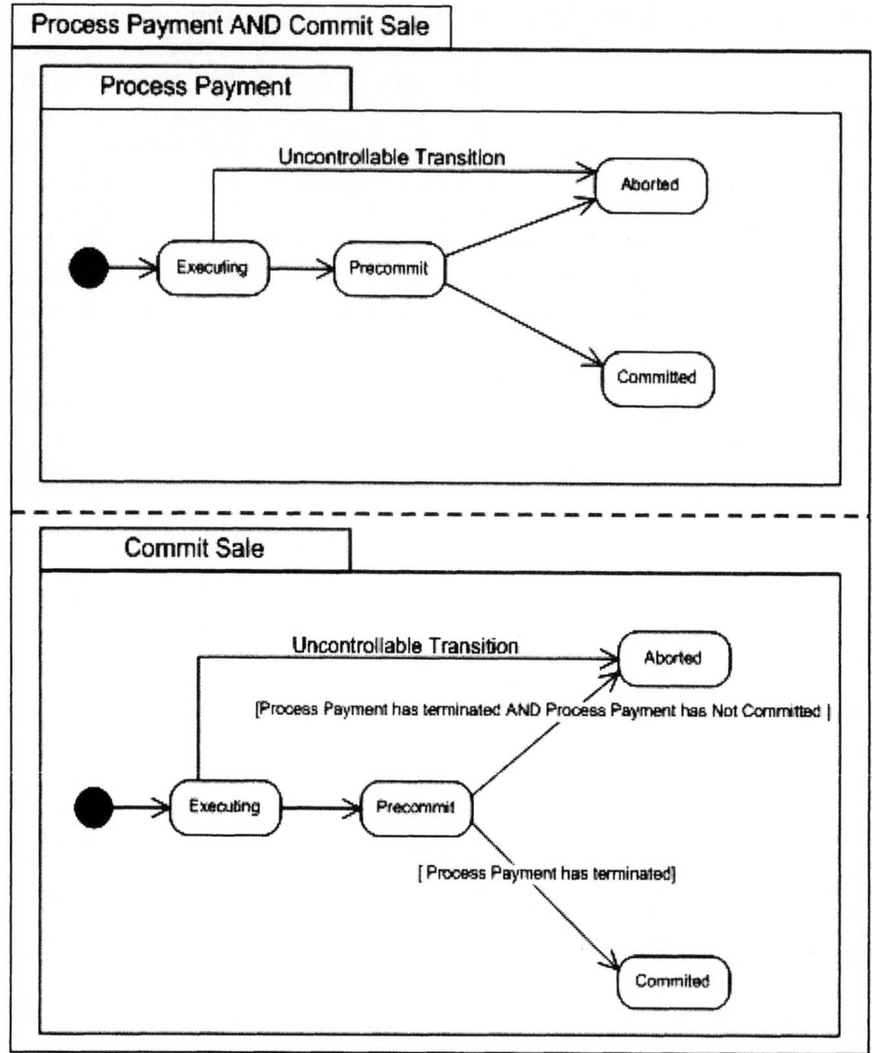

Figure 10.18 State chart for process payment and commit sale tasks with uncontrollable abort transitions.

10.8 SUMMARY

Workflow management systems are large-scale software systems, which may involve business process modeling and coordination, document and image management, and communications required by business-to-customer and business-to-business interactions in addition to other functionalities that support enterprise-wide goals and objectives. A significant part of the functionality of workflow management systems involves interaction with enterprise databases to record, retrieve, and update data resulting from or necessary for the coordinated execution of tasks. In addition, workflow management systems provide a high-level tracking and control system that enforces coordinated and proper task execution by the appropriate resources within the enterprise. Businesses competing in markets with high degrees of control and accountability, such as federally regulated industries, almost always implement some degree of workflow management to control, track, and document daily activities of their enterprise. This chapter introduces the basic concepts associated with workflows and informal and formal process definition models from a control-flow perspective based on intertask dependencies. Workflow management system software are too complex to discuss within the scope of this chapter; however, the interested reader is referred to Kiepuszewski (2003) for a comprehensive review of the capabilities of the popular applications including SAP's SAP R/3 software suite, which includes an integrated workflow component called *SAP R/3 Workflow* and IBM's MQ Series Workflow re-released under the name *Websphere MQ Workflow*. Links to other workflow management software, which are also marketed under the name business process management software by companies such as Fujitsu and Oracle, are included in the bibliography.

REVIEW EXERCISES

10.1 Consider the following state chart diagram. Determine the resultant state(s) and action(s) (if any) based on the given current state, the event(s), and condition(s) indicated in the table.

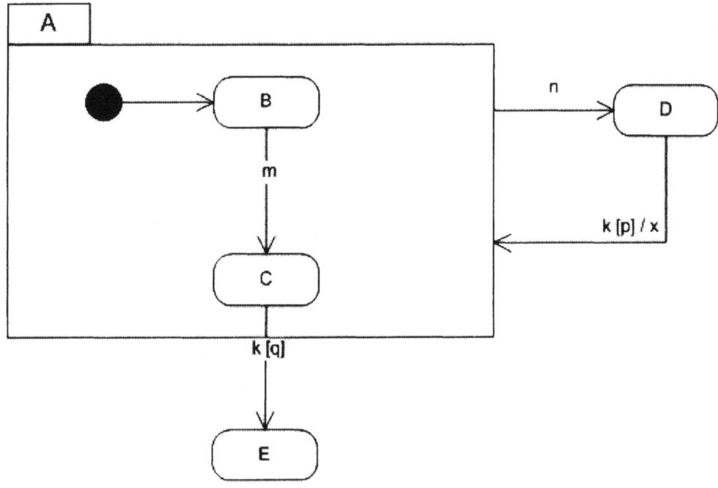

	Current state	True conditions	Events	Action	Resultant state
(a)	B	—	m		
(b)	B	—	n		
(c)	C	p	k		
(d)	D	p	k		
(e)	C	q	k		

10.2 Consider the following state chart diagram. Determine the resultant state(s) and action(s) (if any) based on the given current state, the event(s), and the condition(s) indicated in the following table:

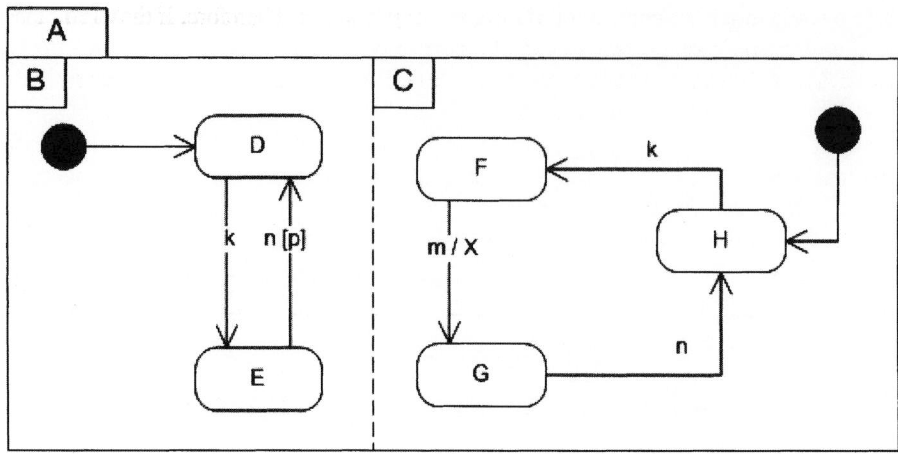

	Current state	True conditions	Events	Action	Resultant state
(a)	D,H	—	k		
(b)	E,F	p	m		
(c)	E,G	—	n		
(d)	E,G	p	n		
(e)	E,H	p	n		

10.3 Consider tasks T_i and T_j, both of which have a two-phase commit task structure. Draw the state charts describing each of the following dependencies between these two tasks:

(a) Task T_j cannot start executing until task T_i commits.

(b) If task T_i aborts then task T_j commits.

(c) If T_i commits, T_j can begin executing after T_i commits.

(d) Task T_i terminates in the aborted state after three unsuccessful executions.

(e) Task T_j is allowed to commit if an external value for parameter T is greater than 5.

(f) T_j cannot begin executing until T_i aborts and if task T_i aborts then task T_j commits.

10.4 Consider the composite state "Process Payment_AND_Commit Sale" described in the illustrative example in Section 10.6. Section 10.7 points out an incorrect dependency specification in the coordination of these tasks and suggests the use of an additional Abort dependency. Illustrate the low-level state chart diagram for this composite state if the abort dependency is specified in addition to the Terminating and Strong Commit dependencies.

10.5 The transfer of funds between bank accounts is accomplished by debiting one account and crediting the other. The workflow for the fund transfer process contains two tasks, Credit and Debit, executing simultaneously. The process requires the following:

1. Debit task cannot start until the Credit task starts (Begin dependency).
2. Debit task cannot commit or abort until Credit task commits or aborts (Terminating dependency).

3. No work should be committed if the Credit task aborts. Therefore, if the Credit task aborts then the Debit task aborts (Abort dependency).

Draw the state chart diagram for the fund transfer process.

CASE STUDIES

10.6 Online Theater Ticket Sale Case(A)

Consider the policy change for the online theater ticket sale application described in Review Exercise 9.3. Based on the described changes:

(a) Describe the set of additional tasks and actions that must be incorporated into the workflow definition.
(b) Determine additional intertask dependencies for the updated online ticket sales workflow and illustrate them with an intertask dependency diagram.
(c) Draw the top-level state chart diagram for the online ticket sales workflow.
(d) Draw the lower-level state chart diagrams for the task defined based on the policy changes.

10.7 University Food Company Online Ordering Case(F)

In an attempt to integrate their online ordering capability into the rest of their production scheduling process, the University Food Company wants to model and analyze the associated activities, which can later be used as a part of its larger efforts to implement a workflow management system. When a customer places an online order with the University Food Company, several activities must take place before the sale is confirmed via displaying a Web page that informs the customer that the order has been accepted. These are as follows:

1. *Process payment.* Based on the customer's credit information on file, which includes the customer's bank account information, the payment for the order is processed. If the payment cannot be processed for reasons such as the available funds are insufficient, the customer is informed and the workflow instance terminates.
2. *Schedule production.* While the customer's payment is being processed, the current production schedule is updated to include the customer's new order.
3. *Select shipper.* Based on the delivery schedules of the shippers that University Food works with, a shipper is selected (if available) that can deliver the order on the selected delivery date. If none of the shippers can deliver the product on the selected date, the order is canceled with the production department, payment is credited to the customer's account, and the customer is informed.
4. *Finalize sale.* If payment for the order is processed and a shipper is available to deliver the order on the selected delivery date, the sale is finalized by informing the customer.

The activity diagram of Figure CS10.1 describes the coordination of the preceding activities. Complete the following exercise to aid the University Food Company in this project.

(a) Describe the set of tasks and actions that must be incorporated into the workflow definition.
(b) Determine the intertask dependencies for the online order process, and illustrate them with an intertask dependency diagram.

(c) Complete the top-level state chart diagram for the online order process.
(d) Complete the lower-level state chart diagrams for the tasks defined.
(e) Provide the final lower-level state chart diagram for the online order process.

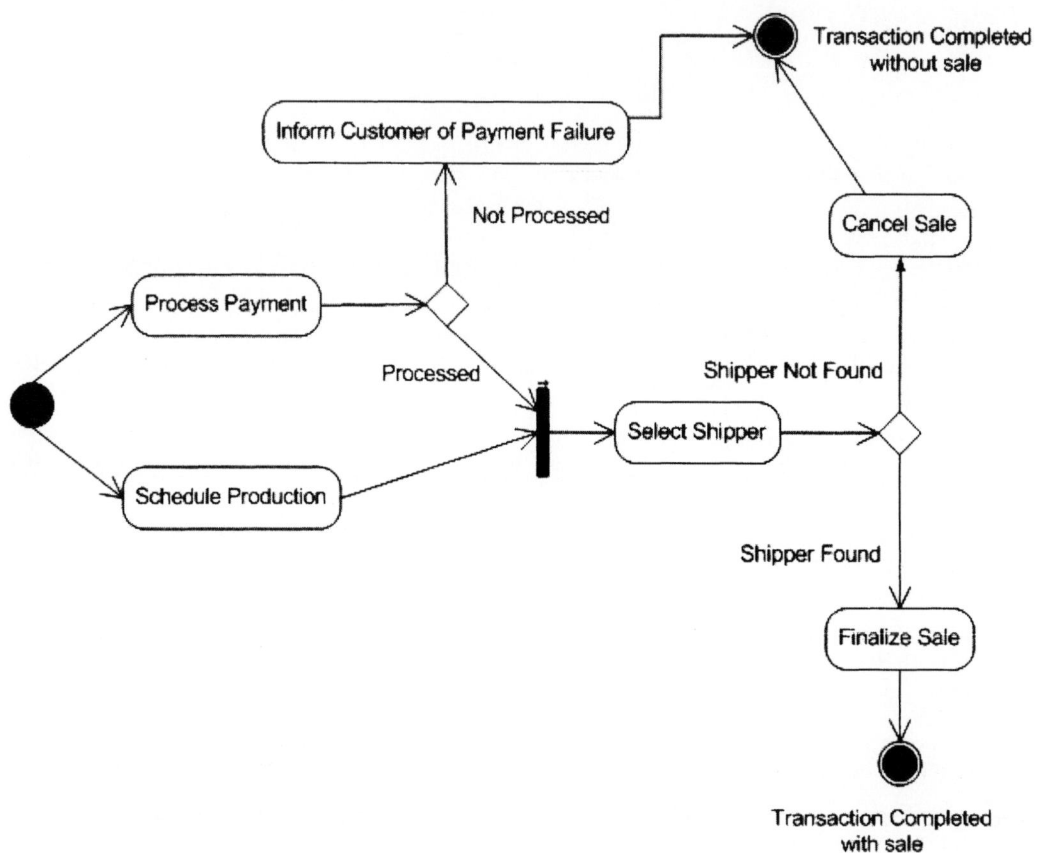

Figure CS10.1 Activity diagram for Case Study 10.7.

Bibliography

Adam, N. R., V. Atluri and W. Huang (1998) Modeling and Analysis of Workflows Using Petri Nets, *Journal of Intelligent Information Systems, Special Issue on Workflow and Process Management*, vol. 10, no. 2, pp. 131–158.

ANSI (1986) *The Database Language SQL*, ANSI Document X3.135–1986. New York: American National Standards Institute.

ANSI (1992) *The Database Language SQL*, ANSI Document X3.135–1992. New York: American National Standards Institute.

Askin, R. G. and J. B. Goldberg (2002) *Design and Analysis of Lean Production Systems*. New York: John Wiley & Sons.

Bae, J., H. Bae, S. H. Kang and Y. Kim (2004) Automatic Control of Workflow Processes Using ECA Rules, *IEEE Transactions on Knowledge and Data Engineering*, vol. 16, no. 8, pp. 1010–1023.

Baker, R. (1990) *Entity Relationship Modeling*. Essex: Addison Wesley Longman.

Beavers, A. N. (2001) *Roadmap to the e-Factory*. Boca Raton, FL: Auerbach Publishers.

Billo, R. E., J. D. Porter and R. J. Puerzer (2006) An Architecture for the Design of Industrial Information Systems, in *Handbook of Industrial and Systems Engineering*, A. Badiru, ed. Boca Raton, FL: CRC Press.

Boucher, T.O. and M.A. Jafari (1992) Design of a Factory Floor Sequence Controller from a High Level System Specification, *Journal of Manufacturing Systems*, vol. 11, no. 6, pp. 401–417.

Boucher, T.O. (1996) *Computer Automation in Manufacturing*. London: Chapman & Hall.

Bowersox, D. J., D. J. Closs and M. B. Cooper (2002) *Supply Chain Logistics Management*. New York: McGraw-Hill.

Bruce, T. (1992) *Designing Quality Databases with IDEF1X Information Models*. New York: Dorset House.

Chen, P. (1976) The entity-relationship model—toward a unified view of data, *ACM Transactions on Database Systems*, vol. 1, no. 1, pp. 9–36.

Chopra, S. and P. Meindl (2004) *Supply Chain Management*, 2nd Edition. Upper Saddle River, NJ: Prentice Hall.

Connolly, T. and C. Begg (1999) *Database Systems*. Essex: Addison Wesley Longman Limited.

Darabi, H. (2006) Finite Automata Modeling and Analysis of Workflow Management Systems, *International Journal of Industrial and Systems Engineering*, vol. 1, no. 3, pp. 388–411.

Date, C. J. (2004) *An Introduction to Database Systems*, 8th Edition. Reading, MA: Addison-Wesley.

Davenport, T. H. and J. E. Short (1990) The New Industrial Engineering: Information Technology and Business Process Redesign, Sloan Management Review, vol. 31, no. 4, pp. 11–27.

DeGarmo, E.P., J.T. Black and R.A. Kohser (2003) *Materials and Processes in Manufacturing*, 9th Edition. New York: John Wiley & Sons.

Dennis, A. (2003) *Systems Analysis and Design*, 2nd Edition. New York: John Wiley & Sons.

Dobson, R. (2003) *Programming Microsoft Office Access 2003*. Redmond, WA: Microsoft Press.

Elmasri, R. and S. B. Navathe (2004) *Fundamentals of Database Systems*, 4th Edition. Reading, MA: Addison-Wesley.

Elsayed, E.A. and T.O. Boucher (1994) *Analysis and Control of Production Systems*, 2nd Edition. Englewood Cliffs, NJ: Prentice Hall.

Farahvash, P. and T. O. Boucher (2004) A Multi-agent Architecture for Control of AGV Systems, *Robotics and Computer-integrated Manufacturing*, vol. 20, no. 6, pp. 473–483.

Garland, J. and R. Anthony (2003) *Large Scale Software Architecture: A Practical Guide using UML*. Chichester: John Wiley & Sons.

Georgakopoulos, M. H. and A. Sheth (1995) An Overview of Workflow Management: From Process Modeling to Workflow Automation Infrastructure, *Distributed and Parallel Databases*, vol. 3, no. 2, pp 119–153.

Gosnell, D.M. (2004) *Beginning Access 2003 VBA*. New York: John Wiley & Sons.

Groff, J. (2002) *SQL: The Complete Reference*, 2nd Edition. New York: McGraw-Hill.

Groover, M. P. (2001) *Automation, Production Systems, and Computer-Integrated Manufacturing* Englewood Cliffs, NJ: Prentice Hall.

Hanna, M.M., R. L. Ahuja and W.L. Winston (2003) *Developing Spreadsheet-Based Decision Support Systems Using VBA for Excel.* (http://www.ise.ufl.edu/it/)

Harel, D. and M. Politi (1998) *Modeling Reactive Systems with Statecharts: The STATEMATE Approach.* New York: McGraw-Hill.

Hay, D. C. (1999) *A Comparison of Data Modeling Techniques*, Essential Strategies Inc. (www.essentialstrategies.com/publications/modeling/compare.htm).

IDEF: Integrated Computer-Aided Manufacturing (ICAM) Architecture, Part II, Volume IV—Functional Modeling Manual (1981) USAF Report no. AFWAL-TR-81-4023. Wright-Patterson AFB, Ohio, June.

Kusiak, A., T. Letsche and A. Zakarian (1997) Data Modeling with IDEF1X, *International Journal of Computer Integrated Manufacturing*, vol. 10, no. 6, pp. 470–486.

Jacobson, I. (1992) *Object-Oriented Software Engineering: A Use Case Driven Approach.* Reading, MA: Addison-Wesley.

Jacobson, I., G. Booch and J. Rumbaugh (1999) *The Unified Software Development Process.* Reading, MA: Addison-Wesley.

Kendall, K. E. and J. E. Kendall, (2002) *Systems Analysis and Design*, 5th Edition. Englewood Cliffs, NJ: Prentice Hall.

Kiepuszewski, B. (2003) *Expressiveness and Suitability of Languages for Control Flow Modeling in Workflows*, Queensland University of Technology Brisbane Australia, Faculty of Information Technology, Ph.D. Thesis. Available via: http://citeseer.ist.psu.edu

Knowledge Based Systems, Inc. (1997) *AI0 User Manual: Automated Functional Modeling for Windows.* College Station, TX: KBSI.

Knowledge Based Systems, Inc. (2003) *Users Manual: Smart ER5, Automated Information Data Modeling and Analysis.* College Station, TX: KBSI.

Kobielus, J. G. (1997) *Workflow Strategies.* Foster City, CA: IDG Books Worldwide.

Kratochvil, M. (2003) *UML Xtra-light: How to Specify Your Software Requirements.* Cambridge: Cambridge University Press.

Ladd, E. and J. O'Donnell (1999), *Using HTML 4, XML, and Java 1.2.* Indianapolis, IN: Que Publishing.

LaRocca, D. (1999) *Teach Yourself SAP R/3 in 24 Hours.* Indianapolis, IN: Sams Publishing.

Laudon, K. C. and J. P. Laudon (2004) *Management Information Systems*, 8th Edition. Englewood Cliffs, NJ: Prentice Hall.

Lewis, P.M., A. Bernstein and M. Kifer (2002) *Database and Transaction Processing.* Reading, MA: Addison-Wesley.

Loos, P. and A.-W. Scheer (1994) Graphical Recipe Management and Scheduling for Process Industries, *Proceedings of Rutgers Conference on Computer Integrated Manufacturing in the Process Industries*, Dept. of Industrial Engineering, Rutgers University.

Lutchen, M. D. (2004) *Managing IT as a Business.* Hoboken, NJ: John Wiley & Sons.

Marca, D.A. and C.L. McGowan (1988) *Structured Analysis and Design Technique.* New York: McGraw-Hill Book Company.

MESA International (1997) *MES Explained: A High Level Vision*, MESA International White Paper. (www.mesa.org)

Microsoft (2003) *Microsoft Office Access 2003 Step by Step.* Redmond, WA: Microsoft Press.

Moynihan, G.P. (1997) The Application of Enterprise Modeling for Aerospace Manufacturing System Integration, *International Journal of Flexible Manufacturing Systems*, vol. 9, no. 2, pp. 195–210.

National Institute of Standards and Technology (1993) *Integration Definition for Information Modeling (IDEF1X)*, FIPS 184, U. S. Department of Commerce.

Perry, J. T. and G. P. Schneider (2004) *Building Accounting Systems using Access 2003*, 3rd Edition. South-Western College Publishers.

Prague, C.N., M.R. Irwin, and J. Reardon (2003) *Access 2003 Bible.* New York: John Wiley & Sons.

Pol, A. A. and R. K. Ahuja (2003) *Developing Web-Enabled Decision Support Systems Using VB. NET and ASP.NET.* (www.ise.ufl.edu/it/)

Potts, S. and M. Kopack (2003) *Web Services.* Indianapolis, IN: Sams Publishing.

Rob, P. and C. Coronel (1995) *Database Systems.* Danvers, MA: Boyd & Fraser Publishing Company.

Romney, M. B. and P. J. Steinbart (2005) *Accounting Information Systems*, 10th Edition. Englewood Cliffs, NJ: Prentice Hall.

Roques, P. (2001) *UML in Practice.* Chichester: John Wiley & Sons.

Rumbaugh, J., I. Jacobson and G. Booch (1999) *The Unified Modeling Language Reference Manual.* Reading, MA: Addison Wesley Longman, Inc.

SAP R/3 Workflow (www.sapgenie.com/workflow/)

Silverston, L., W. H. Inmon, and K. Graziano (1997) *The Data Model Resource Book*. New York: John Wiley & Sons.

Simchi-Levi D., P. Kaminsky, and E. Simchi-Levi (2000) *Designing and Managing the Supply Chain*. New York: McGraw-Hill.

Strong, D. (2004) A Roadmap for Enterprise System Integration, *Computer*, vol. 37, no. 6, pp. 22–29.

Taylor, A.G. and V. Anderson (2004) *Access 2003 Power Programming with VBA*. New York: John Wiley & Sons.

Toh, K. T. K. (1999) Modeling architectures: a review of their application in structured methods for information systems specification, *International Journal of Production Research*, vol. 37, no. 7, pp. 1439–1458.

Turban, E., J. E. Aronson and T. P. Liang (2004) *Decision Support Systems and Intelligent Systems*, 7th Edition. Englewood Cliffs, NJ: Prentice Hall.

Van Der Aalst, W.M.P. (2002) *Workflow Management*. Cambridge, MA: The MIT Press.

Van Der Lans, R. F. (2004) *Introduction to SQL*, 4th Edition. Reading, MA: Addison-Wesley.

Viescas, J.L. (2003) *Microsoft Office Access 2003 Inside Out*. Redmond, WA: Microsoft Press.

Viescas, J.L. (2005) *Building Microsoft Access Applications*. Redmond, WA: Microsoft.

Websphere MQ Workflow. (www-306.ibm.com/software/integration/wmqwf/)

Weiss, G. (1999) *Multiagent Systems*, Cambridge: Cambridge, MA: The MIT Press.

Whitehead, P. (2000) *Active Server Pages 3.0*, Foster City, CA: IDG Books Worldwide, Inc.

Workflow Management Coalition. (www.wfmc.org)

Wu, B. (2000) *Manufacturing and Supply Systems Management—A Unified Framework of Systems Design and Operation*. New York: Springer.

Wu, T. and J. Blackhurst (2006) Information Engineering, in *Handbook of Industrial and Systems Engineering*, A. Badiru, ed. Boca Raton, FL: CRC Press.

Yalçın, A., A. Khemuka, P. Deshpande (2005) Modeling and Control of Workflow Managements Systems based on Discrete Event Systems Theory, *International Journal of Production Research*, vol. 43 no. 20, pp. 4359–4379.

Young, M. J. (2002) *XML Step by Step*. Redmond, WA: Microsoft Press.

Young, R.E. and J. Vesterager (1991) An approach to CIM system development whereby manufacturing people can design and build their own CIM systems, *International Journal of Computer Integrated Manufacturing*, vol. 4, no. 5, pp. 288–299.

Yourdon, E. and L.L. Constantine (1979) *Structured Design: Fundamentals of a Discipline of Computer Program and Systems Design*. Englewood Cliffs, NJ: Prentice Hall.

Zachman, J. A. (1987) A Framework for Information Systems Architecture, *IBM Systems Journal*, vol. 26, no. 3, pp. 276–292.

CASE Tool Web Sites

There are many computer-aided software engineering products that help automate the model building processes discussed in this book. What follows is a listing of a few Web sites that have trial versions of software that support one or all of the model types presented here. A Visio 2003 Lab Manual for creating IDEF, DFD, and UML diagrams is available on the Web site that supports this book.

Artisan Studio, ARTiSAN Software Tools, Inc., 16055 SW Walker Road, #422, Beaverton, OR 97006. www.artisansw.com/—supports UML.

E/R Studio, Embarcadero Technologies, 100 California Street, San Francisco, CA 94111. www.embarcadero.com/downloads/—supports IDEF1X.

AI0, Knowledge Based Systems, Inc., One KBSI Place, 1408 University Drive, East, College Station, TX 77840. www.kbsi.com/software/—supports IDEF0 and IDEF1X.

Visio 2003, Microsoft Corporation, One Microsoft Way, Redmond, WA 98052. www.microsoft.com/office/visio/profinfo/trial.mspx—supports IDEF0, IDEF1X, DFD, UML.

Visual UML, Visual Object Modelers, Inc., 4450 Arapahoe Avenue, Suite 100, Boulder, CO 80303. www.visualobject.com/—supports UML.

Index

Printed and bound by CPI Group (UK) Ltd, Croydon, CR0 4YY
14/05/2025
01871570-0002